The mechanics of the contact between deformable bodies

This book contains the proceedings of the symposium of the International Union of Theoretical and Applied Mechanics on 'The mechanics of the contact between deformable bodies', held in Enschede, The Netherlands, 20–23 August 1974.

It gives a survey of theoretical and fundamental experimental studies on the mechanical aspects of the contact between solid bodies: solid surfaces, area of contact, mechanics of friction in the broadest sense; mathematical analysis of the normal contact and tangential contact (pure sliding, steady and unsteady rolling) between two elastically, plastically or visco-elastically deformable solid bodies on the basis of models of the contact mechanics. Problems involving hydrodynamic lubrication as the main phenomenon are not considered.

The mechanics of the contact between deformable bodies

Proceedings of the symposium
of the International Union of Theoretical and Applied Mechanics (IUTAM)
Enschede, Netherlands, 20–23 August 1974

Edited by
A.D. de Pater and J.J. Kalker

1975
Delft University Press

PROF. DR. IR. ANTON D. DE PATER
Department of Mechanical Engineering
Delft University of Technology
Mekelweg 2
Delft
Netherlands

DR. IR. JOOST J. KALKER
Subdepartment of Mathematics
Delft University of Technology
Julianalaan 132
Delft
Netherlands

Printed in Belgium
ISBN 90 298 2001 2

Contents

Preface VII

List of participants IX

J.J. Kalker/Aspects of contact mechanics 1

K.L. Johnson/Non-Hertzian contact of elastic spheres 26

M. Boucher/Signorini's problem in viscoelasticity 41

J. Dundurs/Properties of elastic bodies in contact 54

D.A. Spence/Similarity considerations for contact between dissimilar elastic bodies 67

A.Ju. Išlinskij/Consideration of the theory of cracks from the point of view of contact problems of the theory of elasticity 77

B.L. Abramjan/Certain asymmetrical contact problems for a half-space 84

G.M.L. Gladwell/Unbonded contact between a circular plate and an elastic foundation 99

J.B. Alblas/On the two-dimensional contact problem of a rigid cylinder, pressed between two elastic layers 110

L.E. Goodman and L.M. Keer/Influence of an elastic layer on the tangential compliance of bodies in contact 127

J. Christoffersen/Small scale plastic flow associated with rolling 152

F. F. Ling/Heat effects in rolling contact 165

J. R. Barber/Thermoelastic contact problems 177

R. A. Burton/An axisymmetric contact patch configuration for two slabs in frictionally heated contact 191

Y. M. Tsai/Dynamic contact stresses produced by impact in elastic plates of finite thickness 206

K. Kawatate/Transition of collision contact force between a visco-elastic half-space and a flat-headed rigid body 221

P. A. Engel/Impact on a worn surface 239

T. G. Johns and A. W. Leissa/The normal contact of arbitrarily shaped multilayered elastic bodies 254

K. P. Singh, B. Paul and W. S. Woodward/Contact stresses for multiply-connected regions – the case of pitted spheres 264

J. F. Archard, R. T. Hunt and R. A. Onions/Stylus profilometry and the analysis of the contact of rough surfaces 282

F. T. Barwell, M. H. Jones and S. D. Probert/ The interaction and lubrication of rough surfaces 304

J. Halling and K. A. Nuri/Contact of rough surfaces of work-hardening materials 330

H. Krause and A. Halim Demirci/Factors influencing the real trend of the coefficient of friction of two elastic bodies rolling over each other in the presence of dry friction 342

A. Schallamach/The frictional contact of rubber 359

L. A. Mitchell/Applications for contact theories in nuclear reactor technology 377

P. R. Nayak/Linearized contact vibration analysis 393

Preface

A recommendation to hold a symposium on contact problems in mechanics, sponsored by the International Union of Theoretical and Applied Mechanics (IUTAM), was made by the Engineering Mechanics Group of the Department of Mechanical Engineering, Delft University of Technology.

In order to outline the scope of the Symposium and select the place and date IUTAM appointed a Study-group of seven members in 1971. Following an interchange of views between these members the Study-group proposed to IUTAM to organise an International Symposium on the Mechanics of the Contact between Deformable Bodies on August 20–23, 1974, at Twente University of Technology, Netherlands, with a number of participants restricted to about 60, the outline of scope being as follows:

Theoretical and fundamental experimental studies on the mechanical aspects of the contact between solid bodies: solid surfaces, area of contact, mechanics of friction in the broadest sense. Mathematical analysis of the normal contact and tangential contact (pure sliding, steady and unsteady rolling) between two (elastically, plastically and viscoelastically) deformable solid bodies on the basis of simplified models of the contact mechanism. Problems involving hydrodynamic lubrication as the main phenomenon should be excluded.

IUTAM accepted this proposal and appointed in December, 1972, a scientific committee consisting of the following members:

N.H. Arutjunjan (USSR),
H. Bufler (BRD),
L.E. Goodman (USA),
A. Ju. Išlinskij (USSR),
K.L. Johnson (UK),
W. Nowacki (Poland),
A.D. de Pater (Chairman, Netherlands) and
D. Tabor (UK).

Their main duty was the selection of lecturers and other participants.

The local arrangements of the Symposium were in the hands of a Local

Organizing Committee (J.J. Kalker, A.D. de Pater, C. de Pater (Chairman), and H.A. Verbeek), which was assisted by a number of members of the staff of Twente University of Technology and by the University office which usually takes care of congresses organized at the campus ('Drienerloo's Organisatie Bureau'). Nearly all the participants were lodged in the student rooms on the campus, whereas the meals were taken in the Student's Union Building. The papers were read in the Applied Mathematics Building.

The 26 papers were presented in seven sessions, each mainly dealing with one special topic of the Symposium theme. Moreover, at the end of the seventh session five brief communications were presented. The manuscripts of the 26 papers were distributed to the participants at the beginning of the Symposium. Nearly all the papers provoked ample discussion. Delft University Press has been found willing to take care of the publication of the present Symposium Proceedings.

The Symposium was considered to be a success, due to the following circumstances: the high standard of the papers, the intensive contacts between the participants, promoted by the excellent facilities of the University campus and the limited number of participants, and the unusually good weather. The Symposium was honoured by the presence of the Rector Magnificus of the University, Prof. P.J. Zandbergen, and the Vice-President of IUTAM, Prof. W.T. Koiter, who both addressed the participants at the opening ceremony.

The help of a number of bodies is gratefully mentioned:

- the International Union of Theoretical and Applied Mechanics granted an amount of U.S.$ 4000,– for covering travel and living expenses of participants;
- N.V. Nederlandse Spoorwegen, Shell Nederland B.V. and N.V. Philips' Gloeilampenfabrieken made gifts of altogether Hfl. 1000,– towards the running costs of the Symposium;
- Delft University of Technology provided adequate clerical and technical help for the preparation of the scientific part of the Symposium and the Proceedings;
- Twente University of Technology assisted in the production of the preprints of a number of papers;
- Drienerloo's Organisatie Bureau took care of lodging and meals and assisted in registration, organizing projection facilities, organizing excursions etc. in a very effective way;
- the Council of Mayor and Alderman of the City of Enschede invited the participants for a reception in the Town Hall of Enschede on the first Symposium day.

Delft, Netherlands
February 1975

A.D. de Pater
J.J. Kalker

List of participants[1]

* ABRAMJAN, B.L., Academy of Sciences of Armenian SSR, Institute of Mechanics, Barekamutsjan Street 24b, 375200 Jerevan, USSR

* ALBLAS, J.B., Technische Hogeschool Eindhoven, Postbus 513, Eindhoven, Netherlands

* ARCHARD, J.F., University of Leicester, Department of Engineering, Leicester LE1 7RH, Great Britain

* BARBER, J.R., University of Newcastle-upon-Tyne, Department of Mechanical Engineering, Newcastle-upon-Tyne NE1 7RU, Great Britain

* BARWELL, F.T., University of Swansea, Singleton Park, Swansea, Wales SA2 8PP, Great Britain

BECKUM, F.P.H. VAN, Technische Hogeschool Twente, Postbus 217, Enschede, Netherlands

BENTALL, R.H., European Space Research Organisation, Domeinweg, Noordwijk, Netherlands

BLOK, H., Laboratorium voor Werktuigonderdelen, Afdeling der Werktuigbouwkunde der TH Delft, Mekelweg 2, Delft, Netherlands

* BOUCHER, M., 75 Voie de Châtenay, 91370 Verrières-le-Buisson, France

BREKELMANS, W.A.M., Technische Hogeschool Eindhoven, Postbus 513, Eindhoven, Netherlands

BUFLER, H., Institut für Mechanik (Bauwesen), Lehrstuhl II, Universität Stuttgart, 7000 Stuttgart 1, Keplerstrasse 11, BRD

* BURTON, R.A., Department of Mechanical Engineering and Astronautical Sciences, Northwestern University, The Technological Institute, Evanston, Illinois 60201, USA

1. Authors are marked with an asterisk.

CAMPEN, D.H. VAN, Technische Hogeschool Twente, Postbus 217, Enschede, Netherlands

CHILDS, T.H.C., University of Bradford, Bradford, Yorkshire BD7 1DP, Great Britain

* CHRISTOFFERSEN, J., Technical University of Denmark, Department of Solid Mechanics, Building 404, DK–2800 Lyngby, Denmark

DAUTZENBERG, J.H., Technische Hogeschool Eindhoven, Postbus 513, Eindhoven, Netherlands

DESOYER, K., Vorstand des II. Instituts für Mechanik der Technische Hochschule Wien, A 1040 Karlsplatz 13, Wien, Österreich

DUBBELDAM, J.W., Division of Research and Development, Civil Engineering Department, Netherlands Railways, Utrecht, Netherlands

* DUNDURS, J., Department of Civil Engineering, Northwestern University, Evanston, Illinois 60201, USA

ECK, H.N. VAN, Technische Hogeschool Twente, Postbus 317, Enschede, Netherlands

* ENGEL, P.A., IBM, System Development Division, P.O. Box 6, Endicott, New York 13760, USA

ENGLAND, A.H., Department of Theoretical Mechanics, The University of Nottingham, University Park, Nottingham NG7 2RD, Great Britain

FIELD, J., Cavendish Laboratory, Madingley Road, Cambridge CB3 0HE, Great Britain

FRÉMOND, M., Laboratoire Central des Ponts et Chaussées, 58 Boulevard Lefèbvre, 75732 Paris 15e, France

* GLADWELL, G.M.L., Department of Civil Engineering, Faculty of Engineering, University of Waterloo, Waterloo, Ontario N2L 3G1, Canada

* GOODMAN, L.E., Department of Engineering Mechanics, University of Minnesota, Minneapolis, Minnesota 55455, USA

GRAND, P. LE, Technische Hogeschool Twente, Postbus 217, Enschede, Netherlands

* HALLING, J., Department of Mechanical Engineering, University of Salford, Salford M5 4WT, Great Britain

HAMILTON, G.M., Department of Applied Physical Sciences, The University of Reading, Whiteknights, Reading, Berks., Great Britain

Herrmann, G., Department of Applied Mechanics, School of Engineering, Stanford, California 94305, USA

* Išlinskij, A. Ju., The Institute of Mechanical Problems, USSR Academy of Sciences, 125040 Leningrad Pr. 7, Moscow A40, USSR

* Johns, T. G., Batelle Columbus Laboratories, 505 King Avenue, Columbus, Ohio 43201, USA

* Johnson, K. L., University of Cambridge, Engineering Department, Trumpington Street, Cambridge CB2 1PZ, Great Britain

* Kalker, J. J., Onderafdeling der Wiskunde, Technische Hogeschool Delft, Julianalaan 132, Delft, Netherlands

* Kawatate, K., Research Institute for Applied Mechanics, Kyushu University, Hakozaki, Fukuoka 812, Japan

* Krause, H., Institut für Fördertechnik und Schienenfahrzeuge, Rheinisch-Westfälische Technische Hochschule Aachen, 51 Aachen, Seffenterweg 8, BRD

Krettek, O., Lehrstuhl und Institut für Fördertechnik und Schienenfahrzeuge, Technische Hochschule Aachen, Seffenterweg 8, Aachen 51, BRD

Kuipers, M., Faculty of Natural Science, University of Groningen, Netherlands

Lachat, J. C., Chef du Département de Résistance des Structures, Cetim BP. 67, 5, Avenue Félix Louat, 60304 Senlis, France

Landheer, D., Technische Hogeschool Eindhoven, Postbus 513, Eindhoven, Netherlands

* Ling, F. F., School of Engineering, Mechanics Division, Rensselaer Polytechnic Institute, Troy, New York 12181, USA

Maugis, D., Laboratoire de Bellevue, 1, Place Aristide Briand, 92190 Meudon, France

Meijers, P., Afdeling der Werktuigbouwkunde, Technische Hogeschool Delft, Mekelweg 2, Netherlands

* Mitchell, L. A., Berkeley Nuclear Laboratories (CEGB), Research and Development Department, Berkeley, Gloucestershire GL13 9PB, Great Britain

Moes, H., Technische Hogeschool Twente, Postbus 217, Enschede, Netherlands

* Nayak, P. R., 346E Sind Coop. Housing Society, Aundh, Poona 7, India

O'CONNOR, J.J., Department of Engineering Science, Park Road, Oxford OX1 3PJ, Great Britain

PACEJKA, H.B., Afdeling der Werktuigbouwkunde, Technische Hogeschool Delft, Mekelweg 2, Delft, Netherlands

PÁCZELT, ISTVÁN, Department of Mechanics, Technical University of Heavy Industry, 3515 Miskolc-Egyetemváros, Hungary

PARLAND, H., Tampere University of Technology PL 527, 33101 Tampere 10, Finland

PATER, A.D. DE, Afdeling der Werktuigbouwkunde, Technische Hogeschool Delft, Mekelweg 2, Delft, Netherlands

PATER, C. DE, Technische Hogeschool Twente, Postbus 217, Enschede, Netherlands

* PAUL, B., University of Pennsylvania, College of Engineering and Applied Science, Philadelphia 19174, 111 Towne Building, USA

PERSSON, B.G.A., Linköping University, Department of Mechanical Engineering, Fack, S 581 88 Linköping, Sweden

ROSSMANITH, H., Vorstand des II. Instituts für Mechanik der Technischen Hochschule Wien, A 1040 Karlsplatz 13, Wien, Österreich

SAVKOOR, A.R., Afdeling der Werktuigbouwkunde, Technische Hogeschool Delft, Mekelweg 2, Delft, Netherlands

SAYIR, M., Eidgenössige Technische Hochschule, Leonhardstrasse 33, CH-8006, Zürich, Switserland

* SCHALLAMACH, A., 73 Longmore Avenue, Barnet, Herts. EN5 1LA, Great Britain

SCHOOFS, A.J.G., Technische Hogeschool Eindhoven, Postbus 513, Eindhoven, Netherlands

* SINGH, K.P., Joseph Oat & Sons Inc., 2500 Broadway, Camden, New Jersey 08104, USA

SOLOMON, L., Université de Poitiers, Laboratoire de Mécanique, 40 Avenue du Recteur-Pineau, 86022-Poitiers, France

* SPENCE, D.A., Department of Science Engineering, University of Oxford, Parks Road, Oxford OX1 3PJ, Great Britain

SPIERING, R.M.E.J., Technische Hogeschool Twente, Postbus 217, Eindhoven, Netherlands

Szabo, B.A., Department of Civil Engineering, Washington University, St. Louis, Missouri 63130, USA

Tabor, D., Cavendish Laboratory, University of Cambridge, Madingley Road, Cambridge CB3 0HE, Great Britain

* Tsai, Y.M., Department of Engineering Science and Mechanics, Iowa State University, Ames, Iowa 50010, USA

Verbeek, H.A., Technische Hogeschool Twente, Postbus 217, Enschede, Netherlands

Vries, R.W. de, Technische Hogeschool Twente, Postbus 217, Enschede, Netherlands

Vroegop, R.R., Technische Hogeschool Twente, Postbus 217, Enschede, Netherlands

Whitelaw, R.L., P.O. Box 223, Blacksburg, Virginia 24060, USA

Willis, J., University of Bath, Great Britain

Wittmeyer, H., European Research Center, SKF Research, Postbus 50, Jutphaas, Netherlands

Zandbergen, P.J., Technische Hogeschool Twente, Postbus 217, Enschede, Netherlands

Aspects of contact mechanics

J. J. Kalker

1. INTRODUCTION

In the present paper two aspects of contact mechanics will be considered. The first concerns the variational theory of contact mechanics, and the second is a review of the theory of contact mechanics. Thus the paper is divided into two parts.

PART 1. VARIATIONAL THEORY

1.1 *Introduction*

The variational theory of contact mechanics was initiated by Signorini[82]* in 1959. It was developed almost exclusively in Italy and France, so that the literature on the subject is in Italian and French and not readily accessible to the Russian and English speaking scientific public. It is my hope that the present introductory lecture will stimulate the reader to study the Italian and French sources, in particular the book by G. Duvaut and J.L. Lions, *Les Inéquations en Mécanique et en Physique*[25]. The variational theory employs the inequalities that determine the contact conditions both with and without friction in a systematic manner.

Up to now, the primal object of the theory has been to obtain existence and uniqueness theorems for the field quantities that determine the mechanical state of the deformable body. You may be under the impression that, e.g., the Kirchhoff uniqueness theorem (see, e.g. Love's *Treatise*[55], pp. 170–178) also covers contact problems, but this is not so, owing to the special boundary conditions that characterise contact problems.

On the other hand, the variational inequalities of the theory often lead to an alternative statement in the form of a minimization problem of functionals. In a numerical implementation these problems can be

* When an author's name is mentioned, reference is made to the alphabetical list of literature at the end of this paper.

discretised, and then lead to problems which can be tackled by means of the well-known techniques of mathematical programming (see, e.g., the book by Kowalik and Osborne[47], and the Colville report[15]).

Also, a book is due to be published by Glowinski, Lions and Tremolières[31] in which the direct numerical implementation of variational inequalities is set forth.

The Franco-Italian school freely makes use of the concepts and methods of functional analysis. There is no space in the present paper to do so; consequently the derivations presented here have a purely formal character.

1.2 *Notations*

Latin subscripts run from 1 to 3. Summation over repeated indices is understood.

$'$	$\partial/\partial t$
$,k$	$\partial/\partial x_k$
$(\alpha, \beta)_\gamma$	$= \int_\gamma (\alpha \cdot \beta)\, d\gamma$; if α, β are vectors, $\alpha \cdot \beta$ denotes the inner product
$a(u, v)$	$= \int_\Omega a_{ijhk}(x) \varepsilon_{ij}(u) \varepsilon_{hk}(v)\, d\Omega$
$A(\tau, \eta)$	$= \int_\Omega A_{ijhk}(x) \tau_{ij} \eta_{hk}\, d\Omega$
N, T	= normal, tangential component (w.r.t. vector n)
$\| \|$	length of vector
$a, a_{ijhk}(x)$	elasticity tensor
A, A_{ijhk}	inverse elasticity tensor
$f, f_h(x)$	body force
$F, F_h(x)$	prescribed surface traction
$h(x, t)$	distance function
$n, n(x)$	outer normal on Ω at x
t	time
$u, u_h(x, t)$	Lagrangean displacement
U	prescribed surface displacement
v	(in part 2) translation vector
w	(in part 2) rotation vector
x, x_k	Cartesian coordinates of the reference state
Γ	boundary of elastic body
Γ_c	part of boundary where contact conditions are prescribed
Γ_F	part of boundary where surface traction is prescribed
Γ_u	part of boundary where displacement is prescribed
Δ	(in part 2) a representative diameter of the contact region
ε	(in part 2) a representative strain
$\varepsilon_{ij}(u)$	$= \frac{1}{2}(u_{i,j} + u_{j,i})$, linearised strain
$\mu(x)$	coefficient of friction
$\rho(x)$	density
$\sigma, \sigma_h(x, t)$	$= \tau_{hk} n_k$, surface traction
τ, τ_{ij}	stress
Ω	space occupied by the elastic body in reference state.

Further notations are defined where they are introduced.

1.3 *The principle of virtual work*

In its classical form, the principle of virtual work states that the work by the impressed and inertial forces on the kinematically admissible virtual displacements must vanish:

$$\delta W_{\text{impressed and inertial forces}} = 0 \tag{1}$$

In this formulation it is implicitly understood that the kinematically admissible virtual displacements are of a *reversible* nature, i.e. that if δq is kinematically admissible, then also $(-\delta q)$ is kinematically admissible. Under these circumstances it is said that all constraints present in the problem are *bilateral.*

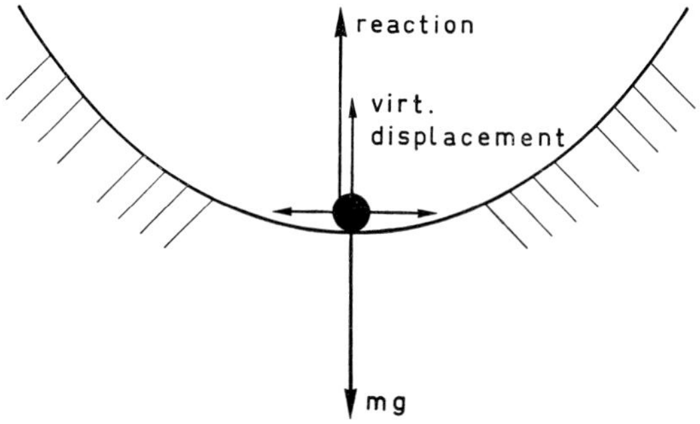

Fig. 1. A particle lying in a bowl

It is, however, quite easy to find problems in which some constraints are not bilateral. As an example, consider the following 'normal contact problem'. A particle is lying in a frictionless bowl, in the presence of gravity (see Fig. 1). The impressed force is mg, acting downwards; the particle may move sidewards and upwards, but not downwards. The vertical displacement is irreversible; a constraint admitting such a virtual displacement is called unilateral. It is also seen that the virtual work by the impressed gravity force is non-positive:

$$\delta W_{\text{unilateral constraint}} \leq 0 \tag{2}$$

This was already pointed out by Fourier (see, e.g., Lanczos[49], pp. 86–87). Of course, when the constraint is replaced by the reaction force, formulation (1) holds again.

Now, there is no objection in utilizing formulation (2) also in the case of bilateral constraints. Indeed, let the generalised forces be given by Q_i, and the virtual displacements by δq_i. Then the principle (2) reads

$$\delta W = \sum_i Q_i \, \delta q_i \leq 0, \; \delta q_i\text{: kinematically admissible virtual displacements} \qquad (3)$$

However, all virtual displacements are reversible, which means that

$$-\delta W = -\sum_i Q_i \, \delta q_i = \sum_i Q_i(-\delta q_i) \leq 0$$

Combination with (3) then yields the form (1).

In the sequel we will use the virtual work postulate in the form (2)–(3).

1.4 *The normal contact problem and the principle of virtual work*

We first write down the principle of virtual work for an elastic body without contact conditions. The problem is

Find u, solution of

$$a_{ijhk}(x)\, \varepsilon_{ij}(u) = \tau_{hk}(u) \qquad \text{in } \Omega \qquad (4)$$

$$\tau_{hk,k}(u) + f_h = \rho u_h'' \qquad \text{in } \Omega \qquad (5)$$

$$\sigma = F, \text{ given} \qquad \text{in } \Gamma_F \qquad (6)$$

$$u = U, \text{ given} \qquad \text{in } \Gamma_u \qquad (7)$$

$$u, u' \text{ satisfy proper initial conditions} \qquad (8)$$

$$\Gamma = \Gamma_u \cup \Gamma_F, \quad \Gamma_u \cap \Gamma_F = \emptyset \qquad (9)$$

This problem is equivalent to the following principle of virtual work:

$$\text{Find } u\text{: } \delta W \equiv -(\rho u'', v-u)_\Omega + (f, v-u)_\Omega - a(u, v-u) + (F, v-u)_{\Gamma_F}$$

$$\leq 0 \; u \in U_{ad}, \; v \in U_{ad}, \; u, u' \text{ satisfy proper initial conditions,} \qquad (10)$$

$$U_{ad} = \{v \,|\, v \text{ vector in } \Omega, \; v = U \text{ on } \Gamma_u\} \qquad (11)$$

For a numerical method of solution for the case of an elastic, work: The class of functions to which u and v belong must also be specified but as we observed in section 1.1 we work only formally and will omit such mentions which are, of course, extremely important. The proper formulation is found in Duvaut-Lions[25], chapter 3.

If the inequality sign in (10) is replaced by an equality, and $(v-u)$ by δu, the classical formulation is recovered. This procedure is justified since $(v-u)$ defined in (11) is indeed reversible.

Now we consider the case that frictionless contact with a rigid support is possible in the subregion Γ_c of Γ. We replace (9) by (9′):

$$\Gamma = \Gamma_u \cup \Gamma_F \cup \Gamma_c, \; \Gamma_u, \Gamma_F, \Gamma_c \text{ disjoint.} \qquad (9')$$

and we add the contact conditions:

$$u_N \leq h(x, t) \qquad \text{in } \Gamma_c \qquad (12)$$

$$\sigma_N \leq 0, \sigma_N(h-u_N) = 0 \qquad \text{in } \Gamma_c \tag{13}$$

$$\sigma_T = 0 \qquad \text{in } \Gamma_c \tag{14}$$

In (12), $h(x, t)$ is the distance of the particle $x \in \Gamma_c$ to the rigid support, measured along the outer normal n at x on Ω. $u = h$ indicates contact, $u_N < h$ indicates a gap between support and body. Thus (12) is a non-penetration condition. $\sigma_N \leq 0$, $\sigma_T = 0$ means that the surface traction is purely compressive, and $\sigma_N(h-u_N) = 0$ signifies that the pressure can only be non-zero when there is contact.

The special problem with $h = 0$ is called the Signorini problem. Its physical significance is that of a perfectly conforming punch with a sharp edge pressed on the body Ω. h not necessarily zero corresponds to a more general normal contact problem.

We will now show that (13) and (14) follow from (12) and the principle of virtual work (10) by simply narrowing U_{ad} down to

$$U_{ad} = \{v \mid v \text{ vector in } \Omega,\ v = U \text{ on } \Gamma_u,\ h - v_N \geq 0 \text{ on } \Gamma_c\} \tag{15}$$

To that end we analyse (10), (15). We integrate the term $-a(u, v-u)$ partially, and find the following expression by utilizing (4)

$$0 \geq \delta W = (\tau_{hk,k} + f_h - \rho u_h'', v_h - u_h)_\Omega - (\sigma, v-u)_{\Gamma_u} + (F - \sigma, v-u)_{\Gamma_F} + \\ - (\sigma_T, v_T - u_T)_{\Gamma_c} - (\sigma_N, v_N - u_N)_{\Gamma_c} \tag{16}$$

sub $u, v \in U_{ad}$ (15); u, u' satisfy proper initial conditions.

All terms but the last contain reversible virtual displacements only; note that (15) does not restrict v_T and u_T, so that by the fundamental lemma of the calculus of variations the equations (5), (6), (14) are satisfied. Equation (7) is automatically satisfied by (15); the form (16) implies that σ is unrestricted in Γ_u.

We obtain for the principle of virtual work:

$$0 \geq -(\sigma_N, v_N - u_N)_{\Gamma_c} = -(\sigma_N, v_N - u_N)_{\Gamma c, u_N < h} - (\sigma_N, v_N - u_N)_{\Gamma c, u_N = h} \tag{17}$$

In the region $\{\Gamma_c, u_N < h\}$ the variation $(v_N - u_N)$ is again reversible, so that there $\sigma_N = 0$. In the region $\{\Gamma_c, u_N = h\}$ we have that $(v_N - u_N) = (v_N - h) \leq 0$. It follows from (17) as in the fundamental lemma that $\sigma_N \leq 0$. So we obtain (13), and the verification is complete.

When in (10) the first (inertial) term vanishes, the problem may be written as a minimum principle:

Find $u \in U_{ad}$ minimizing $I(u) \equiv \frac{1}{2}a(u,u) - (f,u)_\Omega - (F,u)_{\Gamma_F}$

$$U_{ad} = \{v \mid v \text{ vector in } \Omega;\ v = U \text{ on } \Gamma_u,\ h - v_N \geq 0 \text{ in } \Gamma_c\} \tag{18}$$

The classical problem is recovered if the condition '$(h-u_N) \geq 0$ in Γ_c' is omitted and $\Gamma_c = \phi$.

Under certain mild restrictions on F, U, and the elasticity tensor a, Fichera[28] established existence and uniqueness of the static Signorini problem ($h=0$, $u'=u''=0$). When $\Gamma_u=\phi$, i.e. when the force is prescribed over the entire boundary, outside the potential contact area Γ_c, he needs an additional condition on F and f to prevent the body from falling off its support Γ_c.

The minimisation problem may be dualised to yield a complementary principle. We do not embark on details and merely state the result:

$$\underset{\tau K}{\text{maximize}}\ J(\tau) \equiv -\tfrac{1}{2}A(\tau,\tau) + (\sigma, U)_{\Gamma u} + (\sigma_N, h)_{\Gamma c}$$

$$A(\tau,\tau) = \int_\Omega A_{ijhk}\tau_{ij}\tau_{hk}\,\mathrm{d}\Omega,\ A_{ijhk}\tau_{hk} = \varepsilon_{ij}, \tag{19}$$

$$K = \{\tau \mid \tau \text{ tensor in } \Omega;\ \tau_{hk,k} + f_h = 0, \text{ in } \Omega,\ \tau_{hk} = \tau_{kh},$$

$$\sigma = F \text{ on } \Gamma_F,\ \sigma_N \leq 0 \text{ on } \Gamma_c,\ \sigma_T = 0 \text{ on } \Gamma_c\}$$

The classical principle of complementary energy is recovered when $\Gamma_c=\phi$. The complementary principle (19), in combination with the point-load field of Boussinesq-Cerrutti (see Love's book[55], p. 191) was utilized by Kalker and Van Randen[44] to implement numerically the half space normal contact problem.

1.5 *The contact problem with Coulomb friction and the principle of virtual work*

We will consider a problem of elastostatics with friction. We take as frictional law a law corresponding to a shift problem (see Duvaut-Lions[25], chapter 3):

$$\sigma_T(x) = -g(x)\,u_T/|u_N| \text{ in } \Gamma_s = \{x \mid x\in\Gamma_c,\ u_T \neq 0\} \text{ slip area} \tag{20}$$

$$|\sigma_T(x)| \leq g(x) \qquad \text{in } \Gamma_a = \{x \mid x\in\Gamma_c,\ u_T = 0\} \text{ stick area} \tag{21}$$

where

$$g(x) = \mu(x)\,|F_N(x)| \geq 0,\ \mu(x)\text{: coefficient of friction, prescribed in } \Gamma_c, \tag{22}$$

$$\sigma_N(x) = F_N(x) \text{ in } \Gamma_c, \text{ prescribed} \tag{23}$$

The principle of minimum potential energy which is closely connected with the principle of virtual work, is given in (18). We must add a term $(g, |u_T|)_{\Gamma c}$ which represents the potential energy of the shift due to friction. The problem becomes

$$\text{Find } u\in U_{ad} \text{ minimizing } J(u) \equiv \tfrac{1}{2}a(u,u) - (f,u)_\Omega - (F,u)_{\Gamma_F} - \\ - (F_N, u_N)_{\Gamma c} + (g, |u_T|)_{\Gamma c}$$

$$U_{ad} = \{u \mid u \text{ vector in } \Omega;\ u = U \text{ on } \Gamma_u\};\ g = \mu|F_N| \text{ on } \Gamma_c; \\ F,\ F_N \text{ prescribed on } \Gamma_F,\ \Gamma_c \text{ respectively}. \tag{24}$$

We verify whether this problem is actually equivalent to (20), (21), (22), (23). To that end we vary $J(u)$:

$$\delta J(u) = a(u, \delta u) - (f, \delta u)_\Omega - (F, \delta u)_{\Gamma_F} - (F_N, \delta u_N)_{\Gamma_c} + \left(g, \frac{u_T \delta u_T}{|u_T|}\right)_{\Gamma s} +$$

$$+ (g, |\delta u_T|)_{\Gamma a} \quad \text{sub } \delta u = 0 \text{ on } \Gamma_u$$

Partial integration of the first term yields, if τ_{hk} is substituted for $a_{ijhk}\varepsilon_{ij}$, and σ, σ_N for $\tau_{hk}n_k$, $\tau_{hk}n_k n_h$ respectively:

$$\delta J(u) = -(\tau_{hk,k} + f_h, \delta u_h)_\Omega + (\sigma - F, \delta u)_{\Gamma_F} + (\sigma_N - F_N, \delta u_N)_{\Gamma c} +$$

$$+ \left(\sigma_T + g \frac{u_T}{|u_T|} \delta u_T\right)_{\Gamma s} + \int_{\Gamma a} (\sigma_T \delta u_T + g|\delta u_T|)\, \mathrm{d}\Gamma \geq 0$$

δu: $\delta u = 0$ on Γ_u.

From this follows:

$\tau_{hk,k} + f_h = 0$	on Ω (eq. of equilibrium)	
$\sigma = F$	on Γ_F (prescribed surface traction)	
$\sigma_N = F_N$	on Γ_c (prescribed normal traction)	
$\sigma_T = -g u_T/\lvert u_T\rvert$	on Γ_s $(u_T \neq 0)$	Friction
$\sigma_T \delta u_T + g\,\lvert\delta u_T\rvert \geq 0\ \Gamma \delta u_T$	on $\Gamma_a \rightarrow g \geq \lvert\sigma_T\rvert$ on Γ_a	conditions

which completes the verification.

The problem (24) may be dualized to the complementary principle

$$\underset{\tau K}{\text{maximize}} - \tfrac{1}{2} A(\tau, \tau) + (\sigma, U)_{\Gamma u},$$

$$K = \{\tau \mid \tau \text{ tensor in } \Omega;\ \tau_{ij,j} + f_i = 0 \text{ in } \Omega,\ |\sigma_T| \leq g \text{ in } \Gamma_c, \qquad (25)$$

$$\sigma = F \text{ in } \Gamma_F,\ \sigma_N = F_N \text{ in } \Gamma_c\}.$$

Existence and uniqueness of the solution of problem (24) have been established by Duvaut and Lions[25] under proper conditions of regularity on F, a, g, F_N. It is seen that the problem is not a Signorini problem combined with friction: the friction bound g, which is a multiple of the normal load σ_N, is prescribed independently of the frictional field. Existence and uniqueness of the combined, interacting fields constitute an open problem.

The objection may be raised that the friction law of (20), (21) is not at all the true friction law since displacements are acted upon but not velocities. To this it may be answered that for quasistatic problems an incremental form of (24) may be used, viz.

$$\text{Find } w \in U_{ad} \text{ minimizing } J(w) = \tfrac{1}{2} a(u+w, u+w) - (f, w)_\Omega - (F, w)_{\Gamma_F} +$$
$$- (F_N, w_N)_{\Gamma c} + (g, |w_T|)_{\Gamma c},$$

$$U_{ad} = \{w \mid w \text{ vector in } \Omega,\ w = W \text{ on } \Gamma_u\};\ g = \mu|F_N|, \qquad (26)$$

F_N prescribed on Γ_c;

F prescribed on Γ_F; u a fixed vector field in Ω, w: an increment thereon.

Also, for the theory of dynamic frictional contact problems reference is made to Duvaut-Lions[25], who not only treat elastic bodies, but also viscoelastic bodies with simple constitutive relations. These theories are based on the friction law acting on velocities rather than shifts.

PART 2. A SCHEMATIC REVIEW OF THE THEORY

In this second part we shall briefly review the state of the art in the theory of contact mechanics, with special attention to unsolved problems. To that end we made a scheme into which many contact problems fit (see Tables 1–5). The contact problems are divided into those where there is no friction between the contacting bodies (N), and those in which there is friction (F). A special case of the contact problems with friction is that of adhesion, in which the friction is 'infinite' so that no slip occurs in the interface.

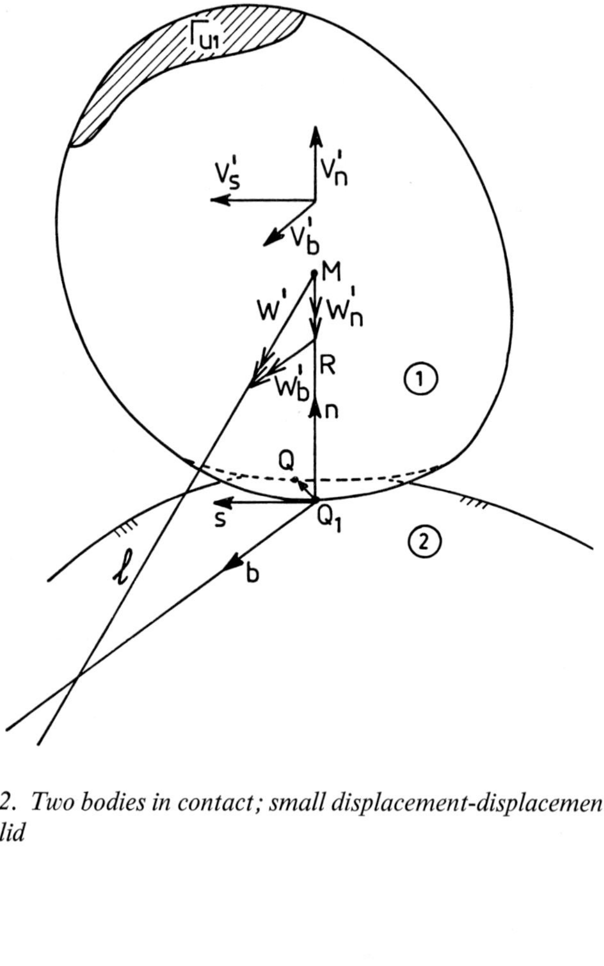

Fig. 2. Two bodies in contact; small displacement-displacement gradient theory is valid

The categories N and F are subdivided into static or quasistatic problems (Q), in which inertial effects are neglected (by far the majority), and into dynamic problems in which inertial effects play a role (D). So we obtain the combinations NQ, ND, FQ, FD. Especially in the case FQ, the quasistatic problems with friction, this subdivision is not specific enough. In order to make a further classification, the possible modes of motion of the bodies are considered.

Consider two bodies 1, 2 in contact which are rigidly supported at the subregions Γ_{ui} ($i = 1, 2$) of their surfaces. The bodies perform a motion to which may be added a rigid motion in such a way that Γ_{u2} remains at rest. At every instant t the motion of Γ_{u1} may be characterised by a rectilinear velocity $v'(t)$ and a vectorial angular velocity $w'(t)$ about an arbitrary point $M(t)$. This point will be chosen in the following way (see Fig. 2).

The bodies touch each other at the time t along the contact area $\Gamma_c(t)$; let $Q(t)$ denote some point of $\Gamma_c(t)$. In the state of deformation which occurs when body 2 is taken away (the reference state of body 1 at time t) Q corresponds to the point $Q_1(t)$ of the surface of body 1. Let n be the unit inner normal to body (1) at Q_1. s is a unit vector orthogonal to n and in the same sense as $w' \times n$, while b is the unit vector $n \times s$.* (n, s, b) thus form a right-handed system of orthonormal vectors. The point $M(t)$ is now defined as the center of curvature of the intersection of the boundary of (1) in the reference state with the (s, n)-plane at the point Q_1, and $R(t)$ is the corresponding radius of curvature which is counted positive when n points at $M(t)$.

The instantaneous motion is now characterised by

$v' = (v'_n, v'_s, v'_b)$	translational velocity;
$w' = (w'_n, 0, w'_b),\ w'_b \geq 0$	vectorial angular velocity about M.

In some cases it may be convenient to work in terms of displacements rather than velocities; then we write

$v = (v_n, v_s, v_b)$	translation;
$w = (w_n, 0, w_b)$	rotation about M.

In the special case of body 1 being a body of revolution, with w' parallel to its axis, $M(t)$ lies on the axis of symmetry. Then (n, s, b) corresponds to the coordinate system which is introduced by several authors.

The following forms of motion are considered. In the first instance we distinguish between steady and non-steady motions. A motion is called steady when the equations governing it may be written in a form into which neither explicit time nor distance traversed enter. This may necessitate the superposition on the system of a non-vanishing motion $\mathscr{M}_2$ of the rigid support Γ_{u2}.

* When $w' = 0$, s is some unit vector orthogonal to n.

Further we need a definition of rolling. Rolling is said to occur when

$$\left.\begin{array}{l} w_b' R = v_s' (1+0(\varepsilon)),\ v_b' = 0(\varepsilon v_s'),\ v_s' \neq 0 \\ \text{No restrictions are placed on } w_n' \text{ nor } v_n' \end{array}\right\} \text{ rolling} \tag{1}$$

$$\left.\begin{array}{l} \varepsilon: \text{a typical strain of the bodies;} \\ \Delta: \text{a typical diameter of the contact area} \end{array}\right\} \tag{2}$$

The first motion we recognize is denoted by '5' in the tables.

a. *Steady rolling* ('5' in Tables). (1) is satisfied, and the motion is steady. All steady motions which are not rolling are steady sliding.

b. *Steady sliding* ('2' in Tables). (1) is not satisfied, and the motion is steady. In terms of surface tractions, steady sliding is characterised by the fact that under Coulomb conditions of friction

$$|\sigma_T| = \mu |\sigma_N|,\ \mu\text{: coefficient of friction} \tag{3}$$

at every point of the contact area. This condition may also be taken as a definition of (not necessarily steady) sliding. In a narrower sense (see Galin [30]) sliding is said to take place when

$$\sigma_T = \mu\sigma_N e,\ e\text{: some constant unit vector orthogonal to } n. \tag{4}$$

The second type of motion we distinguish is the shift. For the description of such a motion use is made of translations and rotations rather than of velocities and angular velocities.

c. *Compression* ('1' in Tables)

$$w = 0,\ v_s = v_b = 0,\ v_n < 0 \tag{5}$$

The Hertz problem falls under this heading.

d. *Impact* ('7' in Tables) is a generalization of compression, in which two bodies are shot at each other, come into contact, and rebound.

$$w' = 0,\ v_s' = v_b' = 0,\ v_n'(t) < 0 \text{ if } t < 0, = 0 \text{ if } t = 0, > 0 \text{ if } t > 0 \tag{6}$$

Hertzian impact is the special case of this motion in which the bodies are elastic half-spaces, the contact is frictionless, and inertial effects are disregarded. An interesting generalisation, as far as I am aware, not treated in the literature, is that

$$\left.\begin{array}{l} w',\ v_s',\ v_b' \text{ unrestricted},\ v_n'(t) < 0 \text{ if } t < 0,\ = 0 \text{ if } t = 0, \\ \qquad > 0 \text{ if } t > 0 \end{array}\right\} \tag{7}$$

Friction in the interface is present

e. *Translation shift* ('4' in Tables) is said to occur when

$$w = 0,\ v_b = 0,\ v_s \neq 0 \tag{8}$$

It becomes sliding in the sense of (4) when $|v_s| \gg \Delta|\varepsilon|$. Translation shift has been treated by Cattaneo[13] under the assumption that $v_n = 0$. Mindlin and Deresiewicz[64] treated the case that v_s and v_n are functions of t. The bodies considered were elastic half-spaces with equal elastic constants.

f. *Rotation shift* ('3' in Tables) is said to occur when

$$v_s = v_b = 0,\ w_b = 0,\ w_n \neq 0 \tag{9}$$

These conditions are usually accompanied by the assumption of rotation symmetry, which entails among other things that Q is the center of the circular contact area. Also, the normal shift will often be taken zero. Such a problem was treated by Lubkin[56].

Rotation shift may become sliding in the sense (3) when $|w_n| \gg |\varepsilon|$; such a problem was treated by Galin[30].

Mindlin[62] treats the combination of e and f under the condition that $v_n = 0$ and that there is full adhesion (no slip in the interface).

g. *Transients* ('6' in Tables). The remaining motions, by far the majority, are termed transients. Transients may be defined by loading and kinematical programmes such as

$$\left.\begin{aligned} v_n' &= A\delta(t-t_1) + BH(t)\cos\omega t; \\ v_s' &= VH(t); \\ w_n' &= 0;\ \acute{w}_b R = V(1+0(\varepsilon)\,H(t)) \end{aligned}\right\} \left\{\begin{array}{l}\text{Compression followed by rolling} \\ \text{with longitudinal creepage} \\ \text{with a time dependent contact} \\ \text{area.}\end{array}\right.$$

$\delta(t)$: Dirac delta function; A, B, V: constants;

$H(t) = 0\ t < 0,\ = 1\ t \geq 0$, Heaviside function.

Such problems have been treated by Kalker[40–43].

After these preliminaries, we study the Tables in detail.

2.1 *The elastic half-plane (see Table 1)*

We begin our survey with the simplest of two-dimensional elastic bodies, viz. the half-plane. Most contributions were made in the period before 1953, when Galin's book, Kontaktnye Zadachi Teorii Uprugosti (Contact Problems of the Theory of Elasticity)[30] appeared. In 1961, a translation of this book was made under the supervision of Sneddon. All kinds of two-dimensional punch problems are treated in the first half of the book. The methods are mostly those presented in Muskhelishvili's works[69,70], which also contain a number of contact problems.

As you are well aware, an unbalanced load on the elastic half-plane is accompanied by an infinite displacement. Of course, we always deal with finite bodies which are approximated locally by the half-plane and the consequence is that the normal approach of the bodies is not found.

Table 1. The elastic half-plane

N – PROBLEMS WITHOUT FRICTION		
NQ – quasistatic problems	*NQ*	All these problems are in principle identical
1. Compression	1	Solutions: see, e.g. Galin
2. Steady-state rolling/sliding	2	Perturbation of Hertz solution for a circle, and in principle for any body: Schwartz-Harper
6. Sliding-rolling transients	6	
7. Impact	7	
ND – dynamic problems	*ND*	
2. Steady-state rolling/sliding	2	Craggs-Roberts
6. Transients	6	OPEN
7. Impact	7	Robertson-Thompson, Persson
F – PROBLEMS WITH FRICTION		
FQ – quasistatic problems	*FQ*	
1. Compression	1	Kalker
2. Steady-state sliding	2	See Galin
3. Rotational shift	3	Inapplicable
4. Translational shift	4	Cattaneo, Mindlin [s]
5. Steady-state rolling	5	Carter [s], Fromm [s], Kalker, Bentall-Johnson
6. Transients	6	Kalker
7. Impact	7	OPEN
FD – dynamic problems	*FD*	
5. Steady state rolling	5	OPEN
7. Impact	7	Persson [a]

s. Symmetric bodies
a. No slip.

Several authors have tried to find ways out of this dilemma, such as Tsu-Tao Loo[94], who considered the circular cylinder. A very elegant way was presented recently by Schwartz and Harper[80], who, in order to find the elastic field in a cylinder, need only the half-plane solution, and the field of a line load on the cylinder: the latter may present a formidable problem when the cylinder is not circular.

ND Since we have a frictionless problem, we only need to consider steady motions, transients, and impact. Steady motions were already considered by Galin[30], where the conclusion was reached that for steel bodies the influence of inertia becomes only noticeable when the velocity exceeds 50 m/sec (180 km/h): a velocity indeed reached by modern high-speed trains. Craggs and Roberts[20] made a study of a cylinder moving over an elastic half-plane and encountered no serious problems up to the Rayleigh velocity. Above that velocity, the problem becomes ill-defined.

Regarding impact, Robertson and Thompson[79] treat the case of a half-plane being indented by a wedge. Only the loading half of impact is considered, and it is assumed that the contact area expands with a constant velocity, which may be below, between and above the propagation velocities of the elastic waves.

Persson[76] considers the case of two plates being shot at each other. He not only considers the frictionless case, but also the case of full adhesion (*FD* 7).

The case of transients is open, but, it would seem, of little interest.

1 *Compression.* When two bodies are pressed together in the presence of friction, not only normal stresses occur in the contact area, but also shear stresses. The correct way of determining those is the 'incremental' argument presented by Goodman in a paper treating the axisymmetric case of a spherical punch on a half-space[32]. The calculation in the two-dimensional case is perfectly straight-forward.

2 For *steady-state sliding* reference is made to Galin[30].

4 *Translation shift.* Although here also the displacement becomes infinite, the stress distribution can be easily found by allowing the axial ratio of the elliptic contact area treated by Cattaneo[13] and Mindlin[62] in their papers to tend to zero. The problem was treated by those authors for symmetric bodies only, that is, for two half-planes with elastic constants. It is well-known (see, e.g. de Pater[74]) that under these conditions the compression does not influence the shear quantities and vice-versa, not only for symmetric half-planes, but for symmetric half-spaces as well.

The asymmetric problem is open. It can be treated by Kalker's method[42] (see *FQ* 6).

5 *Steady rolling* was treated by Carter in 1926[12] and by Fromm in 1927[29]; they considered symmetric bodies. A numerical treatment of the asymmetrical case was given by Bentall and Johnson in 1967[9] and later by Kalker in 1971[42].

6 *Transients.* Transient traction distributions were calculated numerically by Kalker in 1971[42].

7 *Impact* with friction is an open problem.

5 *Steady rolling* with friction and with retention of inertial effects has, as far as I am aware, not been treated correctly in the literature. It may be of interest for high speed trains.

7 One case of *impact* with complete adhesion has been treated by Persson[76] (see *ND*).

We now turn to the elastic half-space.

2.2 *The elastic half-space (see Table 2)*

NQ All quasistatic problems are, in fact, identical to the problem of compression. A number of cases with rotation symmetry may be found in Galin's book[30].

Table 2. The elastic half-space

N – PROBLEM WITHOUT FRICTION			
NQ – quasistatic problems	*NQ*	These problems are in principle identical	
1. Compression	1	Rot. Sym.: Galin	
2. Steady-state rolling/sliding	2	Elliptic C.A.: Hertz, Galin	
		Slender C.A.: Kalker	
6. Sliding-rolling transients	6	General C.A.: Conry-Seireg, Kalker-Van Randen	
7. Impact	7	Conway-Farnham, Singh-Paul	
ND – dynamic problems	*ND*		
2. Steady state rolling/sliding	2	OPEN	
6. Transients	6	OPEN	
7. Impact	7	Rot. Sym: Hunter, Tsai	
F – PROBLEMS WITH FRICTION			
FQ – quasistatic problems	*FQ*		
1. Compression	1	Rot. Sym.: Goodman [a]	
2. Steady-state sliding	2	Elliptical, Circular C.A.: Galin	
3. Rotational shift	3	Rot. Sym.: Galin, Lubkin, Deresiewicz	combined [a, s]: Mindlin (3 and 4)
4. Translational shift	4	Cattaneo [s], Mindlin [s]	
5. Steady-state rolling	5	Kalker [a, s, s]	
6. Transients	6	OPEN	
7. Impact	7	OPEN	
FD – dynamic problems	*FD*		
5. Steady-state rolling	5	OPEN	
7. Impact	7	OPEN	

a. Complete adhesion.
s. Symmetric bodies.

The elliptic contact area was introduced by Hertz in 1881 (see Love[55], p. 191). A number of contact problems with elliptic contact areas may be found in Galin's book.

The slender contact area was treated by Kalker in 1972[43]. It turns out that in slices orthogonal to the center line of the contact area a traction distribution acts which may be found from two-dimensional elasticity theory, but for the determination of the contact width, total force, depth of penetration, which all vary in the axial direction, an integral equation must be solved. Not only normal stresses are considered, but also shear stresses.

For contact areas of unrestricted form, recourse must be had to the numerical methods of Conry-Seireg[16], Kalker-Van Randen[44], Conway-Farnham[17], or Singh-Paul[83].

Only *impact* has been considered in the literature; see Hunter[35, 36] and, more recently, Tsai[91–93]. Hunter finds that for impact velocities, small compared with the propagation velocities of elastic waves in the specimen, a negligible proportion of the original kinetic energy of the particle is transferred to the specimen by collision. Tsai relates the field quantities to the contact time, and finds that in the axisymmetric case the maximum contact radius is given by Hertz's quasistatic theory with great precision; but the contact radius shortly after the onset of contact may differ considerably from Hertz's prediction.

1 *Compression* of a half-space by a rough sphere in the presence of complete adhesion was studied by Goodman[32], who was the first to formulate the correct boundary conditions for this problem. The problem without axisymmetry is open.

2 *Steady sliding*. For an elliptic and a circular contact area reference is made to Galin[30].

3 *Rotation shift* was treated by Lubkin[56] for partial slip in the contact area, and by Galin[30] for complete slip. Axisymmetry was assumed by these authors. A curious circumstance facilitates the calculation, viz. that the compression does not influence the rotation shift and vice versa, even when the bodies are not symmetric. Deresiewicz[21] considered rotation shift transients.

4 *Translation shift* was considered by Cattaneo[13] and Mindlin (1949)[62], for symmetric bodies and a Hertzian normal pressure distribution. Mindlin et al. (1952)[63] generalized Cattaneo's result by considering shift transients, i.e. $w' = 0$, $v_b' = 0$, $v_n' = 0$ and v_s' is arbitrary. Mindlin and Deresiewicz[64] removed the restriction $v_n' = 0$ in 1953 and thus came very close to the impact problem with friction. The case of combined rotation and translation shift was treated by Mindlin in 1949 for symmetric, Hertzian normal contact and full adhesion.

5 *Steady rolling* was considered by Kalker[40] for symmetric bodies both with full adhesion and with finite friction. Hertzian normal contact is assumed. The method is numerical. The complete adhesion results are available in a one-page table which covers all possible cases; see e.g. Kalker[41].

6, 7 Neither frictional impact nor other transients have been treated in the literature; but see the work of Mindlin-Deresiewicz (1953)[64] under *FQ* 4.

No dynamical problems with friction have been treated.

2.3 *The elastic layer (see Table 3)*

NQ All previous results start from the hypothesis that the dimensions of the contact area are small with respect to the dimensions of the body, or that the body may be thought of as being composed of slices in which the elastic field is identical. Here we shall consider the simplest geometry in which the contact width is comparable to the depth of the body, viz. the elastic layer. Contact problems of the elastic layer began to be considered at a comparatively recent date; no trace of the layer problem is found in Galin's book (1953)[30]. The rotation symmetric case, for instance, was solved by Lebedev and Ufliand in 1958[50]. The *NQ* problems are again all identical.

Table 3. The elastic layer

N – PROBLEM WITHOUT FRICTION		
NQ – quasistatic problems	*NQ*	The problems *NQ* are all in principle identical
1. Compression	1	Aleksandrov, Alblas-Kuipers: parabolic punch [p]
2. Steady-state rolling/sliding	2	Rot-Sym.: Ledebev-Ufliand
6. Sliding-rolling transients	6	
7. Impact	7	
ND – dynamic problems	*ND*	
2. Steady-state rolling/sliding	2.	Keer-Sve [p]
6. Transients	6	OPEN
7. Impact	7	Tsai [s]
F – PROBLEMS WITH FRICTION		
FQ – quasistatic problems	*FQ*	
1. Compression	1	Engel-Conway [p]
2. Steady-state sliding	2	Alblas-Kuipers [p]
3. Ratational shift	3	OPEN
4. Translational shift	4	Goodman-Keer [s]
5. Steady-state rolling	5	Bentall-Johnson [f, p]
6. Transients	6	OPEN
7. Impact	7	OPEN
FD – dynamic problems	*FD*	
5. Steady-state rolling	5	OPEN
7. Impact	7	OPEN

p. Plane problem.
f. No net tangential force transmitted.
s. Presented in this Symposium.

The plane problem of a stamp with a parabolic base was treated by Aleksandrov in 1962[5], who solved the problem in the asymptotic case that the layer is thick. Meyers[60], and Alblas and Kuipers (1970)[2,3] solved the problem of the thin layer. Alblas and Kuipers also consider punches with a plane base (1969–1970)[1–3]. As was said before, the case of rotation symmetry was solved by Lebedev and Ufliand[50].

2 *Dynamic problems.* Keer and Sve[46] first solved the plane dynamic problem of a single punch steadily moving without friction over an elastic layer, and later extended their analysis to a periodic array of punches. In the present Symposium, Tsai treated the case of dynamical impact (see pp. 206).

1 Regarding problems with friction, Engel and Conway[26] considered the two-dimensional case that a layer is indented by a cylinder in the presence of adhesion. The finiteness of the cylinder is taken into account. The analysis is numerical.

2 Alblas and Kuipers (1971)[4] considered the two-dimensional case that a rough punch slides slowly over the layer. Both thin layer and thick layer asymptotic solutions are established.

All other frictional layer problems – shift, rolling, impact, either dynamical or quasistatic, are open.

4 In the present Symposium, Goodman and Keer treated the symmetric translation shift (see pp. 127).

5 In 1968, Bentall and Johnson[10] considered plane rolling contact without a net tangential force. They found that up to five slip areas may form which are separated by four locked regions.

2.4 *Viscoelastic contact problems (see Table 4)*

A linear viscoelastic material is a material with linear stress-strain relations which differ from elastic materials in the respect that they are time- and rate-dependent. Viscoelastic materials are closely akin to elastic materials, and it will thus cause no surprise that existence and uniqueness of the solution of problems in viscoelastic contact mechanics can be established. This has indeed been done by Duvaut for the Signorini problem in 1969[24], and for the problem with friction reference is made to the book by Duvaut and Lions[25] for possibly inhomogeneous and anisotropic materials which are, as far as viscoelasticity is concerned, of a fairly simple nature.

Owing to the time-dependence of the stress-strain law in viscoelasticity we must distinguish between the problems of compression and of rolling-sliding even when friction is absent.

1 The field of viscoelastic contact problems was opened by Lee and Radok in 1960[51] when they considered the frictionless contact of a punch with a viscoelastic half-space. They considered only the case that the contact area

Table 4. Viscoelastic contact problems

N – PROBLEMS WITHOUT FRICTION		
NQ – quasistatic problems	*NQ*	
1. Compression	1	Lee, Graham, Ting (3–dimensional)
2. Steady-state rolling/sliding	2	Hunter, Morland: half-plane; Morland: cylinders; Margetson: layer; Alblas: thin layer (all 2-dimensional)
6. Sliding-rolling transients	6	OPEN
7. Impact	7	Kawatate [s]
ND – dynamic problems	*ND*	OPEN
2. Steady-state rolling/sliding		
6. Transients		
7. Impact		
F – PROBLEMS WITH FRICTION		
FQ – quasistatic problems	*F*	OPEN
1. Compression		
2. Steady-state sliding		
3. Rotational shift		
.		
.		
.		
.		
.		
7		

s. Presented in this Symposium.

was growing. Later, Graham, and, independently, Ting[88,89] extended those results to the case that the contact area could both grow and decrease, but only so that the time-dependent contact area $\Gamma_c(t)$ was monotonic in the sense that at any two times t_1 and t_2 ($t_1 > t_1$) either $\Gamma_c(t_1) \subset \Gamma_c(t_2)$ or $\Gamma_c(t_2) \subset \Gamma_c(t_1)$. Then the contact area $\Gamma_c(t)$ appears to coincide with a contact area of the bodies if they had been elastic rather than viscoelastic. These results, which hold for many geometries, were implemented for, e.g., the Hertz problem in viscoelasticity with elliptic contact area.

NQ 2 A second problem which is also simple in nature is that of steady-state rolling-sliding without friction. The technical interest is here to find the resistance to rolling. So far, only the plane problem has been considered. Hunter[36] calculated the solution of the problem of a rigid cylinder rolling over a standard linear half-space in 1961. His results were extended by Morland in 1962 to a rigid cylinder rolling over a viscoelastic half-space with more general constitutive relations[66], and in 1967 Morland considered the case of two finite, identical cylinders rolling over each other[67].

Finally in 1968 Morland reduced the problem of two dissimilar viscoelastic cylinders rolling over each other without friction to the symmetric case[68].

Solutions have also been given for the problem of a rigid cylinder rolling over a viscoelastic layer (Tsai[91–93], Alblas-Kuipers[1–4], Margetson[57]. This is the state of the art in viscoelastic contact mechanics.* Only quasistatic, frictionless contact problems have been considered.

It is seen that a great many problems await solution. We name a few.

1. The introduction of friction in viscoelastic problems would be interesting. One could start with steady-state sliding, or with full adhesion compression, and lead up to steady-state rolling.
2. Rolling-sliding for three-dimensional bodies, quasistatic, without friction. It would seem that Kalker's line contact theory[43] can be used for the purpose of obtaining the solution for a slender contact area.
3. Two-dimensional steady-staie rolling without friction, but with inertia effects would likewise be interesting.

2.5 *Plastic contact problems (see Table 5)*

No proofs of existence-uniqueness of the solutions to plastic contact problems are known to me.

1 Various punch problems have been solved; reference is made to Ling[53]. These problems are all for rigid-perfectly plastic material mostly with two-dimensional geometries, which are solved with the slip line method. Since under those circumstances the plastic deformation is confined to a region near the surface of the rigid-perfectly plastic body, these bodies need not be infinitely deep in order that the 'half-plane' solution holds. Also, in many cases the occurrence of friction does not pose great problems. For a numerical method of solution for the case of an elastic, work-hardening half-space which is indented by a punch reference is made to the paper by Dumas-Baronet[23], while Johnson (1970)[37] has found that indentation pressures in a half-space by variously formed indenters all correlate with the parameter $\{(E/Y)\tan\beta\}$, where E is Young's modulus, G is the yield stress in compression, and β is the angle of inclination of the indenter with respect to the unstressed half-space surface at the edge of the indentation.

2 It is well-known that e/en when there is no outside force or moment acting on a roller rolling over a flat surface these will be dissipation of energy in the rolling bodies which will eventually result in stopping the roller.

This rolling friction cannot be explained by the theory of symmetric elastic bodies rolling over each other. In order to explain it, Tabor postulated in 1955 that the local rate of elastic energy is accompanied by a dissipation of power proportional to this rate[86]. Later, in 1963, Merwin and

* In the present Symposium, Kawatate treated the case of quasistatic impact both without friction and with complete adhesion (see pp. 221).

Table 5. Plastic contact problems

N – PROBLEMS WITHOUT FRICTION	
NQ – quasistatic problems	*NQ*
1. Compression	1 Various punch prob.: see Ling [i], Johnson [e], Dumas-Baronet [e]
2. Steady-state rolling/sliding	2 Merwin-Johnson [p, e], Johnson [p, e, i]
6. Sliding-rolling transients	6 OPEN
7. Impact	7 Conway-Lee-Bayer [e]
ND – dynamic problems	*ND*
2. Steady-state rolling/sliding	2 OPEN
6. Transients	6 OPEN
7. Impact	7 Tsai
F – PROBLEMS WITH FRICTION	
FQ – quasistatic problems	*FQ*
1. Compression	1 See Ling [a]
2. Steady-state sliding	2 see *FQ* 5
3. Rotational shift	3 OPEN
4. Translational shift	4 OPEN
5. Steady-state rolling	5 Collins [i, p]
6. Transients	6 OPEN
7. Impact	7 OPEN
FD – dynamic problems	*FD*
5. Steady-state rolling	5 OPEN
7. Impact	7 OPEN

p. Plane problem.
i. Rigid, perfectly plastic.
e. Elastoplastic.
a. No slip in interface.

Johnson considered rolling loads on a two-dimensional half-space[59] which were so large that around Bielayev's point of maximum shear stress under the surface a plastic region forms. The loads were assumed to move repeatedly over the half-space. They found that with a rolling load below a certain limit shake-down took place, after which there was no more plastic deformation. Above the shake-down limit, which appeared to be largely independent of the yield criterium, the total plastic deformation increases with each passage of the load.

Whereas Johnson's shake-down theory seems to be quite sound, Johnson himself expresses doubt in his 1972 paper[38] that this theory is also valid for much higher contact loads: in fact, he contact problem in the load region between the shake-down load and the onset of full plasticity may be regarded as open; it can perhaps be solved by means of a finite element technique such as Dumas and Baronet utilize[23].

2, 5 When the load becomes so high that plastic flow occurs at the surface of the half-space, the ideally plastic theory of Collins becomes applicable[14]. Collins considers the two-dimensional rolling contact of a cyinder rolling over a thick layer (= half-space) under all possible loading conditions including surface friction. The simplification mentioned in the title of his paper consists of the quite natural assumption that the contact area may be approximated by a straight line.

Johnson, in a recent paper (1972)[38] correlated the three theories of rolling friction which were mentioned above.

7 Quasistatic impact was considered by Conway, Lee and Bayer[18], and
7 Tsai considered axisymmetric impact with inertial effects taken into account[93].

Open problems are for instance:

1. Rolling with loads between shake-down and full plastic deformation.
2. Steady-state sliding-rolling with inertial effects taken into account.
3. Transients between two steady-states of rolling-sliding considered by Collins[14].

2.6 *Other problems*

Much recent work is on elastic contact problems for bodies with geometries which are neither half-spaces nor elastic layers, especially by Soviet scientists. In this connection we think of the work of e.g. Bondareva on elastic spheres[11]; of Parlas and Michalopoulos on the axisymmetric contact problem for a half-space with a cylindrical hole[73]; of the work done on the Melan problem of a half-space on which elastic stiffeners are welded, by Arutiunian and Mkhitarian[6], whose paper contains numerous references.

Further, a number of papers have appeared on the subject of contact problems involving shells, e.g. Updike and Kalnins[96].

Problems of this kind may lead to the contact analysis of automobile tyres.

Also, mention should be made of thermoelastic problems, to which a session of this symposium is dedicated, of the very interesting work on asperity contact and the work on edge effects. These last three fields have been recently reviewed in the book by Ling on *Surface Mechanics*[53].

Finally, no work at all seems to have been done on the subject of contact problems for bodies made of materials permitting large deformations. The first task at hand seems to be the provision of existence and uniqueness theorems for the Signorini problem and the problem of friction by means of the variational theory, a problem mentioned in the book by Duvaut and Lions[25].

REFERENCES

1. Alblas, J.B. and Kuipers, M.; 'Contact Problems of a Rectangular Block on an Elastic Layer of Finite Thickness', *Acta Mech.*, **8**, 1969, pp. 133–145; *ibid.*, **9**, 1970, pp. 1–12.
2. Alblas, J.B. and Kuipers, M.; 'The Contact Problem of a Rigid Cylinder Rolling over a Thin Viscoelastic Layer', *Int. J. Engng. Sci.*, **8**, 1970, pp. 363–380.
3. Alblas, J.B. and Kuipers, M.; 'On the Two-Dimensional Problem of a Cylindrical Stamp Pressed into a Thin Elastic Layer', *Acta Mech.*, **9**, 1970, pp. 292–311.
4. Alblas, J.B. and Kuipers, M.; 'The Two-Dimensional Contact Problem of a Rough Stamp Sliding Slowly on an Elastic Layer', *Int. J. Solids, Structures*, **7**, 1971, pp. 99–109, and pp. 225–237.
5. Aleksandrov, 'On the Approximate Solution of a Type of Integral equation', *PMM*, **26**, 1962, pp. 934–943.
6. Arutiunian, N.Kh. and Mkhitarian, S.M., 'Periodic Contact Problem for a Half-Plane with Elastic Laps', *PMM*, **33**, 1969, pp. 813–843.
7. Baronet, C.N., see G. Dumas [23].
8. Bayer, R.G., see H.D. Conway [18].
9. Bentall, R.H. and Johnson, K.L., 'Slip in the Rolling Contact of Two Dissimilar Elastic Rollers', *Int. J. Mech. Sci.*, **9**, 1967, pp. 389–404.
10. Bentall, R.H. and Johnson, K.L., 'An Elastic Strip in Plane Rolling Contact', *Int. J. Mech. Sci.*, **10**, 1968, pp. 637–663.
11. Bondareva, V.F., 'Contact Problems for the Elastic Sphere', *PMM*, **35**, 1971, pp. 62–70.
12. Carter, F.C, 'On the Action of a Locomotive Driving Wheel', *Proc. Roy. Soc.*, *A***112**, 1926, pp. 151–157.
13. Cattaneo, C., 'Sul Contratto di Due Corpi Elastici: Distribuzione Locale degli Sforzi', *Rend. Accad. Lincei*, Ser. 6, **27**, 1938, pp. 342–348, 434–436, 474–478.
14. Collins, I.F., 'A Simplified Analysis of the Rolling of a Cylinder on a Rigid, Perfectly Plastic Half-Space', *Int. J. Mech. Sci,*. **14**, 1972, pp. 1–14.
15. Colville, A.R., 'A Comparative Study on Non-Linear Programming Codes', IBM New York Scientific Center Report No. 320–2949, June 1968.
16. Cony, T.F. and Seireg, A., 'A Mathematical Programming Method for Design of Elastic Bodies in Contact', *J. Appl. Mech.*, **38**, 1971, pp. 387–392.
17. Conway, H.D. and Farnham, K.A., 'The Relationship between Load and Penetration for a Rigid, Flat-Ended Punch of Arbitrary Cross-Section', *Int. J. Engng. Sci.*, **6**, 1968, pp. 489–496.
18. Conway, H.D., Lee, H.C. and Bayer, R.G., 'The Impact between a Rigid Sphere and a Thin Layer', *J. Appl. Mech.*, **37**, 1970, pp. 159–162.
19. Conway, H.D., see also P.A. Engel [26].
20. Craggs, J.W. and Roberts, A.M., 'On the Motion of a Heavy Cylinder over the Surface of an Elastic Solid', *J. Appl. Mech.*, **34**, 1967, pp. 207–209.
21. Deresiewicz, H., 'Contact of Elastic Spheres under an Oscillating Torsional Couple', *J. Appl. Mech.*, **21**, 1954, pp. 52–56.
22. Deresiewicz, H., see also R.D. Mindlin [64].
23. Dumas, G. and Baronet, C.N., 'Elastoplastic Indentation of a Half-Space by an Infinitely Long Rigid Circular Cylinder', *Int. J. Mech. Sci.*, **13**, 1971, pp. 519–530.
24. Duvaut, G., 'Le Problème de Signorini en Viscoélasticité Linéaire', *C.R. Acad. Sci. Paris*, **268**, 1969, pp. 1044–1046.
25. Duvaut, G. and Lions, J.L., *Les Inéquations en Mécanique et en Physique*, Dunod, Paris, 1972.
26. Engel, P.A. and Conway, H.D., 'Contact Stress Analysis for an Elastic Cylinder Indenting a Slab in the Presence of Friction', *Int. J. Mech. Sci.*, **13**, 1971, pp. 391–402.
27. Farnham, K.A., see H.D. Conway [17].

28. Fichera, G., 'Problemi Elastostatici con Vincoli Unilaterali : il Problema di Signorini con Ambigue Condizioni al Contorno', *Mem. Accad. Naz. Lincei*, Ser. 8, **7**, 1964, pp. 91–140.
29. Fromm, H., 'Calculation of the Slipping in the Case of Rolling Deformable Bars,' *ZAMM V7N1* (in German), 1927.
30. Galin, L.A., *Kontaktnye Zadachi Teorii Uprugosti* (In Russian), Gostekhizdat, 1953. Transl. by Mrs. H. Moss [I.N. Sneddon (ed.)], *Contact Problems in the Theory of Elasticity*, North Carolina State College, Raleigh (N.C.), 1961.
31. Glowinski, R., Lions, J.L. and Tremolières, R., *Approximation Numérique des Solutions des Inéquations en Mécanique et en Physique*, Dunod, Paris (in preparation).
32. Goodman, L.E., 'Contact Stress Analysis of Normally Loaded Rough Spheres', *J. Appl. Mech.*, **29**, 1962, pp. 515–522.
33. Harper, E.Y., see J. Schwartz [80].
34. Hertz, H., see A.E.H. Love's book [55], pp. 193–208.
35. Hunter, S.C., 'Energy Absorbed by Elastic Waves during Impact', *J. Mech. Phys. Solids*, **5**, 1957, pp. 162**8**171.
36. Hunter, S.C., 'The Rolling Contact of a Rigid Cylinder with a Viscoelastic Half-Space', *J. Appl. Mech.*, **28**, 1961, pp. 611-617.
37. Johnson, K.L., 'The Correlation of Indentation Experiments', *J. Mech. Phys. Solids*, **18**, 1970, pp. 119–120.
38. Johnson, K.L., 'Rolling Resistance of a Rigid Cylinder on an Elastic ,Plastic Surface'. *Int. J. Mech. Sci.*, **14**, 1972, pp. 145–148.
39. Johnson, K.L., see also R.H. Bentall [10]; J.E. Merwin [59].
40. Kalker, J.J., 'Rolling with Slip and Spin in the Presence of Dry Friction', *Wear*, **9**, 1966, pp. 20–38.
41. Kalker, J.J., 'The Tangential Force Transmitted by Two Elastic Bodies Rolling over Each Other with Pure Creepage', *Wear*, **11**, 1968, pp. 421–430.
42. Kalker, J.J., 'A Minimum Principle for the Law of Dry Friction with Application to Elastic Cylinders in Rolling Contact', *J. Appl. Mech.*, **38**, 1971, pp. 875–887.
43. Kalker, J.J. 'On Elastic Line Contact', *J. Appl. Mech.*, **39**, 1972, pp. 1125–1132.
44. Kalker, J.J. and van Randen, Y., 'A Minimum Principle for Frictionless Elastic Contact with Application to Non-Hertzian Half-Space Contact Problems', *J. Engng. Math.*, **6**, 1972, pp. 193–206.
45. Kalnins, A., see D.P. Updike [96].
46. Keer, L.M. and Sve, C., 'Indentation of an Elastic Layer by an Array of Punches Moving with Steady Velocity', *J. Appl. Mech.*, **38**, 1971, pp. 92–98.
47. Kowalik, J. and Osborne, M.R., *Methods for Unconstrained Optimization Methods*, American Elsevier, New York, 1968.
48. Kuipers, M., see J.B. Alblas [1–4].
49. Lanczos, C., *The Variational Principles of Mechanics*, 4th ed., Univ. of Toronto, 1970.
50. Lebedev, N.N. and Ufliand, Ia.S., 'Axisymmetric Contact Problem for an Elastic Layer', *PMM*, **22**, 1958, pp. 320–326.
51. Lee, E.H. and Radok, J.R.M., 'The Contact Problem for Viscoelastic Bodies', *J. Appl. Mech.*, **27**, 1960, pp. 438–444.
52. Lee, H.C., see H.D. Conway [18].
53. Ling, F.F., *Surface Mechanics*, Wiley and Sons, New York, 1973.
54. Lions, J.L., see G. Duvaut [25]; R. Glowinski [31].
55. Love, A.E.H., *A Treatise on the Mathematical Theory of Elasticity*, 4th ed., Cambridge University Press, 1952.
56. Lubkin, J.L., 'The Torsion of Elastic Spheres in Contact', *J. Appl. Mech.*, **18**, 1951, pp. 183–186.
57. Margetson, J., 'Rolling Contact of a Smooth, Viscoelastic Strip between Rotating, Rigid Cylinders;, *Int. J. Mech. Sci.*, **13**, 1971, pp. 207–215.
58. Mason, W.P., see R.D. Mindlin [63].

59. Merwin, J.E. and Johnson, K.L., 'An Analysis of Plastic Deformation in Rolling Contact', *Proc. Inst. Mech. Engrs.*, **177**, 1963, pp. 676ff.
60. Meyers, P., 'The Contact Problem of a Rigid Cylinder on an Elastic Layer', *Appl. Sci. Res.*, **18**, 1968, pp. 353–383.
61. Michalopoulos, C.D., see S.C. Parlas [73].
62. Mindlin, R.D., 'Compliance of Elastic Bodies in Contact', *J. Appl. Mech.*, **16**, 1949, pp. 249–268.
63. Mindlin, R.D., Mason, W.P. Osmar, T.F. and Deresiewicz, H., 'Effects of an Oscillating Tangential Force on the Contact Surfaces of Elastic Spheres', *Proc. 1st Nat. Congr. Appl. Mech.*, USA, 1952, pp. 203–207.
64. Mindlin, R.D. and Deresiewicz, H., 'Elastic Spheres in Contact under Varying Oblique Forces', *J. Appl. Mech.*, **20**, 1953, pp. 327–344.
65. Mkhitarian, S.M., see N.Kh. Arutiunian [6].
66. Morland, L.W., 'A Plane Problem of Rolling Contact in Linear Viscoelasticity Theory', *J. Appl. Mech.*, 29, 1962, pp. 345–361.
67. Morland, L.W., 'Exact Solutions for Rolling Contact between Viscoelastic Cylinders', *Q. J. Mech. Appl. Math.*, **20**, 1967, pp. 74ff.
68. Morland, L.W., 'Rolling Contact between Dissimilar Viscoelastic Cylinders', *Q. Appl. Math.*, **25**, 1967–1968, pp. 363–376.
69. Muskhelishvili, N.I., *Singular Integral Equations* (Transl. by J.R.M. Radok), Noordhoff, Groningen, 1953.
70. Muskhelishvili, N.I., *Some Basic Problems of the Mathematical Theory of Elasticity* (Transl. by J.R.M. Radok), Noordhoff, Groningen, 1953.
71. Osborne, M.R., see J. Kowalik [47].
72. Osmar, T.F., see R.D. Mindlin [63].
73. Parlas, S.C. and Michalopoulos, C.D., 'Axisymmetric Contact Problem for an Elastic Half-Space with a Cylindrical Hole;, *Int. J. Engng. Sci.*, **10**, 1972, pp. 699–707.
74. de Pater, A.D., 'On the Reciprocal Pressure between Two Elastic Bodies in Contact', in J.B. Bidwell (ed.), *Rolling Contact Phenomena*, Elsevier, Amsterdam, 1964, p. 33.
75. Paul, B., see K.P. Singh [83].
76. Persson, Aa., 'Shock pressure in Oblique Impact – a Theoretical Analysis', *J. Appl. Mech.*, **41**, 1974, pp. 124–130.
77. Radok, J.R.M., see E.H. Lee [51]; N.I. Muskhelishvili [69, 70].
78. Randen, Y. van, see J.J. Kalker [44].
79. Robertson, A.R. and Thompson, J.C., 'Transient Stresses in an Elastic Half-Space Resulting from the Frictionless Indentation of Rigid Wedge-Shaped Die', *ZAMM*, **54**, 1974, pp. 139–144.
80. Schwartz, J. and Harper, E.Y., 'On the Relative Approach of Two-Dimensional Elastic Bodies in Contact', *Int. J. Solids, Structures*, **7**, 1971, pp. 1613–1626.
81. Seireg, A., see T.F. Conry [16].
82. Signorini, A., 'Questioni di Elastostatica Linearizzata e Semilinearizzata', *Rend. di Matem. e delle Sur Appl.*, **18**, 1959.
83. Singh, K.P. and Paul, B., 'Numerical Solution of Non-Hertzian Elastic Contact Problems', *J. Appl. Mech.*, **41**, 1974, pp. 484–490.
84. Sneddon, I.N., see L.A. Galin [30].
85. Sve, C., see L.M. Keer [46].
86. Tabor, D., 'The Mechanism of Rolling Friction; II. The Elastic Range', *Proc. Roy. Soc. London*, Ser. A., **229** 1955, pp. 198–220.
87. Thompson, J.C., see A.R. Robertson [79].
88. Ting, T.C.T., 'The Contact Stresses between a Rigid Indenter and a Viscoelastic Half-Space', *J. Appl. Mech.*, **33**, 1966, pp. 845-854.
89. Ting, T.C.T., 'Contact Problems in the Linear Theory of Viscoelasticity', *J. Appl. Mech.*, **35**, 1968, pp. 248–254.
90. Tremolières, R., see R. Glowinski [31].

91. Tsai, Y. M., 'Stress Waves Produced by Impact on the Surface of a Plastic Medium', *J. Franklin Inst.*, **285**, 1968, pp. 204–221.
92. Tsai, Y. M., 'Stress Distribution in Elastic and Viscoelastic Plates Subjected to Symmetrical Rigid Indentations', *Q. App. Match.*, **27**, 1969, pp. 371–380.
93. Tsai, Y. M., 'Dynamic Contact Stresses Produced by the Impact of an Axisymmetric Projectile on an Elastic Half-Space', *Int. J. Solids, Structures*, **7**, 1971, pp. 543–448.
94. Tsu-Tao Loo, 'Effect of Curvature on the Hertz Theory for Two Circular Cylinders in Contact', *J. Appl. Mech.*, **25**, 1958, pp. 122–124.
95. Ufliand, Ia. S., see N. N. Lebedev [50].
96. Uptike, D. P. and Kalnins, A., 'Axisymmetric Behaviour of an Elastic, Spherical Shell Compressed between Rigid Plates', *J. Appl. Mech.*, **37**, 1970, pp. 635–640.

Non-Hertzian contact of elastic spheres

K. L. Johnson

1. INTRODUCTION

The Hertz theory of elastic contact has stood the test of time with remarkable endurance. In addition to being a landmark in applied mechanics, it remains the primary tool of the engineer and physicist in predicting the area of contact and stresses between elastic bodies pressed together. Deviations from the Hertz theory are generally small; nevertheless in special circumstances physical conditions which were ignored by Hertz can be important. Some extensions of the theory to non-Hertzian conditions will be reviewed in this paper.

As a starting point we note the relevant restrictions in the Hertz theory:

1. the bodies are homogeneous and isotropic;
2. their surfaces are topographically smooth and continuous;
3. their profiles may be represented by a second order surface;
4. the stresses and displacements may be deduced from the small-strain theory of elasticity applied to a linear elastic half-space;
5. the surfaces are frictionless;
6. the surface tractions are due to contact forces only – adhesive forces are ignored.

It is usual to justify assumptions 3 and 4 by requiring that the size of the contact patch be small compared with the radii of curvature of the bodies.

We shall now consider modifications to the theory which are necessary if some of these restrictions are relaxed.

Firstly, we note in passing that attempts have appeared in the literature to include terms higher than second order in the representation of the profiles of contacting spheres, whilst retaining small-strain theory applied to a half-space. This procedure is generally unsatisfactory since it introduces corrections which are of the same order of magnitude as the errors incurred in half-space small-strain theory. A notable exception occurs in the case of a sphere in contact with a hollow spherical cavity which is *closely conforming*. In this special case it is justifiable to use small-strain theory, provided that the 'half-space approximation' is abandoned and the stresses and deformations are found by methods appropriate to a sphere and a spherical cavity. Such an analysis has been carried out by Goodman and Keer[1].

2. INTERFACIAL FRICTION

Real surfaces are never frictionless, so that an obvious practical extension of the Hertz theory is to remove restriction 5 and to consider the influence of friction at the contact interface. In the normal contact of spheres friction introduces a first order correction only when the two bodies have dissimilar elastic constants (whether or not their radii are equal). Under the action of their mutual contact pressure the two surfaces undergo radially inward tangential displacements whose magnitudes, in the absence of friction, would be proportional to their respective values of the quantity $(1-2\nu)/G$, where G is the shear modulus and ν is Poisson's ratio. With dissimilar materials the displacements will be different for the two surfaces and this 'slip' will be resisted by interfacial friction, which will act radially outwards on the more compliant surface and inwards on the more rigid one.

In principle slip could be prevented entirely by a sufficiently high coefficient of friction μ at the interface. This is the fully 'adhesive' or 'no-slip' solution. At the other extreme, if the coefficient of friction is vanishingly small, slip will take place over the whole contact area and the tangential traction will be radial and equal to $\pm\mu p$ everywhere (where $p=$ normal pressure). This is the 'complete slip' solution. The true state of affairs comprises a central circular region of adhesion surrounded by an annulus of slip; this case will be referred to as one of 'partial slip'.

The behaviour is governed by the two non-dimensional parameters:

$$\kappa \equiv \frac{[(1-2\nu_1)/G_1]-[(1-2\nu_2)/G_2]}{[(1+\nu_1)/G_1]+[(1+\nu_2)/G_2]}, \text{ and } \mu$$

These problems were first studied by the Russian school for the case of a flat-ended rigid punch pressed into an elastic half-space (Mossakovski[2]). In that case the contact area does not change with increase in load, but during the contact of elastic spheres the contact grows with load such that tangential displacements of points on the two surfaces, which are unrestricted whilst they lie outside the contact, become 'frozen' when they are overtaken by the expanding contact area. This feature of the problem led Goodman[3] to seek a solution by a method in which the load and contact area were allowed to grow incrementally. In his analysis Goodman assumed complete adhesion and neglected the (small) influence of the tangential traction upon the Hertz distribution of normal pressure. A complete solution was found incrementally by Mossakovski[4].

From the self-similarity of solutions for any value of the contact radius a, Spence[5] was able to show that the relative tangential displacements of points within a region of no-slip must vary in direct proportion to r^2, from which he obtained Mossakovski's result directly.

These 'no-slip' solutions exhibit an infinite frictional traction at the edge of the contact area which must, in reality, be relieved by slip.

The true situation, in which a central region of no-slip is surrounded by an annulus of slip, has been studied by Spence[6, 7]. He has shown that the radius c of the adhesion region is the same for all bodies whose profiles are of polynomial form, including a flat-ended rigid punch. Having specified the value of c the frictional traction and the normal pressure were evaluated numerically.

Spence's solution applies to monotonic loading. If, at any point, the load is reduced, it follows that slip must immediately take place at the edge of the contact circle in the opposite sense to that during loading, otherwise an infinite tangential traction would develop. Thus the state of contact stress depends upon the complete history of loading. To the writer's knowledge, this aspect of the problem has not been investigated.

For practical values of the elasticity parameter κ (< 0.4) and the coefficient of friction μ (< 0.5) the influence of frictional tractions upon the compliance and load bearing capacity of most contacting bodies is small, so that the use of the Hertz theory (neglecting friction) is justifiable. An exception arises in the case of brittle solids which fracture as a result of the radial tensile stress surrounding the contact area. This component of stress in the absence of friction is given by

$$\sigma_{rr} = (1 - 2\nu) P/2\pi r^2, \qquad r \geqslant a \tag{1}$$

where P is the load and r the radial distance from the centre of the contact. It is particulary sensitive to frictional tractions at the interface.

The effect of interfacial friction upon the radial tensile stress has been analysed by Johnson, O'Connor and Woodward[8] assuming

a. no-slip (Goodman's theory) and
b. complete slip (Spence's work on partial slip was not available at the time).

In both the extreme cases considered the effect is the same; the radially inward motion of the more compliant surface is opposed by interfacial friction, thereby reducing the radial tensile stress outside the contact area compared with equation (1). Furthermore the point of maximum stress is displaced from the edge of the contact circle to a greater radius (see Fig. 1). On the more rigid surface, however, the frictional traction is directed inwards, so that the radial tension is augmented. In this case the maximum tension remains at $r = a$. Thus, when two dissimilar bodies are pressed together, if one or other fails by brittle fracture, we should expect the fracture load to be influenced by the frictional conditions at the interface and by the relative compliance of the two solids.

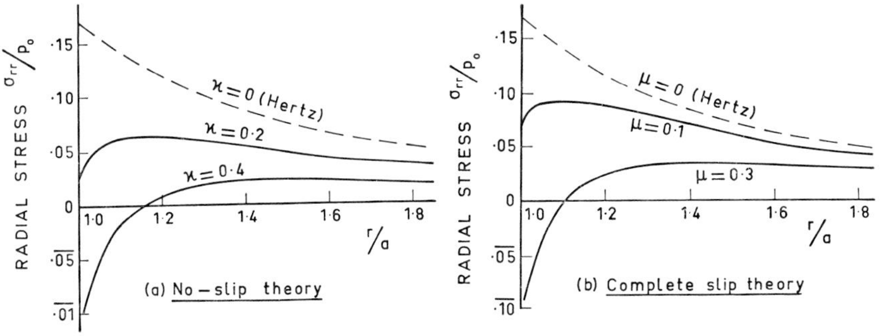

Fig. 1. Influence of frictional traction of radial stress outside the contact circle

To test this hypothesis comparative experiments were carried out using
a. steel spheres and
b. glass spheres
to indent flat plate glass specimens[8]. The radii of the spheres were chosen to obtain the same contact pressures at the same loads. Typical results for the contact pressure at fracture are shown in Fig. 2.

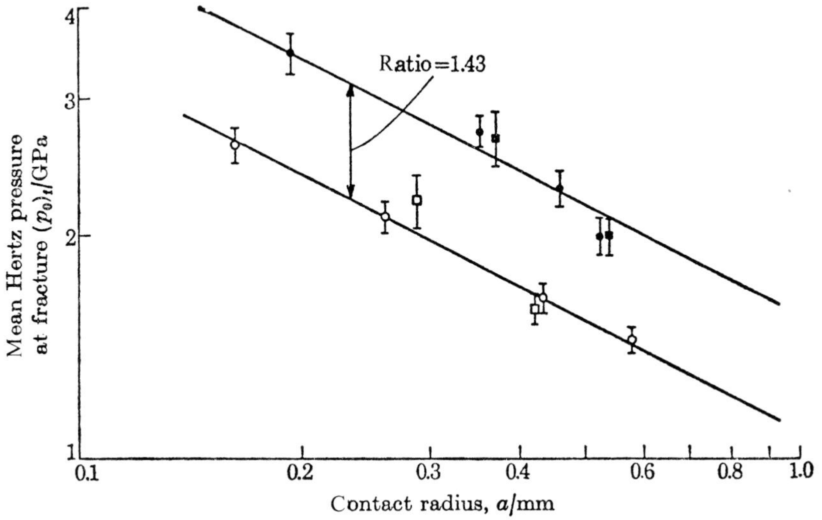

Fig. 2. Mean values and standard deviations of the maximum Hertz pressure at fracture $(p_0)_f$
As received: ○ glass indenters; ● steel indenters.
Abraded and etched plate: □ glass indenters; ■ steel indenters.
Lines of slope $-\frac{1}{2}$ represent Auerbach's law.

With glass indenters the contacting bodies are elastically similar ($\kappa = 0$) and interfacial friction effects do not arise. With steel indenters the glass specimens are more compliant than the indenter, so that friction reduces the radial tensile stress and thereby increases the contact pressure at which fracture occurs. The observed increase of about 40 per cent is consistent with the theoretical reduction in radial stress at the point of fracture shown in Fig. 1.

3. ADHESION OF SMOOTH SPHERES

It is well known that clean, smooth surfaces pressed into intimate contact may in some circumstances adhere strongly together as a result of intermolecular forces[9]. The Hertz theory neglects such forces and, indeed, measurable adhesion between non-conforming elastic bodies is seldom observed in practice. An exception to this common experience has recently been demonstrated in the Cavendish Laboratory at Cambridge using optically smooth rubber spheres of very low elastic modulus in contact with a smooth glass surface. The combination of smooth surfaces and a compliant solid permitted intimate contact between the rubber and the glass surfaces to be achieved throughout the whole of the apparent contact area. Under a compressive load the sphere made contact with the glass over a circular area noticeably greater than that predicted by Hertz and a measurable tensile force was required to separate the two surfaces. A very thin layer of fluid containing a wetting agent between the surfaces destroyed the adhesion and the deformation of the sphere reverted to the Hertzian.

These observations led to an extension of the Hertz theory for spheres to include the influence of adhesive forces[10]. The magnitude of the adhesive forces is high and their range of action is very small (of order 5 Å).

The law of force against separation is not known precisely, so that it is convenient to express the adhesive action in terms of a *surface energy*, γ, which is defined as the work required to separate unit area of the adhered surfaces. This approach proved fruitful in Griffith's analysis of the growth of a brittle crack. The same reasoning has been used in this problem.

If two elastic spheres of radii R_1 and R_2 are pressed into contact by a force P_1, according to Hertz they make contact on a circle of radius a_1 given by

$$a_1^3 = RP_1/K \tag{2}$$

where

$$R = R_1 R_2/(R_1+R_2),\ K = 4E'/3$$

and

$$1/E' = \{(1-\nu_1^2)/E_1 + \{(1-\nu_2^2)/E_2\}.$$

If now the load is reduced to P_0, whilst the surfaces maintain contact by adhesion over the same contact radius a_1, the resulting distribution of surface traction is found by subtracting the stress under a flat cylindrical punch from the Hertz distribution[11], viz.:

$$p(r) = \frac{3P_1}{2\pi a_1^2}\{1-r^2/a_1^2\}^{\frac{1}{2}} - \frac{P_1-P_0}{2\pi a_1^2}\{1-r^2/a_1^2\}^{-\frac{1}{2}} \qquad (3)$$

This traction is tensile(−) at the edge of the contact and compressive(+) in the centre, as shown by curve B in Fig. 3(b). It is straightforward to

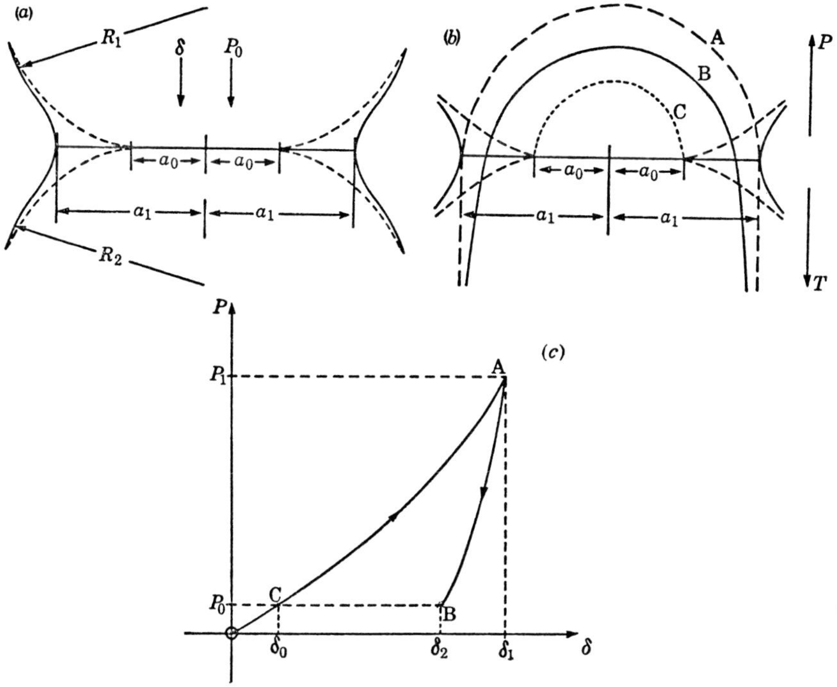

Fig. 3. The contact between two elastic solids both in the presence (contact radius a_1) and absence (contact radius a_0) of surface forces

(*a*) Shows the contact between two convex bodies of radii R_1 and R_2 under a normal load of P_0; δ is the elastic displacement. (*b*) Indicates the distribution of stress in the contacting spherical surfaces. When surfaces are maintained in contact over an enlarged area by surface forces, the stresses between the surfaces are tensile (T) at the edge of the contact and only remain compressive (P) in the centre. Distribution A is the Hertz stress with $a = a_1$ and $P = P_1$; distribution B the actual stress (Johnson, 1958) with $a = a_0$ and $P = P_0$. (*c*) Represents the load-displacement relation for the contracting surfaces.

calculate the elastic strain energy U_E associated with this state of stress by following the loading path in Fig. 3(c) from 0 to A and from A to B, whence

$$U_E = \frac{1}{15} K^{-2/3} R^{1/3} [P_1^{5/3} + 15 P_0^2 P_1^{-1/3} + 5 P_0 P_1^{2/3}] \qquad (4)$$

where P_1 is related to a_1 by equation (2). The surface energy associated with a contact area of radius a_1 is:

$$U_s = -\gamma\pi a_1^2 = -\gamma\pi(P_1 R/K)^{2/3} \qquad (5)$$

We now follow Griffith in stating that the equilibrium value of a_1, for any given value of P_0, occurs when the variation of elastic strain energy with a_1 just matches the variation of surface energy, i.e. when $d(V_E + V_s)/da_1 = 0$. This condition gives

$$Ka_1^3/R = P_1 = P_0 + 3\gamma\pi R + \{(6\gamma\pi RP_0 + (3\gamma\pi R)^2)\}^{\frac{1}{2}} \qquad (6)$$

If we now write $P_0 = P$ and $a_1 = a$, an expression is obtained for the equilibrium value of a under the action of a load P in the presence of adhesive forces, viz.:

$$a^3 = \frac{R}{K}\left[P + 3\gamma\pi R + 36\gamma\pi RP + (3\gamma\pi R)^2\}^{\frac{1}{2}}\right] \qquad (7)$$

When the surface energy is zero, equation (7) reduces to the simple Hertz relationship. It also follows that an equilibrium contact area can be

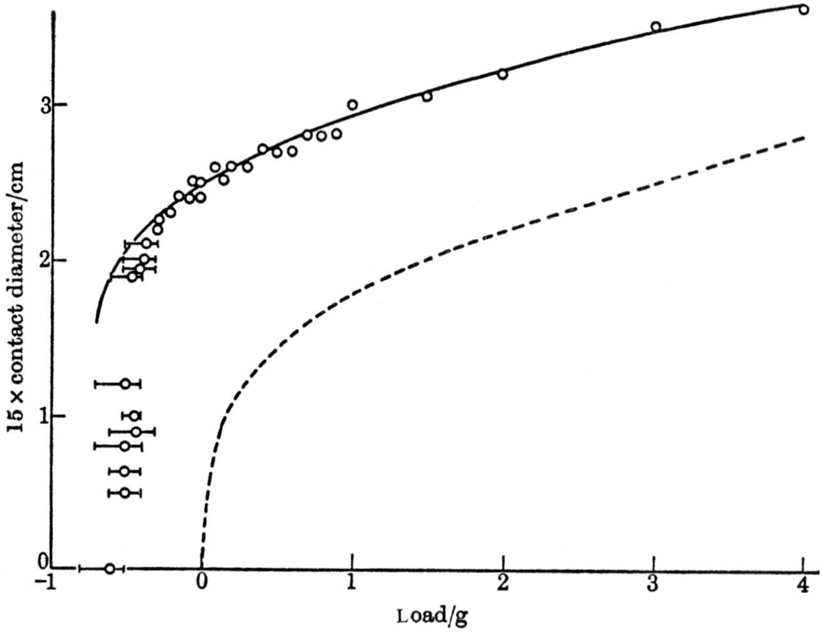

Fig. 4. Results for rubber sphere (R = 2.2 cm) in dry contact with rubber flat under small positive and negative loads

○ Contact results

– – – Hertz theory

——— Modified theory.

maintained even though the force is tensile (−), up to a maximum value of

$$P = -\frac{3}{2}\gamma\pi R \tag{8}$$

This is the force required to separate the spheres.

Observations of the contact size of a smooth rubber sphere in contact with glass are compared with the predictions of equation (7) in Fig. 4.

4. ROUGH SURFACES

Real surfaces are never smooth as Hertz assumed (cleaved mica is an almost unique exception). When two surfaces are pressed into contact they touch at the tips of the surface irregularities, so that the real area of contact is only a fraction of the nominal area. In the case of elastic spheres true contact is not made continuously over the circular area envisaged by Hertz, but through an archipelago of small discrete islands roughly clustered within a circular region. The true contact pressure is, therefore, also discontinuous, being very high within the islands of real contact and falling to zero in between. However, if the scale of the surface asperities is small compared with the bulk scale of the bodies themselves, so that there are a very large number of islands of real contact, it is possible to conceive of the discontinuous pressure distribution being 'smoothed out' into a continuous one. For this purpose it is usual to start from some statistical model of a nominally 'flat' rough surface.

Several such models have been proposed. Some are based upon individual model asperities which are assumed to deform elastically or plastically, e.g. Zhuravlev[12], Archard[13], Ling[14], Greenwood and Williamson[15]; others are based on the random nature of the surface profile, e.g. Whitehouse and Archard[16] and Nayak[17]. For the purpose of this paper it is convenient to use the Greenwood and Williamson model[15]. The asperities are taken to have spherical caps of uniform radius β, whose heights above a mean datum have a statistical distribution $\phi(z)$ and which deform elastically and independently according to the Hertz theory.

Thus the force P_a required to compress an individual asperity by an amount w_a is given by

$$p_a = \frac{4}{3} E' \beta^{\frac{1}{2}} w_a^{3/2} \tag{9}$$

If such a rough surface is in contact with a smooth surface at a separation d, the effective (i.e. smoothed out) pressure between them is:

$$p = \frac{4}{3} NE' \beta^{\frac{1}{2}} \int_d^{\infty} (z-d)^{3/2}\, \phi(z)\, \mathrm{d}z \tag{10}$$

where N is the number of asperities per unit area.

The geometry of rough elastic spheres in contact is shown diagrammatically in Fig. 5. For convenience the roughness is represented on one

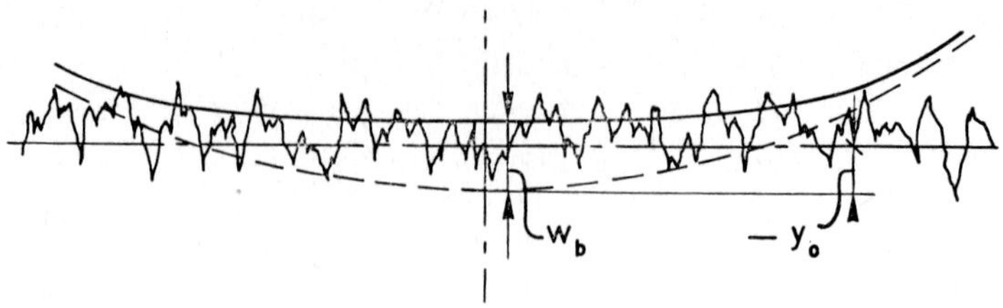

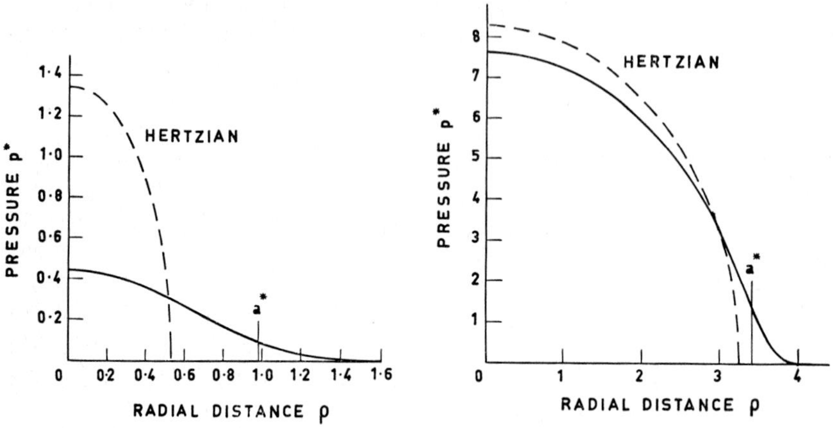

Fig. 5. Comparison of pressure distributions with Hertzian theory

a. At low loads pressures are much lower than Hertzian and spread over a very much larger area.

b. At high loads calculated and Hertzian pressures agrees well.

surface only (taken to be nominally flat) and the bulk elastic deformation w_b is represented on the smooth sphere. Thus the separation is given by

$$d = y_0 + r^2/2R + w_b \tag{11}$$

The distribution of asperity heights for a freshly prepared rough surface is usually taken to be Gaussian, but, as Greenwood and Williamson showed, the main features of the theory can be demonstrated in a simple way by representing the 'upper tail' of the Gaussian curve by an exponential distribution function $\phi(z) = (1/\sigma)\exp(-d/\sigma)$. If the function, with d taken from equation (11), is substituted in equation (10) and the variable changed from z to w, we get

$$p(r) = \pi^{\frac{1}{2}}(N\beta\sigma)\, E'(\sigma/\beta)^{\frac{1}{2}} \exp(-y_0/\sigma) \exp(-w/\sigma) \exp(-r^2/2R\sigma) \tag{12}$$

In addition it is necessary to relate the bulk elastic deformation of the two spheres $w_b(r)$ to the effective pressure $p(r)$. This is properly done through the equations of an elastic half-space loaded by an arbitrary axially-symmetric distribution of normal pressure. Such calculations have been carried out using an elaborate iterative numerical technique by Greenwood and Tripp[18] and by Mikic and Roca[19]. They find that the effect of surface roughness is to reduce the effective pressure at the centre and to spread it more widely, so that the archipelago of real contact extends beyond the Hertz contact circle, as shown in Fig. 5.

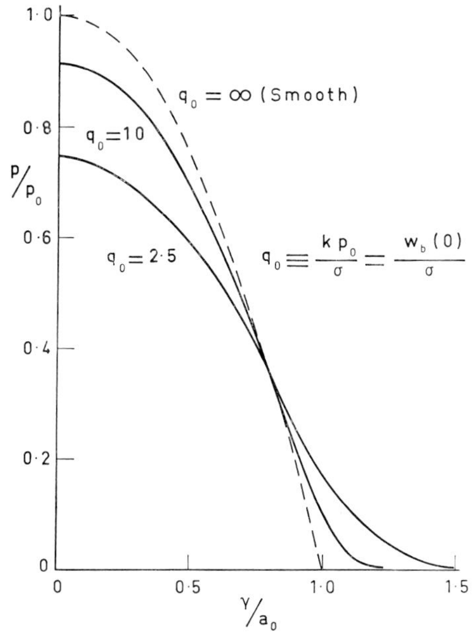

Fig. 6. Pressure distributions for rough spheres in contact; simple elastic foundation analysis

This behaviour can be demonstrated in a straightforward approximate way if the spheres are assumed to deform in bulk as a simple elastic foundation of modulus k, i.e.

$$w_b = kp \tag{13}$$

combining equations (12) and (13), and writing $q = kp/\sigma$, $q_0 = kp_0/\sigma$, $\rho^2 = r^2/2R\sigma$ and $A = (k/\sigma)\pi^{\frac{1}{2}}(N\beta\sigma)E'(\sigma/\beta)^{\frac{1}{2}}\exp(-y_0/\sigma)$, gives

$$q\,e^q = A\,e^{-\rho^2},$$

from which

$$\rho^2 = \ln(q_0/q) + (q_0 - q) \tag{14}$$

The total load

$$P = \pi \int_0^\infty p \, \mathrm{d}(r^2)$$

i.e.

$$\frac{kP}{2\pi R\sigma^2} = \int_0^\infty q \, \mathrm{d}(\rho^2)$$

$$= q_0(\tfrac{1}{2}q_0 + 1) \tag{15}$$

The non-dimensional parameter q_0 can be interpreted as the ratio of the bulk compression at $r = 0$ to the surface roughness σ. The value of this single parameter specifies the pressure distribution through equation (14) and the total load through equation (15). The influence of increasing the surface roughness σ (i.e. decreasing q_0) upon the pressure distribution whilst maintaining the total load P constant is shown in Fig. 6. With smooth surfaces, the consequence of the simplifying assumption embodied in equation (13) is to give a parabolic rather than a semi-elliptical pressure distribution. As the roughness is increased the effective contact pressure spreads over an increasing area in the manner predicted by the complete numerical solutions.

5. ADHESION OF ROUGH SURFACES

In Section 3 the adhesion of smooth spheres was considered and the analysis was satisfactorily supported by experiments with smooth, compliant rubber spheres. When the surface of the rubber or the glass was roughened the measured adhesion decreased drastically, to an extent which could not be accounted for solely by the reduction in real area of contact. In fact, a modest roughness caused the adhesion to fall to an immeasurable value. Careful experiments with metal spheres having the best attainable surface finish revealed no measurable adhesion.

To explain these observations we shall combine the work described in the two previous sections to develop a theory of adhesion of rough surfaces. In the first instance we shall consider nominally 'flat' surfaces. We shall follow Greenwood and Williamson in modelling the rough surface by independent spherical, elastic asperities with a random height distribution but, instead of deforming according to Hertz, they will be assumed to deform according to the modified theory presented in Section 3[10] which includes the effect of adhesion.

The compression of a smooth sphere loaded by P in the presence of adhesive force (denoted by δ_2 in Fig. 3(c)) is given by

$$w = \frac{1}{3K^{2/3}R^{1/3}} \frac{P_1 + 2P}{P_1^{1/3}} \tag{16}$$

The equilibrium value of P_1 is given by equation (6). Writing $P_c = 3\gamma\pi R/2$ from equation (8), we obtain the equilibrium relationship between compression and force in the non-dimensional form:

$$\frac{w}{w_c} = \frac{3(P/P_c) + 2 + 2(1+P/P_c)^{\frac{1}{2}}}{3^{2/3}\{(P/P_c) + 2 + 2(1+P/P_c)^{\frac{1}{2}}\}^{1/3}} \tag{17}$$

where

$$w_c = \left(\frac{P_c^2}{3K^3R}\right)^{1/3} = \left(\frac{3\pi^2\gamma^2 R}{4K^2}\right)^{1/3}$$

Inverting equation (17) we may write

$$P/P_c = F(w/w_c) \tag{17a}$$

The load/compression curve expressed by equation (17) is plotted in Fig. 7. It corresponds to the load/contact radius curve given by equation (7), part of which is plotted in Fig. 4. In the absence of adhesion it reverts to

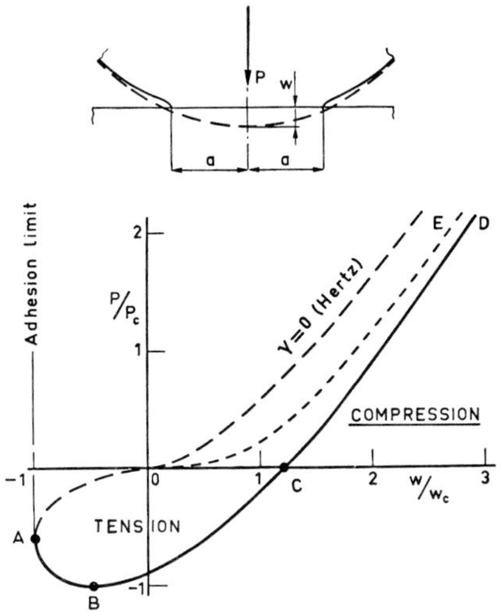

Fig. 7. Force/compression curve for an elastic sphere with adhesion

the Hertz relation shown by a chain line. The effect of adhesive forces is to introduce a regime in which the force is tensile and the surface is pulled out from its undeformed level. The maximum tensile force is denoted by $-P_c$ and the maximum tensile displacement is denoted by $-w_c$. Stable equilibrium is only possible on the branch of the curve ABCD which is drawn as a full line.

We now turn to the Greenwood and Williamson analysis of rough surfaces, outlined in Section 4, and imagine that a nominally flat rough surface, whose asperities deform according to equation (17), is brought into contact with a rigid flat surface. The separation is then increased again by an amount not less than w_c. We wish to find the total adhesive force per unit area acting between the two surfaces. Asperities which are still compressed by an amount greater than OC in Fig. 7 will exert a compressive force whilst those which are extended or compressed by an amount less than OC will exert a tensile (i.e. adhesive force). When an asperity is extended by w_c the adhesion is broken and it no longer remains in contact. The total effective asperity pressure is thus:

$$p = N \int_{d-w_c}^{\infty} P_c \, F(w/w_c) \, \phi(z) \, dz \tag{18}$$

If, once again, for simplicity, we take an exponential distribution of asperity heights, equation (18) for the effective pressure becomes:

$$p = N \exp\{-(d-w_c)/\sigma\} \, P_c(w_c/\sigma) \int_{-1}^{\infty} F(w/w_c) \exp\{-(w+w_c)/w_c\} \, d(w/w_c) \tag{19}$$

The number of asperities per unit area in contact is given by

$$n = N \exp\{-(d-w_c)/\sigma\}$$

Putting $(1+w/w_c) = x$ and $\sigma/w_c = \alpha$, gives

$$p = nP_c \int_0^{\infty} F(x) \exp(-x/\alpha) \, d(x/\alpha) \tag{20}$$

The parameter

$$\alpha = \sigma/w_c = \left(\frac{4\sigma^3 K^2}{3\pi^2 \gamma^2 R}\right)^{1/3}$$

may be described as an *adhesion index*. The integral in equation (20), $I(\alpha)$, has been evaluated numerically and is shown in Fig. 8. It falls from -1.0 when $\alpha = 0$ and becomes positive for values of $\alpha > 1.6$.

When there is no dispersion in the asperity heights ($\alpha = 0$) the effective pressure is $-nP_c$. This result is to be expected since it corresponds to a total adhesion made up of n asperities per unit area, each with a pull-off force P_c. However, when the asperity heights are dispersed with a standard deviation σ, the high asperities exert a compressive force which opposes the tension carried by the lower ones. At a critical value of the adhesion index ($\alpha_c = 1.6$) the compressive force is sufficient to break the tensile junctions so that no overall adhesion can be developed. This result explains why adhesion is difficult to achieve between rough surfaces, particularly with materials which have a high elastic modulus (large K).

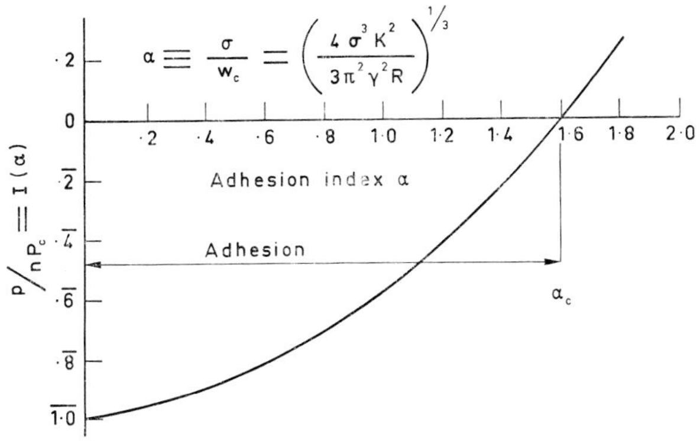

Fig. 8. Adhesion of rough surfaces (adhesion ceases when $\alpha > 1.6$)

So far the theory is restricted to nominally flat surfaces. The next logical step is to extend it to curved surfaces, e.g. elastic spheres, as in the previous section. Unfortunately the form of equation (20) is not suitable for this purpose, since it implies that the sign of the pressure p is independent of the separation between the surfaces, whereas we would expect p to become positive (compressive) at sufficiently small separations. The anomaly lies in the use of the exponential distribution function, and is removed if a more realistic (e.g. Gaussian) distribution is used.

Calculations along these lines are at present being carried out by Dr. K. Fuller at Cambridge and will be reported in due course.

ACKNOWLEDGEMENTS

Much of the work referred to in this paper arose through the stimulus of close cooperation with Professor D. Tabor and his colleagues in the Cavendish Laboratory at Cambridge. The author also acknowledges with gratitude the help from his colleague Dr. J.A. Greenwood, both in discussion of the problems and in the preparation of this paper.

REFERENCES

1. L.E. Goodman and L.M. Keer, 'The Contact Stress Problem for an Elastic Sphere Indenting an Elastic Cavity', *Int. J. Solids & Struct.*, **1**, 1965, pp. 407ff.
2. V.I. Mossakovski, 'The Fundamental General Problem of the Theory of Elasticity for a Half-Space with a Circular Curve Determining the Boundary Conditions', *PMM*, **18**, 1954.
3. L.E. Goodman, 'Contact Stress Analysis of Normally Loaded Rough Spheres', *J. App. Mech.*, **29**, 1954. pp. 515ff.

4. V.I. Mossakovski, 'Compression of Elastic Bodies under Conditions of Adhesion', *PMM*, **27**, 1963, pp. 418ff.
5. D.A. Spence, 'Self-Similar Solutions to Adhesive Contact Problems', *Proc. Roy. Soc.*, **A305**, 1968, pp. 55ff.
6. D.A. Spence, 'An Eigenvalue Problem for Elastic Contact with Finite Friction', *Proc. Camb. Phil. Soc.*, **73**, 1973, pp. 249ff.
7. D.A. Spence, 'The Hertz Contact Problem with Finite Friction', U. of Wisconsin Math. Res. Centre Report No. 1209, 1972.
8. K.L. Johnson, J.J. O'Connor and A.C. Woodward, 'The Effect of Indenter Elasticity on the Hertzian Fracture of Brittle Materials', *Proc. Roy. Soc.*, **A334**, 1973, pp. 95ff.
9. F.P. Bowdon and D. Tabor, *Friction and Lubrication of Solids*, Vol. II, Oxford University Press, 1964, Chapter v.
10. K.L. Johnson, K. Kendall and A.D. Roberts, 'Surface Energy and the Contact of Elastic Solids', *Proc. Roy. Soc.*, **A324**, 1971, 301 pp.ff.
11. K.L. Johnson, 'A Note on the Adhesion of Elastic Solids', *Brit. J. App. Phys.*, **9**, 1958, pp. 199ff.
12. V.A. Zhuravlev, 'The Theoretical Justification of Amonton's Law for the Friction of Unlubricated Surface', *Zh. Tekh. Fiz.*, **10**, 1940, pp. 1447ff.
13. J.F. Archard, 'Elastic Deformation and the Laws of Friction', *Proc. Roy. Soc.*, **A243**, 1957, pp. 190ff.
14. F.F. Ling, 'On Asperity Distributions of Metallic Surfaces', *J. App. Phys.*, **29**, 1958, pp. 1168ff.
15. J.A. Greenwood and J.B.P. Williamson, 'Contact of Nominally Flat Surfaces', *Proc. Roy. Soc.*, **A295**, 1966, pp. 300ff.
16. D.J. Whitehouse and J.F. Archard, 'The Properties of Random Surfaces of Significance in their Contact', *Proc. Roy. Soc.*, **A316**, 1970, pp. 97ff.
17. P.R. Nayak, 'Random Process Model of Rough Surfaces', *J. Lub. Tech., Trans. ASME*, **93** (F), 1971, pp. 398ff.
18. J.A. Greenwood and J.H. Tripp. 'The Elastic Contact of Rough Spheres', *J. App. Mech., Trans. ASME*, **89***E*, 1967, pp. 153ff.
19. B.B. Mikic and R.T. Roca, 'A Solution to the Contact of Two Rough Spherical Surfaces', *J. App. Mech., Trans. ASME*, Ser. E, (to be published).

Signorini's problem in viscoelasticity

M. Boucher

1. TERMS OF THE PROBLEM

Hypothesis and notations

We consider a body which occupies, in its natural configuration, the open and bounded region $\Omega^{(1)}$ of $\mathbb{R}^3$ and whose boundary is called $\partial\Omega^{(1)}$. It is resting on a half-space denoted $\Omega^{(2)}$. We have the following definitions:

$$\begin{aligned}
\Omega^{(2)} &= \{x = (x_1, x_2, x_3) \in \mathbb{R}^3, x_3 < 0\} \\
\partial\Omega^{(2)} &= \mathbb{R}^2 \times \{0\} \\
\Sigma &= \partial\Omega^{(1)} \cap \partial\Omega^{(2)} \\
\Sigma^{(1)} &= \partial\Omega^{(1)} - \Sigma \\
\Sigma^{(2)} &= \partial\Omega^{(2)} - \Sigma
\end{aligned}$$

We act upon $\Omega^{(\alpha)}$ ($\alpha = 1, 2$) by body forces $\mathbf{f}^{(\alpha)}$ and surface forces $\mathbf{F}^{(\alpha)}$ along $\Sigma^{(\alpha)}$.

In a first step, we take elastic materials and we have the constitutive equations[1]

$$\alpha = 1, 2 \quad \sigma_{ij}^{(\alpha)}(x) = a_{ijkl}^{(\alpha)}(x)\, \varepsilon_{kl}(\mathbf{u}^{(\alpha)})$$

where $u_i^{(\alpha)}$ are the components of the displacements, $\sigma_{ij}^{(\alpha)}$ those of the stress tensor and $\varepsilon_{ij}(\mathbf{u}^{(\alpha)})$ those of the linearised strain tensor in $\Omega^{(\alpha)}$:

$$2\varepsilon_{ij}(\mathbf{u}^{(\alpha)}) = u_{i,j}^{(\alpha)} + u_{j,i}^{(\alpha)}$$

The elastic coefficients $a_{ijkl}^{(\alpha)}$ defined on $\Omega^{(\alpha)}$ satisfy the inequality:

$$\alpha = 1, 2 \quad a_{ijkl}^{(\alpha)} \varepsilon_{ij} \varepsilon_{kl} \geqslant \lambda \varepsilon_{ij} \varepsilon_{ij} \tag{1)–(2}$$

where λ is a positive constant, and the classical identities:

$$\alpha = 1, 2 \quad a_{ijkl}^{(\alpha)} = a_{jikl}^{(\alpha)} = a_{klij}^{(\alpha)}$$

Equations and inequalities

Our purpose is to find the displacements $\mathbf{u}^{(\alpha)}$ defined on $\Omega^{(\alpha)}$. For that, we have the following equations and inequalities.
Constitutive equations

$$\alpha = 1, 2 \quad \sigma_{ij}^{(\alpha)}(x) = a_{ijkl}^{(\alpha)}(x)\, \varepsilon_{kl}(\mathbf{u}^{(\alpha)}) \quad \text{in } \Omega^{(\alpha)} \tag{(3)–(4)}$$

Equilibrium equations

$$\alpha = 1, 2 \quad \sigma_{ij,j}^{(\alpha)} + f_i^{(\alpha)} = 0 \quad \text{in } \Omega^{(\alpha)} \tag{(5)–(6)}$$

Boundary conditions on $\Sigma^{(\alpha)}$

$$\alpha = 1, 2 \quad \sigma_{ij}^{(\alpha)} n_j^{(\alpha)} = F_i^{(\alpha)} \tag{(7)–(8)}$$

where the $n_j^{(\alpha)}$ are the components of the unit vector normal to $\Sigma^{(\alpha)}$ directed towards the outside of $\Omega^{(\alpha)}$.

Along Σ, we have two kinds of conditions depending on whether the contact is broken or not. If, in a point of Σ, the body leaves its support, we have:

$$u_3^{(1)} - u_3^{(2)} > 0$$

In such a case, there is no stress:

$$i = 1, 2, 3 \quad \sigma_{i3}^{(1)} = \sigma_{i3}^{(2)} = 0$$

If there is still contact, we have:

$$u_3^{(1)} - u_3^{(2)} = 0$$

As we have a frictionless contact, the stress must satisfy:

$$\begin{aligned} &\sigma_{13}^{(1)} = \sigma_{23}^{(1)} = \sigma_{13}^{(2)} = \sigma_{23}^{(2)} = 0 \\ &\sigma_{33}^{(1)} = \sigma_{33}^{(2)} < 0 \end{aligned}$$

In short we can write

$$u_3^{(1)} - u_3^{(2)} \geqslant 0 \quad \begin{cases} u_3^{(1)} - u_3^{(2)} > 0 & \sigma_{i3}^{(1)} = \sigma_{i3}^{(2)} = 0 \qquad (9) \\ u_3^{(1)} - u_3^{(2)} = 0 & \begin{cases} \sigma_T^{(1)} = \sigma_T^{(2)} = 0 \\ \sigma_{33}^{(1)} = \sigma_{33}^{(2)} < 0 \end{cases} \qquad (10) \end{cases}$$

where $\sigma_T^{(\alpha)}$ represents the tangential stress.

Weak formulation

As it is usual, we study the weak statement of the problem. For that we introduce some definitions:

$$v = (\mathbf{v}^{(1)}, \mathbf{v}^{(2)})$$

$$a^{(\alpha)}(\mathbf{v}^{(\alpha)}, \mathbf{w}^{(\alpha)}) = \int_{\Omega^{(\alpha)}} a_{ijkl}^{(\alpha)}\, \varepsilon_{ij}(\mathbf{v}^{(\alpha)})\, \varepsilon_{kl}(\mathbf{w}^{(\alpha)})\, \mathrm{d}x \tag{(11)}$$

$$L^{(\alpha)}(\mathbf{v}^{(\alpha)}) = \int_{\Omega^{(\alpha)}} \mathbf{f}^{(\alpha)} \cdot \mathbf{v}^{(\alpha)}\, dx + \int_{\Sigma^{(\alpha)}} \mathbf{F}^{(\alpha)} \cdot \mathbf{v}^{(\alpha)}\, d\Gamma \tag{12}$$

$$a(v, w) = a^{(1)}(\mathbf{v}^{(1)}, \mathbf{w}^{(1)}) + a^{(2)}(\mathbf{v}^{(2)}, \mathbf{w}^{(2)}) \tag{13}$$

$$L(v) = L^{(1)}(\mathbf{v}^{(1)}) + L^{(2)}(\mathbf{v}^{(2)}) \tag{14}$$

$$K = \{v = (\mathbf{v}^{(1)}, \mathbf{v}^{(2)}) \mid v_3^{(1)} - v_3^{(2)} \geqslant 0 \text{ along } \Sigma\} \tag{15}$$

It is easy to show[2] that the weak formulation of the problem is:

Find $u \in K$ such that
$a(u, v-u) \geqslant L(v-u)$ for $v \in K$

As the bilinear form $a(u, v)$ is symmetrical, it is well known that we have another formulation which is equivalent to the former:

Find $u \in K$ which minimizes the functional
$J(v) = 1/2\, a(v, v) - L(v)$

Some remarks

We want to sum up some remarks which are developed in [2] and [3]. Let $\Lambda^{(1)}$ the set of the bilateral rigid displacements of $\Omega^{(1)}$, i.e.,

$$\Lambda^{(1)} = \{\boldsymbol{\rho}^{(1)} \mid \boldsymbol{\rho}^{(1)} = \mathbf{a} + \mathbf{b} \Lambda \mathbf{x}\}$$

a and **b** are constant vectors such that

$$a_3 = 0,\ b_1 = b_2 = 0$$

If there is a solution of our problem, the linear form $L^{(1)}$ must verify:

$$L^{(1)}(\boldsymbol{\rho}^{(1)}) = 0 \quad \text{for} \quad \boldsymbol{\rho}^{(1)} \in \Lambda^{(1)} \tag{16}$$

Physically, this means that the system of body forces $\mathbf{f}^{(1)}$ and surface forces $\mathbf{F}^{(1)}$ is equivalent to a single force orthogonal to the plane (x_1, x_2).

Let $\Delta^{(1)}$ be the set of the rigid displacements of $\Omega^{(1)}$ which are compatible with the contact on $\Omega^{(2)}$, i.e.,

$$\Delta^{(1)} = \{\boldsymbol{\rho}^{(1)} \mid \boldsymbol{\rho}^{(1)} = \mathbf{a} + \mathbf{b} \Lambda \mathbf{x}\}$$

a and **b** are constant vectors such that

$$a_3 + b_1 x_2 - b_2 x_1 \geqslant 0 \quad \text{for} \quad (x_1, x_2) \in \Sigma$$

If there is a solution of our problem, the linear form $L^{(1)}$ satisfies the following inequality:

$$L^1(\boldsymbol{\rho}^{(1)}) \leqslant 0 \quad \text{for} \quad \boldsymbol{\rho}^{(1)} \in \Delta^{(1)} \tag{17}$$

Physically, this means that the single force we put in place previously is directed downwards and the intersection of its central axis with $\partial\Omega^{(2)}$ is a point of $C(\Sigma)$, where $C(\Sigma)$ denotes the convex envelope of Σ.

For the proof of the existence and uniqueness theorem in elasticity, we need some results in functional analysis. The first of them is theorem 5.1 of [4]. Let us write it here.

Basic theorem

Let W be a Hilbert space whose norm is denoted $\| \ \|_W$. We assume the following hypotheses:

1. $\| \ \|_W$ is equivalent to p_0+p_1 where
   p_0 is a semi-norm on W,
   p_1 is a pre-Hilbertian norm on W.
2. The space $Y=\{w \in W \mid p_1(w)=0\}$ has finite dimension.
3. There exists a constant c such that

$$\operatorname*{Inf}_{y \in Y} p_0(w-y) \leqslant c p_1(w)$$

4. Let $a(v, w)$ be a continuous bilinear form on W which is semi-coercive, i.e.

$$a(w,w) \geqslant \gamma p_1(w) \quad \text{for} \quad w \in W$$

   where γ is a positive number.
5. Let K be a closed convex containing $\{0\}$.
6. Let L be an element of W' such that

$$L(y) < 0 \quad \text{for} \quad y \in Y \cap K - \{0\}.$$

Then, there exists at least one solution of the following problem:

Find $u \in K$ such that
$$a(u, v-w) \geqslant L(v-u) \quad \text{for} \quad v \in K$$

Functional spaces on $\Omega^{(1)}$

On $\Omega^{(1)}$ we utilize the Sobolev's spaces and we are going to indicate the principal properties used later on and that we can find in many books, e.g. [5].

$\mathscr{D}(\Omega^{(1)})$ is the space of infinitely differentiable functions with compact support in $\Omega^{(1)}$ with its natural topology. Its topological dual, the space of distributions on $\Omega^{(1)}$, is denoted $\mathscr{D}'(\Omega^{(1)})$.

Let $H^1(\Omega^{(1)})$ be the space of functions belonging to $L^2(\Omega^{(1)})$ such that their first partial derivatives in $\mathscr{D}'(\Omega^{(1)})$ are in $L^2(\Omega^{(1)})$.

$$H^1(\Omega^{(1)}) = \{f \mid f \in L^2(\Omega^{(1)}), f_{,i} \in L^2(\Omega^{(1)}) \quad i=1, 2, 3)\}$$

With the norm

$$\|f\|_{H^1(\Omega^{(1)})} = \left[\|f\|^2_{L^2(\Omega^{(1)})} + \sum_{i=1}^{3} \|f_{,i}\|^2_{L^2(\Omega^{(1)})} \right]^{1/2}$$

$H^1(\Omega^{(1)})$ is a Hilbert space whose scalar product is denoted $(f, g)_{H^1(\Omega^{(1)})}$.

We suppose that $\Omega^{(1)}$ is smooth enough so that we denote $H^{1/2}(\partial\Omega^{(1)})$ the space of the trace of functions of $H^1(\Omega^{(1)})$ on $\partial\Omega^{(1)}$ with the natural topology. The trace operator is denoted γ.

Later on, we shall use the following properties:

- The imbedding of $H^1(\Omega^{(1)})$ in $L^2(\Omega^{(1)})$ is compact.
- There is a continuous imbedding of $H^{1/2}(\partial\Omega^{(1)})$ into $L^2(\partial\Omega^{(1)})$.

We define the space

$$\mathscr{H}^1(\Omega^{(1)}) = [H^1(\Omega^{(1)})]^3$$

Thanks to the Korn's inequality, we can use the following norm

$$\|\mathbf{u}^{(1)}\|_{\mathscr{H}^1(\Omega^{(1)})} = \left[\int_{\Omega^{(1)}} u_i u_i \, \mathrm{d}x + \int_{\Omega^{(1)}} \varepsilon_{ij}(\mathbf{u}^{(1)}) \, \varepsilon_{ij}(\mathbf{u}^{(1)}) \, \mathrm{d}x \right]^{1/2} \tag{18}$$

instead of the Euclidian one.

Functional spaces on $\Omega^{(2)}$

We denote $K^1(\Omega^{(2)})$ the space obtained by the closure of $\mathscr{D}(\bar{\Omega}^{(2)})$ provided with the topology defined by the norm

$$\|f\|_{K^1(\Omega^{(2)})} = \left[\sum_{i=1}^{3} \|f_{,i}\|^2_{L^2(\Omega^{(2)})} \right]^{1/2}$$

$K^1(\Omega^{(2)})$ is a Hilbert space whose scalar product is denoted $(f, g)_{K^1(\Omega^{(2)})}$.

We denote $K^{1/2}(\partial\Omega^{(2)})$ the space of the trace of functions of $K^1(\Omega^{(2)})$ on $\partial\Omega^{(2)}$ with the natural topology. In [6] the following properties are proved:

- $K^1(\Omega^{(2)})$ is the space of the functions of $L^6(\Omega^{(2)})$ whose first partial derivatives in $\mathscr{D}'(\Omega^{(2)})$ belong to $L^2(\Omega^{(2)})$.
- There is a continuous imbedding of $K^{1/2}(\partial\Omega^{(2)})$ into $L^4(\partial\Omega^{(2)})$.

Now we consider the product space:

$$\mathscr{K}^1(\Omega^{(2)}) = [K^1(\Omega^{(2)})]^3$$

The classical product norm is

$$|\mathbf{u}^{(2)}|_{\mathscr{K}^1(\Omega^{(2)})} = \left[\int_{\Omega^{(2)}} u^{(2)}_{i,j} u^{(2)}_{i,j} \, \mathrm{d}x \right]^{1/2} \tag{19}$$

We do not use this norm but the following one:

$$\|\mathbf{u}^{(2)}\|_{\mathscr{H}^1(\Omega^{(2)})} = \left[\int_{\Omega^{(2)}} \varepsilon_{ij}(\mathbf{u}^{(2)}) \, \varepsilon_{ij}(\mathbf{u}^{(2)}) \, \mathrm{d}x \right]^{1/2} \tag{20}$$

This is possible because the norms (19) and (20) are equivalent. This assertion is proved in [7] and reproduced in [2]. We sum up the two points of the proof:

- At first, the equivalence is easy to show when $\mathbb{R}^3$ is considered instead of $\Omega^{(2)}$ thanks to the Fourier transform.
- Afterwards, it is proved that there exists a linear and continuous operator which extends every function of the closure of $[\mathscr{D}(\bar{\Omega}^{(2)})]^3$ provided with the topology defined by the norm (20) in a function of $\mathscr{K}^1(\mathbb{R}^3)$.

These two results lead to the equivalence of the norms (19) and (20).

3. SOLUTION OF THE ELASTIC PROBLEM

We use the basic theorem and therefore we are going to look up, one after the other, all the conditions which compose it.

Condition 1

Let W be the following space

$$W = V^{(1)} \times V^{(2)}$$

with $V^{(1)} = \mathscr{H}^1(\Omega^{(1)})/\Lambda^{(1)}$, the quotient space of $\mathscr{H}^1(\Omega^{(1)})$ by $\Lambda^{(1)}$ and $V^{(2)} = \mathscr{K}^1(\Omega^{(2)})$.

If $\mathbf{v}^{(1)}$ is an element of $\mathscr{H}^1(\Omega^{(1)})$, we denote $v^{(1)}$ its range in $V^{(1)}$. The norm on $V^{(1)}$ is given by

$$|v^{(1)}| = \operatorname*{Inf}_{\boldsymbol{\rho}^{(1)} \in \Lambda^{(1)}} \|\mathbf{v}^{(1)} + \boldsymbol{\rho}^{(1)}\|_{\mathscr{H}^1(\Omega^{(1)})} \tag{21}$$

Lemma

The norm (21) is equivalent to the following one:

$$\|v^{(1)}\|_{V^{(1)}} = [v_3{}^{(1)}\|^2_{L^2(\Omega^{(1)})} + \varepsilon(\mathbf{v}^{(1)})]^{1/2} \tag{22}$$

where $\varepsilon(\mathbf{v}^{(1)})$ denotes the expression

$$\varepsilon(\mathbf{v}^{(1)}) = \int_{\Omega^{(1)}} \varepsilon_{ij}(\mathbf{v}^{(1)})\, \varepsilon_{ij}(\mathbf{v}^{(1)})\, dx$$

The proof can be found in [2].

We provide W with the norm

$$\|w\|_W = \|(w^{(1)}, \mathbf{w}^{(2)})\|_W = [\|w^{(2)}\|^2_{V^{(1)}} + \|\mathbf{w}^{(2)}\|^2_{V^{(2)}}]^{1/2}$$

and we define p_0 and p_1 by

$$p_0(w) = [\tfrac{1}{2}\|\mathbf{w}^{(2)}\|^2_{V^{(2)}} + \|w_3^{(1)}\|^2_{L^2(\Omega^{(1)})} + \tfrac{1}{2}\varepsilon(\mathbf{w}^{(1)})]^{1/2}$$

$$p_1(w) = [\tfrac{1}{2}\|\mathbf{w}^{(2)}\|^2_{V^{(2)}} + \tfrac{1}{2}\varepsilon(\mathbf{w}^{(1)})]^{1/2}$$

It is easy to see that these expressions are independent of the element $\mathbf{w}^{(1)}$ chosen in the class $w^{(1)}$. Furthermore, p_0 is a pre-Hilbert norm on W.

Condition 2

If $\mathscr{R}^{(1)}$ denotes the space of the rigid displacements defined on $\Omega^{(1)}$, Y is the space:

$$Y = \mathscr{R}^{(1)}/\Lambda^{(1)} \times \{0\}$$

Y has finite dimension: dim. $Y = 3$.

Condition 3

We must find a constant c such that

$$c \operatorname*{Inf}_{y \in Y} p_0(w-y) \leqslant p_1(w)$$

If P denotes the projector of W on Y with respect to the pre-Hilbert structure on W defined by the norm p_0, we can write this condition in the following form:

$$\text{Find } c \text{ such that } cp_0(w-\rho w) \leqslant p_1(w)$$

If we assume that c does not exist and if we change w for $w/p_0(w-Pw)$, we can suppose the existence of a sequence w_n such that

$$p_1(w_n) \leqslant 1/n \quad \text{and} \quad p_0(w_n - Pw_n) = 1$$

But

$$Pw_n = P(w_n^{(1)}, w_n^{(2)}) = (Hw_n^{(1)}, 0)$$

where H is the projector of $V^{(1)}$ on $\mathscr{R}^{(1)}/\Lambda^{(1)}$ with respect to the pre-Hilbert structure on $V^{(1)}$ defined by the norm

$$[\|w_3^{(1)}\|^2_{L^2(\Omega^{(1)})} + \tfrac{1}{2}\varepsilon(\mathbf{w}^{(1)})]^{1/2}$$

Consequently, we have the following hypothesis:

$$\|\mathbf{w}_n^{(2)}\|_{V^{(2)}} \to 0 \quad \|w_n^{(1)} - Hw_n^{(1)}\|_{V^{(1)}} \text{ is bounded} \quad \varepsilon(\mathbf{w}_n^{(1}) \to 0$$

Setting $v_n^{(1)} = w_n^{(1)} - Pw_n^{(1)}$, the sequence $\|v_n^{(1)}\|_V{}^{(1)}$ is bounded and $\varepsilon(\mathbf{v}_n^{(1)}) \to 0$. As $V^{(1)}$ is a Hilbert space, there exists a subsequence $v_\Gamma^{(1)}$ such that $v_\Gamma^{(1)} \to v^{(1)}$ weakly in $V^{(1)}$. The functional ε being lower semicontinuous, we have $\varepsilon(v^{(1)}) = 0$ and therefore $v^{(1)}$ is an element of $\mathscr{R}^{(1)}/\Lambda^{(1)}$. But $v_\Gamma^{(1)}$ belongs to $[\mathscr{R}^{(1)}/\Lambda^{(1)}]^\perp$ and so does $v^{(1)}$. Consequently, we have $v^{(1)} = 0$.

As $v_{\Gamma 3}^{(1)} \to 0$ weakly in $H^1(\Omega^{(1)})$ and the imbedding of $H^1(\Omega^{(1)})$ into $L^2(\Omega^{(1)})$ is compact, $v_{\Gamma 3}^{(1)} \to 0$ strongly in $L^2(\Omega^{(1)})$. We have obtained a contradiction with the fact that

$$p_0^2(w_n - Pw_n) = \tfrac{1}{2}\|\mathbf{w}_n^{(2)}\|_{V^{(2)}}^2 + \|v_{n3}^{(1)}\|_{L^2(\Omega^{(1)})}^2 + \varepsilon(\mathbf{v}_n^{(1)}) = 1$$

Condition 4

We assume that the elastic coefficients $a_{ijkl}^{(\alpha)}$ belong to $L^\infty(\Omega^{(\alpha)})$. In that way, the bilinear form (11) is defined and continuous on $\mathscr{H}^1(\Omega^{(1)})$ for $\alpha = 1$ and on $\mathscr{K}^1(\Omega^{(2)})$ for $\alpha = 2$. When $\alpha = 1$, it is easy to show that the continuous bilinear form $a^{(1)}(\mathbf{v}^{(1)}, \mathbf{w}^{(1)})$ defines a bilinear and continuous form on $V^{(1)}$ in the following manner:

$$a^{(1)}(v^{(1)}, w^{(1)}) = a^{(1)}(\mathbf{v}^{(1)}, \mathbf{w}^{(1)})$$

If we do not regard the little change in the notation, we take the bilinear form (13) and, thanks to the inequalities (1)–(2), it is semi-coercive.

Condition 5

We must prove that the following definition

$$K = \{v = (v^{(1)}, \mathbf{v}^{(2)}) \mid v_3^{(1)} - v_3^{(2)} \geqslant 0 \quad \text{along} \quad \Sigma\} \tag{23}$$

is correct. Let γ_3 be trace operator $\mathbf{v}^{(1)} \to v_3^{(1)}|_\Sigma$ whose domain is $\mathscr{H}^1(\Omega^{(1)})$ and range $H^{1/2}(\Sigma)$. γ_3 is a linear and continuous map which vanishes on $\Lambda^{(1)}$. Therefore, it defines a linear and continuous map whose domain is $V^{(1)}$ and range $H^{1/2}(\Sigma)$. If we take the properties of $H^{1/2}(\Sigma)$ and $K^{1/2}(\Sigma)$ into account, (23) defines a closed convex cone containing $\{0\}$.

Condition 6

We take $\mathbf{f}^{(1)}$ in $[L^2(\Omega^{(1)})]^3$, $\mathbf{F}^{(1)}$ in $[L^2(\Sigma^{(1)})]^3$, $\mathbf{f}^{(2)}$ in $[L^{6/5}(\Sigma^{(2)})]^3$ and $\mathbf{F}^{(2)}$ in $[L^{4/3}(\Sigma^{(2)})]^3$. We assume that the linear form $L^{(1)}(\mathbf{v}^{(1)})$ satisfies the equality (16). It is a physical condition which is necessary to get equilibrium. $L^{(1)}(\mathbf{v}^{(1)})$, vanishing on $\Lambda^{(1)}$, defines a linear and continuous form on $V^{(1)}$.

$$L^{(1)}(v^{(1)}) = L^{(1)}(\mathbf{v}^{(1)})$$

Again, if we do not regard the little change in the notation, we take the linear form (14).

We have

$$Y \cap K = \Delta^{(1)}/\Lambda^{(1)} \times \{0\}$$

We assume that $L^{(1)}$ satisfies the following condition which is stronger than the condition (16):

$$L^{(1)}(v^{(1)}) < 0 \quad \text{for} \quad v^{(1)} \in \Delta^{(1)}/\Lambda^{(1)} - \{0\}$$

i.e.,

$$\int_{\Omega^{(1)}} \mathbf{f}^{(1)} \cdot \mathbf{v}^{(1)} \, \mathrm{d}x + \int_{\Sigma^{(1)}} \mathbf{F}^{(1)} \cdot \mathbf{v}^{(1)} \, \mathrm{d}\Gamma < 0 \quad \text{for} \quad \mathbf{v}^{(1)} \in \Delta^{(1)} - \Lambda^{(1)} \tag{24}$$

Now, we can assert:

Theorem 1
Under all the previous hypotheses, the elastic problem has at least one solution.

Theorem 2
There is one and only one solution of the elastic problem.

The proof is given in [2]. In short, we can say that, if there are two solutions, the stress along Σ is an element of $H^{-1/2}(\Sigma)$, dual of $H^{1/2}(\Sigma)$, whose support is contained in a straight line.

In [5], it is proved that such an element of $H^{-1/2}(\Sigma)$ is necessary 0. This is impossible because the condition (24) leads to the following one:

$$\int_{\Omega^{(1)}} f_3^{(1)} \, \mathrm{d}x + \int_{\Sigma^{(1)}} F_3^{(1)} \, \mathrm{d}\Gamma < 0$$

In the same time, it is proved that the sufficient condition (24) is also necessary for the existence of a solution.

4. SOLUTION OF THE VISCOELASTIC PROBLEM

Terms of the problem

Here, we have the following constitutive equations:

$$\alpha = 1,2 \quad \sigma_{ij}^{(\alpha)}(x,t) = a_{ijkl}^{(\alpha)}(x)\, \varepsilon_{kl}[\mathbf{u}^{(\alpha)}(x,t)] + \int_0^t b_{ijkl}^{(\alpha)}(t-\mathscr{T}, x) \times \\ \times \varepsilon_{kl}[\mathbf{u}^{(\alpha)}(x,\mathscr{T})] \, \mathrm{d}\mathscr{T}$$

The elastic coefficients $a_{ijkl}^{(\alpha)}$ satisfy (1)–(2) and the conditions of symmetry. The relaxation coefficients $b_{ijkl}^{(\alpha)}$ have also the same properties of symmetry.

We assume that for $t < 0$ the system is at rest. For $t \geqslant 0$, we act upon $\Omega^{(\alpha)}$ $(\alpha = 1, 2)$ by body forces $\mathbf{f}^{(\alpha)}(x, t)$ and surface forces $\mathbf{F}^{(\alpha)}(x, t)$ along $\Sigma^{(\alpha)}$.

We consider a quasi-static evolution so that we have the same equations and inequalities as in the elastic problem except for the constitutive equations and the intervention of the time. The weak formulation leads to:

Find $u(t) = [\mathbf{u}^{(1)}(t), \mathbf{u}^{(2)}(t)]$, $t \in I =]0, T[$

where T is finite, such that

$$\sum_{\alpha=1}^{2} \int_{\Omega^{(\alpha)}} \sigma_{ij}[\mathbf{u}^{(\alpha)}(t)]\, \varepsilon_{ij}[\mathbf{v}^{(\alpha)}-\mathbf{u}^{(\alpha)}(t)]\, \mathrm{d}x \geqslant \sum_{\alpha=1}^{2} \int_{\Omega^{(\alpha)}} \mathbf{f}^{(\alpha)} \cdot$$

$$\cdot [\mathbf{v}^{(\alpha)}-\mathbf{u}^{(\alpha)}(t)]\, \mathrm{d}x + \sum_{\alpha=1}^{2} \int_{\Sigma^{(\alpha)}} \mathbf{F}^{(\alpha)} \cdot [\mathbf{v}^{(\alpha)}-\mathbf{u}^{\alpha}(t)]\, \mathrm{d}\Gamma$$

$$\text{for} \quad v = (\mathbf{v}^{(1)}, \mathbf{v}^{(2)}) \in K$$

where K is the convex (16).

We introduced the subsequent definitions:

$$a^{(\alpha)}[\mathbf{u}^{(\alpha)}(t), \mathbf{v}^{(\alpha)}] = \int_{\Omega^{(\alpha)}} a_{ijkl}^{(\alpha)}\, \varepsilon_{ij}[\mathbf{u}^{(\alpha)}(t)]\, \varepsilon_{kl}(\mathbf{v}^{(\alpha)})\, \mathrm{d}x$$

$$a[u(t), v] = \sum_{\alpha=1}^{2} a^{(\alpha)}[\mathbf{u}^{(\alpha)}(t), \mathbf{v}^{(\alpha)}]$$

$$b^{(\alpha)}[t, \mathbf{u}^{(\alpha)}(\mathscr{T}), \mathbf{v}^{(\alpha)}] = \int_{\Omega^{(\alpha)}} b_{ijkl}^{(\alpha)}(t, x)\, \varepsilon_{ij}[\mathbf{u}^{(\alpha)}(\mathscr{T})]\, \varepsilon_{kl}(\mathbf{v}^{(\alpha)})\, \mathrm{d}x$$

$$b[t, u(\mathscr{T}), v] = \sum_{\alpha=1}^{2} b^{(\alpha)}[t, \mathbf{u}^{(\alpha)}(\mathscr{T}), \mathbf{v}^{(\alpha)}]$$

$$L^{(\alpha)}[\mathbf{u}^{(\alpha)}(t)] = \int_{\Omega^{(\alpha)}} \mathbf{f}^{(\alpha)} \cdot \mathbf{u}^{(\alpha)}(t)\, \mathrm{d}x + \int_{\Sigma^{(\alpha)}} \mathbf{F}^{(\alpha)} \cdot \mathbf{u}^{(\alpha)}(t)\, \mathrm{d}\Gamma$$

$$L[u(t)] = \sum_{\alpha=1}^{2} L^{(\alpha)}[\mathbf{u}^{(\alpha)}(t)]$$

Thus, the problem is to find $u(t)$ such that for $t \in I$ we have:

$$u(t) \in K \tag{25}$$

$$a[u(t), v-u(t)] + \int_0^t b[t-\mathscr{T}, u(\mathscr{T}), v-u(t)]\, \mathrm{d}\mathscr{T} \geqslant L[v-u(t)] \quad \text{for} \quad v \in K \tag{26}$$

Functional spaces

For $p \geqslant 1$ and E, Hilbert space whose norm is denoted $\| \ \|_E$, $L^p(I, E)$ represents the space of strongly measurable functions f from $I \to E$ such that

$$\left[\int_0^T \|f(t)\|_E^P\, \mathrm{d}t\right]^{1/P} = \|f\|_{L^P(I,E)} < +\infty \quad p \neq \infty$$

$$\operatorname*{ess.\,sup.}_{t \in I} \|f\|_E = \|f\|_{L^\infty(I,E)} < +\infty$$

We assume the following conditions

$$f_i^{(1)} \in L^\infty[I, L^2(\Omega^{(1)})] \qquad F_i^{(1)} \in L^\infty[I, L^2(\Sigma^{(1)})]$$

$$f^{(2)} \in L^\infty[I, L^{6/5}(\Omega^{(2)})] \qquad F_i^{(2)} \in L^\infty[I, L^{4/3}(\Sigma^{(2)})]$$

$$a_{ijkl}^{(\alpha)} \in L^\infty(\Omega^{(\alpha)}) \qquad b_{ijkl}^{(\alpha)} \in L^\infty(I \times \Omega^{(\alpha)})$$

and the strong Signorini's hypothesis: for almost every $t \in I$, $L(\boldsymbol{\rho}^{(1)}, 0) < 0$ for $\boldsymbol{\rho}^{(1)} \in \Delta^{(1)} - \Lambda^{(1)}$.

Then, we must find u in $L^2(I, W)$ verifying for almost every $t \in I$ (25) and (26).

Existence of a solution

We use a method of approximation. The viscoelastic problem is the limit, in a sense that we shall define, of a sequence of elastic problems.

Lemma 1

Let w be a function of $L^2(I, W)$ and let us fix the time t. We consider the following expression:

$$\phi_w(v) = L(v) - \int_0^t b(t-\mathscr{T}, w(\mathscr{T}), v]\, d\mathscr{T}$$

We state that ϕ_w is a linear and continuous form on W.

The proof is obvious. Moreover, we have the inequality

$$\|\phi_w\|_{W'} \leqslant \|L\|_{W'} + \beta \int_0^t \|w(\mathscr{T})\|_W\, d\mathscr{T}$$

where $\| \ \|_{W'}$ is the norm in the dual of W and β is a number which depends only on the norm of $b_{ijkl}^{(\alpha)}$ in $L^\infty(I \times \Omega^{(\alpha)})$.

Lemma 2

The linear form ϕ_w satisfies the strong Signorini's condition. It is also obvious because an element of $Y \cap K - \{0\}$ is reduced to $v = (\boldsymbol{\rho}^{(1)}, 0)$ and so we have

$$\phi_w(v) = L^{(1)}(\boldsymbol{\rho}^{(1)})$$

Now, let us consider the following problem: t being fixed in I and w being a given element of $L^2(I, W)$, find $u \in K$ such that

$$a(u, v-u) \geqslant \phi_w(v-u) \quad \text{for} \quad v \in K$$

Lemma 3

The previous problem has a unique solution.

Thanks to lemmas 1 and 2, we come back to the elastic problem with the linear form ϕ_w instead of L.

Thus, we have a map

$$(t, w) \rightarrow u = (Tw)(t)$$

Lemma 4

w being fixed in $L^2(I, W)$, the map $t \to u(t)$ from $I \to W$ belongs to $L^\infty(I, W)$ [and consequently to $L^2(I, W)$].

It is a consequence of some inequalities proved in [2]. Thanks to lemma 3, we can put in place a sequence in the following manner:

$$u_0 = 0,\ u_1 = (Tu_0)(t),\ \dots\ u_n(t) = (Tu_{n-1})(t)$$

Lemma 5

The sequence (u_n) is bounded in $L^\infty(I, W)$ [and consequently in $L^2(I, W)$]. The proof uses the same inequalities as in lemma 4.

$L^2(I, W)$ is a Hilbert space and $L^\infty(I, W)$ is identified with the dual of $L^1(I, W)$. So, there is a subsequence, still denoted (u_n), which converges to an element u of $L^2(I, W) \cap L^\infty(I, W)$ in the weak topology of $L^2(I, W)$ and the weak star topology of $L^\infty(I, W)$.

The sequence (u_n) verifies:

$$a[u_n(t), v-u_n(t)] + \int_0^t b[t-\mathscr{T},\ u_{n-1}(\mathscr{T}),\ v-u_n(t)]\ \mathrm{d}\mathscr{T} \geqslant L[v-u_n(t)] \quad \text{for} \quad v \in K \qquad (25)$$

$$u_n(t) \in K \qquad (26)$$

We can write (25) as follows:

$$\begin{aligned} &a[u_n(t), v] + \int_0^t b[t-\mathscr{T},\ u_{n-1}(\mathscr{T}),\ v]\ \mathrm{d}\mathscr{T} - L[v-u_n(t)] \\ &\qquad \geqslant a[u_n(t), u_n(t)] + \int_0^t b[t-\mathscr{T}, u_{n-1}(\mathscr{T}), u_n(t)]\ \mathrm{d}\mathscr{T} \end{aligned} \quad \text{for} \quad v \in K \qquad (27)$$

Proceeding to the limit, we obtain

$$\begin{aligned} &a[u(t), v] + \int_0^t b[t-\mathscr{T}, u(\mathscr{T}), v]\ \mathrm{d}\mathscr{T} - L[v-u(t)] \\ &\geqslant \liminf_{n \to \infty} \left\{ a[u_n(t), u_n(t)] + \int_0^t b[t-\mathscr{T}, u_{n-1}(\mathscr{T}), u_n(t)]\ \mathrm{d}\mathscr{T} \right\} \\ &\geqslant \liminf_{n \to \infty} a[u_n(t), u_n(t)] + \liminf_{n \to \infty} \int_0^t b[t-\mathscr{T}, u_{n-1}(\mathscr{T}), u_n(t)]\ \mathrm{d}\mathscr{T} \\ &\geqslant a[u(t), u(t)] + \liminf_{n \to \infty} \int_0^t b[t-\mathscr{T}, u_{n-1}(\mathscr{T}), u_n(t)]\ \mathrm{d}\mathscr{T} \end{aligned} \qquad (28)$$

To finish, we need two lemmas whose proofs can be read in [2].

Lemma 6

The sequence $p_1(u_n)$ is a Cauchy sequence in $L^\infty(I)$.

Lemma 7
We have

$$\lim_{n \to \infty} \int_0^t b[t-\mathscr{T}, u_{n-1}(\mathscr{T}), u_n(t)] \, d\mathscr{T} = \int_0^t b[t-\mathscr{T}, u(\mathscr{T}), u(t)] \, d\mathscr{T}$$

Now, we can assert:

Theorem 1
There is a solution of the viscoelastic problem. It is an obvious consequence of the inequality (28) and the lemma 7.

Theorem 2
There is uniqueness of the solution. The proof is the same one as in the elastic problem.

BIBLIOGRAPHY

1. P. Germain, *Cours de Mécanique des Milieux Continus*, Part 1, Masson, 1973.
2. M. Boucher, Thèse de 3ᵉ cycle, Université Paris 6.
3. G. Fichera, Problemi Elastostatici con Vincoli Unilaterali il Problema di Signorini con Ambigue Condizioni al Contorno, *Memorie. Atti della Accademia Nazionale dei Lincei*, Series 8, Vol. 7, 1963.
4. J.L. Lions and G. Stampacchia, Variational Inequalities, *Communications on Pure and Applied Mathematics*, 20, 1967, p. 493.
5. J.L. Lions and E. Magenes, *Problèmes aux Limites Non Homogenes*, Part I, Dunod, Paris, 1968.
6. J. Barros-Neto, Inhomogeneous Boundary Value Problem in a Half-Space, *Annali della Scuola Normale Superiore di Pisa*, Series 3, Vol. 19, 1965, p. 331.
7. J.L. Lions and G. Duvaut, American edition (to be published).

Properties of elastic bodies in contact*

J. Dundurs

1. CLASSIFICATION OF CONTACT PROBLEMS

It is of an advantage to classify contacts between elastic bodies by comparing the extent of contact C in the deformed configuration with the initial contact C^0 in the load-free state. The initial contact between two bodies may be a point, a curve, a surface, or a combination of these elements. If the bodies are individually supported, it also is possible that C^0 is empty and that the bodies start to touch only after the loads have reached a certain level. While the initial contact C^0 is determined by the geometric features of the bodies and the support conditions, the contact C in the deformed state generally depends on the nature of applied loads, the level of loading and also on the elastic constants of the materials.

One should recognize that contact problems may also arise in the context of a single body, if one part of its surface can touch another part. Such self-contact is possible, for instance, in a slit hollow cylinder. There is, however, no essential difference between the contact of several bodies and self-contact.

The classification of contacts is based on the criterion of whether or not new surface points come into contact as the bodies deform. Adopting a terminology suitable for linear elastostatics in which boundary conditions are satisfied on the original geometry, C is not contained in C^0 ($C \not\subset C^0$) in the first case. Such contacts are called *advancing*. It may be noted that contacts, which upon loading shrink in some places but expand in others, are by definition advancing. If the initial contact extends over a surface, it may happen on the contrary that the contact shrinks everywhere. In such case, C is contained in C^0 ($C \subset C^0$), and the contact is called *receding*. The special case of a receding contact when C coincides with C^0 ($C = C^0$) has been called a stationary contact[1], but this term will presently be used in a different context.

* The results reported here were obtained in the course of research supported by the National Science Foundation of the United States.

Receding contacts are by no means as unusual as it at first may seem. In a well built machine or structure with many carefully fitted parts, it is likely that there are more receding than advancing contacts because of gaps opening between the individual parts, as they distort under loads. Simple examples of receding contact can also be recognized in every branch of solid mechanics: a plate on unilateral simple supports, a layer resting on a subgrade, an unbonded inclusion in a matrix, a cylindrical shell fitted inside another shell, fractured mass of rock, two beams or columns placed side by side, etc., where in each case the bodies may partially separate as they deform. Although strictly speaking every contact is advancing because of unavoidable surface roughness, its effect is in most cases confined to a boundary layer influencing little the deformations on a global scale. Exceptions may be expected, however, when heat is conducted through the contact surface.

The importance of the preceding classification lies in the fact that it is possible to arrive at some general properties of receding contacts. In contrast, over-all statements about advancing contacts seem to be restricted to mere scaling laws involving the initial gap between the bodies and the elastic constants of the materials[1].

Unfortunately it is not possible in every given case to decide, without solving the posed problem, whether the contact between the bodies is advancing or receding. For given geometries of the bodies, the type of contact depends on the nature of the applied loads and the elastic constants. In many situations, the type of contact can nevertheless be predicted by inspection, and it becomes possible to take advantage of the classification in analysis or experiments.

2. PROPERTIES OF RECEDING CONTACTS

Consider two bodies with initial contact C^0, as for example shown in Fig. 1 with C^0 extending over a surface. Denoting by n_i the unit outer normal and by u_i displacement, the normal component of the displacement is $U = u_i n_i$.

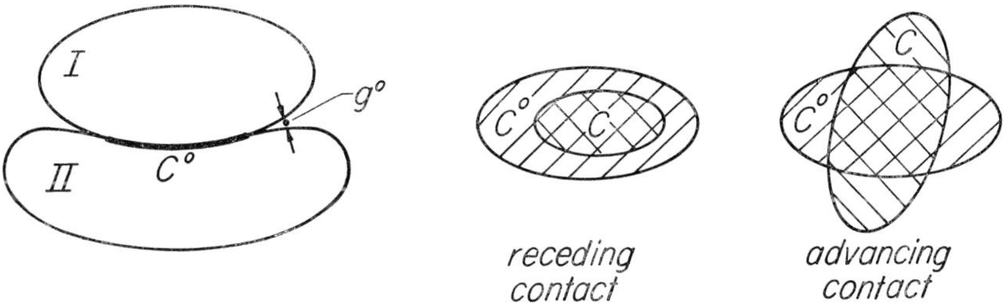

Fig. 1

The gap between the bodies in the deformed configuration is

$$g = g^0 - (U^{\mathrm{I}} + U^{\mathrm{II}}) \geq 0, \tag{1}$$

where g^0 is the gap in the load-free state. If the contact problem is considered within the framework of linear elasticity, so that strains, rotations and displacements are small and the boundary conditions are imposed on the original geometry, the contact C can extend only over those parts of the surfaces that initially have adjacent tangent planes and are separated by a gap g^0 which is small in comparison to the overall dimensions involved. Consequently the normals n_i^{I} and n_i^{II} are nearly parallel in the vicinity of $C^0 \cup C$, and it makes no difference whether g^0 and g are measured along n_i^{I}, n_i^{II} or some intermediate direction.

In the three regions that can be recognized in Fig. 1, we have

$$\begin{array}{llll} C \cap C^0: & g^0 = 0, & g = 0, & U^{\mathrm{I}} + U^{\mathrm{II}} = 0, \\ C \;- C \cap C^0: & g^0 > 0, & g = 0, & U^{\mathrm{I}} + U^{\mathrm{II}} = g^0, \\ C^0 - C \cap C^0: & g^0 = 0, & g > 0, & U^{\mathrm{I}} + U^{\mathrm{II}} < 0. \end{array} \tag{2}$$

If the contact is receding, the region $C - C \cap C^0$ is empty.

The other quantities, which besides the normal displacement are needed for the boundary conditions in the contact region, are the normal component of traction

$$N = t_i n_i = \sigma_{ij} n_i n_j, \tag{3}$$

and the shearing traction

$$s_i = t_i - N n_i = (\delta_{ij} - n_i n_j)\, \sigma_{jk} n_k. \tag{4}$$

In the absence of inelastic strain, Hooke's law gives stress in terms of displacement as

$$\sigma_{ij} = c_{ijkl} \partial_k u_l, \tag{5}$$

where c_{ijkl} is the modulus of the material, and ∂_k denotes differentiation with respect to x_k. Substituting the last expression into the previous two,

$$N = n_i n_j c_{ijkl} \partial_k u_l, \tag{6}$$

$$s_i = (\delta_{ij} - n_i n_j)\, n_k c_{jklm} \partial_l u_m. \tag{7}$$

It suffices to consider only two bodies, as extension of the results to a larger number or self-contact is immediate. The field equations that the displacements of the two bodies must satisfy are

$$c_{ijkl}^{\mathrm{I}} \partial_j \partial_k u_l^{\mathrm{I}} = -f_i^{\mathrm{I}}, \tag{8}$$

$$c_{ijkl}^{\mathrm{II}} \partial_j \partial_k u_l^{\mathrm{II}} = -f_i^{\mathrm{II}}, \tag{9}$$

where f_i denotes the intensity of body forces.

The boundary conditions on C for a *frictionless* contact are

$$n_i^{\mathrm{I}} u_i^{\mathrm{I}} + n_i^{\mathrm{II}} u_i^{\mathrm{II}} = \begin{cases} 0, & C \cap C^0, \\ g^0, & C - C \cap C^0, \end{cases} \tag{10}$$

$$n_i^{\mathrm{I}} n_j^{\mathrm{I}} c_{ijkl}^{\mathrm{I}} \partial_k u_l^{\mathrm{I}} - n_i^{\mathrm{II}} n_j^{\mathrm{II}} c_{ijkl}^{\mathrm{II}} \partial_k u_l^{\mathrm{II}} = 0, \tag{11}$$

$$(\delta_{ij} - n_i^{\mathrm{I}} n_j^{\mathrm{I}})\, n_k^{\mathrm{I}} c_{jklm}^{\mathrm{I}} \partial_l u_m^{\mathrm{I}} = 0, \tag{12}$$

$$(\delta_{ij} - n_i^{\mathrm{II}} n_j^{\mathrm{II}})\, n_k^{\mathrm{II}} c_{jklm}^{\mathrm{II}} \partial_l u_m^{\mathrm{II}} = 0. \tag{13}$$

These express the requirements that the normal components of the displacements are continuous after the gap between the bodies has been closed, that Newton's third law is obeyed, and that there are no frictional forces. Further, (10) to (13) must be supplemented with the conditions that the contact tractions are neither tensile ($N^{\mathrm{I}} = N^{\mathrm{II}} \leq 0$), nor unbounded except when one of the bodies has a discontinuous tangent plane. In the latter case, the nature of the proper singularities must be established by means of some special considerations.

Suppose that the surfaces of the bodies outside C are subjected to prescribed displacements υ_i or tractions τ_i. The boundary conditions are

$$u_i^{\mathrm{I}} = \upsilon_i^{\mathrm{I}}, \tag{14}$$

$$u_i^{\mathrm{II}} = \upsilon_i^{\mathrm{II}} \tag{15}$$

on those parts of the surfaces where displacements are specified, and

$$c_{ijkl}^{\mathrm{I}} n_j^{\mathrm{I}} \partial_k u_l^{\mathrm{I}} = \tau_i^{\mathrm{I}}, \tag{16}$$

$$c_{ijkl}^{\mathrm{II}} n_j^{\mathrm{II}} \partial_k u_l^{\mathrm{II}} = \tau_i^{\mathrm{II}} \tag{17}$$

where tractions are given. As C is generally unknown and its determination is part of the contact problem, care must be exercised in specifying the surface loads to avoid contradictory situations where the loaded regions would fall inside C. It also must be assumed that the applied loads are such that the global equilibrium conditions can be satisfied for each of the bodies in absence of friction. The last question can be settled using the equilibrium conditions of a rigid body.

It may be noted that the field equations and boundary conditions are made inhomogeneous by body forces, surface loads and the initial gap between the bodies. All of these quantities can be viewed as forcing terms, and there is no essential difference between the surface loads and the initial gap. If the contact problem also involved inelastic strain e_{ij}^*, so that Hooke's law became

$$\sigma_{ij} = c_{ijkl}(\partial_k u_l - e_{kl}^*), \tag{18}$$

additional forcing terms would appear in the field equations (8) and (9), and the boundary conditions (11), (12), (13), (16) and (17).

Let us assume that the problem on basis of (8) to (17) and the restrictions on the nature of the contact tractions is well posed, and that there is a unique displacement field specified by u_i^{I} and u_i^{II}. The extent of contact for given u_i^{I} and u_i^{II} is determined simply as the locus of

$$g^0 - (U^{\mathrm{I}} + U^{\mathrm{II}}) = 0. \tag{19}$$

The first set of properties of a receding contact follows from considering the displacements ku_i^{I} and ku_i^{II} $(0 < k < 1)$.* It is seen by inspection that (11), (12) and (13) are satisfied identically, and (8), (9), (14), (15), (16) and (17) will be satisfied if the body force intensities are changed to kf_i^{I} and kf_i^{II}, the prescribed displacements to kv_i^{I} and kv_i^{II}, and the specified surface tractions to $k\tau_i^{\mathrm{I}}$ and $k\tau_i^{\mathrm{II}}$. The remaining condition (10) requires that g^0 be changed to kg^0 on $C - C \cap C^0$. If this were done, (19) would determine the same contact C as for u_i^{I} and u_i^{II}, and no other disconnected contacts could develop outside C because of (1). If the contact is receding, however, the region $C - C \cap C^0$ is empty, (10) is satisfied by ku_i^{I} and ku_i^{II} identically, and the geometries of the bodies need not be slightly altered by changing g^0. This fact brings out the fundamental difference between receding and advancing contacts: Eq. (10) is homogeneous for receding but inhomogeneous for advancing contacts.

The conclusions about receding contacts are immediate:

1. *The displacement, strain and stress are proportional to the level of loading.*
2. *The contact C is independent of the level of loading or remains stationary.*
3. *Except for the special case of $C = C^0$, the change from C^0 to C is discontinuous.*

It should be noted that, although there is a proportionality between the applied loads and the induced elastic fields, the problem of a receding contact is not linear in the sense of addition. Thus, the results for two systems of loads cannot be superposed, unless it happens by coincidence that both load systems give identical extents of receding contact. On the other hand, superposition can be made to work in the solution of certain specific problems involving receding contact[2].

Scaling laws for receding contacts that are perhaps of less interest in analysis than experiments involve the elastic constants. Suppose that the moduli of the materials are changed from c_{ijkl} to mc_{ijkl}. The following conclusions can then be reached essentially by inspection of (8) to (17) and (19):

1. If v_i^{I} and v_i^{II} are changed to mv_i^{I} and mv_i^{II} but f_i^{I}, f_i^{II} and τ_i^{I}, τ_i^{II} left the same, the contact C and stress σ_{ij} are not altered and u_i^{I}, u_i^{II} change to $(1/m)u^{\mathrm{I}}$, $(1/m)u^{\mathrm{II}}$.
2. If $f_i^{\mathrm{I}}, f_i^{\mathrm{II}}$ are changed to mf_i^{I}, mf_i^{II}, and τ_i^{I}, τ_i^{II} to $m\tau_i^{\mathrm{I}}$, $m\tau_i^{\mathrm{II}}$, but v_i^{I} and v_i^{II}

* As will be discussed later, the restriction $k<1$ is connected with the possibility of a multiphase contact. If no multiphase contacts are involved, this restriction can be dropped.

left the same, the contact C and displacements u_i^{I}, u_i^{II} are left unaltered, and the stresses change from σ_{ij}^{I}, $\sigma_{ij}^{\mathrm{II}}$ to $m\sigma_{ij}^{\mathrm{I}}$, $m\sigma_{ij}^{\mathrm{II}}$.

3. SPECIAL PROPERTIES OF RECEDING CONTACTS IN PLANE PROBLEMS

Theory of elasticity is not rich in conclusions that are both general and physically unexpected. A notable exception is the result in plane elasticity due to J.H. Michel[3] concerning the dependence of stress on the elastic constants, which has been of untold importance particularly in photo-elasticity. It asserts that the in-plane components of stress do not depend on the elastic constants of an isotropic body, provided the body is subjected only to prescribed surface tractions and no net forces are transmitted through any closed contours, or

$$\oint t_i \, \mathrm{d}s = \oint \sigma_{ij} n_j \, \mathrm{d}s = 0, \quad (i,j = 1,2). \tag{20}$$

Since dimensional analysis predicts that, for specified surface tractions, the stress can depend only on Poisson's ratio, the Michel result constitutes essentially a reduction by one in the dependence of stress on the properties of the material.

A reduction in the dependence of stress on the elastic constants can also be achieved in plane problems involving a receding contact between *isotropic* bodies. It is convenient to base the derivations on the complex-variable formulation of plane elasticity[4], in which the displacement components u_x, u_y are related to the complex potentials $\varphi(z)$, $\psi(z)$ through

$$2\mu(u_x + iu_y) = \varkappa\varphi - z\overline{\varphi}' - \overline{\psi}. \tag{21}$$

In (21), μ denotes the shear modulus, $\varkappa = 3-4\nu$ for plane strain and $\varkappa = (3-\nu)/(1+\nu)$ for generalized plane stress, where ν is Poisson's ratio. If there are no body forces, the formulas giving stress components in terms of the complex potentials contain no elastic constants. Consequently the dependence of stress on the elastic constants is of the same form as that of the potentials. The elastic constants can enter the complex potentials through the requirement of single valued displacement and the boundary conditions. The first possibility can be precluded by insisting that (20) be satisfied. If the contacting bodies are loaded by prescribed surface tractions, and no displacement conditions of the type (14) and (15) are imposed, the elastic constants enter the potentials only through condition (10). Therefore, it suffices to examine this condition in detail.

The normal component of displacement is given by

$$4\mu U = \mathrm{e}^{-i\theta}(\varkappa\varphi - \bar{z}\varphi' - \overline{\psi}) + \mathrm{e}^{i\theta}(\varkappa\overline{\varphi} - \bar{z}\varphi' - \psi), \tag{22}$$

where θ is the angle from the x-axis to the outer normal. The differential relation

$$d(\varphi+z\bar{\varphi}'+\bar{\psi}) = i(t_x+it_y)\,ds \tag{23}$$

is also needed in the derivation. In (23), the arc-coordinate must be chosen so that the material is on the left when moving in the direction of increasing s along the boundary.

The boundary condition (10) for a *receding* contact becomes in terms of complex potentials

$$\begin{aligned}\Gamma[&\exp(-i\theta_1)\,(\varkappa_1\varphi_1-z\bar{\varphi}_1'-\bar{\psi}_1)+\exp(i\theta_1)\,(\varkappa_1\bar{\varphi}_1-\bar{z}\varphi_1'-\psi_1)]+\\&+\exp(-i\theta_2)\,(\varkappa_2\varphi_2-z\bar{\varphi}_2'-\bar{\psi}_2)+\exp(i\theta_2)\,(\varkappa_2\bar{\varphi}_2-\bar{z}\varphi_2'-\psi_2)=0.\end{aligned} \tag{24}$$

where subscripts 1 and 2 refer to the two bodies I and II, and $\Gamma=\mu_2/\mu_1$. At this point, the boundary condition contains the three constants Γ, $\varkappa_1$ and $\varkappa_2$. Newton's third law requires that

$$d(\varphi_1+z\bar{\varphi}_1'+\bar{\psi}_1) = d(\varphi_2+z\bar{\varphi}_2'+\bar{\psi}_2). \tag{25}$$

Due to the fact that an arbitrary complex constant γ can be added to $\varphi(z)$ and $\varkappa\bar{\gamma}$ to $\psi(z)$ without affecting the displacement, it is easy to show that (25) yields

$$\varphi_1+z\bar{\varphi}_1'+\bar{\psi}_1 = \varphi_2+z\bar{\varphi}_2'+\bar{\psi}_2, \tag{26}$$

provided (20) is satisfied on all closed contours lying in the region occupied by *both* bodies. Noting that $\exp(i\theta_2)=-\exp(i\theta_1)$, it follows from (26) that

$$\exp(-i\theta_1)\,(\varphi_1+z\bar{\varphi}_1'+\bar{\psi}_1)+\exp(-i\theta_2)\,(\varphi_2+z\bar{\varphi}_2'+\bar{\psi}_2)=0, \tag{27}$$

$$\exp(i\theta_1)\,(\bar{\varphi}_1+\bar{z}\varphi_1'+\psi_1)+\exp(i\theta_2)\,(\bar{\varphi}_2+\bar{z}\varphi_2'+\psi_2)=0. \tag{28}$$

Finally forming the linear combination $(24)+\frac{1}{2}(\Gamma+1)[(27)+(28)]$ of the preceding equations and dividing by $\Gamma(\varkappa_1+1)+\varkappa_2+1$, the result is

$$\begin{aligned}&(2+\alpha+\beta)\,[\exp(-i\theta_1)\,\varphi_1+\exp(i\theta_1)\,\bar{\varphi}_1]-\\&-(\alpha-\beta)\,[\exp(-i\theta_1)\,(z\bar{\varphi}_1'+\bar{\psi}_1)+\exp(i\theta_1)\,(\bar{z}\varphi_1'+\psi_1)]+\\&+(2-\alpha-\beta)\,[\exp(-i\theta_2)\,\varphi_2+\exp(i\theta_2)\,\bar{\varphi}_2]+\\&+(\alpha-\beta)\,[\exp(-i\theta_2)\,(z\bar{\varphi}_2'+\bar{\psi}_2)+\exp(i\theta_2)\,(\bar{z}\varphi_2'+\psi_2)]=0,\end{aligned} \tag{29}$$

where

$$\alpha=\frac{\Gamma(\varkappa_1+1)-(\varkappa_2+1)}{\Gamma(\varkappa_1+1)+\varkappa_2+1},\qquad \beta=\frac{\Gamma(\varkappa_1-1)-(\varkappa_2-1)}{\Gamma(\varkappa_1+1)+\varkappa_2+1}. \tag{30}$$

The constants α and β are measures for the mismatches in the uniaxial and voluminal compliances of two elastic materials[5,6]. It may be noted that $-1\le\alpha\le+1$, $-\frac{1}{2}\le\beta\le\frac{1}{2}$, that $\alpha=\beta=0$ for identical materials, and that α and β simply assume opposite signs upon interchange of the two materials. The last fact is clearly reflected in the structure of (29).

The immediate conclusion from (29) is that the extent of contact and the stresses in two contacting bodies depend only on the two constants α and β, provided the contact is receding, the bodies are loaded by specified surface tractions, and (20) is satisfied on all contours in the union of the regions occupied by the bodies, including the contact in the deformed configuration. Because of the last restriction, the result holds, for example, in case a shown in Fig. 2, but not in case b.

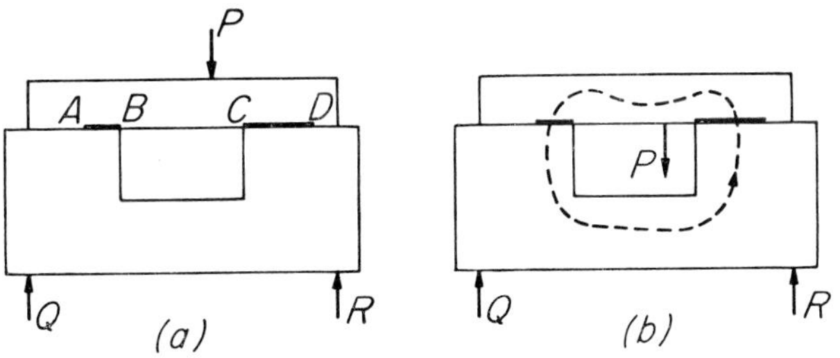

Fig. 2

It may be recalled[6] for the purposes of generalization that the stress in a composite body made of two *adhering* isotropic phases also depends only on α and β under similar conditions. It is then easy to arrive at the following results under the same restrictions as listed in the previous paragraph:

1. *In an assemblage of possibly inhomogeneous bodies involving two elastic phases, the extents of contact and the stresses depend only on the two constants α and β containing the properties of the materials.*
2. *If the bodies in the assemblage are homogeneous and made of identical materials, neither the extent of contact nor the stresses depend on the elastic constants.*

A further reduction in the dependence on the elastic constants can be attained for bodies in *contact along a straight boundary*. Placing the x-axis on this boundary, (10) becomes

$$u_y^{\mathrm{I}} = u_y^{\mathrm{II}}. \tag{31}$$

This condition can be replaced for a wide class of problems with

$$\frac{\partial u_y^{\mathrm{I}}}{\partial x} = \frac{\partial u_y^{\mathrm{II}}}{\partial x}. \tag{32}$$

Recalling that

$$4i\mu\,\frac{\partial u_y}{\partial x} = (\varkappa+1)\,(\varphi'-\bar{\varphi}') + \bar{z}\varphi''-z\bar{\varphi}''+\psi'-\bar{\psi}\,'. \tag{33}$$

$$2i\sigma_{xy} = \bar{z}\varphi'' - z\bar{\varphi}'' + \psi' - \bar{\psi}', \tag{34}$$

(32) leads immediately to

$$(1+\alpha)(\varphi_1' - \bar{\varphi}_1') = (1-\alpha)(\varphi_2' - \bar{\varphi}_2') \tag{35}$$

for a *frictionless* receding contact. It is seen that (35) contains only one parameter formed from the elastic constants, and that this boundary condition becomes independent of all elastic constants not only when the materials are identical, but also when one of the bodies is rigid ($\alpha = +1$ or -1).

The last reduction in the dependence on the elastic constants hinges on (32) being equivalent to (31). Eq. (32) ensures that the slopes of the deformed boundaries are equal at all points along the extent of contact. If one of the bodies is not constrained against rigid body motion and the contact is over a single interval, the bodies can be brought together and the normal displacements matched by translating one of them with respect to the other. In such case, (32) is indeed fully equivalent to (31). If, on the contrary, the contact consists of several disjoint parts, it is generally not possible to do so, and (32) cannot replace (31). Thus for example, when the bodies in Fig. 2a are put together along the segment AB, there is likely to be a gap or overlap on the second contact interval CD in spite of matching slopes.

More generally, conditions of the type (32) are not equivalent to those of (31) in an assemblage of bodies if it is possible to draw a closed curve that, first, does not lie wholly in one of the bodies and, second, does not cross a continuous contact interval more than once. Assuming that it is not possible to draw such a closed curve, and that all contacts are frictionless, receding and lie straight boundaries, the following conclusions can be reached for an assemblage of bodies subjected to specified surface tractions:

1. *If the assemblage consists of homogeneous bodies made of two materials, the extents of contact and the stresses depend on the elastic constants only through the single parameter* α.
2. *If the assemblage consists of homogeneous elastic bodies made of the same material and rigid bodies, the extents of contact and the stresses do not depend on the elastic constants.*

It may be noted that no restrictions need be placed on the shape of those parts of the boundaries which are not in contact.

Finally it is possible to combine the conclusion for contacts along straight boundaries involving elastic and rigid bodies (α dependence) with the previous result for inhomogeneous bodies made of two phases, but contact allowed on curved boundaries (α, β dependence). The result valid *under appropriate restrictions* asserts that, *in an assemblage of bodies consisting of inhomogeneous elastic members and rigid members, the extents of contact and the stresses depend only on the constants* α *and* β. *If the elastic members are homogeneous and are all of the same material, neither the extents of contact nor the stresses depend on the elastic constants.* The restrictions are:

1. The assemblage of bodies is subjected to specified surface tractions.
2. The elastic members involve at most two different materials.
3. All contacts are frictionless and receding.
4. The contacts between the elastic and rigid bodies are along straight boundaries.
5. Eq. (20) is satisfied in the union of the regions occupied by the bodies, including the contact C in the deformed state (those parts of C^0 where gaps have opened are viewed as cuts).
6. It is not possible to draw a closed curve in this union of regions, such that it does not lie wholly in one of the bodies, passes into a rigid member, and does not intersect a continuous contact interval between an elastic and rigid member more than once.

The reduction in the dependence on the elastic constants can also be traced for assemblages involving more than two elastic materials, but there is noting that could not be anticipated from the previous results.

4. MULTIPHASE CONTACTS

In order to develop some further ideas, consider the four different contact problems indicated in Fig. 3, in which a slab resting on a subgrade is used for purposes of illustration.

In the first case, the contact between the slab and the subgrade is undoubtedly receding and remains such regardless of the magnitude of the applied force P. Consequently, the length of contact c_1 shown in Fig. 3a is stationary or independent of P ($P > 0$), and there is a discontinuous change from the initial contact length c_0 to c_1. Only one phase in the development of contact between the bodies can be recognized in this case.

There are two phases in the deformation process connected with the next problem (Fig. 3b). This case illustrates the point that a receding contact can change into an advancing contact. The receding contact of length c_1 exists only up to some magnitude P' of the applied force, because new contacts start to develop at A and B for sufficiently high levels of loading. Thus the contact is receding during the first ($0 < P < P'$) but advancing during the second ($P > P'$) phase.

In the third case shown (Fig. 3c), the contact is advancing, but it develops in several phases as the load is increased. Furthermore, if the gap between the bottom of the slab and the subgrade is small and the slab is sufficiently long in comparison to its thickness, the contact during the third phase ($P'' < P < P'''$) is stationary ($c_2 = \text{const}$). The reason for this is that, once the humps on the bottom of the slab lift off the subgrade, there is no difference with the case shown in Fig. 3a. During the last phase when there is also contact at A and B, c_2 ceases to be constant.

The last example indicated in Fig. 3d shows that it also is possible to have receding contact change to advancing contact, which in turn becomes stationary at some level of loading ($P > P''$).

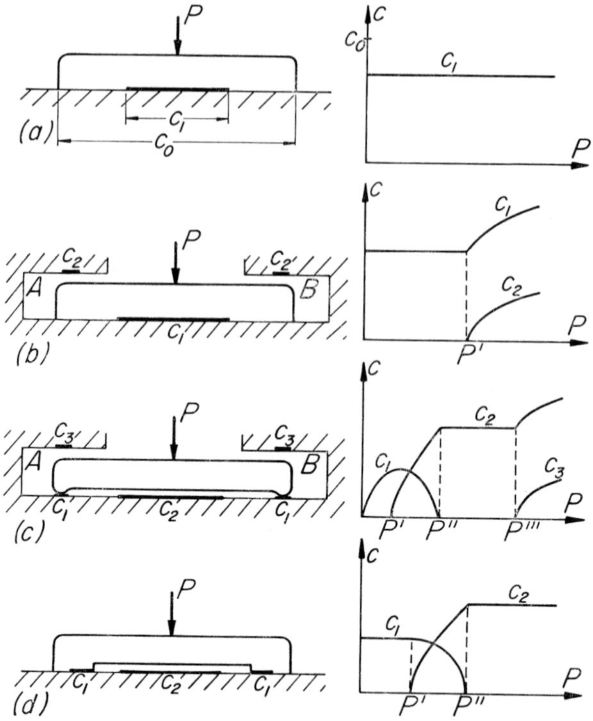

Fig. 3

Not many general statements can be made about multiphase contacts. The only exceptions are those that derive from the properties of receding contacts:

1. A receding contact cannot be preceded by an advancing contact. In other words, if $C \subset C^0$ at some level of loading (test load), $C \subset C^0$ at all lower load levels. This conclusion follows from (1) by considering the possibility of the gap closing somewhere outside C^0 as the load is lowered, and assuming existence and uniqueness of the elastic fields. In regions where $g < g^0$ at the test load, $U^{\mathrm{I}} + U^{\mathrm{II}} > 0$. If the load is reduced, $U^{\mathrm{I}} + U^{\mathrm{II}}$ decreases and the gap widens. Consequently no new contacts can develop in these regions. On those parts of the boundaries where $g > g^0$ at the test load, the gap narrows as the load is reduced, but it cannot become smaller than g^0.
2. If the contact is receding during the first phase, new contacts can eventually develop *only outside* C^0. This is due to the fact that in the

region $C^0 - C$, $g^0 = 0$ and $U^{\mathrm{I}} + U^{\mathrm{II}} < 0$, and that $U^{\mathrm{I}} + U^{\mathrm{II}}$ decreases monotonically with increasing level of loading during the phase of receding contact.

3. If the contacts at two load levels are identical, the contact is stationary at intermediate levels of loading. This result also follows from (1) by reasoning along similar lines.
4. If an advancing contact has a stationary phase, it possesses all properties of a receding contact during this phase.

5. DYNAMIC CONTACT PROBLEMS

The distinction between receding and advancing contacts retains its significance also in dynamic contact problems. However, general statements again are possible only for receding contacts, or the situations in which the contact C in the deformed configuration is at all times contained in the contact C^0 associated with the undeformed state. No suitable solutions of dynamic contact problems are available to guide thoughts on this subject, but the physically more transparent cases, such as a layer in contact with a half space, or a matrix containing loose inclusions may be suggested for the purpose of visualization.

The field equations for dynamic problems involving two bodies are

$$c^{\mathrm{I}}_{ijkl} \partial_j \partial_k u^{\mathrm{I}}_l - \rho^{\mathrm{I}} \ddot{u}^{\mathrm{I}}_i = -f^{\mathrm{I}}_i, \tag{36}$$

$$c^{\mathrm{II}}_{ijkl} \partial_j \partial_k u^{\mathrm{II}}_l - \rho^{\mathrm{II}} \ddot{u}^{\mathrm{II}}_i = -f^{\mathrm{II}}_i, \tag{37}$$

which replace (8) and (9) for the static case. The boundary conditions (10) to (17) remain intact, but the statement of the problem must be supplemented with initial conditions at time $t = 0$:

$$u^{\mathrm{I}}_i(t=0) = u^{0\mathrm{I}}_i, \qquad u^{\mathrm{II}}_i(t=0) = u^{0\mathrm{II}}_i, \tag{38}$$

$$\dot{u}^{\mathrm{I}}_i(t=0) = \dot{u}^{0\mathrm{I}}_i, \qquad \dot{u}^{\mathrm{II}}_i(t=0) = \dot{u}^{0\mathrm{II}}_i. \tag{39}$$

The right sides of the initial conditions constitute forcing terms in the sense that they make the relations involved inhomogeneous. It is assumed that (36) to (39) and (10) to (17) lead to a well posed problem which has a unique solution.

Similarly to the equilibrium case, the properties of the elastic fields emerge when one assumes that there are displacements u^{I}_i and u^{II}_i satisfying (36) to (39) and (10) to (17), and inquires as to what conditions are satisfied by the displacements ku^{I} and ku^{II}. The answer is immediate: if the contact is receding the scaled displacements yield the same contact and satisfy the field equations, boundary conditions and initial conditions in which all forcing terms are scaled by the factor k.

There is a large number of special situations that could be considered, but elaborations on the general conclusion are straightforward and only one case will be mentioned. Suppose that the bodies are initially ($t = 0$) undeformed and at rest, and are subsequently subjected to some time-dependent surface loads. An example of this might be the layer which is contact with a half-space and is loaded by a set of downward forces during $t > 0$. A very complicated contact pattern could be anticipated, possibly with many zones of separation. The conclusion is, however, that the history of contact $C(t)$, while determined by the spatial and temporal distribution of the applied forces, is independent of the general level of loading. Furthermore, the elastic fields are proportional to this level. Thus, doubling the forces doubles the displacements, strains and stresses, but but leaves $C(t)$ unaffected.

REFERENCES

1. J. Dundurs and M. Stippes, 'Role of Elastic Constants in Certain Contact Problems', *J. Appl. Mech.*, **37**, 1970, pp. 965-970.
2. K.C. Tsai, J. Dundurs and L.M. Keer, 'Elastic Layer Pressed Against a Half-Space', *J. Appl. Mech.*, **41**, 1974, pp. 703-707.
3. J.H. Michell, 'On the Direct Determination of Stress in an Elastic Solid, with Application to the Theory of Plates', *Proc. Lond. Math. Soc.*, **31**, 1899, pp. 100-124.
4. N.I. Muskhelishvili, *Some Basic Problems of the Mathematical Theory of Elasticity*, 3rd ed., transl. by J.R. M. Radok, P. Noordhoff, Groningen, 1953.
5. J. Dundurs, 'Discussion', *J. Appl. Mech.*, **36**, 1969, pp. 650-652.
6. J. Dundurs, 'Some Properties of Elastic Stresses in a Composite', in A.C. Eringen (ed.), *Recent Advances in Engineering Science*, Gordon and Breach, New York, 1970, pp. 203–216.

Similarity considerations for contact between dissimilar elastic bodies

D. A. Spence

1. INTRODUCTION

We consider the problem of indentation of a linearly-elastic half space by an axisymmetric elastic punch. From dimensional considerations we infer in section 2 the form of the integral kernel in the relationship between contact stresses and displacements, without evaluating its precise form. The dependence of the kernel on the elastic constants – in particular on the ratio $\gamma=(1-2\nu)/(2-2\nu)$ – in fact is such that the problem of elastic-elastic contact between bodies having different constants can be reduced to that of rigid-elastic contact with a modified value of γ. This is shown in section 3. We then consider in detail in section 4 the boundary value problem posed by a 'power law' indentor under adhesive conditions, and infer the functional form of the radial displacement from similarity arguments. It is then possible to apply the results of section 2 to obtain the stress distributions for such indentors by quadrature from those for flat indentors, as shown in section 5. The frictional problem is outlined in section 6.

2. GENERAL EQUATIONS

For a uniform half space $z>0$, write the surface stresses as $(\sigma_{zz})_{z=0}=\sigma_1$, $(\sigma_{rz})_{z=0}=\sigma_2$ and displacements $(u_z)_{z=0}=u_1$, $(u_r)_{z=0}=u_2$.

Axisymmetric case

Consider a stress distribution $\sigma_j(s)$ over the annulus s, $s+\mathrm{d}s$, and suppose the ith component of the corresponding displacement at radius r is $\mathrm{d}u_i(r)$ $(i,j=1,2)$. Then in the limit $\mathrm{d}s\to 0$, the dimensionless ratio

$$\frac{\mu\,\mathrm{d}u_i(r)}{\sigma_j(s)\,\mathrm{d}s}$$

(where μ is the shear modulus) is for a given material, a function only of the dimensionless ratio r/s, namely $(1-\nu)k_{ij}(r/s)$ say. Of course k_{ij} also depends on the Poisson's ratio ν of the material. Integrating over all such elements gives

$$\left(\frac{\mu}{1-\nu}\right) u_i(r) = \int_0^a k_{ij}\left(\frac{r}{s}\right) \sigma_j(s)\, \mathrm{d}s \tag{2.1}$$

with summation over the repeated suffix, where a is the contact radius.

Reciprocal theorem

If u_i, σ_i is one virtual state of strain, $\tilde{u}_i$, $\tilde{\sigma}_i$ another, the theorem states that

$$\int u_i \tilde{\sigma}_i\, \mathrm{d}S = \int \tilde{u}_i \sigma_i\, \mathrm{d}S$$

Therefore

$$\int k_{ij}\left(\frac{r}{s}\right) \sigma_j(s)\, \tilde{\sigma}_i(r)\, r\mathrm{d}r\mathrm{d}s = \int k_{ij}\left(\frac{r}{s}\right) \tilde{\sigma}_j(s)\, \sigma_i(r)\, r\mathrm{d}r\mathrm{d}s$$

Interchange of r and s, and of i and j, in the right hand integral gives

$$\int k_{ji}\left(\frac{s}{r}\right) \tilde{\sigma}_i(r)\, \sigma_j(s)\, s\mathrm{d}s\,\mathrm{d}r$$

Therefore

$$rk_{ij}\left(\frac{r}{s}\right) = sk_{ji}\left(\frac{s}{r}\right) \tag{2.2}$$

Forms of the contact equations

They may be derived in many ways, e.g. by use of transform methods[1] to solve the biharmonic equation, as is necessary to treat a layer of finite thickness, and from the complex variable approach in two dimensions[2]. For half spaces, the Green's function for point loading gives a direct derivation: Landau and Lifshitz[3], equations (8.19), p. 29, can be written

$$\left(\frac{\mu}{1-\nu}\right)\begin{pmatrix} u_x \\ u_y \\ u_z \end{pmatrix} = \frac{1}{2\pi r}\begin{bmatrix} X + \left(\dfrac{\nu}{1-\nu}\right)\left(\dfrac{xT}{r} - \dfrac{\gamma xZ}{r}\right) \\ Y + \left(\dfrac{\nu}{1-\nu}\right)\left(\dfrac{yT}{r} - \dfrac{\gamma yZ}{r}\right) \\ Z + \gamma T \end{bmatrix} \tag{2.3}$$

$\mu =$ shear modulus $= \frac{1}{2}E/(1+\nu)$, $\gamma = (1-2\nu)/(2-2\nu)$, X, Y, Z are point loads at the origin, and $T=(xX+yY)/r$ ($r=\sqrt{(x^2+y^2)}$). This is of the form

$$\left(\frac{\mu}{1-\nu}\right) \underset{\sim}{u}(x) = G(\underset{\sim}{x}-\underset{\sim}{\xi})\, \underset{\sim}{F}(\underset{\sim}{\xi})$$

so that the displacement for a distribution of stress $\underset{\sim}{\sigma}(\underset{\sim}{x})$ is of the form

$$\left(\frac{\mu}{1-\nu}\right) \underset{\sim}{u}(\underset{\sim}{x}) = \int G(\underset{\sim}{x}-\underset{\sim}{\xi})\, \underset{\sim}{\sigma}(\underset{\sim}{\xi})\, \mathrm{d}S(\underset{\sim}{\xi})$$

Integration over an axisymmetric distribution gives the expressions

$$\left(\frac{\mu}{1-\nu}\right)\begin{pmatrix} u_z \\ u_r \end{pmatrix} = \begin{pmatrix} -\int k_{11}\left(\frac{r}{s}\right)\sigma_{zz}(s)\,\mathrm{d}s + \gamma\int k_{12}\left(\frac{r}{s}\right)\sigma_{rz}(s)\,\mathrm{d}s \\ \gamma\int k_{21}\left(\frac{r}{s}\right)\sigma_{zz}(s)\,\mathrm{d}s - \int k_{22}\left(\frac{r}{s}\right)\sigma_{rz}(s)\,\mathrm{d}s \end{pmatrix} \tag{2.4}$$

where the diagonal elements are

$$\left.\begin{aligned} k_{11}(\lambda) &= \frac{2}{\pi}K(\lambda) \\ k_{22}(\lambda) &= \frac{2}{\pi\lambda}[K(\lambda)-E(\lambda)] \end{aligned}\right\} \quad (\lambda<1) \tag{2.5}$$

(K, E are the complete elliptic integrals of first and second kind)

The values for $\lambda>1$ are given by the reciprocal theorem (2.2):

$$k_{ij}(\lambda) = \frac{1}{\lambda}k_{ji}\left(\frac{1}{\lambda}\right).$$

The off-diagonal elements are

$$k_{12}(\lambda) = \begin{cases} 0 \\ 1 \end{cases}, \quad k_{21}(\lambda) = \begin{cases} 1/\lambda & (\lambda>1) \\ 0 & (\lambda<1) \end{cases} \tag{2.6}$$

k_{11} and k_{22} are logarithmically singular at $\lambda=1$.
Hence (2.4) reduces to

$$\left(\frac{\mu}{1-\nu}\right)\begin{pmatrix} u_z \\ u_r \end{pmatrix} = \begin{pmatrix} -\int_0^a k_{11}\left(\frac{r}{s}\right)\sigma_{zz}(s)\,\mathrm{d}s + \gamma\int_0^a \sigma_{rz}(s)\,\mathrm{d}s \\ \frac{\gamma}{r}\int_0^r s\sigma_{zz}(s)\,\mathrm{d}s - \int_0^a k_{22}\left(\frac{r}{s}\right)\sigma_{rz}(s)\,\mathrm{d}s \end{pmatrix} \tag{2.7}$$

These expressions were previously derived by Noble and Spence[4] from the transform solution of Sneddon.

If now we write ρ for r in (2.7), and apply the operators

$$A_{11} = \frac{d}{dr}\int_0^r \frac{\rho\, d\rho}{\sqrt{(r^2-\rho^2)}}, \qquad A_{22} = \frac{d}{dr}\int_0^r \frac{r\, d\rho}{\sqrt{(r^2-\rho^2)}} \tag{2.8}$$

to the first and second equations, they become

$$-\int_r^a \frac{s\sigma_{zz}(s)\, ds}{\sqrt{(s^2-r^2)}} + \gamma\left[\int_0^a \sigma_{rz}(s)\, ds - r\int_0^r \frac{\sigma_{rz}(s)\, ds}{\sqrt{(r^2-s^2)}}\right] = \left(\frac{\mu}{1-\nu}\right) u_z^*(r) \tag{2.9a}$$

$$\gamma\int_0^r \frac{s\sigma_{zz}(s)\, ds}{\sqrt{(r^2-s^2)}} - r\int_r^a \frac{\sigma_{rz}(s)\, ds}{\sqrt{(s^2-r^2)}} = \left(\frac{\mu}{1-\nu}\right) u_r^*(r) \tag{2.9b}$$

respectively, where $u_z^*(r) = \dfrac{d}{dr}\displaystyle\int_0^r \frac{su_z(s)\, ds}{\sqrt{(r^2-s^2)}}$, $u_r^*(r) = \displaystyle\int_0^r \frac{[su_r(s)]'\, ds}{\sqrt{(r^2-s^2)}}$

Symbolically, equation (2.7) may be represented as

$$\begin{pmatrix} K_{11} & \gamma K_{12} \\ K_{21} & K_{22} \end{pmatrix}\begin{pmatrix} \sigma_1 \\ \sigma_2 \end{pmatrix} = \left(\frac{\mu}{1-\nu}\right)\begin{pmatrix} u_1 \\ u_2 \end{pmatrix} \tag{2.10}$$

or, more briefly, as

$$K\underset{\sim}{\sigma} = \left(\frac{\mu}{1-\nu}\right)\underset{\sim}{u} \tag{2.11}$$

and the above operations represent multiplication by a matrix

$$A = \begin{pmatrix} A_{11} & 0 \\ 0 & A_{22} \end{pmatrix}$$

to give the results (2.9) in the matrix-operation form

$$\begin{pmatrix} L_{11} & \gamma L_{12} \\ \gamma L_{21} & L_{22} \end{pmatrix}\begin{pmatrix} \sigma_1 \\ \sigma_2 \end{pmatrix} = \left(\frac{\mu}{1-\nu}\right)\begin{pmatrix} A_{11}\, u_1 \\ A_{22}\, u_2 \end{pmatrix} \equiv \left(\frac{\mu}{1-\nu}\right)\begin{pmatrix} u_1^* \\ u_2^* \end{pmatrix} \text{ say} \tag{2.12}$$

It is, likewise, possible to invert the equations (2.10) to the form

$$\begin{pmatrix} \sigma_1 \\ \sigma_2 \end{pmatrix} = 4\mu\left(\frac{1-\nu}{3-4\nu}\right)\begin{pmatrix} M_{11} & \gamma M_{12} \\ \gamma M_{21} & M_{22} \end{pmatrix}\begin{pmatrix} u_1 \\ u_2 \end{pmatrix} \tag{2.13}$$

The M operators are given in [4], equations 2.13 and 2.14.

Two dimensional case

Exactly analogous results are obtainable. We mention in particular the equations corresponding to (2.7), which are now

$$\left(\frac{\mu}{1-\nu}\right)\begin{pmatrix} v'(x) \\ u'(x) \end{pmatrix} = \begin{pmatrix} \dfrac{1}{\pi}\displaystyle\int_{-a}^{a} \frac{\sigma(t)\,\mathrm{d}t}{t-x} - \gamma\tau(x) \\ \gamma\sigma(x) + \dfrac{1}{\pi}\displaystyle\int_{-a}^{a} \frac{\tau(t)\,\mathrm{d}t}{t-x} \end{pmatrix} \qquad \begin{matrix}(2.14\mathrm{a}) \\ (2.14\mathrm{b})\end{matrix}$$

as given by Muskhelishvili, and the corresponding Abel equations for the case when $v(=u_z)$ is even and $u(=u_x)$ odd with respect to x:

$$\begin{pmatrix} \displaystyle\int_x^a \frac{t\sigma(t)\,\mathrm{d}t}{\sqrt{(t^2-x^2)}} - \gamma\int_0^x \frac{t\tau(t)\,\mathrm{d}t}{\sqrt{(t^2-x^2)}} \\ \gamma\displaystyle\int_0^x \frac{\sigma(t)\,\mathrm{d}t}{\sqrt{(x^2-t^2)}} + \int_x^a \frac{\tau(t)\,\mathrm{d}t}{\sqrt{(t^2-x^2)}} \end{pmatrix} = \left(\frac{\mu}{1-\nu}\right)\int_0^x \frac{\mathrm{d}t}{\sqrt{(x^2-t^2)}}\begin{pmatrix} tv'(t) \\ u'(t) \end{pmatrix}$$

3. EQUIVALENCE OF BOUNDARY VALUE PROBLEMS FOR RIGID-ELASTIC AND FOR ELASTIC-ELASTIC CASE

Consider the contact between two materials, referred to by suffixes 1 and 2. Static equilibrium requires that

$$(\sigma_{zz})_1 = (\sigma_{zz})_2 = \bar{\sigma}, \quad (\sigma_{rz})_1 = -(\sigma_{rz})_2 = \bar{\tau} \quad \text{say} \tag{3.1}$$

Then

$$(u_z)_\alpha = \frac{1-\nu_\alpha}{\mu_\alpha} K_{11}\bar{\sigma} \pm \frac{1-2\nu_\alpha}{2\mu_\alpha} K_{12}\bar{\tau} \tag{3.2}$$

$$(u_r)_\alpha = \frac{1-2\nu_\alpha}{2\mu_\alpha} K_{21}\bar{\sigma} \pm \frac{1-\nu_\alpha}{\mu_\alpha} K_{22}\bar{\tau}, \ \alpha=1,2, \tag{3.3}$$

(+ for $\alpha=1$, − for $\alpha=2$) where K_{ij} are the integral operators defined by (2.10). Likewise, the combined displacements

$$\bar{w} = (u_z)_1 + (u_z)_2 \tag{3.4}$$

$$\bar{u} = (u_r)_1 + (u_r)_2$$

are given by

$$\begin{pmatrix} \bar{w} \\ \bar{u} \end{pmatrix} = \left(\frac{1-\nu_1}{\mu_1} + \frac{1-\nu_2}{\mu_2}\right)\begin{pmatrix} K_{11} & \bar{\gamma}K_{12} \\ \bar{\gamma}K_{21} & K_{22} \end{pmatrix}\begin{pmatrix} \bar{\sigma} \\ \bar{\tau} \end{pmatrix} \tag{3.5}$$

where

$$\bar{\gamma} = \left(\frac{1-2\nu_1}{2\mu_1} - \frac{1-2\nu_2}{2\mu_2}\right)\Big/\left(\frac{1-\nu_1}{\mu_1} + \frac{1-\nu_2}{\mu_2}\right)$$

Thus the functional dependence of the barred displacements on the barred stresses is the same as that given by (2.11), with appropriately modified physical constants.

4. THE GENERAL MIXED BOUNDARY VALUE PROBLEM

The problem may be defined as that of solving the integral equation (2.11) for $\underset{\sim}{\sigma}$ and $\underset{\sim}{u}$ on some parts of the boundary $z=0$ (not, in general, the same regions for $\underset{\sim}{\sigma}$ as for $\underset{\sim}{u}$). The boundary conditions most often encountered are:
i. *Classical* boundary conditions: u_1 prescribed, $\sigma_2 = 0$.
ii. *Adhesive:* u_1 prescribed, $(\partial u_2)/(\partial a) = 0$ $(r < a)$.
For a flat punch, a is fixed and we therefore have $u_2 = 0$. For a punch whose shape is given by a power law, when the applied force is P, the contact radius $a(P)$, and maximum indentation $\delta(a)$, let us write $x = r/a$ and seek solutions of the form

$$\sigma_{zz} = \left(\frac{\mu}{1-\nu}\right)\frac{\delta}{a}\, p(x),$$

$$\sigma_{rz} = \left(\frac{\mu}{1-\nu}\right)\frac{\delta}{a}\, q(x)$$

Such solutions can exist only if the surface displacements take the form

$$u_z = \delta(P)\, w(x), \qquad u_r = \delta(P)\, u(x) \tag{4.1}$$

and will then be given by solving the coupled equations

$$K(\gamma)\begin{pmatrix} p \\ q \end{pmatrix} = \begin{pmatrix} w \\ u \end{pmatrix} \tag{4.2}$$

In fact $u_z = \delta(P) - Br^n$ say ($n = 2$ for Hertzian case).
So

$$w(x) = 1 - \frac{Ba^n}{\delta(a)}\, x^n \tag{4.3}$$

This is of the required form if

$$\delta = Aa^n \tag{4.4}$$

where A depends on γ but not on a.
Then

$$u_r(r, a) = Aa^n\, u(x) \tag{4.5}$$

and since $(\partial u_r)/(\partial a) = 0$, this must be a function of r only. This can be the case only if

$$u_r = Cr^n \tag{4.6}$$

Thus the adhesive problem can be formulated in terms of two unknown constants A and C, on which it depends linearly, and which are found finally from the condition of vanishing stress at $r=a$. This solution was obtained in the author's earlier paper[5] using the solution of an integral equation of Wiener-Hopf type[6].

However a property of the kernel makes a simpler solution possible, by reduction to the flat punch problem, as follows.

5. REDUCTION TO 'FLAT PUNCH' PROBLEM

The equations are

$$K\begin{pmatrix} p \\ q \end{pmatrix} = \begin{pmatrix} 1-Ax^n \\ Cx^n \end{pmatrix} \tag{5.1}$$

where in dimensionless form

$$K_{ij}p_j = \int_0^1 k_{ij}\left(\frac{x}{y}\right) p_j(y)\,\mathrm{d}y$$

Now differentiating with respect to x and integrating by parts gives

$$\begin{aligned} x\frac{\mathrm{d}}{\mathrm{d}x}(K_{ij}p_j) &= \int_0^1 \frac{x}{y} k'_{ij}\left(\frac{x}{y}\right) p_j(y)\,\mathrm{d}y \\ &= -\left[k_{ij}\left(\frac{x}{y}\right) y p_j(y)\right]_0^1 \\ &\quad + \int_0^1 k_{ij}\left(\frac{x}{y}\right)(y p_j(y))'\,\mathrm{d}y \end{aligned}$$

and since $p_j(1)=0$ (this defines the area of contact) we have

$$x\frac{\mathrm{d}}{\mathrm{d}x}K_{ij}p_j = K_{ij}(xp_j)' \tag{5.2}$$

Now apply the operator $1-\dfrac{x}{n}\dfrac{\mathrm{d}}{\mathrm{d}x}$ to 5.1):

$$\left(1-\frac{x}{n}\frac{\mathrm{d}}{\mathrm{d}x}\right)Kp = \begin{pmatrix} 1 \\ 0 \end{pmatrix} = K\underset{\sim}{p}^{(0)} \quad \text{say} \tag{5.3}$$

Therefore, by (4.8),

$$\underset{\sim}{p} - \frac{1}{n}(x\underset{\sim}{p})' = \underset{\sim}{p}^{(0)} \tag{5.4}$$

and integration with respect to x gives the solution $\underset{\sim}{p} = \underset{\sim}{p}^{(n)}$ say which vanishes at $x = 1$ as

$$\underset{\sim}{p}^{(n)}(x) = nx^{n-1} \int_x^1 y^{-n}\, \underset{\sim}{p}^{(0)}(y)\, \mathrm{d}y \tag{5.5}$$

Note that in the Hertz ($n = 2$) case, for frictionless indentation, since these arguments apply *a fortiori*

$$p_1^{(0)} = \frac{2}{\pi}(1-x^2)^{-1/2} \quad \text{(Boussinesq)}$$

and

$$p_1^{(2)} = \frac{4}{\pi}(1-x^2)^{1/2} \quad \text{(Hertz)}$$

For a more general indentor of shape $z = g(r)$, for which

$$w(r, a) = w_0(a) - g(r) \tag{5.6}$$

we can express the solution in terms of that for a flat punch, which we suppose known, as follows.

Write the normal stress and normal displacement for the flat adhesive punch as

$$\begin{aligned} \sigma^{(0)}(r, a) &= \left(\frac{\mu}{1-\nu}\right)\frac{\delta}{a}\, p^{(0)}\left(\frac{r}{a}\right) \\ w(r, a) &= \delta f_0\left(\frac{r}{a}\right) \end{aligned} \tag{5.7}$$

say, where $p^{(0)} = 0$ for $r/a > 1$, $f_0 = 1$ for $r/a < 1$.

Then when incremental loading takes place, at each stage when the normal displacement within the contact area increases by $\mathrm{d}w_0$, the increment over the whole surface is $\mathrm{d}w_0(a) f_0(r/a)$.

Therefore for $r < a$,

$$w_0(a) - g(r) = \int_0^a w_0'(s) f_0\left(\frac{r}{s}\right) \mathrm{d}s$$

whence on integration by parts, we obtain a Volterra equation for w_0:

$$w_0(r) - \int_0^r w_0'(s) f_0\left(\frac{r}{s}\right) \mathrm{d}s = g(r) \tag{5.8}$$

In particular, if

$$g(r) = \Sigma B_n r^n, \text{ then } w_0(a) = \Sigma A_n a^n \tag{5.9}$$

where

$$A_n = B_n \Big/ \left[1 - n \int_1^\infty t^{-n-1} f_0(t)\, \mathrm{d}t \right] \tag{5.10}$$

(A result of this form was first obtained by Segedin[9] for the smooth dunch.) Similarly, the pressure and shear stress distributions are given by

$$\sigma_i(r, a) = \left(\frac{\mu}{1-\nu}\right) \int_r^a p_i^{(0)} \left(\frac{r}{s}\right) w_0'(s) \frac{\mathrm{d}s}{s} \; (i = 1, 2) \tag{5.11}$$

and for a polynomial indentor this becomes

$$\sigma_i(r, a) = \left(\frac{\mu}{1-\nu}\right) \Sigma A_n\, a^{n-1}\, p_i^{(n)} \left(\frac{r}{s}\right) (i = 1, 2) \tag{5.12}$$

The functions f_0, $p_i^{(0)}$ are given in Spence [5]. The incremental argument leading to (5.8) is due to Mossakovski[10].

6. FRICTIONAL IDENTATION WITH PARTIAL SLIP UNDER MONOTONIC LOADING

We here consider the problem where, as in section 4, (ii) above,

$$u_z = \delta - Br^n \tag{6.1}$$

but the contact circle $r < a$ is now divided into two parts.

1. An adhesive inner circle $r < ac$, where

$$\frac{\partial u_r}{\partial a} = 0, \quad \eta p_1 - p_2 > 0 \tag{6.2}$$

where η is the Coulomb coefficient of friction.

2. A frictional annulus $ac < r < a$, in which

$$\frac{\partial u_r}{\partial a} < 0, \; \eta p_1 - p_2 = 0 \tag{6.3}$$

Here c is an eigenvalue, initially unknown, which must depend on η and ν. This has been treated in Spence[7] (two dimensional case) and [8] (axi-symmetric) but was first posed for a flat indentor by Galin[11]. The same arguments as in the fully adhesive case show that in region (1) $u_r = Cr^n$, so application of the same operator gives

$$\left(1-\frac{r}{n}\frac{\mathrm{d}}{\mathrm{d}r}\right)\begin{pmatrix}u_1\\u_2\end{pmatrix}=\begin{pmatrix}\delta\\u_2^{(0)}\end{pmatrix} \tag{6.4}$$

where $u_2^{(0)}=0$ for $0<\rho<c$.

These displacements are of the same form as those for the frictional problem for a flat punch, so we deduce as before that $p_i^{(n)}(p)$ are related to $p_i^{(0)}(p)$ by (5.5); and that the required inequalities (6.2) and (6.3) are satisfied by the $p_i^{(n)}$, since

$$\eta p_1^{(n)} - p_2^{(n)} = nx^{n-1}\int_x^1 [\eta p_1^{(0)}(t) - p_2^{(0)}(t)]\, t^{-n}\,\mathrm{d}t$$

This expression vanishes for $x>c$, and is positive for $x<c$, since the integrand is so, and it follows that *the slip radius c is the same for the curved as for the flat indentor.* A similar argument shows that the condition on u_2 is also satisfied. The conclusion can moreover, be extended to indentors given by the polynomial law (5.9), since we then have

$$\eta\sigma_1-\sigma_2=\left(\frac{\mu}{1-\nu}\right)\Sigma A_n\, a^{n-1}(\eta p_1^{(n)}-q_1^{(n)})$$

which is zero or positive accordingly as $r \gtrless ac$ for all a.

REFERENCES

1. I.N. Sneddon, *Fourier Transforms*, McGraw Hill, New York.
2. N.I. Muskhelishvili, *Some Basic Problems of the Mathematical Theory of Elasticity*, Eng. transl. by J.R.M. Radok, Noordhoff, Groningen, 1953.
3. L.D. Landau and E.M. Lifshitz, *Theory of Elasticity*, Pergamon, London, 1959.
4. B. Noble and D.A. Spence, *Formulation of Two Dimensional and Axisymmetric Boundary Value Problems*, U. Wisconsin, Math. Res. Center, TR 1089, 1971.
5. D.A. Spence, *Proc. Roy. Soc.*, **A305**, 1968a, pp. 55–80.
6. D.A. Spence, *Proc. Roy. Soc.*, **A305**, 1968b, pp. 81–92.
7. D.A. Spence, *Proc. Camb. Phil. Soc.*, **73**, 1973, pp. 249–268.
8. D.A. Spence, *The Hertz Contact Problem with Finite Friction*, Math. Research Center, U. Wisconsin, TR 1209 (presented at IUTAM congress, Moscow, 1972, to appear in J. Elasticity, 1975).
9. C.M. Segedin, *Mathematika*, **4**, 1957, pp. 156–161.
10. V.I. Mossakovski, *PMM*, 27, 1963, p. 418.
11. L.A. Galin, *PMM*, 9, 1945, pp. 413–424. Translated by Mrs. H. Moss [I.N. Sneddon (ed.)], *Contact Problems in the Theory of Elasticity*, Nort Carolina State Coll. Dep. Math. Report, Raleigh (N.C.), 1961.

Consideration of the theory of cracks from the point of view of contact problems of the theory of elasticity

A. Ju. Išlinskij

The problem of the development of cracks in elastic bodies has much in common with the contact problems of the theory of elasticity. We will present one of them: that which has served S.A. Hristianovič[1] as an assumption model for the establishment of a theory of rupture of oil strata (Fig. 1).

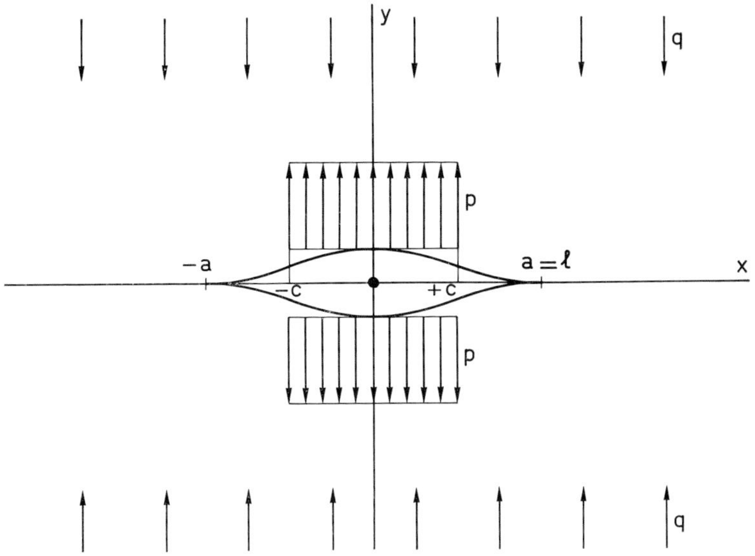

Fig. 1. Model for the establishment of a theory of rupture of oil strata

Two elastic half-planes are pressed against each other (without any friction) by distributed loads of constant intensity $-q$, which are orthogonal to their common boundary (the axis of x) and which apply far away from that axis. Two equal and opposite forces P and P' tend to break the contact of these half-planes. The force P is orientated in the positive direction of the axis y and is the resultant of a uniformly distributed load p which applies on the part of the boundary $-c<x<c$, $y=0$ of the upper half-plane.* The force P' has an analogous significance; the load p', the resultant of which is P', applies at the lower half-plane. It is obvious that

$$2pc = P = P' = 2p'c\,.$$

Let as a result of the action of the forces P and P' the contact of the elastic half-planes be broken on the segment of the boundary $-a<x<a$, where a gap appears, the length of which $(2a)$ is a quantity which has to be determined.

In order to find the length of the gap we shall consider the plane problem of the theory of elasticity on the equilibrium, e.g., of the upper half-plane. In agreement with the considerations presented above, we have the following boundary conditions at the boundary of this half-plane:

$$\sigma_x \to 0, \quad \tau_{xy} \to 0, \quad \sigma_y \to -q \quad \text{for} \quad y \to \infty\,,$$

$$\sigma_y = 0, \quad \tau_{xy} = 0 \quad \text{for} \quad -a<x<-c \quad \text{and}$$

$$\text{for} \quad c<x<a\,,$$

$$\sigma_y = -p, \tau_{xy} = 0 \quad \text{for} \quad -c<x<c\,,$$

and, moreover, $\tau_{xy}=0$ and $v=0$ everywhere on the part of the boundary of both half-planes, where they are in contact, i.e. for $x>a$, $y=0$. Here, as is usual, σ_x, σ_y and τ_{xy} are the components of the stress tensor and u and v are the components in the x- and y-directions of the displacement of an arbitrary point of the elastic half-plane.

By using well-known methods of solution of the plane problem of the theory of elasticity (see, e.g., [2]) the functions u and v of the variables x and y can be found from the functions σ_x, σ_y and τ_{xy} which satisfy both the equilibrium equations and the boundary conditions which we just mentioned. It is found that on the segment where there is no contact, i.e. for $|x|<a$, $y=0$, the displacement of the points of the boundary of the upper elastic half-plane in the vertical direction is given by the following formula:

$$v = v(x,0) = \frac{\lambda+2\mu}{2\mu(\lambda+\mu)}\left(\frac{2}{\pi}\,p\,\arcsin\frac{c}{\pi} - q\right)\sqrt{a^2-x^2} +$$

$$+\frac{p}{\pi}\left(c\ln\left|\frac{\sqrt{a^2-x^2}+\sqrt{a^2-c^2}}{\sqrt{a^2-x^2}-\sqrt{a^2-c^2}}\right| -\right.$$

* The following argument undergoes unessential modifications when the forces P and P' are regarded as concentrated. The calculations even simplify somewhat. But then essential difficulties in the calculation of the variation of the potential energy of the half-plane arise.

$$- x \ln \left| \frac{c\sqrt{a^2 - x^2} + x\sqrt{a^2 - c^2}}{c\sqrt{a^2 - x^2} - x\sqrt{a^2 - c^2}} \right| \Bigg), \tag{1}$$

where λ and μ are Lamé's constants. The expression for the distributed load maintaining the contact between the two half-planes in the domain where there is contact, reads

$$\sigma_y = \left(\frac{2}{\pi} p \arcsin \frac{c}{a} - q \right) \frac{x}{\sqrt{x^2 - a^2}} +$$

$$+ \frac{p}{\pi} \left[\arcsin \frac{a^2 - cx}{a(x - c)} - \arcsin \frac{a^2 + cx}{a(x + c)} \right], \tag{2}$$

which holds, of course, only for $x > a$.

Let us denote by l the root in the interval (c, ∞) of the trigonometric equation

$$\frac{2}{\pi} p \arcsin \frac{c}{a} - q = 0 \tag{3}$$

in which the quantity a is unknown. Under the condition $p > q$ the equation (3) will always have exactly one such a root.

It is not difficult to show that the half-length a of the gap which is formed as a result of the forces P and P' which tend to separate the half-planes, are exactly equal to the root l of the equation (3).

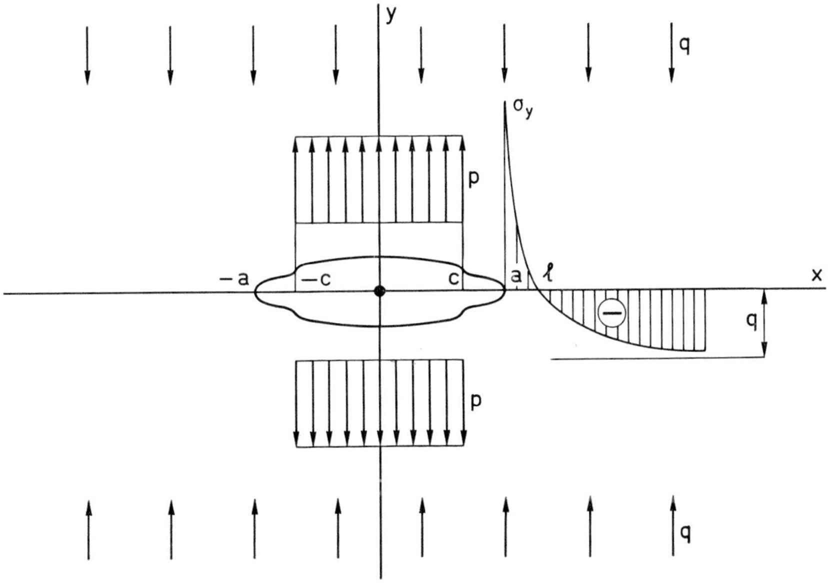

Fig. 2. The case $a < l$

Indeed, let $a < l$. In this case the gap has the shape of a stretched oval (Fig. 2). The tangent to the contour of the gap is vertical in its extreme points $x = \pm a$, $y = 0$. In the immediate proximity of the ends of the gap the stress σ_y in the points of the axis x becomes positive. This means that the elastic half-planes must exert a tensile traction on each other in order to maintain contact. In the present problem, as in the classical statement of contact problems of the theory of elasticity, a tensile traction between contracting bodies is not admitted. Thus, the case $a < l$ is impossible.

Let, on the other hand, $a > l$. Now everywhere in the contact area, i.e. for $|x| > a$ and $y = 0$, the stress σ_y is negative and, consequently, the elastic half-planes are pressed against each other. But in this case the function $v = v(x, 0)$, i.e. the displacement of the boundary of the upper elastic half-plane, is positive near the ends of the gap and negative in its central part. In consequence of this the contour of the gap acquires the shape of a curve which intersects itself (Fig. 3), as if the elastic half-planes had

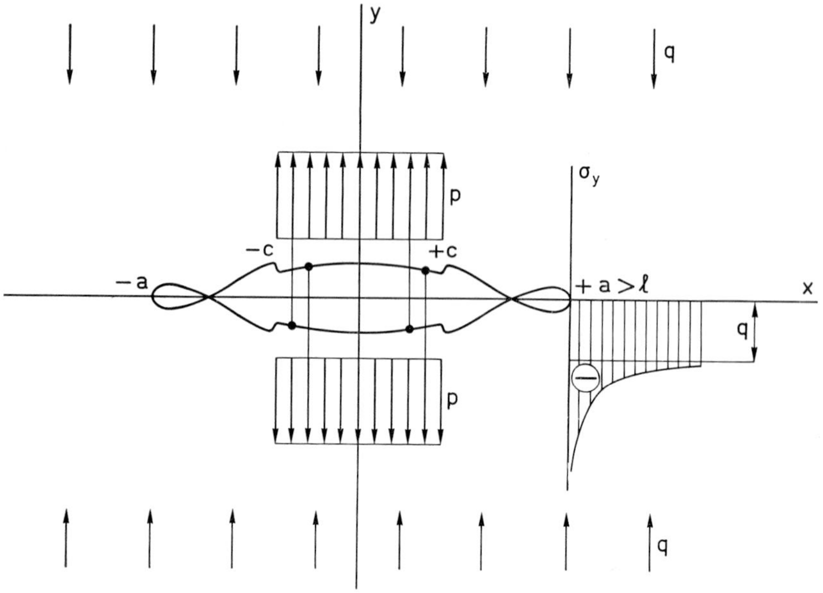

Fig. 3. The case $a > l$

penetrated one into the other. This is, of course, impossible and the case $a > l$ also is eliminated.

In such a way the solution of the plane problem of the theory of elasticity which we just have found, leads for $a \neq l$ either to a physically impossible stress state or to an impossible displacement field.

We are left with the only possible case $l = a$. Here the formulae (3) and (4) reduce to*

$$v(x,0) = \frac{\lambda+2\mu}{2\mu(\lambda+\mu)}\frac{p}{\pi}\left(c \ln\left|\frac{\sqrt{l^2-x^2}+\sqrt{l^2-c^2}}{\sqrt{l^2-x^2}-\sqrt{l^2-c^2}}\right| - \right.$$
$$\left. - x \ln\left|\frac{c\sqrt{l^2-x^2}+x\sqrt{l^2-c^2}}{c\sqrt{l^2-x^2}-x\sqrt{l^2-c^2}}\right|\right) \tag{4}$$

and

$$\sigma_y(x,0) = \frac{p}{\pi}\left[\arc\sin\frac{a^2-cx}{a(x-c)} - \arc\sin\frac{a^2+cx}{a(x+c)}\right]. \tag{5}$$

Now the ends of the gap smoothly converge to each other. The tangent to the contour of the gap in the points $x=\pm l$, $y=0$ is directed along the axis of x. The stress σ_y and, consequently, the distributed pressure of one elastic half-plane upon the other is in the just mentioned points equal to zero and increases gradually, tending to the negative value $-q$ as the distance from the gap increases. The length of the gap increases with the growth of the cleaving forces P and P'. The equations of the theory of elasticity and the boundary conditions are satisfied, and, consequently, the problem is solved in the correct way.

We can arrive at the same value of the gap length when we draw up the condition of boundedness of the stress σ_y in the neighbourhood of its end points $x=\pm a$, $y=0$ or, alternatively, the condition that in these points the tangent to the contour of the gap is directed along the axis x.

In the above, the problem on the determination of the gap length was solved without utilising such requirements. On the contrary, they have been satisfied without imposing any auxiliary conditions, resulting from these requirements, on the solution. The basic condition for the choice of the gap length turned out to be the requirement of the physical feasibility of the solution – absence of mutual penetration of the elastic half-planes and absence of their adhesion in the whole contact area.

As a matter of fact, on the requirements of feasibility of the stress state and absence of mutual penetration of half-planes one of the simplest crack theories can be based, viz. the one which has been proposed by S.A. Hristianovič and G.I. Barenblatt and has been developed by the latter[3]. In this theory it is assumed that under the action of the same loading P and P' as in the contact problem considered above (Fig. 1), the material tears along the axis x, forming a crack. At the ends of the crack appear adhesive forces

* We mention that in the points $x = \pm c$ the tangent to the graph of the function $v = v(x, 0)$ is always vertical. This property possesses all solutions of problems of the theory of elasticity in places of step-wise change of intensity of the distributed load at the boundary. We observe, however, that the derivative $\partial v/\partial x$ of the function has only a logarithmic singularity in the neighbourhood of these points.

$q(\xi)$, operating at a certain distance Δ (Fig. 4). Loads which press the elastic half-planes together and which are situated far away from the axis x, are absent. The length of the crack which is required, can be found by means of considerations, analogous to those which we encountered above in the

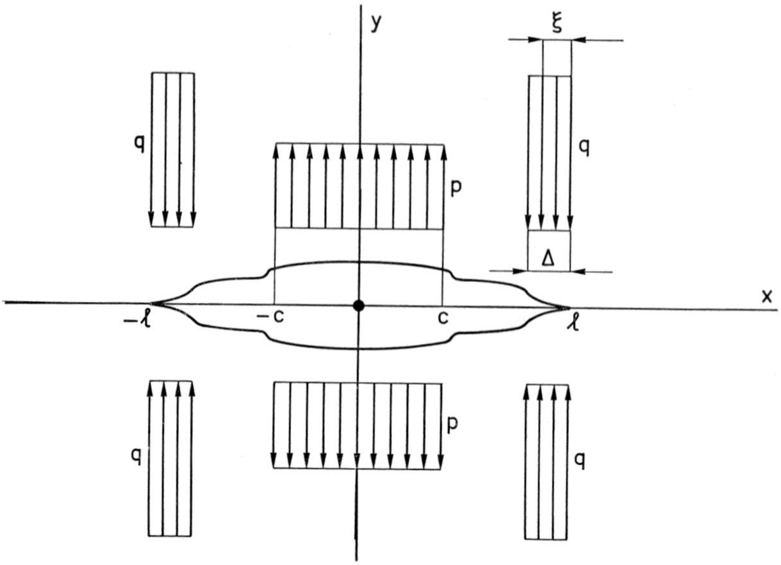

Fig. 4. Model for the establishment of the crack theory of Hristianovič and Barenblatt

determination of the gap length at the contact of two half-planes. To be exact, the boundaries of the crack cannot be so that mutual penetration of the material on both sides of the crack occurs, and the material cannot stand unboundedly large tensile stresses: in particular, the stresses σ_y at the axis x in the neighbourhood of the end of the crack must be finite.

And here again, as a final remark the conclusion can be drawn that the ends of the crack smoothly close down and that infinitely large stresses in the material are absent. The presence of infinitely large stresses is an obvious deficiency of the well-known theory of crack propagation of Griffith[4]. After all, the observation of Irwin[5] about the possibility to apply the Griffith theory on the formation of cracks in an elasto-plastic material is presented without a mathematical statement and solution of the corresponding problem of mechanics.

REFERENCES

1. JuP. Želtov, S.A., Hristianovič, 'On the Mechanism of a Hydraulical Rupture of an Oil Stratum' (in Russian), *Transactions Ac. of Sc. USSR Dept. of Technical Sc.*, 1955, **5**, pp. 3–41.
2. N.I. Mushelišvili, *Some Basic Problems of the Mathematical Theory of Elasticity* (in Russian), 'Nauka', Moscow 1966 (also English translation, published by Noordhoff, Groningen, 1953).
3. G.I. Barenblatt, 'On the equilibrium cracks due to brittle fracture' (in Russian), *Prikl. Mat i Meh.*, **23**, 1959, no. 3–5.
4. A.A. Griffith, 'The Phenomenon of Rupture and Flow in Solids', *Phil. Trans. Roy. Soc.*, **A221**, 1920, pp. 163–198.
5. G.R. Irwin, *Fracturing of Metals*, ASM, Cleveland, 1948, pp. 147–166 (Chapter 'Fracture dynamics').

Certain asymmetrical contact problems for semi-space

B. L. Abramjan

SUMMARY

The paper deals with effective solutions for two asymmetric contact problems in the linear theory of elasticity for the semi-space.

In considering the problems the semi-space is assumed to have at least one plane with respect to which the deformation is symmetric. Further, the harmonic functions of P.F. Papkovich-H. Neuber are used to find solutions which are sought for in a cylindrical system of coordinates.

In the first problem a thin rigid round-shaped plate, coupled with the semi-space surface, is displaced for a short length, δ, under the action of a horizontally directed force. Only tangential stresses are assumed to develop under the plate.

The solution to this problem is attained by means of Hankel's integrals. The determination of the arbitrary functions of integration is reduced to dual integral equations, involving Bessel functions. Using the solutions to these equations to determine the displacements and stresses on the semi-space surface, closed expressions are obtained.

In the second problem a round cylindrical rigid punch, coupled through its flat butt end with the semi-space surface, is displaced along the semi-space surface under the action of an external load. The external load is assumed to be applied to the punch at a certain height from the semi-space surface as a force directed horizontally along the axis OX. In this problem certain normal stresses are assumed to develop under the punch.

The solution to this problem is reduced to a set of integral equations of the second kind.

Further, the laws of distribution of normal and tangential stresses under the punch are determined as well as the values of horizontal displacement of the punch along the semi-space surface and the angle of its revolution around the axis, perpendicular to the direction of the external force action.

It is assumed that a homogeneous elastic solid of revolution is deformed symmetrically with respect to one or several of its axial sections under the action of external forces.

Axis OZ is directed along the axis of revolution. If the plane XOZ is one of the axial sections, with respect to which deformation is symmetric, then the coordinate of the cylindrical system φ is counted from the axis OX.

In this case the components of elastic displacements $u_r(r, \varphi, z)$, $u_\varphi(r, \varphi, z)$ and $u_z(r, \varphi, z)$ may be represented as the Fourier series

$$
\begin{aligned}
u_r &= \sum_{k=0}^{\infty} u_k(r,z) \cos \gamma_k \varphi \\
u_\varphi &= \sum_{k=1}^{\infty} v_k(r,z) \sin \gamma_k \varphi \\
u_z &= \sum_{k=0}^{\infty} w_k(r,z) \cos \gamma_k \varphi
\end{aligned}
\qquad (0.1)
$$

γ_k depends upon the number of axial sections which are the planes of deformation symmetry.

In the case where such planes are two in number, we have $\gamma_k = k$.

In a similar statement certain asymmetric spatial problems were considered in [1–4]. To solve the problems in such a statement we use the harmonic functions of P.F. Papkovich-H. Neuber[5–6].

Let the harmonic functions $\phi_i(r, \varphi, z)$ $(i = 1, 2, 3, 0)$ be presented in the form

$$
\begin{aligned}
\phi_1 &= \sum_{k=0}^{\infty} \varphi_{1k}(r,z) \cos(\gamma_k + 1)\, \varphi \\
\phi_2 &= \sum_{k=0}^{\infty} \varphi_{1k}(r,z) \sin(\gamma_k + 1)\, \varphi \\
\phi_i &= \sum_{k=0}^{\infty} \varphi_{ik}(r,z) \cos \gamma_k \varphi \quad (i = 3, 0)
\end{aligned}
\qquad (0.2)
$$

where the functions $\varphi_{ik}(r, z)$ satisfy the equations

$$
\begin{aligned}
&\nabla^2_{\gamma_k+1} \varphi_{1k} = 0, \quad \nabla^2_{\gamma_k} \varphi_{3k} = 0, \quad \nabla^2_{\gamma_k} \varphi_{0k} = 0 \\
&\nabla^2_\alpha = \frac{\partial^2}{\partial r^2} + \frac{1}{r}\frac{\partial}{\partial r} + \frac{\partial^2}{\partial z^2} - \frac{\alpha^2}{r^2}
\end{aligned}
\qquad (0.3)
$$

Using expressions (0.1) and (0.2) as well as the formulas of P.F. Papkovich, we obtain

$$
u_k(r,z) = \varphi_{1k} - \frac{1}{4(1-\nu)} \frac{\partial \Omega_k}{\partial r} \quad (k = 0, 1, \ldots)
$$

$$v_k(r,z) = \varphi_{1k} + \frac{1}{4(1-\nu)}\frac{k\Omega_k}{r} \quad (k=1,2,\ldots) \tag{0.4}$$

$$w_k(r,z) = \varphi_{3k} - \frac{1}{4(1-\nu)}\frac{\partial\Omega_k}{\partial z} \quad (k=0,1,\ldots)$$

where $\Omega_k = r\varphi_{1k}(r, z)+z\varphi_{3k}(r, z)+\varphi_{0k}(r, z)$, and ν is Poisson's coefficient.

Using expressions (0.1), (0.4) and familiar formulas, the stress components may expressed by the functions $\varphi_{ik}(r, z)$.

1. A CONTACT PROBLEM FOR SEMI-SPACE WITH A THIN RIGID PLATE OF ROUND SHAPE ON ITS SURFACE

Let us assume that on the surface of semi-space, under complete coupling, there is a thin plate of round shape. The plate is assumed not to be deformed in its plane and not to offer any resistance to bending.

We take the origin of the coordinate system in the centre of the plate and assume that under the action of external load, applied to the plate, the plate is drawn over the semi-space surface along the axis OX and displaced for a certain short length, δ, which is to be determined later.

The assumption made with respect to the plate enable us to presume that all the points of the semi-space surface under the plate are displaced along the axis OX for a length, δ, and that no normal stresses develop at these points on the surface. The semi-space surface beyond the plate is considered to be free loads.

For simplicity we assume also that under the plate the tangential stresses τ_{xz} alone are different from zero.

A closed solution may be constructed for such a simplified problem.

The boundary conditions on the semi-space surface are given by the relations:

$$\begin{aligned} u_x|_{z=0} &= \delta \quad \text{inside the domain of the circle} \quad r \leqslant a \\ \tau_{xz}|_{z=0} &= 0 \quad \text{outside the domain of the circle} \quad r > a \\ \left.\begin{aligned} \tau_{yz}|_{z=0} &= 0 \\ \delta_z|_{z=0} &= 0 \end{aligned}\right\} & \quad \text{throughout the surface} \end{aligned} \tag{1.1}$$

Conditions (1.1) allow to seek the solution to the problem in the cylindrical system of coordinates in the form:

$$\begin{aligned} u_r(r,\varphi,z) &= u_1(r,z)\cos\varphi \\ u_\varphi(r,\varphi,z) &= v_1(r,z)\sin\varphi \\ u_z(r,\varphi,z) &= w_1(r,z)\cos\varphi \end{aligned} \tag{1.2}$$

where u_1, v_1 and w_1 are determined by the relations

$$u_1 = \varphi_{11}(r, z) - \frac{1}{4(1-\nu)} \frac{\partial \Omega_1}{\partial r} \tag{1.3}$$

$$v_1 = \varphi_{11}(r, z) + \frac{1}{4(1-\nu)} \frac{\Omega_1}{r}$$

$$w_1 = \varphi_{31}(r, z) - \frac{1}{4(1-\nu)} \frac{\partial \Omega_1}{\partial z}$$

$$\Omega_1(r, z) = r\varphi_{11} + z\varphi_{31} + \varphi_{01}$$

$$\nabla_2^2 \varphi_{11} = 0, \quad \nabla_1^2 \varphi_{31} = 0, \quad \nabla_1^2 \varphi_{01} = 0 \tag{1.4}$$

By means of (1.2) we obtain for the stresses τ_{rz}, $\tau_{z\varphi}$ and σ_z the expressions

$$\tau_{rz}(r, \varphi, z) = \tau_{rz}^{(1)}(r, z) \cos \varphi$$

$$\tau_{z\varphi}(r, \varphi, z) = \tau_{z\varphi}^{(1)}(r, z) \sin \varphi \tag{1.5}$$

$$\sigma_z(r, \varphi, z) = \sigma_z^{(1)}(r, z) \cos \varphi$$

where

$$\tau_{rz}^{(1)} = G \left[\frac{\partial \varphi_{31}}{\partial r} + \frac{\partial \varphi_{11}}{\partial z} - \frac{1}{2(1-\nu)} \frac{\partial^2 \Omega_1}{\partial r \partial z} \right]$$

$$\tau_{z\varphi}^{(1)} = G \left[\frac{\partial \varphi_{11}}{\partial z} - \frac{\varphi_{31}}{r} + \frac{1}{2(1-\nu)} \frac{1}{r} \frac{\partial \Omega_1}{\partial z} \right] \tag{1.6}$$

$$\sigma_z^{(1)} = 2G \left[\frac{\partial \varphi_{31}}{\partial z} - \frac{1}{4(1-\nu)} \frac{\partial^2 \Omega_1}{\partial z^2} + \frac{\nu}{2(1-\nu)} \left(\frac{\partial \varphi_{11}}{\partial r} + \frac{2\varphi_{11}}{r} + \frac{\partial \varphi_{31}}{\partial z} \right) \right]$$

in which G denotes the rigidity modulus and ν the Poisson ratio of the semi-space.

The solutions to equations (1.4) are taken in the form of Hankel's integral

$$\varphi_{11}(r, z) = \int_0^\infty A(\lambda) \, e^{-\lambda z} J_2(\lambda r) \, d\lambda$$

$$\varphi_{31}(r, z) = \int_0^\infty B(\lambda) \, e^{-\lambda z} J_1(\lambda r) \, d\lambda \tag{1.7}$$

$$\varphi_{01}(r, z) = \int_0^\infty C(\lambda) \, e^{-\lambda z} J_1(\lambda r) \, d\lambda$$

It should be noted here that for the function $r\varphi_{11}(r, z)$ the following representation also takes place:

$$r\varphi_{11}(r, z) = -z \int_0^\infty A(\lambda) \, e^{-\lambda z} J_1(\lambda r) \, d\lambda \tag{1.8}$$

Considering these expressions, according to (1.3) and (1.6) we obtain

$$u_1 = \int_0^\infty A(\lambda)\, e^{-\lambda z}\, J_2(\lambda r)\, d\lambda - \frac{1}{4(1-\nu)} \times$$
$$\times \int_0^\infty \lambda [(B-A)\, z + C]\, e^{-\lambda z}\, J_1'(\lambda r)\, d\lambda \tag{1.9}$$

$$v_1 = \int_0^\infty A(\lambda)\, e^{-\lambda z}\, J_2(\lambda r)\, d\lambda + \frac{1}{4(1-\nu)} \times$$
$$\times \int_0^\infty [(B-A)\, z + C]\, e^{-\lambda z}\, \frac{J_1(\lambda r)}{r}\, d\lambda \tag{1.10}$$

$$w_1 = \int_0^\infty B(\lambda)\, e^{-\lambda z}\, J_1(\lambda r)\, d\lambda - \frac{1}{4(1-\nu)} \times$$
$$\times \int_0^\infty [(B-A)\,(1-\lambda z) - \lambda C]\, e^{-\lambda z}\, J_1(\lambda r)\, d\lambda \tag{1.11}$$

$$\tau_{rz}^{(1)} = G \left\{ \int_0^\infty \lambda e^{-\lambda z}\, J_1'(\lambda r) \left[B - \frac{(B-A)\,(1-\lambda z) - \lambda C}{2(1-\nu)} \right] d\lambda - \right.$$
$$\left. - \int_0^\infty \lambda\, A(\lambda)\, e^{-\lambda z}\, J_2(\lambda r)\, d\lambda \right\} \tag{1.12}$$

$$\tau_{z\varphi}^{(1)} = G \left\{ \int_0^\infty e^{-\lambda z}\, \frac{J_1(\lambda r)}{r} \left[\frac{(B-A)\,(1-\lambda z) - \lambda C}{2(1-\nu)} - B \right] d\lambda - \right.$$
$$\left. - \int_0^\infty \lambda\, A(\lambda)\, e^{-\lambda z}\, J_2(\lambda r)\, d\lambda \right\} \tag{1.13}$$

$$\sigma_z^{(1)} = 2G \left\{ - \int_0^\infty \lambda\, B(\lambda)\, e^{-\lambda z}\, J_1(\lambda r)\, d\lambda - \right.$$
$$\left. - \frac{1}{4(1-\nu)} \int_0^\infty \lambda e^{-\lambda z}\, J_1(\lambda r)\, [(A-B)\,(2-2\nu-\lambda z) + \lambda C]\, d\lambda \right\} \tag{1.14}$$

By means of the notation used, boundary conditions (1.1) may be reduced to the following form:

$$\begin{aligned} &u_1(r,0) - v_1(r,0) = 2\delta && (r \leqslant a) \\ &\tau_{rz}^{(1)}(r,0) = 0 && (r > a) \\ &\left.\begin{aligned} &\sigma_z^{(1)}(r,0) = 0 \\ &\tau_{rz}^{(1)}(r,0) + \tau_{z\varphi}^{(1)}(r,0) = 0 \end{aligned}\right\} && (0 < r < \infty) \end{aligned} \tag{1.15}$$

Satisfying these conditions, we obtain the following equations for the determination of arbitrary functions of integration:

$$\int_0^\infty \lambda\, C(\lambda)\, J_0(\lambda r)\, d\lambda = -8(1-\nu)\,\delta \quad (r \leqslant a) \tag{1.16}$$

$$\int_0^\infty \lambda[(1-2\nu)\, B(\lambda) + A(\lambda) + \lambda C(\lambda)]\, J_0(\lambda r)\, d\lambda = 0 \quad (r > a)$$

$$(1-2\nu)\, B(\lambda) + (5-4\nu)\, A(\lambda) + \lambda C(\lambda) = 0 \tag{1.17}$$

$$2(1-\nu)\,[A(\lambda) + B(\lambda)] + \lambda C(\lambda) = 0 \tag{1.18}$$

Excluding by means of relations (1.17) and (1.18) the functions $A(\lambda)$ and $B(\lambda)$ from the second equation of (1.16), we obtain dual integral equations of the form:

$$\int_0^\infty \lambda C(\lambda)\, J_0(\lambda r)\, d\lambda = -8(1-\nu)\,\delta \quad (r \leqslant a)$$

$$\int_0^\infty \lambda^2\, C(\lambda)\, J_0(\lambda r)\, d\lambda = 0 \quad (r > a) \tag{1.19}$$

Such equations were considered in the studies of I. Busbridge[7], E. Titchmarsh[8], I.N. Sneddon[9] and other researchers.

The solution to equations (1.19) is of the form

$$\lambda C(\lambda) = -\frac{16(1-\nu)\,\delta}{\pi\lambda} \sin \lambda a \tag{1.20}$$

Then we find also

$$A(\lambda) = \frac{4\delta \sin \lambda a}{\pi(2-\nu)\lambda}$$

$$B(\lambda) = \frac{4(3-2\nu)\,\delta \sin \lambda a}{\pi(2-\nu)\lambda} \tag{1.21}$$

Substituting these values into the formulas for displacements and stresses, we obtain the following expressions:

$$u_r(r,\varphi,0) = \begin{cases} \delta \cos\varphi & (r \leqslant a) \\ \dfrac{2\delta \cos\varphi}{\pi}\left(\arcsin \dfrac{a}{r} + \dfrac{\nu}{2-\nu}\dfrac{a}{r^2}\sqrt{r^2-a^2}\right) & (r \geqslant a) \end{cases} \tag{1.22}$$

$$u_\varphi(r,\varphi,0) = \begin{cases} -\delta \sin\varphi & (r \leqslant a) \\ -\dfrac{2\delta \sin\varphi}{\pi}\left(\arcsin \dfrac{a}{r} - \dfrac{\nu}{2-\nu}\dfrac{a}{r^2}\sqrt{r^2-a^2}\right) & (r \leqslant a) \end{cases} \tag{1.23}$$

$$u_z(r,\varphi,0) = \begin{cases} \dfrac{2\delta(1-2\nu)}{\pi(2-\nu)} \dfrac{r\cos\varphi}{a+\sqrt{a^2-r^2}} & (r \leqslant a) \\ \dfrac{2\delta(1-2\nu)}{\pi(2-\nu)} \dfrac{a\cos\varphi}{r} & (r \geqslant a) \end{cases} \tag{1.24}$$

$$\tau_{rz}(r,\varphi,0) = \begin{cases} -\dfrac{4\delta G}{\pi(2-\nu)} \dfrac{\cos\varphi}{\sqrt{a^2-r^2}} & (r < a) \\ 0 & (r > a) \end{cases} \tag{1.25}$$

$$\tau_{z\varphi}(r,\varphi,0) = \begin{cases} \dfrac{4\delta G}{\pi(2-\nu)} \dfrac{\sin\varphi}{\sqrt{a^2-r^2}} & (r < a) \\ 0 & (r > a) \end{cases} \tag{1.26}$$

When obtaining these expressions, the values of the well known integrals are used[10]:

$$\int_0^\infty \frac{\sin\lambda a\, J_1(\lambda r)\, \mathrm{d}\lambda}{\lambda^2} = \begin{cases} \dfrac{\pi r}{4} & (\tau \leqslant a) \\ \dfrac{1}{2}\left(r \arcsin\dfrac{a}{r} + \dfrac{a}{r}\sqrt{r^2-a^2}\right) & (r \geqslant a) \end{cases} \tag{1.27}$$

$$\int_0^\infty \frac{\sin\lambda a\, J_1(\lambda r)\, \mathrm{d}\lambda}{\lambda} = \begin{cases} \dfrac{a}{r} & (a \leqslant r) \\ \dfrac{r}{a+\sqrt{a^2-r^2}} & (a \geqslant r) \end{cases} \tag{1.28}$$

$$\int_0^\infty \sin\lambda a\, J_2(\lambda r)\, \mathrm{d}\lambda = \begin{cases} \dfrac{2a}{r^2} & (a < r) \\ -\dfrac{r^2}{\sqrt{a^2-r^2}\,(a+\sqrt{a^2-r^2})^2} & (a < r) \end{cases} \tag{1.29}$$

If the plate is drawn along the axis OX by the force P, then working out the equation of equilibrium of forces and stresses acting upon the plate, we obtain

$$-P = \iint_\Omega \tau_{xz}|_{z=0}\, \mathrm{d}x\mathrm{d}y = \int_0^{2\pi} \mathrm{d}\varphi \int_0^a [\tau_{rz}(r,\varphi,0)\cos\varphi - \\ - \tau_{z\varphi}(r,\varphi,0)\sin\varphi]\, r\, \mathrm{d}r = -\frac{8\delta G a}{2-\nu} \tag{1.30}$$

where Ω is the domain of the plate contact with semi-space. That is

$$\delta = \frac{P(2-\nu)}{8\,Ga} \tag{1.31}$$

Thus, the value of displacement of the plate coupled with the surface of elastic semi-space, is directly proportional to the acting force P and inversely proportional to the modulus of elasticity of semi-space and to radius of the plate.

2. A CONTACT PROBLEM FOR SEMI-SPACE WITH A RIGID PUNCH HAVING A ROUND FLAT BASE ON THE SURFACE

We assume that a rigid punch with a round flat base, coupled with the semi-space surface, under the action of a horizontally directed force P, applied to the punch at a height 'h' from the semi-space surface, is displaced in the direction of the action of force at a length δ.

We take the origin of the coordinate system in the centre of the circular domain of contact on the semi-space surface which occupies the space $z \geqslant 0$. The axis OX is guided along the direction of action of the force P.

Under the action of force P the punch revolves around the axis OY and the points lying under the punch undergo the axial displacements u_z due to revolution of the punch butt end.

Thus it is assumed that tangential stresses τ_{xz} and normal stresses σ_z develop under the punch which are balanced by the force P and by the moment of this force with respect to the axis OY.

$$\iint_{\Omega} \tau_{xz}|_{z=0} \, \mathrm{d}\Omega = -P, \quad \iint_{\Omega} x\sigma_z|_{z=0} \, \mathrm{d}\Omega = hP \tag{2.1}$$

where Ω is the domain of the punch contact with semi-space.

As far as the tangential stresses τ_{yz} are concerned, as in section 1, we assume that under the punch they may be taken equal to zero to simplify the problem.

Outside the punch the semi-space surface is considered to be free from stresses.

Similar problems, using different methods, were considered by A.I. Lurie[11], I.J. Staerman[12], L.A. Galin[13], J.S. Ufland[14], J.J. Kalker[15], G.M.L. Gladwell[19]* et al.

To solve the problem formulated, we again will use a cylindrical system of coordinates.

Using notations (1.2) and (1.5) and the condition of freedom from the stresses τ_{yz} throughout the semi-space surface, we will have

$$\tau_{rz}^{(1)}(r,0) + \tau_{z\varphi}^{(1)}(r,0) = 0 \qquad (0 < r < \infty) \tag{2.2}$$

Using (2.3), we will also have

$$\tau_{xz}|_{z=0} = \tau_{rz}^{(1)}(r,0) \tag{2.3}$$

* It should be mentioned that Gladwell's work was unknown to author when he prepared the present paper.

Let us introduce the notation

$$\tau_{rz}^{(1)}(r,0) = \begin{cases} \dfrac{G}{4(1-\nu)} \tau(r) & (r < a) \\ 0 & (r > a) \end{cases} \tag{2.4}$$

$$\sigma_z^{(1)}(r,0) = \begin{cases} -\dfrac{G}{2(1-\nu)} \rho(r) & (r < a) \\ 0 & (r > a) \end{cases} \tag{2.5}$$

Then relation (2.1) may be written in the form

$$\int_0^a r\,\tau(r)\,\mathrm{d}r = -\frac{2(1-\nu)\,P}{G\pi} \tag{2.6}$$

$$\int_0^a r^2\,p(r)\,\mathrm{d}r = -\frac{2(1-\nu)\,hP}{G\pi} \tag{2.7}$$

Apart from conditions (2.3), (2.4) and (2.5) we will also have the conditions

$$u_1(r,0) = -\,v_1(r,0) = \delta \qquad (r \leqslant a) \tag{2.8}$$

and

$$w_1(r,0) = \varkappa r \qquad (r \leqslant a) \tag{2.9}$$

where $\varkappa$ is the angle of revolution of the punch around the axis OY.

The functions $\tau(r)$, $\rho(r)$ and values δ and $\varkappa$ are to be determined.

To solve this problem the stresses and displacements should be found by means of the functions $\varphi_{i1}(r, z)$ $(i = 1, 3, 0)$ which are taken in the form of (1.7).

Using expressions (1.9)–(1.14) and satisfying boundary conditions, we will obtain the relations

$$\int_0^\infty \lambda C(\lambda)\,J_0(\lambda r)\,\mathrm{d}\lambda = -8(1-\nu)\,\delta \qquad (r \leqslant a) \tag{2.10}$$

$$\int_0^\infty \lambda[(1-2\nu)\,B(\lambda) + A(\lambda) + \lambda C(\lambda)]\,J_0(\lambda r)\,\mathrm{d}\lambda = \begin{cases} \tau(r) & (r < a) \\ 0 & (r > a) \end{cases} \tag{2.11}$$

$$\int_0^\infty [(3-4\nu)\,B(\lambda) + A(\lambda) + \lambda C(\lambda)]\,J_1(\lambda r)\,\mathrm{d}\lambda = 4(1-\nu)\,\varkappa r \qquad (r \leqslant a) \tag{2.12}$$

$$\int_0^\infty \lambda\{2(1-\nu)\,[A(\lambda) + B(\lambda)] + \lambda C(\lambda)\}\,J_1(\lambda r)\,\mathrm{d}\lambda = \begin{cases} p(r) & (r < a) \\ 0 & (r > a) \end{cases}$$

$$(1-2\nu)\,B(\lambda) + (5-4\nu)\,A(\lambda) + \lambda C(\lambda) = 0 \tag{2.14}$$

From (2.11) and (2.13) we find

$$(1-2\nu)\, B(\lambda) + A(\lambda) + \lambda C(\lambda) = \int_0^a r\tau(r)\, J_0(\lambda r)\, dr \tag{2.15}$$

$$2(1-\nu)\, [A(\lambda) + B(\lambda)] + \lambda C(\lambda) = \int_0^a r\, p(r)\, J_1(\lambda r)\, dr \tag{2.16}$$

Solving equations (2.14)–(2.16) for the integration functions we will have

$$A(\lambda) = -\frac{1}{4(1-\nu)} \int_0^a r\, \tau(r)\, J_0(\lambda r)\, dr \tag{2.17}$$

$$B(\lambda) = \int_0^a r\, p(r)\, J_1(\lambda r)\, dr - \frac{3-2\nu}{4(1-\nu)} \int_0^a r\, \tau(r)\, J_0(\lambda r)\, dr \tag{2.18}$$

$$\lambda C(\lambda) = (2-\nu) \int_0^a r\, \tau(r)\, J_0(\lambda r)\, dr - (1-2\nu) \int_0^a r\, p(r)\, J_1(\lambda r)\, dr \tag{2.19}$$

Substituting expressions (2.17)–(2.19) into equations (2.10) and (2.12), we will obtain a set of integral equations for determining the unknown functions $\tau(r)$ and $\rho(r)$.

By such a method the solution of dual equations in the paper by B. Noble[16] was reduced to the integral equation of Fredholm of the first kind.

Substituting (2.19) into equation (2.10), we will obtain

$$(2-\nu) \int_0^\infty J_0(\lambda z)\, d\lambda \int_0^a r\, \tau(r)\, J_0(\lambda r)\, dr -$$
$$- (1-2\nu) \int_0^\infty J_0(\lambda z)\, d\lambda \int_0^a r\, p(r)\, J_1(\lambda r)\, dr = -8(1-\nu)\, \delta$$
$$(z \leqslant a) \tag{2.20}$$

Changing here the sequence of integration and using the values

$$\int_0^\infty J_0(\lambda z)\, J_1(\lambda r)\, d\lambda = \begin{cases} 0 & z > r \\ \dfrac{1}{r} & z < r \end{cases} \tag{2.21}$$

$$\int_0^\infty J_0(\lambda z)\, J_0(\lambda r)\, d\lambda = \frac{2}{\pi} \int_0^{\min(r,z)} \frac{dx}{\sqrt{(r^2-x^2)\,(z^2-x^2)}} \tag{2.22}$$

from (2.20) we will obtain

$$\frac{2}{\pi}(2-\nu) \int_0^a r\, \tau(r)\, dr \int_0^{\min(r,z)} \frac{dx}{\sqrt{(r^2-x^2)\,(z^2-x^2)}} -$$
$$- (1-2\nu) \int_z^a p(r)\, dr = -8(1-\nu)\, \delta \qquad (z \leqslant a)$$

or

$$\int_0^z r\tau(r)\,\mathrm{d}r \int_0^r \frac{\mathrm{d}x}{\sqrt{(r^2-x^2)\,(z^2-x^2)}} + \int_z^a r\tau(r)\,\mathrm{d}r \int_0^z \frac{\mathrm{d}x}{\sqrt{(r^2-x^2)(z^2-x^2)}}$$
$$= \frac{(1-2\nu)\,\pi}{2(2-\ )} \int_z^a p(r)\,\mathrm{d}r - \frac{4\ 1-\nu)\,\delta\pi}{2-\nu} \tag{2.23}$$

Following here the method of Copson[18] and changing the sequence of integration on the left side of equation (2.23) we will obtain an integral equation of the form

$$\int_0^z \frac{\mathrm{d}x}{\sqrt{z^2-x^2}} \int_x^a \frac{r\tau(r)\,\mathrm{d}r}{\sqrt{r^2-x^2}} = \frac{(1-2\nu)\,\pi}{2(2-\nu)} \int_z^a p(r)\,\mathrm{d}r - \frac{4(1-\nu)\,\delta\pi}{2-\nu} = \phi(z)$$

or the equations of Abel

$$\int_x^a \frac{r\tau(r)\,\mathrm{d}r}{\sqrt{r^2-x^2}} = F(x), \qquad \int_0^z \frac{F(x)\,\mathrm{d}x}{\sqrt{z^2-x^2}} = \phi(z) \tag{2.24}$$

if the right sides of equations (2.24) are assumed to be known.

By means of the solutions of these equations

$$F(x) = \frac{2}{\pi}\frac{\mathrm{d}}{\mathrm{d}x} \int_0^x \frac{z\phi(z)\,\mathrm{d}z}{\sqrt{x^2-z^2}}, \qquad r\tau(r) = -\frac{2}{\pi}\frac{\mathrm{d}}{\mathrm{d}r} \int_r^a \frac{xF(x)\,\mathrm{d}x}{\sqrt{x^2-r^2}} \tag{2.25}$$

we will obtain

$$r\tau(r) = -\frac{4}{\pi^2}\frac{\mathrm{d}}{\mathrm{d}r} \int_r^a \frac{x\,\mathrm{d}x}{\sqrt{x^2-r^2}} \frac{\mathrm{d}}{\mathrm{d}x} \int_0^x \frac{z\,\mathrm{d}z}{\sqrt{x^2-z^2}} \times$$
$$\times \left[\frac{(1-2\nu)\,\pi}{2(2-\nu)} \int_z^a p(t)\,\mathrm{d}t - \frac{4(1-\nu)\,\delta\pi}{2-\nu} \right] \tag{2.26}$$

or

$$\tau(r) = -\frac{16(1-\nu)\,\delta}{(2-\nu)\,\pi} \frac{1}{\sqrt{a^2-r^2}} +$$
$$+ \frac{2(1-2\nu)}{(2-\nu)\,\pi} \left[\frac{1}{\sqrt{a^2-r^2}} \int_0^a p(t)\,\mathrm{d}t + \right.$$
$$\left. + \frac{1}{r}\frac{\mathrm{d}}{\mathrm{d}r} \int_r^a \frac{x^2\,\mathrm{d}x}{\sqrt{x^2-r^2}} \int_0^x \frac{p(t)\,\mathrm{d}t}{\sqrt{x^2-t^2}} \right] \tag{2.27}$$

By means of expressions (2.17)–(2.19) we will have

$$(3-4\nu)\,B(\lambda) + A(\lambda) + \lambda C(\lambda) =$$

$$= 2(1-\nu)\int_0^a r\,p(r)\,J_1(\lambda r)\,\mathrm{d}r - \frac{1-2\nu}{2}\int_0^a r\,\tau(r)\,J_0(\lambda r)\,\mathrm{d}r \tag{2.28}$$

Substituting (2.28) into equation (2.12), we will have

$$4(1-\nu)\int_0^\infty J_1(\lambda z)\,\mathrm{d}\lambda\int_0^a r\,p(r)\,J_1(\lambda r)\,\mathrm{d}r -$$
$$-(1-2\nu)\int_0^\infty J_1(\lambda z)\,\mathrm{d}\lambda\int_0^a r\tau(r)\,J_0(\lambda r)\,\mathrm{d}r = 8(1-\nu)\,\varkappa z$$
$$(z \leqslant a) \tag{2.29}$$

Changing here the sequence of integration and using values (2.21) and

$$\int_0^\infty J_1(\lambda z)\,J_1(\lambda r)\,\mathrm{d}\lambda = \frac{2}{\pi r z}\int_0^{\min(r,z)} \frac{x^2\mathrm{d}x}{\sqrt{(r^2-x^2)(z^2-x^2)}} \tag{2.30}$$

from (2.29) we will obtain

$$\frac{8(1-\nu)}{\pi z}\int_0^a p(r)\,\mathrm{d}r\int_0^{\min(r,z)} \frac{x^2\,\mathrm{d}x}{\sqrt{(r^2-x^2)(z^2-x^2)}} =$$
$$= \frac{1-2\nu}{z}\int_0^z r\tau(r)\,\mathrm{d}r + 8(1-\nu)\,\varkappa z \qquad (z \leqslant a)$$

or

$$\int_0^z p(r)\,\mathrm{d}r\int_0^r \frac{x^2\,\mathrm{d}x}{\sqrt{(r^2-x^2)(z^2-x^2)}} + \int_z^a p(r)\,\mathrm{d}r\int_0^z \frac{x^2\,\mathrm{d}x}{\sqrt{(r^2-x^2)(z^2-x^2)}} =$$
$$= \frac{(1-2\nu)\,\pi}{8(1-\nu)}\int_0^z r\tau(r)\,\mathrm{d}r + \pi\varkappa z^2 \tag{2.31}$$

Similarly, as it is done for equation (2.23), from (2.31) we will obtain the equation

$$\int_0^z \frac{x^2\,\mathrm{d}x}{\sqrt{z^2-x^2}}\int_x^a \frac{p(r)\,\mathrm{d}r}{\sqrt{r^2-x^2}} = \frac{(1-2\nu)\,\pi}{8(1-\nu)}\int_0^z \tau r(r)\,\mathrm{d}r + \pi\varkappa z^2 \tag{2.32}$$

from which we will find

$$p(r) = -\frac{1-2\nu}{2(1-\nu)\,\pi}\frac{\mathrm{d}}{\mathrm{d}r}\int_r^a \frac{\mathrm{d}x}{x\sqrt{x^2-r^2}}\frac{\mathrm{d}}{\mathrm{d}x}\int_0^x \frac{z\,\mathrm{d}z}{\sqrt{x^2-z^2}}\int_0^z t\tau(t)\,\mathrm{d}t -$$
$$-\frac{4\varkappa}{\pi}\frac{\mathrm{d}}{\mathrm{d}r}\int_r^a \frac{\mathrm{d}x}{x\sqrt{x^2-r^2}}\frac{\mathrm{d}}{\mathrm{d}x}\int_0^x \frac{z^3\,\mathrm{d}z}{\sqrt{x^2-z^2}} \tag{2.33}$$

or

$$p(r) = \frac{8\varkappa}{\pi} \frac{r}{\sqrt{a^2-r^2}} - \frac{1-2\nu}{2(1-\nu)\pi} \frac{\mathrm{d}}{\mathrm{d}r} \int_r^a \frac{\mathrm{d}x}{\sqrt{x^2-r^2}} \int_0^x \frac{t\,\tau(t)\,\mathrm{d}t}{\sqrt{x^2-t^2}} \tag{2.34}$$

The length of horizontal displacement of the punch over the semi-space surface and the angle of the punch revolution around the axis, perpendicular to the direction of action of the external force P, are determined by means of equations (2.6) and (2.7) of equilibrium of forces and stresses, acting upon the punch, and of moments, revolving the punch around the axis OY.

Substituting (2.27) into (2.6) and making certain transformations, we will obtain

$$\delta = \frac{(2-\nu)P}{8\,Ga} + \frac{1-2\nu}{8(1-\nu)\,a} \int_0^a p(t)\,[a - \sqrt{a^2-t^2}]\,\mathrm{d}t \tag{2.35}$$

Substituting (2.34) into (2.7) and here also making transformation, including a change in the sequence of integration, we will obtain

$$\varkappa = -\frac{3(1-\nu)\,hP}{8\,Ga^3} - \frac{3(1-2\nu)}{16(1-\nu)\,a^3} \int_0^a t\sqrt{a^2-t^2}\,\tau(t)\,\mathrm{d}t \tag{2.36}$$

Substituting values δ and $\varkappa$ into equations (2.27) and (2.34), we will have

$$\tau(r) = -\frac{2(1-\nu)\,P}{Ga\pi\sqrt{a^2-r^2}} + \frac{2(1-2\nu)}{\pi(2-\nu)} \left[\frac{1}{a\sqrt{a^2-r^2}} \int_0^a p(t)\sqrt{a^2-t^2}\,\mathrm{d}t + \right.$$

$$\left. + \frac{1}{r}\frac{\mathrm{d}}{\mathrm{d}r} \int_r^a \frac{x^2\,\mathrm{d}x}{\sqrt{x^2-r^2}} \int_0^x \frac{p(t)\,\mathrm{d}t}{\sqrt{x^2-t^2}}\right] \tag{2.37}$$

$$p(r) = -\frac{3(1-\nu)\,hP}{\pi Ga^3} \frac{r}{\sqrt{a^2-r^2}} + \frac{1-2\nu}{2\pi(1-\nu)} \left[\frac{3r}{a^3\sqrt{a^2-r^2}} \times \right.$$

$$\left. \times \int_0^a t\sqrt{a^2-t^2}\,\tau(t)\,\mathrm{d}t - \frac{\mathrm{d}}{\mathrm{d}r} \int_r^a \frac{\mathrm{d}x}{\sqrt{x^2-r^2}} \int_0^x \frac{t\,\tau(t)\,\mathrm{d}t}{\sqrt{x^2-t^2}}\right] \tag{2.38}$$

The functions $\tau(r)$ and $\rho(r)$, which characterize tangential and normal stresses under the punch, are determined from a set of integral equations of form (2.37) and (2.38).

Examining relations (2.37) and (2.38) and (2.36) and considering that according to the notation adopted the $\tau(r)$ and $\rho(r)$ are negative functions, the following conclusions may be drawn.

The tangential stress under the punch on the contour of the contact domain has a singularity of the form $(a^2-r^2)^{-1/2}$. The action of the moment revolving the punch partly increases the tangential stresses under the punch.

The normal stress under the punch at the boundary of contact domain has a singularity of the same form $(a^2 - r^2)^{-1/2}$, but in the centre of contact domain and on the diametrical line, perpendicular to the direction of action of external force, this stress is equal to zero.

Increasing of tangential stresses under the punch results in partial intensification of normal stresses.

The magnitude of the punch displacement towards the action of force decreases with increase in normal stress under the punch.

The punch revolution around the axis, perpendicular to the direction of action of force, is directly proportional to the value of tilting moment but as the tangential stresses under the punch increase, it decreases.

These conclusions agree with similar conclusions presented in the paper by J.S. Ufland[14].

REFERENCES

1. K.V. Solianik-Krassa, 'On the Problem of Elastic Equilibrium of Solids Revolution', *Dokl. Akad. Nauk SSSR*, **114**, 1, 1957. pp. 49–52 (Russian).
2. R. Muki, 'Asymmetric Problems of the Theory of Elasticity for a Semi-infinite Solid and a Thick Plate', I.N. Sneddon and R. Hill (eds.), in: *Progress in Solid Mechanics*, Vol. I, North-Holland Publish. Company, Amsterdam, 1960, pp. 401–439.
3. B.L. Abramian, 'On one Method of Solution to Certain Spatial Problems in the Theory of Elasticity', *Dokl. Akad. Nauk Arm. SSR*, **35**, 4, 1962, pp. 151–159 (Russian).
4. Westman, 'Asymmetric Boundary Problems of a Mixed Type for an Elastic Semi-space', *J. Appl. Mech.* (*Transactions of ASME*), **ser. E**, **32**, 2, 1965, pp. 178–185.
5. P.F. Papkovich, 'Expression of the General Integral of Principal Equations of the Theory of Elasticity by Harmonic Functions', *Izv. Akad. Nauk SSSR Otd. Mat. Est. Nauk*, 32, 10, 1932, pp. 1425–1335 (Russian).
6. H. Neuber, 'Ein neuer Ansatz zur Lösung räumlicher Probleme der Elastizitätstheorie. Der Hohlkegel unter Einzellast als Beispiel', *ZAMM*, **14**, 4, 1934, pp. 203–212.
7. I. Busbridge, 'Dual Integral Equations', *Proc. London Math. Soc.*, 2 ser. ,44, 2207, 1938, pp. 115–129.
8. E. Titchmarsh, *Introduction to the Theory of Fourier Integrals*, Moscow, 1948 (Russian)
9. I.N. Sneddon, 'The Elementary Solution of Dual Integral Equations', *Proc. Glasgow Math. Assoc.*, **4**, Part. 3, 1960, pp. 108–110.
10. G.N. Watson, *The Theory of Bessel Functions*, Part I, Moscow, 1949 (Russian).
11. A.I. Lurie, *Spatial Problems of the Theory of Elasticity*, Gostekhizdat, Moscow, 1955 (Russian).
12. I.J. Staerman, *A Contact Problem of the Theory of Elasticity*, Gostekhizdat, Moscow-Leningrad, 1949 (Russian).
13. L.A. Galin, *Contact Problems of the Theory of Elasticity*, Gostekhizdat, Moscow, 1953 (Russian).
14. J.S. Ufland, *Integral Transformation in the Problems of the Theory of Elasticity*, Izdatel'stvo Akad. Nauk SSSR, Leningrad, 1967 (Russian).
15. J.J. Kalker, 'Elastic Contact on Line', *J. Appl. Mech.* (*Transactions of ASME*), **ser. E**, **39**, **4**, 1972, pp. 270–278.
16. B. Noble, 'Certain Dual Integral Equations', *Journ. of Math. and Phys.*, **37**, 2, 1958, pp. 128–136.
17. B. Noble, *Method Winer-Hopf*, Izd. IL, Moscow, 1962 (Russian).

18. E.T. Copson, 'On the Problem of the Electrified Disk', *Proc. Edinb. Math. Soc.*, **3**, **8**, 1947, pp. 14–19.
19. G.M.L. Gladwell, 'A Contact Problem for a Circular Cylindrical Punch in Adhesive Contact with an Elastic Half-Space. The Case of Rocking and Translation Parallel to the Plane', *Int. J. Engng. Sci.*, **7** ,1969, pp. 295–307.

Unbonded contact between a circular plate and an elastic foundation

G. M. L. Gladwell

1. INTRODUCTION

A review of the literature on the problem of a circular plate in bonded or unbonded contact with an elastic half-space was given in a recent paper by Gladwell and Iyer[1]. That paper also presented a (rather inelegant) solution and some results for the title problem. Here it is sufficient to discuss a few papers which deal with problems closely related to the title problem. Weitsman[2] considered an *infinite*, *weightless* plate lying in unbonded contact on an elastic half-space. He formulated the problem as an integral equation in the radial derivative of the contact pressure $p(r)$. Because the kernel $K(r, s)$ of the equation was symmetric and had a logarithmic singularity at $r = s$ he was able to use a general result of Noble and Hussain[3] and assume a solution of the form

$$p'(r) = B(c)\,(c^2 - r^2)^{-1/2} + f(r)$$

where $f(r)$ was continuous at the edge $r = c$ of the contact region. He then used a variational process to find c, taking $f(r) = A(c)(c^2 - r^2)^{1/2}$ and choosing c so that $B(c) = 0$. Some errors in his later calculations have been corrected by Iyer[4].

Although the important numerical constant that Weitsman obtains for the infinite weightless plate is very close to the one obtained in this paper, the correctness of his basic assumption, that $B(c) = 0$, can be seriously questioned. It would appear, both from *a priori* considerations and from actual values of the pressure obtained by solving the integral equation numerically, that the contact pressure has the form $p(r) \propto (c^2 - r^2)^{1/2}$ near $r = c$. This would give rise to a singularity $(c^2 - r^2)^{-1/2}$ in the derivative – a singularity that Weitsman specifically excludes. On purely mathematical grounds it would seem that the only restriction on the pressure at $r = c$ is that is zero; it will then behave like $(c^2 - r^2)^{1/2}$ near $r = c$. It could be argued on *physical* grounds that thin plate theory can be applied only when the pressure varies slowly as a function of r. In that case the form of $p(r)$ described above would be excluded and $p'(r)$ would have to be finite at

$r = c$, and possibly would be zero. The question will not be fully elucidated until the unbonded contact problem has been solved for a finite circular slab resting on an elastic foundation.

Weitsman's analysis gives only an approximate solution, and even the single-term approximation that he used led to some very tedious algebraic manipulations. The method is not well suited to computation. The formulation used by Keer, Dundurs and Tsai[5] is more suitable for computation. They consider the unbonded contact between an *infinite layer* of thickness h and an elastic half-space. For both the plane and the axisymmetric problems they formulate the problem as a Fredholm integral equation of the second kind in a transform of the contact pressure. It is this type of formulation that is used in the present paper.

In Section 2 a relation between displacement and load is obtained for the plate in integral form. In Section 3 the contact problem is considered. In Section 4 expressions are given for the case of uniform loading, before the results are discussed in Section 5.

2. THE PLATE

Consider a plate of radius a and thickness h lying on a half-space, and loaded symmetrically by a pressure $q(r)$, as shown in Fig. 1. The equation

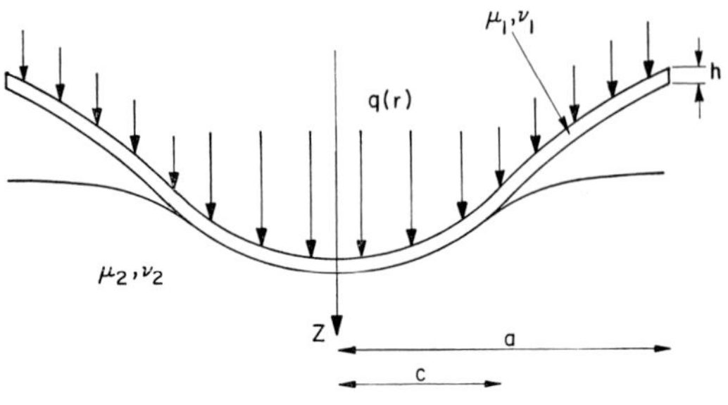

Fig. 1. Plate lying on an elastic half-space

governing the plate deflection $w(r)$ is

$$\nabla^4 w(r) = \{q(r) - p(r)\}/D = \phi(r) \tag{2.1}$$

where $p(r)$ is the contact pressure,

$$D = \mu_1 h^3/\{12(1-\nu_1)\} \tag{2.2}$$

and μ_1, ν_1 are the shear modulus and Poisson's ratio of the plate. The boundary conditions are

$$w'(0) = 0 \tag{2.3}$$

$$\nabla^2 w(r) = (1-\nu_1)\, w'(r)/r \quad \text{at} \quad r = a \tag{2.4}$$

$$\mathrm{d}\{\nabla^2 w(r)\}/\mathrm{d}r = 0 \quad \text{at} \quad r = a \tag{2.5}$$

where the last two state that the edge of the plate is free. The equilibrium condition for the plate is

$$P + W = 2\pi \int_0^c r\, p(r)\, \mathrm{d}r = 2\pi \int_0^a r\, q(r)\, \mathrm{d}r \tag{2.6}$$

where P is the applied load and W is the weight of the plate.

The plate equation (2.1) may be inverted. Since

$$\nabla^2 = \frac{1}{r}\frac{\mathrm{d}}{\mathrm{d}r}\left(r\frac{\mathrm{d}}{\mathrm{d}r}\right)$$

the equation may be integrated once to give

$$r\frac{\mathrm{d}}{\mathrm{d}r}\nabla^2 w(r) = -\int_r^a s\phi(s)\, \mathrm{d}s$$

Equation (2.5) shows that no arbitrary constant is involved. A second integration and use of equation (2.4) gives

$$\nabla^2 w(r) = \int_r^a \frac{\mathrm{d}t}{t} \int_t^a s\phi(s)\, \mathrm{d}s + (1-\nu_1)\frac{w'(a)}{a}$$

while a further integration and use of equation (2.3) gives

$$r w'(r) = \int_0^r u\, \mathrm{d}u \int_u^a \frac{\mathrm{d}t}{t} \int_t^a s\phi(s)\, \mathrm{d}s + (1-\nu_1)\frac{w'(a)\, r^2}{2a} \tag{2.7}$$

which yields

$$\frac{(1+\nu_1)}{2} a w'(a) = \int_0^a u\, \mathrm{d}u \int_u^a \frac{\mathrm{d}t}{t} \int_t^a s\, \phi(s)\, \mathrm{d}s \tag{2.8}$$

Now interchange the orders of the integration, first of s and t, then of s and u. The result is

$$w'(r) = \int_0^a f_1(r, s)\, s\, \phi(s)\, \mathrm{d}s \tag{2.9}$$

where

$$f_1(r, s) = \begin{cases} \dfrac{1}{4}\left(\kappa r \dfrac{s^2}{a^2} + \dfrac{s^2}{r}\right) & s < r \\[2ex] \dfrac{1}{4}\left(\kappa r \dfrac{s^2}{a^2} + r + 2r \ln\left(\dfrac{s}{r}\right)\right) & s \geq r \end{cases} \tag{2.10}$$

Equation (2.9) may be integrated once more to yield

$$w(r) - w(0) = \int_0^a s f_2(r, s)\, \phi(s)\, ds \tag{2.11}$$

where

$$f_2(r, s) = \begin{cases} \dfrac{1}{8}\left(\dfrac{\kappa r^2 s^2}{a^2} + 2s^2 \ln\left(\dfrac{r}{s}\right) + 2s^2\right) & s < r \\ \dfrac{1}{8}\left(\dfrac{\kappa r^2 s^2}{a^2} + 2r^2 \ln\left(\dfrac{s}{r}\right) + 2r^2\right) & s \geq r \end{cases} \tag{2.12}$$

and $\kappa = (1 - \nu_1)/(1 + \nu_1)$.

3. THE HALF-SPACE

It is well known (Sneddon[4]) that on the boundary of the half-space the normal deflection $u_z \equiv v(r)$ and pressure $p(r)$ may be written

$$v(r) = \frac{1 - \nu_2}{\mu_2} \int_0^\infty e(\lambda)\, J_0(\lambda r)\, d\lambda \tag{3.1}$$

$$p(r) = \int_0^\infty \lambda\, e(\lambda)\, J_0(\lambda r)\, d\lambda \tag{3.2}$$

where μ_2, ν_2 are the shear modulus and Poisson's ratio of the half-space. It will now be assumed that the plate is in contact with the half-space for $0 \leq r \leq c$, and out of contact with it for $r > c$. Thus it is assumed that the loading $q(r)$ is such that the plate does not bend back after it has lost contact at $r = c$. The contact conditions are therefore

$$p(r) > 0, \quad 0 \leq r < c; \quad p(r) = 0 \quad r \geq c \tag{3.3}$$

$$v(r) = w(r) \quad 0 \leq r \leq c \tag{3.4}$$

Put

$$\begin{aligned} e(\lambda) &= \int_0^c g(t) \cos(\lambda t)\, dt \\ &= g(c) \frac{\sin(\lambda c)}{\lambda} - \int_0^c g'(t) \frac{\sin(\lambda t)\, dt}{\lambda} \end{aligned} \tag{3.5}$$

and make use of the integral

$$\int_0^\infty \sin(\lambda t)\, J_0(\lambda r)\, d\lambda = (t^2 - r^2)^{-1/2}\, H(t - r).$$

Then

$$p(r) = \left\{\frac{g(c)}{(c^2-r^2)^{1/2}} - \int_r^c \frac{g'(t)\,\mathrm{d}t}{(t^2-r^2)^{1/2}}\right\} H(c-r) \tag{3.6}$$

so that $p(r)=0$ when $r>c$, in accordance with (3.3). This equation shows that if $g(c)\neq 0$ the contact pressure will have a square root singularity at the boundary $r=c$ of the contact region. If the contact is truly unbonded and $c<a$ then such a singularity is not allowable, so that

$$g(c) = 0 \qquad \text{when} \qquad c<a \tag{3.7}$$

and this implies that $p(c)=0$. The corresponding normal displacement is

$$v(r) = \frac{1-\nu_2}{\mu_2}\int_0^{\min(c,r)} \frac{g(t)\,\mathrm{d}t}{(r^2-t^2)^{1/2}} \tag{3.8}$$

This equation may be inverted to give $g(t)$ in terms of $v(r)$; thus

$$g(t) = \frac{2}{\pi}\left(\frac{\mu_2}{1-\nu_2}\right)\frac{\mathrm{d}}{\mathrm{d}t}\int_0^t \frac{x\,v(x)\,\mathrm{d}t}{(t^2-x^2)^{1/2}}$$

which may be rewritten

$$g(t)-g(0) = \frac{2\mu_2}{\pi(1-\nu_2)}\,t\int_0^t \frac{v'(x)\,\mathrm{d}x}{(t^2-x^2)^{1/2}}\ (t\le c) \tag{3.9}$$

It is now necessary to combine equations (2.9) and (3.8) by substituting $w'(x)$ for $v'(x)$. It is found that

$$g(t)-g(0) = \frac{\mu_2}{2\pi(1-\nu_2)}\int_0^a h(t,s)\,s\,\phi(s)\,\mathrm{d}s \tag{3.10}$$

where

$$h(t,s) = \begin{cases} \kappa\dfrac{s^2t^2}{a^2} + 3t(t-\sqrt{t^2-s^2}) + (s^2+2t^2)\ln\left[\dfrac{t+\sqrt{t^2-s^2}}{2t}\right] \\ \qquad + s^2\ln\left(\dfrac{2t}{s}\right) \quad (s<t) \\ \kappa\dfrac{s^2t^2}{a^2} + 2t^2+2t^2\ln\left(\dfrac{s}{2t}\right) \quad (s\ge t) \end{cases} \tag{3.11}$$

Now put

$$D\,\phi(r) = q(r) + \int_r^c \frac{g'(t)\,\mathrm{d}t}{(t^2-r^2)^{1/2}}\,H(c-r); \tag{3.12}$$

the result is

$$g(t) - g(0) = n \left\{ \int_0^a h(t,s)\, s\, q(s)\, \mathrm{d}s - \int_0^c H_1(t,u)\, g(u)\, \mathrm{d}u \right\} \tag{3.13}$$

where

$$H_1(t,u) = \frac{2\kappa t^2 u^2}{a^2} + 2tu \ln \left| \frac{t+u}{t-u} \right| + t^2 \ln \left| \frac{t^2-u^2}{t^2} \right| + u^2 \ln \left| \frac{t^2-u^2}{u^2} \right| \tag{3.14}$$

$$H_1(t,u) = H_1(u,t) \tag{3.15}$$

$$n = \mu_2 / \{2\pi (1-\nu_2) D\} \tag{3.16}$$

Equation (3.13) is an integral equation for the unknown function $g(t)$; it will now be manipulated into a more convenient form. The equilibrium condition (2.6) and equations (3.2) and (3.5) give

$$P+W = 2\pi \int_0^c r\, p(r)\, \mathrm{d}r = 2\pi\, e(0) = 2\pi \int_0^c g(u)\, \mathrm{d}u \tag{3.17}$$

and equation (3.13) becomes

$$g\ t) - g(0) = n \int_0^c H_2(t,u)\, g(u)\, \mathrm{d}u \tag{3.18}$$

where

$$H_2(t,u) = -H_1(t,u) + \frac{2\pi}{P+W} \int_0^a h(t,s)\, s\, q(s)\, \mathrm{d}s\,. \tag{3.19}$$

Since $g(c) = 0$ equation (3.18) may be written as

$$g(t) = n \int_0^c H_3(t,u)\, g(u)\, \mathrm{d}u \tag{3.20}$$

where

$$H_3(t,u) = H_2(t,u) - H_2(c,u) \tag{3.21}$$

Equation (3.20) is an integral equation containing two unknowns, $g(t)$ and c. It is more convenient however to consider c as fixed, especially as c appears in the kernel $H(t, u)$, and find the eigenvalue n corresponding to it. The eigenvalue must be chosen so that the pressure $p(r)$, related to $g(t)$ through equations (3.2) and (3.5), satisfies the condition (3.3), that it be positive.

4. CONTACT UNDER UNIFORM LOAD

Assume for the sake of definiteness that the external load $q(r)$ is made up of two parts, a uniform load P distributed over a circle of radius b, and the weight W evenly distributed over the whole plate. Then the parameters in the problem are the lengths a, b, c and h; variables t, u, r; displacements v, w; function $g(t)$ and pressure $p(r)$; loads P, W; and the parameter n. Introduce the following dimensionless parameters:

$$(\alpha, \beta, \gamma, x, y, \rho) = (a, b, c, t, u, r)/h$$

$$(v^*(\rho),\ w^*(\rho)) = \frac{\mu_2}{1-\nu_2} \cdot \frac{2\pi}{P+W} \cdot h \cdot (v(r), w(r))$$

$$f(y) = \frac{2\pi h}{P+W} \cdot g(u), \qquad p^*(\rho) = \frac{2\pi h^2}{P+W} \cdot p(r)$$

$$(P^*, W^*) = (P, W)/(P+W).$$

Notice then that $P^* + W^* \equiv 1$.

For the parameter n occurring with the elastic half-space one may introduce the dimensionless quantity

$$\eta = \frac{3}{\pi n h^3} = \frac{\mu_1(1-\nu_2)}{\mu_2(1-\nu_2)} \tag{4.1}$$

Keer, Dundurs and Tsai[5] use the mismatch parameter (which they designate α)

$$\lambda = \frac{\mu_2(1-\nu_1) - \mu_1(1-\nu_2)}{\mu_2(1-\nu_1) + \mu_1(1-\nu_2)} \tag{4.2}$$

With this notation

$$\eta = \frac{1-\lambda}{1+\lambda} \quad \text{or} \quad \lambda = \frac{1-\eta}{1+\eta} \tag{4.3}$$

The functions $f(y)$ and $p^*(\rho)$ now satisfy

$$\int_0^\gamma p^*(\rho)\, d\rho = \int_0^\gamma f(y)\, dy = 1 \tag{4.4}$$

and equations (3.6), (3.8) for the half-space show that

$$p^*(\rho) = -\int_\rho^\gamma \frac{f'(x)\, dx}{(x^2-\rho^2)^{1/2}} H(\gamma-\rho) \tag{4.5}$$

$$v^*(\rho) = \int_0^{\min(\gamma,\rho)} \frac{f(y)\, dy}{(\rho^2-y^2)^{1/2}} \tag{4.6}$$

Equation (4.6) gives the centre displacement as

$$v^*(0) = w^*(0) = \frac{\pi}{2} f(0) \tag{4.7}$$

After some algebra it is found that the dimensionless plate displacement is given by

$$w^*(\rho) - w^*(0) = \frac{3}{2\eta}\left\{-\int_0^{\gamma} f(y)\, k_i(\rho, y)\, dy + P^* f_j(\rho, \beta) + \right.$$
$$\left. + W^* f_3(\rho, \alpha)\right\} \qquad (4.8)$$

where

$$i = \begin{cases} 1, & y \le \rho \\ 2, & y > \rho \end{cases}, \qquad j = \begin{cases} 3, & \rho < \beta \\ 4, & \rho \ge \beta \end{cases}$$

and

$$k_1(\rho, y) = \frac{\kappa \rho^2 y^2}{\alpha^2} + 3y^2 + 2y^2 \ln\left(\frac{\rho}{2y}\right)$$

$$k_2(\rho, y) - k_1(\rho, y) = -3y(y^2-\rho^2)^{1/2} +$$
$$+ (2y^2 + \rho^2) \ln\left[\frac{y + (y^2-\rho^2)^{1/2}}{\rho}\right]$$

$$f_3(x, y) = \frac{\kappa x^2 y^2}{4\alpha^2} + \frac{x^4}{8y^2} + x^2 \ln\left(\frac{y}{x}\right) + \frac{x^2}{2}$$

$$f_4(x, y) = \frac{\kappa x^2 y^2}{4\alpha^2} + \frac{5y^2}{8} + \frac{y^2}{2} \ln\left(\frac{x}{y}\right)$$

The integral equation for $f(x)$ is

$$f(x) = \frac{3}{\pi\eta}\int_0^{\gamma} K_3(x, y)\, f(y)\, dy \qquad (4.9)$$

where

$$K_3(x, y) = K_2(x, y) - K_2(\gamma, y)$$
$$K_2(x, y) = -H_3(hx, hy)/h^2 + P^* f_k(x, \beta) + W^* f_5(x, \alpha)$$

and

$$f_5(x, \beta) = 2\left\{\frac{\kappa\beta^2 x^2}{4\alpha^2} + x^2 + x^2 \ln\left(\frac{\beta}{2x}\right) + \frac{x^4}{6\beta^2}\right\}$$

$$f_6(x, \beta) - f_5(x, \beta) = -2\left\{\frac{x(x^2-\beta^2)^{3/2}}{6\beta^2} + \frac{5x}{4}(x^2-\beta^2)^{1/2} - \right.$$
$$\left. - \frac{(4x^2+\beta^2)}{4} \ln\left[\frac{x + (x^2-\beta^2)^{1/2}}{\beta}\right]\right\}$$

and

$$k = \begin{cases} 5 & x < \beta \\ 6 & x \geq \beta \end{cases}.$$

The point loading case is obtained by putting f_4 and f_6 identically zero.

5. RESULTS

Dimensional analysis shows that for an infinite, weightless plate, for which there are only two lengths involved, namely contact radius c and thickness h, the relation

$$nc^3 = \text{constant}$$

must hold. This relation gives rise to

$$(c/h)^3 = k\eta$$

where k is a dimensionless constant. Fig. 2 shows $\eta^{-1}(c/h)^3$ plotted for various *finite* weightless plates as a function of (c/h). When a/h is large then $\eta^{-1}(c/h)^3$ is almost constant. The value obtained for $(c/h) = 2$, $(a/h) = 50$ was 4.15, compared to Weitsman's 4.18.

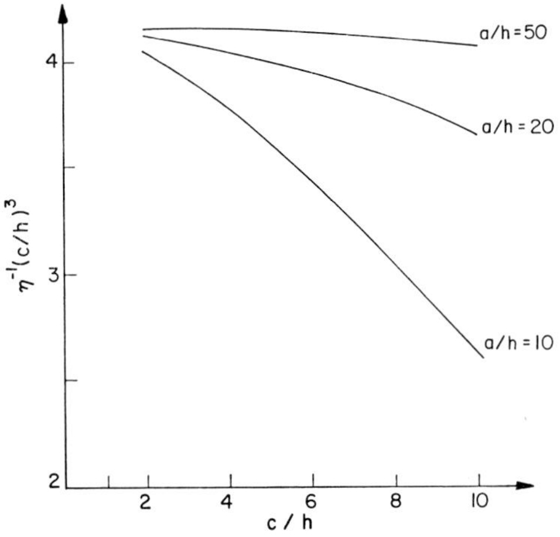

Fig. 2. Dimensionless elastic constant plotted against contact radius

It should be noted that in the problem of a beam in unbonded contact with a half-plane, considered by Weitsman in [6], his assumption that $p'(y) \propto (c^2 - y^2)^{1/2}$ near $y = c$ leads him to a completely wrong numerical value for the dimensionless eigenvalue ratio. This will be discussed in a further paper.

When the plate is weightless it is found that for given values of radius a, load radius b and contact radius c there is only one value of η for which the contact pressure $p(r)$ is positive for all r in the range $0 \leq r < c$. However, when the plate is allowed to have weight the situation is more complicated. Fig. 3 shows the eigenvalue η plotted as a function of dimensionless weight for a fixed value of $a/h = 10$, point loading ($b = 0$) and different values of c/h.

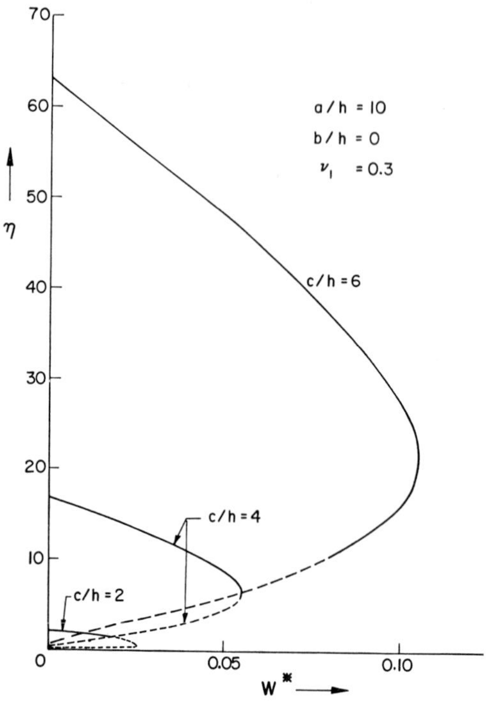

Fig. 3. Variation of elastic constant rates with dimensionless plate weight

Once an eigenvalue η has been found, three conditions must be satisfied:

1. The eigenvalue must be positive.
2. The contact pressure must be positive throughout [0, c].
3. The plate must remain above the half-space for values of r in [c, a].

For $W^* = 0$ it is found that although there are two or more positive eigenvalues, only one satisfies conditions 2 and 3. As W^* is increased the eigenvalues change as shown, the acceptable one on the full line, and the other on the dashed line. The two eigenvalues approach each other and at some stage the secondary one may become acceptable. A further increase in W^* then leads to breakdown either because the eigenvalues become complex conjugates or because one or both of conditions 2 and 3 are violated.

In Fig. 3 condition 3 is violated on the curve $c/h = 2$ somewhere between $W^* = 0.02$ and $W^* = 0.025$. On the curve $c/h = 4$ the eigenvalues become complex between 0.055 and 0.056, while for $c/h = 6$ they become complex between 0.104 and 0.105.

The conclusion to be drawn from Fig. 3 is that, for given values of a, b, c h there is a limiting value of W^* above which the given contact configuration is not possible, for any combination of elastic constants of the plate and half-space. This conclusion holds for loading other than point loading; the curve of η against W^* for loading over a circle of radius $b/h = 1$ is very similar to that shown in Fig. 3.

It should be noted that there are cases – see $c/h = 2$ on Fig. 3 – where condition 3 is broken before the eigenvalues become complex. In this case there is a second region of contact between plate and half-space and the analysis given above on longer applies.

ACKNOWLEDGMENT

The author gratefully acknowledges the help provided by Shimon Coen in the computation. This work was supported by the National Research Council of Canada.

REFERENCES

1. G.M.L. Gladwell and K.R.P. Iyer, 'On the Unbonded Contact between a Circular Plate and an Elastic Foundation', *J. Elasticity*, 1974.
2. Y. Weitsman, 'On the Unbonded Contact between Plates and an Elastic Half-Space', *J. Appl. Mech.*, **36**, 1969, pp. 198–202.
3. B. Noble and M. Hussain, 'A Variational Method for Inclusion and Indentation Problems', US Army Mathematics Research Centre, University of Wisconsin, Report 812, 1967.
4. K.R.P. Iyer, *The Unbonded Contact Problem*, Ph.D. Thesis, University of Waterloo, Waterloo, Ontario, 1973.
5. L.M. Keer, J. Dundurs and K.C. Tsai, 'Problems Involving a Receding Contact between a Layer and a Half-Space', *J. Appl. Mech.*, **39**, 1972, pp. 1115–1120.
6. Y. Weitsman, 'A Tensionless Contact between a Beam and an Elastic Half-Space,' *Int. J. Engng. Sci.*, **10**, 1972, pp. 73–81.

On the two-dimensional contact problem of a rigid cylinder, pressed between two elastic layers

J. B. Alblas

1. INTRODUCTION

This paper may be considered as a generalization and continuation of a paper[1] by the present author, concerning the deformation of two elastic half-spaces, pressed against each other with a cylindrical inclusion between them. Here we investigate the corresponding problem for the case that the half-spaces are replaced by two elastic layers. The layers are supported by rigid bases, along which they may slide without friction.

This problem being complicated, it is desirable to consider first a more limited version of it. In this paper we only consider the two-dimensional problem of two layers, loaded in plane strain. Further we confine our discussions to the symmetrical case, in which the layers have equal widths and equal constants of elasticity. It must be noted that the more general non-symmetrical problem is not a simple generalization of the one, treated here. But the mathematical methods for the treatment of both problems are similar.

Because we have to deal with a two-dimensional problem we only have to consider a cross-section, perpendicular to the axis of the cylinder. Therefore, we often shall describe the half-spaces as half-planes and the layers as strips, as has also be done in [1].

A basic part of the problem consists of the determination of the values of the length of the contact regions with the cylinder and the length of the regions of separation, for given values of the displacement of the supports (or the pressure at infinity), the width of the cylinder and the thickness of the layers. After the solution of this part of the problem the surface pressure and the normal displacement at the surface may be obtained. With (one of) these quantities the distributions of displacements and stresses throughout the layers can be calculated.

The solution of the corresponding fundamental problem, concerning the contact of two half-spaces may be found in closed form (cf. [1]). Its surface pressure and normal displacement are presented in the form of elliptic functions, Jacobi's Zeta-function and Heumann's Lambda-function, respectively. The relations between the pressure at infinity, the width of the cylinder and the lengths of contact and separation regions are given by two simple transcendental equations.

For the layer the surface pressure and normal displacement are represented in the form of convergent series expansions in powers of the reciprocal thickness parameter, i.e. the ratio of the values of the layer thickness and the sum of the lengths of contact and separation regions. The series only converge for values of this parameter which are greater than two. If this parameter is much smaller than two, another type of asymptotics has to be applied, which is not the object of this study. It may be expected that for this case the stress distribution near the cylinder will not differ much from that of the simple stamp problem.

The terms in the series expansions may be calculated step by step, using for each term the results obtained in the preceding ones. The half-space solution corresponds with the zero'th term.

Numerical results are given for several values of the width of the inclusion and the thickness of the layers. They refer to the contact and separation lengths and to the pressure distribution and normal displacement at the surface.

The results are of special interest for layers of soft (i.e. small shear modulus) materials, e.g. rubbery materials. For hard materials the inclusion must be very thin, otherwise plastic flow or fracture will occur.

2. STATEMENT OF THE PROBLEM

We consider two isotropic homogeneous elastic layers of the same material, occupying the regions of space $-\infty < x < \infty$; $-\infty < z < \infty$; $b < y < 0$ and $-b < y < 0$, where (x, y, z) is a right handed Cartesian coordinate system (cf. Fig. 1). Between the layers a smooth rigid cylinder of infinite

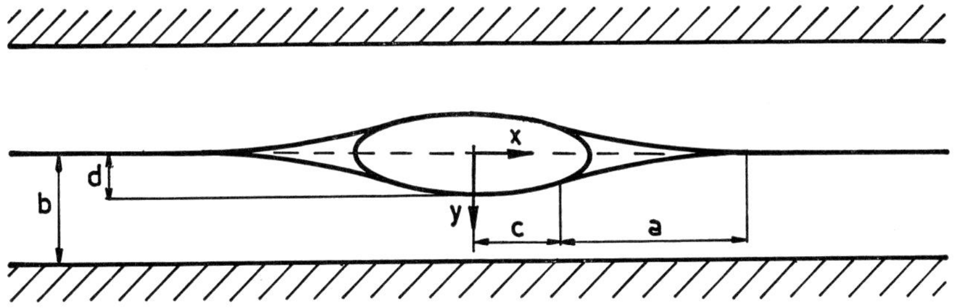

Fig. 1. Geometry of the problem

extension is included. The cross-section of the cylinder is bounded by two parabolic arcs with rounded corners. The layers are pressed by a uniform displacement at $y = \pm b$, where they are in contact with smooth rigid bases.

Within linear elastostatics the problem to be solved is formulated by the system of equations

$$\left.\begin{aligned} \nabla^2 u + \frac{1}{1-2\nu}(u_{,x}+v_{,y})_{,x} &= 0, \\ \nabla^2 v + \frac{1}{1-2\nu}(u_{,x}+v_{,y})_{,y} &= 0, \end{aligned}\right\} \tag{2.1}$$

where u and v are the displacements in the x and y directions, respectively, ∇^2 is the plane Laplacian Operator and ν is Poisson's ratio. Because of symmetry we may confine our discussion to the displacements and stresses in one strip, for which we take $y > 0$.

The boundary conditions of the problem under discussion are

$$\left.\begin{aligned} y = 0,\ & -\infty < x < \infty,\ \tau_{xy} = 0, \\ ,\ & -c < x < c,\ v = d - \frac{x^2}{2R}, \\ ,\ & |x| > c+a,\ v = 0 \\ ,\ & c < |x| < a,\ \overline{p(x)} \equiv -\sigma_y = 0, \\ y = b,\ & -\infty < x < \infty,\ \tau_{xy} = 0, \\ & ,\ v = -d_1, \end{aligned}\right\} \tag{2.2}$$

where τ_{xy} and σ_y are shear stress and normal stress, respectively, $2d$ is the width of the cylinder, $1/R$ is the curvature at the vertex of the parabola, c is the unknown length of the interval of contact with the cylinder, while the region of contact between the strips is given by $|x| > c+a$, with a also an unknown length. The displacement of the support at $y = b$ is equal to d_1. The pressure at infinitely is given by $p^{(0)}$. It appears to be preferable to replace in the formulae the parameter a by ρ, given by

$$\rho = a+c. \tag{2.3}$$

For $|x| < c$ and $|x| > \rho$, the pressure $\overline{p(x)}$ has to satisfy the inequality

$$\overline{p(x)} \geq 0. \tag{2.4}$$

3. DECOMPOSITION OF THE PROBLEM

We assume the solution S of the problem to be split, according to

$$S = S_1 + S_2, \tag{3.1}$$

where S_1 is an elementary solution, given by

$$u = \frac{\nu}{2\mu} p^{(0)} x, \qquad v = -\frac{1-\nu}{2\mu} p^{(0)} y, \tag{3.2}$$

with μ the shear modulus. The boundary conditions for the residual problem S_2 may be stated as

$$\left.\begin{array}{lll} y=0, & -\infty < x < \infty, & \tau_{xy} = 0, \\ , & -c < x < c, & v = d - \dfrac{x^2}{2R}, \\ , & |x| > \rho, & v = 0, \\ , & c < |x| < a, & p(x) = -p^{(0)}, \\ y = b, & -\infty < x < \infty, & \tau_{xy} = v = 0, \end{array}\right\} \tag{3.3}$$

where $p(x)$ is the pressure in the residual problem, vanishing at infinity. The pressure $p^{(0)}$ is determined by the displacement d_1 at $y=b$, according to (3.2)

$$p^{(0)} = \frac{2\mu}{1-\nu} \frac{d_1}{b}. \tag{3.4}$$

Of course we have $d_1 > 0$, otherwise there is no contact at infinity.

4. THE INTEGRAL EQUATIONS

We solve the residual problem (2.1), (3.3) by the method of Fourier integrals. Denoting the transformed quantities by a bar, we find for $y=0$

$$\frac{1-\nu}{\mu} \bar{p} = \frac{s(\operatorname{sh} 2sb + 2sb)}{(\operatorname{ch} 2sb - 1)} \bar{v}, \tag{4.1}$$

where s is the transform parameter. To simplify the equations we introduce the non-dimensional pressure $\tilde{p}$ by

$$\tilde{p} = \frac{1-\nu}{\mu} p. \tag{4.2}$$

In the following we shall omit the tilde. We define the function $K(sb)$ by

$$K(sb) = sb \left[\frac{\operatorname{sh} 2sb + 2sb - \operatorname{ch} 2sb + 1}{\operatorname{ch} 2sb - 1} \right]. \tag{4.3}$$

Inversion of (4.1) yields with (4.2) and (4.3)

$$p(x) = \sqrt{\frac{2}{\pi}} \int_0^\infty \frac{\bar{v}(s)}{b} sb \cos sx \, ds + \sqrt{\frac{2}{\pi}} \int_0^\infty \frac{\bar{v}(s)}{b} K(sb) \cos sx \, ds. \tag{4.4}$$

To meet the boundary condition (3.3_3) we write $\bar{v}(s)$ in the form

$$\bar{v}(s) = \int_0^\rho g(t)\, J_0(st)\, \mathrm{d}t\,, \tag{4.5}$$

where $J_0(st)$ denotes Bessel's function of the first kind of order zero and $g(t)$ is an auxiliary function. Assuming that we may change the order of integration, we find that $v(x)$ can be expressed as

$$v(x) = \sqrt{\frac{2}{\pi}} \int_x^\rho g(t) \frac{\mathrm{d}t}{\sqrt{t^2 - x^2}}\,, \quad \text{for} \quad 0 < x < \rho\,, \tag{4.6}$$

while

$$v(x) = 0\,, \quad \text{for} \quad x > \rho\,. \tag{4.7}$$

Equation (4.6) may be considered as an integral equation of Abel's type for $g(t)$. The solution is

$$g(t) = -\sqrt{\frac{2}{\pi}}\, t \int_t^\rho \frac{v'(x)}{\sqrt{x^2 - t^2}}\, \mathrm{d}x\,, \quad 0 < t < \rho\,, \tag{4.8}$$

where the prime denotes the derivative.

With (4.5) we write (4.4) in the form

$$p(x) = \sqrt{\frac{2}{\pi}} \frac{\mathrm{d}}{\mathrm{d}x} \int_0^x \frac{g(t)\, \mathrm{d}t}{\sqrt{x^2 - t^2}} + \sqrt{\frac{2}{\pi}} \int_0^\infty \frac{\bar{v}(s)}{b} K(sb) \cos sx\, \mathrm{d}s\,,$$
$$0 < x < \rho\,. \tag{4.9}$$

This equation may also be solved for $g(t)$. There results

$$g(t) = \sqrt{\frac{2}{\pi}}\, t \int_0^t \frac{p(x)\, \mathrm{d}x}{\sqrt{t^2 - x^2}} - \sqrt{\frac{2}{\pi}}\, t \int_0^t \frac{F(x)}{\sqrt{t^2 - x^2}}\, \mathrm{d}x\,, \quad 0 < t < \rho\,, \tag{4.10}$$

with

$$F(x) = \sqrt{\frac{2}{\pi}} \int_0^\infty \bar{v}(s) \frac{K(sb)}{b} \cos sx\, \mathrm{d}s\,. \tag{4.11}$$

From (4.8) and (4.10) we can eliminate $g(t)$ to obtain the basic integral equation

$$-\int_t^\rho \frac{v'(x)\, dx}{\sqrt{x^2 - t^2}} = \int_0^t \frac{p(x)\, \mathrm{d}x}{\sqrt{t^2 - x^2}} - \int_0^t \frac{F(x)\, \mathrm{d}x}{\sqrt{t^2 - x^2}}\,, \quad 0 < t < \rho\,. \tag{4.12}$$

Another from for this equation will be obtained, if we multiply it with $1/(\sqrt{t^2 - u^2})$ and integrate over t from u to ρ. We find

$$v'(x) = \frac{1}{\pi} \frac{1}{\sqrt{\rho^2 - x^2}} \int_{-\rho}^\rho p(u) \frac{\sqrt{\rho^2 - u^2}}{u - x}\, \mathrm{d}u$$

$$-\frac{1}{\pi}\frac{1}{\sqrt{\rho^2-x^2}}\int_{-\rho}^{\rho} F(u)\frac{\sqrt{\rho^2-u^2}}{u-x}\,du, \quad |x|<\rho. \tag{4.13}$$

We note that (4.13) is a singular integral equation for the unknowns $p(x)$ on $(-c, c)$ and $v(x)$ on $(-\rho, -c)$ and (c, ρ), while c and ρ are unknown parameters. The solution is determined by the conditions that $v'(x)$ has to be continuous at $|x| = c$ and $|x| = \rho$.

5. REDUCTION TO A SET OF INTEGRAL EQUATIONS

We first rewrite $F(u)$, according to (4.11), in the following form

$$\begin{aligned} F(u) &= \frac{1}{\pi}\int_0^\infty v(x)\,dx \int_0^\infty \frac{K(sb)}{b}\{\cos s(u+x) + \cos s(u-x)\}\,ds \\ &= \frac{1}{\pi}\int_{-\infty}^\infty v(x)\,S(u-x)\,dx = \frac{1}{\pi}\int_{-\rho}^{\rho} v(x)\,S(u-x)\,dx, \end{aligned} \tag{5.1}$$

with

$$\begin{aligned} S(t) &= \int_0^\infty \cos st\, s\, ds \left\{\frac{sb}{\mathrm{sh}^2\, sb} + \frac{e^{-sb}}{\mathrm{sh}\, sb}\right\} \\ &= \frac{\pi^2}{4b^2}\,\frac{2\left(\frac{\pi t}{2b}\right)\mathrm{ch}\left(\frac{\pi t}{2b}\right) - 3\,\mathrm{sh}\left(\frac{\pi t}{2b}\right) + \frac{4b^2}{\pi^2 t^2}\,\mathrm{sh}^3\left(\frac{\pi t}{2b}\right)}{\mathrm{sh}^3\left(\frac{\pi t}{2b}\right)}. \end{aligned} \tag{5.2}$$

We expand this function in a convergent series. We have

$$\frac{2t\,\mathrm{ch}\,t - 3\,\mathrm{sh}\,t + \frac{1}{t^2}\,\mathrm{sh}^3 t}{\mathrm{sh}^3 t} = \sum_{m=0}^{\infty} \alpha_m t^{2m}, \tag{5.3}$$

which converges for $|t| < \pi/2$. We have evaluated the first five coefficients in the expansion (5.3). The results are presented in Table 1.

Table 1. The α-coefficients

α_0	α_1	α_2	α_3	α_4
1	$-\frac{1}{3}$	$\frac{2}{27}$	$-\frac{1}{75}$	$\frac{2}{945}$

To proceed we write the integral equation (4.13) in non-dimensional form.

We introduce

$$x = \rho\tilde{x}, \quad u = \rho\tilde{u}, \quad v = \rho\tilde{v}, \quad k = c/\rho, \quad q = b/\rho \tag{5.4}$$

and (4.13) takes the form

$$v'(x) = \frac{1}{\pi}\frac{1}{\sqrt{1-x^2}}\left\{\int_{-1}^{1} p(u)\frac{\sqrt{1-u^2}}{u-x}\,du - \int_{-1}^{1} F(u)\frac{\sqrt{1-u^2}}{u-x}\,du\right\},$$
$$|x| < 1. \tag{5.5}$$

In (5.5) we have dropped the tildes. The parameter q is called the thickness parameter.

The next step consists of the introduction of the actual pressure $\overline{p(u)}$ by

$$p(u) = \overline{p(u)} - p^{(0)}. \tag{5.6}$$

Now (5.5) becomes

$$v'(x) - p^{(0)}\frac{x}{\sqrt{1-x^2}} = \frac{1}{\pi}\frac{1}{\sqrt{1-x^2}}\int_{-k}^{k} \overline{p(u)}\frac{\sqrt{1-u^2}}{u-x}\,du -$$
$$-\frac{1}{\pi}\frac{1}{\sqrt{1-x^2}}\int_{-1}^{1} F(u)\frac{\sqrt{1-u^2}}{u-x}\,du, \quad |x| < 1, \tag{5.7}$$

where the boundary condition (3.3_4) has been used.

We put

$$\left.\begin{aligned} v(x) &= \sum_{l=0}^{\infty}\frac{v_l(x)}{q^{2l}}, \\ \overline{p(x)} &= \sum_{l=0}^{\infty}\frac{\overline{p_l(x)}}{q^{2l}}, \\ p^{(0)} &= \sum_{l=0}^{\infty}\frac{p_l^{(0)}}{q^{2l}}, \\ F(u) &= \sum_{l=1}^{\infty}\frac{F_l(u)}{q^{2l}}, \end{aligned}\right\} \quad q > 2, \tag{5.8}$$

and introduce these expansions into (5.7). Putting the corresponding coefficients of q^{-2l} equal to zero, we find the system

$$v_l'(x) - p_l^{(0)}\frac{x}{\sqrt{1-x^2}} + \frac{1}{\pi}\frac{1}{\sqrt{1-x^2}}\int_{-1}^{1} F_l(u)\frac{\sqrt{1-u^2}}{u-x}\,du$$
$$= \frac{1}{\pi}\frac{1}{\sqrt{1-x^2}}\int_{-k}^{k} \overline{p_l(u)}\frac{\sqrt{1-u^2}}{u-x}\,du, \quad |x| < 1,$$
$$(l = 0, 1, 2 \ldots, F_0 = 0). \tag{5.9}$$

An important feature of this system is that it can be solved step by step.

From (5.1) to (5.4) and (5.8) we derive

$$F_l(u) = \frac{\pi}{4} \sum_{m=0}^{l-1} \alpha_m \left(\frac{\pi}{2}\right)^{2m} \int_{-1}^{1} v_{l-m-1}(x)\,(x-u)^{2m}\,dx, \quad (l = 1,2,3,\dots). \tag{5.10}$$

By expanding $(x-u)^{2m}$ we obtain for $F_l(u)$

$$F_l(u) = \frac{\pi}{4} \sum_{n=0}^{l-1} C_{ln} u^{2n}, \tag{5.11}$$

with

$$C_{ln} = -\sum_{m=n}^{l-1} \alpha_m \left(\frac{\pi}{2}\right)^{2m} \binom{2m}{2n} \frac{1}{2m-2n+1} \int_{-1}^{1} v'_{l-m-1}(x)\, x^{2m-2n+1}\,dx. \tag{5.12}$$

Introduction of (5.11) into (5.9) leads to

$$\begin{aligned}\int_{-1}^{1} F_l(u) \frac{\sqrt{1-u^2}}{u-x}\,du &= \frac{\pi}{4} \sum_{n=0}^{l-1} C_{ln} \int_{-1}^{1} u^{2n} \frac{\sqrt{1-u^2}}{u-x}\,du \\ &= \frac{\pi}{4} \sum_{n=0}^{l-1} a_{ln} x^{2n+1}, \quad |x| < 1,\end{aligned} \tag{5.13}$$

where the coefficients a_{ln} are given by

$$a_{ln} = \sum_{p=n}^{l-1} C_{lp} \frac{\Gamma(p-n-\frac{1}{2})\,\Gamma(\frac{3}{2})}{\Gamma(p-n+1)} = -\pi \sum_{p=n}^{l-1} C_{lp} (-1)^{p-n} \binom{1/2}{p-n} \tag{5.14}$$

For the derivation of (5.14) we have used

$$\begin{aligned}I_n &= \int_{-1}^{1} \frac{u^{2n}\sqrt{1-u^2}}{u-x}\,du = \sum_{p=0}^{n} \frac{\Gamma(p-\frac{1}{2})\,\Gamma(\frac{3}{2})}{\Gamma(p+1)} x^{2n-2p+1} \\ &= -\pi \sum_{p=0}^{n} \binom{1/2}{p} (-1)^p x^{2n-2p+1}, \quad |x| < 1.\end{aligned} \tag{5.15}$$

With (5.13) the system of integral equations (5.9) becomes

$$\begin{aligned}&v'_l(x) - p_l^{(0)} \frac{x}{\sqrt{1-x^2}} + \frac{1}{4} \frac{1}{\sqrt{1-x^2}} \sum_{p=0}^{l-1} a_{lp} x^{2p+1} \\ &= \frac{1}{\pi} \frac{1}{\sqrt{1-x^2}} \int_{-k}^{k} p_l(u) \frac{\sqrt{1-u^2}}{u-x}\,du, \quad |x| < 1,\ (l = 0,1,2,\dots).\end{aligned} \quad (5.16$$

6. SOLUTION OF THE INTEGRAL EQUATIONS

We first treat the case $l = 0$. Then the contribution of $F(u)$ disappears, we have to do with the half-plane problem. Applying (5.16) to the interval

$|x| < k$, we have the equation

$$-\frac{x\rho}{R} - \frac{p_0^{(0)}x}{\sqrt{1-x^2}} = \frac{1}{\pi}\frac{1}{\sqrt{1-x^2}}\int_{-k}^{k} \overline{p_0(u)}\,\frac{\sqrt{1-u^2}}{u-x}\,du\,, \quad |x| < k\,, \tag{6.1}$$

with the solution

$$\overline{p_0(x)} = \frac{1}{\pi}\frac{\rho}{R}\sqrt{(1-x^2)\,(k^2-x^2)}\int_{-k}^{k}\frac{t\,dt}{\sqrt{(k^2-t^2)\,(1-t^2)}}\,\frac{1}{t-x}\,, \quad |x| < k\,, \tag{6.2}$$

while the condition that $v_0'(x)$ must be continuous at $x = 1$ yields

$$p_0^{(0)} = \frac{1}{\pi}\frac{\rho}{R}\int_{-k}^{k}\frac{t^2\,dt}{\sqrt{(1-t^2)\,(k^2-t^2)}}\,. \tag{6.3}$$

Applying (5.16) to the interval $k < x < 1$, we find $v_0'(x)$ with the aid of (6.3). We obtain

$$v_0'(x) = \frac{1}{\pi}\frac{\rho}{R}\sqrt{(1-x^2)\,(x^2-k^2)}\int_{-k}^{k}\frac{t\,dt}{\sqrt{(1-t^2)\,(k^2-t^2)}}\,\frac{1}{t-x}\,, \quad k < x < 1\,. \tag{6.4}$$

For $l > 0$ we take

$$v_l'(x) = 0, \quad |x| < k, \quad l = 1, 2, \ldots. \tag{6.5}$$

while we have

$$\overline{p_l(x)} = 0, \quad k < |x| < 1, \quad l = 1, 2, \ldots. \tag{6.6}$$

We now find

$$\overline{p_l(x)} = -\frac{1}{\pi}\sqrt{\frac{k^2-x^2}{1-x^2}}\int_{-k}^{k}\frac{\left(-p_l^{(0)}\,t + \frac{1}{4}\sum_{p=0}^{l-1} a_{lp}t^{2p+1}\right)}{\sqrt{k^2-t^2}\,(t-x)}\,dt\,, \quad |x| < k, \quad l = 1, 2, \ldots \tag{6.7}$$

and

$$v_l'(x) = p_l^{(0)}\frac{x}{\sqrt{1-x^2}} - \frac{1}{4}\frac{1}{\sqrt{1-x^2}}\sum_{p=0}^{l-1} a_{lp}x^{2p+1} - \frac{1}{\pi^2}\frac{1}{\sqrt{1-x^2}}\int_{-k}^{k}\frac{\sqrt{k^2-u^2}}{u-x}\,du \times \int_{-k}^{k}\frac{\left(-p_l^{(0)}\,t + \frac{1}{4}\sum_{p=0}^{l-1} a_{lp}t^{2p+1}\right)}{\sqrt{k^2-t^2}\,(t-u)}\,dt\,, \quad k < |x| < 1, \quad l = 1, 2, \ldots. \tag{6.8}$$

We have the following identities

$$\int_{-k}^{k} \frac{\left(-p_l^{(0)} t + \frac{1}{4} \sum_{p=0}^{l-1} a_{lp} t^{2p+1}\right)}{\sqrt{k^2-t^2}\,(t-x)} \, dt$$

$$= -\pi \left[\frac{\left(-p_l^{(0)} x + \frac{1}{4} \sum_{p=0}^{l-1} a_{lp} x^{2p+1}\right)}{\sqrt{x^2-k^2}} - \sum_{p=0}^{l-1} b_{lp} x^{2p} \right],$$

$$k < |x| < 1, \quad l = 2, 3, \ldots \tag{6.9}$$

and

$$\int_{-k}^{k} \frac{\left(-p_l^{(0)} t + \frac{1}{4} \sum_{p=0}^{l-1} a_{lp} t^{2p+1}\right)}{\sqrt{k^2-t^2}\,(t-x)} \, dt$$

$$= +\pi \sum_{p=0}^{l-1} b_{lp} x^{2p}, \quad |x| < k, \quad l = 2, 3, \ldots . \tag{6.10}$$

In (6.9) and (6.10) b_{lp} are the first coefficients of expansion of the integrands for large $|x|$.

From (6.7) and (6.10) we obtain

$$\overline{p_l(x)} = -\sqrt{\frac{k^2-x^2}{1-x^2}} \sum_{p=0}^{l-1} b_{lp} x^{2p}, \quad |x| < k, \quad l = 2, 3, \ldots . \tag{6.11}$$

With (6.9) the expression (6.8) for $v_l'(x)$ becomes

$$v_l'(x) = -\sqrt{\frac{x^2-k^2}{1-x^2}} \sum_{p=0}^{l-1} b_{lp} x^{2p}, \quad k < |x| < 1, \quad l = 2, 3, \ldots . \tag{6.12}$$

As $v_l'(x)$ is bounded if x tends to 1, we have

$$\sum_{p=0}^{l-1} b_{lp} = 0, \quad l = 2, 3, \ldots, \tag{6.13}$$

a condition which determines $p_l^{(0)}$, because

$$b_{l0} = -p_l^{(0)} + \tfrac{1}{4} a_{l0} + \tfrac{1}{8} k^2 a_{l1}, \quad l = 2, 3, \ldots . \tag{6.14}$$

while b_{lp}, $p \neq 0$ does not depend on $p_l^{(0)}$. It follows from (6.12) and (6.13) that $v_l'(1)$ is equal to zero.

The problem being formally solved, it is possible to write some of the results in the form of tabulated functions. We have for $l = 0$

$$p_0^{(0)} = \frac{2}{\pi}\frac{\rho}{R}(K-E), \tag{7.1}$$

$$v_0'(x) = -x\frac{\rho}{R}, \quad |x| < k, \tag{7.2}$$

$$v_0'(x) = -x\frac{\rho}{R}(1-\Lambda_0(\theta, k)), \quad k < |x| < 1, \tag{7.3}$$

$$\overline{p_0(x)} = \frac{2}{\pi}\frac{\rho}{R}\left[K\sqrt{(1-x^2)(k^2-x^2)} + KxZ\left(\sin^{-1}\frac{x}{k}, k\right)\right], \quad |x| < k. \tag{7.4}$$

In (7.1) to (7.4) K and E denote the complete elliptic integrals of the first and second kind, respectively, Λ_0 is Heumann's Lambda-function [2] and Z is the Zeta-function of Jacobi [2]. The parameter k, given by

$$k = c/\rho \tag{7.5}$$

is the modulus of the elliptic functions, while θ is given by

$$\theta = \sin^{-1}\frac{1}{x}\sqrt{\frac{x^2-k^2}{1-k^2}}. \tag{7.6}$$

For $l = 1$ we easily derive

$$p_1^{(0)} = \frac{a_{10}}{4} = -\frac{\pi}{4}\alpha_0\int_{-1}^{1} v_0(x)\,\mathrm{d}x \tag{7.7}$$

and

$$v_1'(x) = 0; \quad \overline{p_1(x)} = 0, \quad 0 < |x| < 1. \tag{7.8}$$

For $l = 2$ we find

$$p_2^{(0)} = \frac{a_{20}}{4} + \frac{a_{21}}{4} + \frac{a_{21}}{8}k^2, \tag{7.9}$$

$$\overline{p_2(x)} = \frac{a_{21}}{4}\sqrt{(1-x^2)(k^2-x^2)}, \quad |x| < k, \tag{7.10}$$

$$\overline{p_2(x)} = 0 \quad , \quad k < |x| < 1, \tag{7.11}$$

$$v_2'(x) = 0 \quad , \quad |x| < k, \tag{7.12}$$

$$v_2'(x) = \frac{a_{21}}{4}\sqrt{(1-x^2)(x^2-k^2)}, \quad k < |x| < 1. \tag{7.13}$$

For a given $p^{(0)}$ we have derived expressions for $p(x) = \overline{p(x)} - p^{(0)}$ and $v'(x)$. These expressions, however, contain the unknown parameters c and ρ, which have to be determined.

We have

$$v(1) - v(0) = \int_0^1 v'(x)\,\mathrm{d}x = -\frac{d}{\rho}. \tag{8.1}$$

From (5.8_1) we obtain

$$v'(x) = \sum_{l=0}^{\infty} \frac{v_l'(x)}{q^{2l}}, \tag{8.2}$$

which expression is introduced into (8.1). There results

$$-\frac{d}{\rho} = \sum_{l=0}^{\infty} \frac{1}{q^{2l}} \int_0^1 v_l'(x)\,\mathrm{d}x = \int_0^1 v_0'(x)\,\mathrm{d}x + \sum_{l=2}^{\infty} \frac{1}{q^{2l}} \int_0^1 v_l'(x)\,\mathrm{d}x. \tag{8.3}$$

It is possible to evaluate the integrals in (8.3) in closed form. We find

$$-\frac{d}{\rho} = -\frac{1}{\pi}\frac{\rho}{R}\{E'(E-K) + k^2 KK'\} + \sum_{l=2}^{\infty} \frac{1}{q^{2l}} \sum_{p=1}^{l-1} b_{lp} \int_k^1 \sqrt{(x^2-k^2)(1-x^2)} \left(\frac{1-x^{2p}}{1-x^2}\right) \mathrm{d}x. \tag{8.4}$$

In (8.4), K' and E' are the associated complete elliptic integrals. The coefficients b_{lp} and a_{lp} are proportional to ρ/R, so we may write

$$b_{lp} = \rho/R\ \bar{b}_{lp}, \quad a_{lp} = \rho/R\ \bar{a}_{lp}. \tag{8.5}$$

With (8.5_1), (8.4) becomes

$$\frac{\pi d R}{\rho^2} = \{E'(E-K) + k^2 KK'\} - \pi \sum_{l=2}^{\infty} \frac{1}{q^{2l}} \sum_{p=1}^{l-1} \bar{b}_{lp} \int_k^1 \sqrt{(x^2-k^2)(1-x^2)} \left(\frac{1-x^{2p}}{1-x^2}\right) \mathrm{d}x. \tag{8.6}$$

Equation (8.6) may be considered as an equation for $k = c/\rho$. A second equation for c and ρ comes from the expansion (5.8_3). According to (6.13) and (6.14) we have

$$\frac{R p_l^{(0)}}{\rho} = \tfrac{1}{4}\bar{a}_{l0} + \tfrac{1}{8}k^2 \bar{a}_{l1} + \sum_{p=1}^{l-1} \bar{b}_{lp}. \tag{8.7}$$

The solution of the problem will be approximated by truncating the series (5.8) at a certain value of l. If we truncate at $l=0$, we find the solution for the half-plane problem as

$$\left.\begin{aligned} \frac{Rp^{(0)}}{\rho} &= \frac{2}{\pi}(K-E), \\ \frac{\pi Rd}{\rho^2} &= E'(E-K)+k^2KK'. \end{aligned}\right\} \tag{8.8}$$

If we truncate at $l=1$ we have

$$\left.\begin{aligned} \frac{Rp^{(0)}}{\rho} &= \frac{2}{\pi}(K-E)-\frac{\pi}{4}\frac{R}{\rho}\frac{\alpha_0}{q^2}\int_{-1}^{1} v_0(x)\,dx, \\ \frac{\pi Rd}{\rho^2} &= E'(E-K)+k^2KK'. \end{aligned}\right\} \tag{8.9}$$

Note that the value of k from (8.8) is not equal to that of (8.9) for the same $p^{(0)}$.

Truncating at $l=2$ yields

$$\left.\begin{aligned} \frac{Rp^{(0)}}{\rho} &= \frac{2}{\pi}(K-E)-\frac{\pi}{4}\frac{R}{\rho}\frac{\alpha_0}{q^2}\int_{-1}^{1} v_0(x)\,dx \\ &\quad +\frac{1}{4q^4}(\bar{a}_{20}+\bar{a}_{21}+\tfrac{1}{2}\bar{a}_{21}k^2), \\ \frac{\pi dR}{\rho^2} &= E'(E-K)+k^2KK'-\frac{\pi}{q^4}\bar{b}_{21}\int_{k}^{1}\sqrt{(1-x^2)(x^2-k^2)}\,dx. \end{aligned}\right\} \tag{8.10}$$

In (8.10) $\bar{a}_{20}$, $\bar{a}_{21}$ and $\bar{b}_{21}$ are given by (cf. (5.14), (5.12))

$$\left.\begin{aligned} \bar{a}_{20} &= \frac{R}{\rho}\left[-\pi C_{20}+\frac{\pi}{2}C_{21}\right], \\ \bar{a}_{21} &= \frac{R}{\rho}[-\pi C_{21}], \\ \bar{b}_{21} &= \tfrac{1}{4}\bar{a}_{21} = -\frac{R}{\rho}\frac{\pi}{4}C_{21}, \\ C_{20} &= -\alpha_1\frac{\pi^2}{12}\int_{-1}^{1} v_0'(x)\,x^3\,dx, \\ C_{21} &= -\alpha_1\frac{\pi^2}{4}\int_{-1}^{1} v_0'(x)\,x\,dx. \end{aligned}\right\} \tag{8.11}$$

Now we have

$$\alpha_0 = 1, \quad \alpha_1 = -\tfrac{1}{3} \tag{8.12}$$

$$\int_{-1}^{1} v_0(x)\,dx = -\int_{-1}^{1} v_0(x)\,x\,dx = \frac{1}{3}\frac{\rho}{R} \times$$

$$\times\,[E(1+k^2) - K(1-k^2)], \tag{8.13}$$

$$\int_{k}^{1} \sqrt{(1-x^2)\,(x^2-k^2)}\,dx = \tfrac{1}{3}[(1+k^2)\,E' - 2k^2 K'], \tag{8.14}$$

$$-\int_{-1}^{1} v_0'(x)\,x^3\,dx = \frac{1}{20}\frac{\rho}{R} \times$$

$$\times\,[E(3k^4 + 2k^2 + 3) - K(1-k^2)\,(k^2+3)]. \tag{8.15}$$

With (8.12) to (8.15) the approximation (8.10) takes the form

$$\frac{Rp^{(0)}}{\rho} = \frac{2}{\pi}(K-E) - \frac{\pi}{12q^2}\{E(1+k^2) - K(1-k^2)\} +$$

$$+ \frac{\pi 3}{288}\frac{1}{q^4}[E(1+k^2)^2 - K(1-k^4) +$$

$$+ \tfrac{1}{10}E(3k^4 + 2k^2 + 3) - \tfrac{1}{10}K(1-k^2)\,(3+k^2)], \tag{8.16}$$

$$\frac{\pi dR}{\rho^2} = E'(E-K) + k^2 KK' -$$

$$- \frac{\pi^4}{432}\frac{1}{q^4}[\{E'(1+k^2) - 2k^2 K'\}\,\{E(1+k^2) - K(1-k^2\}]. \tag{8.17}$$

From (8.16) and (8.17) we determine $c/\sqrt{Rd}$ and $\rho/\sqrt{Rd}$ for given $p^{(0)}\sqrt{R/d}$ and $b/\sqrt{Rd}$, if the strip is thick enough to neglect errors of the order q^{-6}.

9. DISCUSSION AND RESULTS

We have solved the complete problem if we have determined c and ρ and with these quantities the surface pressure, according to (5.8), (7.4), (7.7) and (6.11), or the surface displacement according to (5.8), (7.2), (7.3), (7.8) and (6.12). We note that the surface pressure for $|x| > \rho$ may easily be found from (4.8) if we take $g(t) = 0$ for $t > \rho$. The surface quantities being known, it is a matter of straightforward calculation to obtain the stress- and displacement distributions in the whole body.

We have to make a few remarks. The theory is essentially limited to the case of thick strips, which we define by

$$b > \sqrt{Rd}, \quad \text{for} \quad d \leqslant R. \tag{9.1}$$

If the strip is thin the elementary contact problem deviates from that of the thick strip (cf. [3]), and with the boundary conditions (2.2), mutual contact of the strips is then not possible. If the problem is modified by applying a pressure at $y = \pm b$, instead of a prescribed displacement, the thin strip may be treated with another kind of asymptotics, e.g. based on plate or beam theory.

For the thick strip we only have solved the case

$$\rho < b/2. \tag{9.2}$$

This is not a serious limitation however, because it may be expected, that deviations from the stress distribution near the stamp according to the elementary contact problem occur only for this case.

Further we have approximated our exact solution by truncating the system (5.16) at $l = 2$. Although the results may simply be improved by taking some more equations, the convergence of the series is such that this seems superfluous for most cases. It appears that deviations from the half-plane problem at least for the surface quantities, are relatively small.

For the numerical calculations we have introduced the non-dimensional quantities

$$\frac{\rho}{\sqrt{Rd}}, \ \frac{b}{\sqrt{Rd}}, \ \frac{c}{\sqrt{Rd}}, \ p^{(0)} \sqrt{R/d}. \tag{9.3}$$

Note that the fundamental parameters k and q are unaltered. In Fig. 2 we have plotted $p^{(0)}\sqrt{R/d}$ and $\rho/\sqrt{Rd}$ as functions of k for several values of $b/\sqrt{Rd}$, including ∞. In Fig. 3 we show the same quantities as functions of $c/\sqrt{Rd}$ while in Fig. 4 $p^{(0)}\sqrt{R/d}$ is given as a function of $\rho/\sqrt{Rd}$. In Fig. 5 we present the surface pressure $\overline{p(x)}$, in Fig. 6 the displacement derivative $v'(x)$ and in Fig. 7 the displacement $v(x)/R$ as functions of x/R. All these surface quantities are calculated for $k = 0{,}5$, $d/R = 0{,}1$, $b/R = 2$, $\rho/R = 0{,}6986$. For the determination of the surface displacement we have used the formula

$$\begin{aligned} \frac{\pi R v(x)}{\rho} &= Kx \sqrt{(1-x^2)(x^2-k^2)} + (E-K-Kx^2)\, E(\theta, k') - \\ &\quad - \{(E-K)\, x^2 - k^2 K\}\, F(\theta, k') - \\ &\quad - \frac{\pi^4}{432} \frac{1}{q^4} [\{E(1+k^2) - K(1-k^2)\}\, \{1+k^2)\, E(\theta, k') - \\ &\quad - 2k^2\, F(\theta, k') + x \sqrt{(1-x^2\, (x^2-k^2)\}}], \quad k < |x| < 1, \end{aligned} \tag{9.4}$$

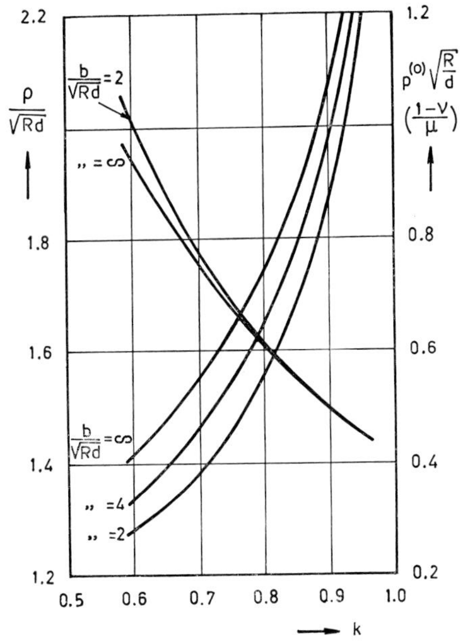

. 2. Relations between $p^{(0)}$, ρ and *k*

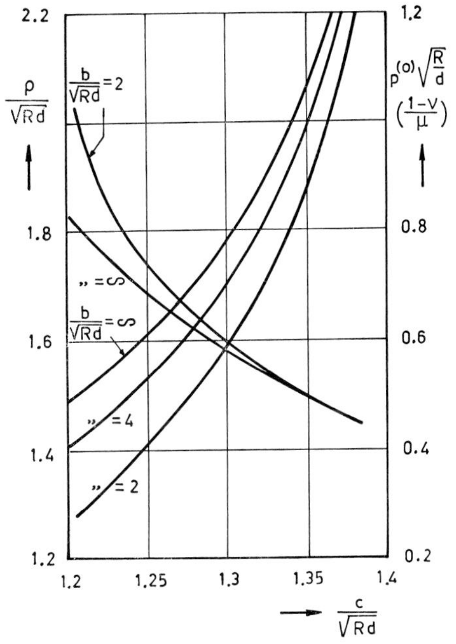

Fig. 3. Relations between $p^{(0)}$, ρ and c

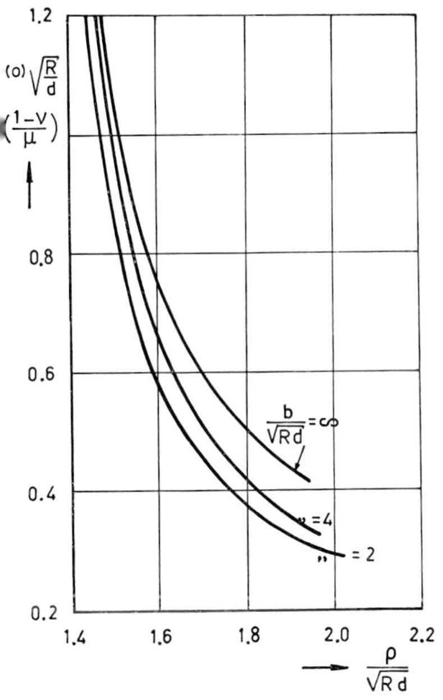

g. 4. Relations between $p^{(0)}$ and ρ

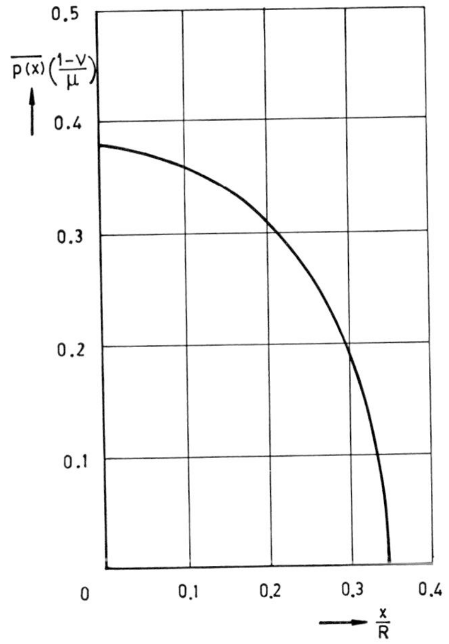

Fig. 5. Surface pressure $\overline{p(x)}$

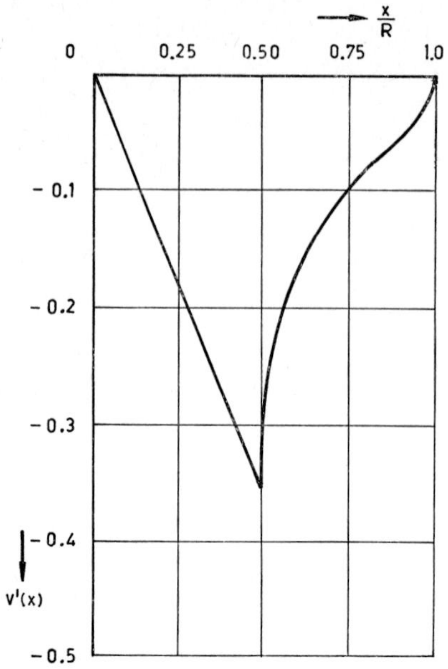

Fig. 6. Normal displacement derivative at surface

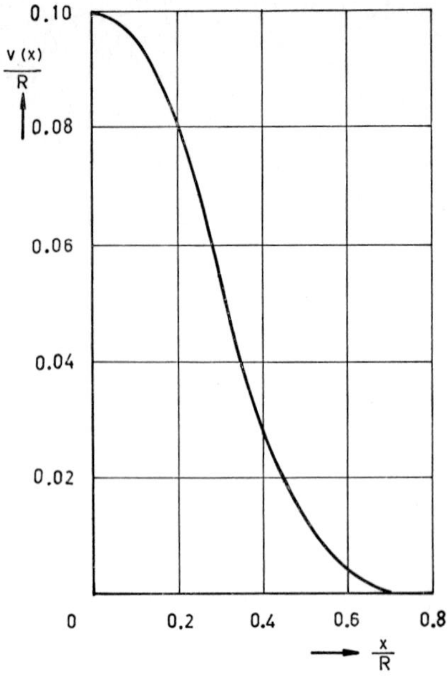

Fig. 7. Normal displacement at surface

which has been obtained by integrating (7.2), (7.3) and (7.13). In (9.4) θ is given by

$$\theta = \sin^{-1} \sqrt{\frac{1-x^2}{1-k^2}}, \tag{9.5}$$

while $F(\theta, k')$ and $E(\theta, k')$ are the incomplete elliptic integrals of the first and second kind, respectively. k' is the complementary modulus.

REFERENCES

1. J.B. Alblas, 'On the Two-Dimensional Contact Problem of a Rigid Cylinder, Pressed between Two Elastic Half-Planes', *Mech. Res. Comm.*, **1**, 1974, pp. 15–20.
2. P.F. Byrd and M.D. Friedman, *Handbook of Elliptic Integrals for Engineers and Scientists*, Springer, Berlin etc., 1971.
3. J.B. Alblas and M. Kuipers, 'On the Two-Dimensional Problem of a Cylindrical Stamp, Pressed into a Thin Elastic Layer', *Acta Mechanica*, **9**, 1970, pp. 292–311.

Influence of an elastic layer on the tangential compliance of bodies in contact

L. E. Goodman and L. M. Keer

Dedicated to Raymond David Mindlin
Magna opera magistri nostri, exquisita in omnes voluntates ejus.

1. INTRODUCTION

The state of stress that arises when two deformable bodies are pressed together by forces normal to the common tangent plane at the point of initial contact is of great technological interest. For isotropic, linearly elastic, smooth homogeneous solids having dimensions large compared with those of the final contact region the state of deformation so created was analyzed by Hertz[1] at a relatively early stage in the history of continuum mechanics. Cattaneo[2] and, in 1949, Mindlin[3], using the same model, treated the case in which the bodies are rough rather than smooth and are subjected to tangential as well as to normal force. Mindlin's seminal work showed that, even for tangential force too small to produce overall motion, slip must occur at the boundary of the contact region. The effect of this slip on compliance and the dimensions of the slip region were worked out for the case in which both bodies are spheres. The physical correctness of Mindlin's analysis and its reconciliation with the Bowden-Tabor theory of friction has been established in the first instance by K. L. Johnson[4] and verified by him and by others[5–10, 23]. In the present paper the Mindlin analysis is extended to the case in which the deformation takes place in elastic surface layers (one in each of the solids) of arbitrary thickness, bonded to a rigid substrate.

The analysis is of theoretical as well as practical interest because it exposes the existence of an asymmetrical effect that is associated with change in material stiffness. This *bona fide* layer effect does not appear in the valuable study that Tu and Gazis have made[11] of a smooth plate pressed between spheres. That is because only normal boundary tractions are considered in [11]. Neither does it appear in the literature dealing with three-dimensional stress analysis of a layered half-space. An extensive survey of this literature has been published by Lysmer and Duncan[12] as part of a review of the use of elastic theory for computation of stresses and deflections in soils. For the most part the boundary conditions considered in the three-dimensional studies reported in [12] correspond to uniform normal traction distributed over a circular surface region.

2. NORMAL LOADING

We first satisfy the field equations of the linear theory of elasticity in the region $0 \leq z \leq h$ subject to the boundary conditions

$$z = 0, \quad \tau_{zr} = \tau_{z\theta} = 0, \qquad 0 \leq r < \infty \tag{1a}$$

$$u_z = \sum_{n=0}^{\infty} w_n(r) \cos n\theta, \qquad 0 \leq r \leq a \tag{1b}$$

$$\tau_{zz} = 0, \qquad a < r < \infty \tag{1c}$$

$$z = h, \quad u_r = u_\theta = u_z = 0, \quad 0 \leq r < \infty . \tag{1d}$$

For the purposes of notation the subscript or superscript n affixed to any of the field quantities will denote its Fourier component as in equation (1 b).

All stress components decay to zero as r increases without limit. The normal contact problem of rigid spheres coated with an adherent elastic layer can be obtained from this stress field by appropriate specification of the functions $w_n(r)$. In the end, of course, it will only be necessary to use the symmetrical case, $n = 0$, which has been studied previously by Hayes, et al.[13].

Following Green and Zerna[14], we satisfy the field equations of the linear theory of elasticity by expressing displacements and stresses in terms of harmonic functions $F(r, \theta, z)$ and $G(r, \theta, z)$ through the relations

$$2\mu u_r = \frac{\partial F}{\partial r} + z\frac{\partial G}{\partial r}, \quad 2\mu u_\theta = \frac{1}{r}\frac{\partial F}{\partial \theta} + \frac{z}{r}\frac{\partial G}{\partial \theta} \tag{2a, b}$$

$$2\mu u_z = \frac{\partial F}{\partial z} - (3-4v)\, G + z\frac{\partial G}{\partial z} \tag{2c}$$

$$\tau_{zr} = \frac{\partial^2 F}{\partial r \partial z} - (1-2v)\frac{\partial G}{\partial r} + z\frac{\partial^2 G}{\partial r \partial z} \tag{2d}$$

$$\tau_{z\theta} = \frac{1}{r}\frac{\partial^2 F}{\partial \theta \partial z} - (1-2v)\frac{1}{r}\frac{\partial G}{\partial \theta} + \frac{z}{r}\frac{\partial^2 G}{\partial \theta \partial z} \tag{2e}$$

$$\tau_{zz} = \frac{\partial^2 F}{\partial z^2} - 2(1-v)\frac{\partial G}{\partial z} + z\frac{\partial^2 G}{\partial z^2}. \tag{2f}$$

We take

$$F = \sum_{n=0}^{\infty} F_n(r, z) \cos n\theta, \quad G = \sum_{n=0}^{\infty} G_n(r, z) \cos n\theta \tag{3}$$

with

$$F_n(r,z) = \int_0^\infty E_n(\xi)\, J_n(\xi r)\, q^{-1}\, \xi^{-1}\{[\beta-(3-4\nu)\,\mathrm{sh}\,\beta\,\mathrm{ch}\,\beta] \times$$
$$\times (1-2\nu)\,\mathrm{sh}(\xi z) - [\beta^2-(1-2\nu)(3-4\nu)\,\mathrm{sh}^2\beta]\,\mathrm{ch}(\xi z)\}\,\mathrm{d}\xi \tag{4a}$$

$$G_n(r,z) = \int_0^\infty E_n(\xi)\, J_n(\xi r)\, q^{-1}\{[\beta-(3-4\nu)\,\mathrm{sh}\,\beta\,\mathrm{ch}\,\beta]\,\mathrm{ch}(\xi z) +$$
$$+ [2(1-\nu)+(3-4\nu)\,\mathrm{sh}^2\beta]\,\mathrm{sh}(\xi z)\}\,\mathrm{d}\xi \tag{4b}$$

$$q = q(\beta,\nu) = \beta^2 + 4(1-\nu)^2 + (3-4\nu)\,\mathrm{sh}^2\beta \tag{5}$$

where $\beta = \xi h$. It is easily verified that the choices for $F_n(r,z)$ and $G_n(r,z)$ as given in equations (4) will when used with the appropriate equations (2) automatically satisfy boundary conditions (1a) and (1d).

If we now write

$$\tau_{zz}(r,\theta,z) = \sum_{n=0}^{\infty} \tau_{zz}^{(n)}(r,z)\cos n\theta \tag{6a}$$

$$u_z(r,\theta,z) = \sum_{n=0}^{\infty} u_z^{(n)}(r,z)\cos n\theta \tag{6b}$$

it follows from (4) that

$$\tau_{zz}^{(n)} = -\int_0^\infty E_n(\xi)\, J_n(\xi r)\, q^{-1}\,\xi\{q\,\mathrm{ch}(\xi z) +$$
$$+ [\beta-(3-4\nu)\,\mathrm{sh}\,\beta\,\mathrm{ch}\,\beta]\,\mathrm{sh}(\xi z) -$$
$$- [2(1-\nu)+(3-4\nu)\,\mathrm{sh}^2\beta]\,(\xi z)\,\mathrm{sh}(\xi z) - (\xi z)\,\mathrm{ch}(\xi z)\}\,\mathrm{d}\xi \tag{7a}$$

$$2\mu u_z^{(n)} = -\int_0^\infty E_n(\xi)\, J_n(\xi r)\, q^{-1}\,\{[\beta-(3-4\nu)\,\mathrm{sh}\,\beta\,\mathrm{ch}\,\beta] \times$$
$$\times 2(1-\nu)\,\mathrm{ch}(\xi z) + [\beta^2 + 2\{1-\nu)(3-4\nu)\,\mathrm{ch}^2\beta]\,\mathrm{sh}(\xi z) -$$
$$- [\beta-(3-4\nu)\,\mathrm{sh}\,\beta\,\mathrm{ch}\,\beta]\,(\xi z)\,\mathrm{ch}(\xi z) -$$
$$- [2(1-\nu)+(3-4\nu)\,\mathrm{sh}^2\beta]\,(\xi z)\,\mathrm{sh}(\xi z)\}\,\mathrm{d}\xi\,. \tag{7b}$$

In order to satisfy boundary conditions (1b,c), we must have

$$\int_0^\infty E_n(\xi)\, J_n(\xi r)\,\mathrm{d}\xi - \int_0^\infty E_n(\xi)\, L(\beta,\nu)\, J_n(\xi r)\,\mathrm{d}\xi = \frac{\mu}{1-\nu}\, w_n(r) \qquad 0 \le r \le a \tag{8a}$$

$$\int_0^\infty E_n(\xi)\, J_n(\xi r)\,\xi\,\mathrm{d}\xi = 0, \qquad a < r < \infty \tag{8b}$$

with

$$qL(\beta,\nu) = \beta+\beta^2+4(1-\nu)^2 - (3-4\nu)\,\mathrm{e}^{-\beta}\,\mathrm{sh}\,\beta\,. \tag{8c}$$

To solve this integral equation we follow a technique due to Keer[15] and Westmann[16] and based on work of Sneddon[17] and of Copson[18]. Let

$$E_n(\xi) = \xi^{1/2} \int_0^a t^{1/2} \phi_n(t) J_{n-1/2}(\xi t) \, dt . \tag{9}$$

Then (8b) is satisfied automatically.* In fact

$$\tau_{zz}^{(n)}(r,0) = \sqrt{\frac{2}{\pi}} \left\{ \frac{-a^{-n} r^n}{(a^2-r^2)^{\frac{1}{2}}} \phi_n(a) + r^n \int_r^a \frac{[t^{-n} \phi_n(t)]'}{(t^2-r^2)^{\frac{1}{2}}} dt \right\} \qquad 0 \leq r \leq a \tag{10a}$$

$$= 0, \qquad a < r < \infty \tag{10b}$$

The prime symbol indicates differentiation with respect to t. If the pressure is to be continuous and is to vanish at the edge of the contact region, $r = a$, it is necessary that

$$\phi_n(a) = 0 . \tag{11}$$

To find the equation governing ϕ_n substitute (9) in (8a). After inverting the order of integration so as to carry out one integration with respect to ξ multiply by $(s^2-r^2)^{-1/2} r^{n+1}$ and integrate with respect to r over $0 \leq r \leq s$. The integration with respect to r can then be carried out.† Then on differentiating with respect to s we have

$$\phi_n(s) - s^{1/2} \int_0^a t^{1/2} \phi_n(t) \, dt \int_0^\infty L(\beta, v) \, \xi \, J_{n-1/2}(\xi s) \, J_{n-1/2}(\xi t) \, d\xi$$

$$= \sqrt{\frac{2}{\pi}} \frac{\mu}{1-v} s^{-n} \frac{d}{ds} \int_0^s \frac{w_n(r) \, r^{n-1}}{(s^2-r^2)^{1/2}} dr . \tag{12}$$

We also require an expression for the resultant normal force.

$$P_z = \int_0^{2\pi} \int_0^a -\tau_{zz}(r,\theta,0) \, r \, dr d\theta \tag{13a}$$

$$= 4 \sqrt{\frac{\pi}{2}} \int_0^a \phi_0(t) \, dt . \tag{13b}$$

Consider now two spheres with rigid cores to which are bonded concentric elastic surface layers. When these bodies are compressed by forces directed along the line connecting their centers they are brought into contact over a circular region whose radius, a, is small compared with the smaller of the radii, R_1, R_2, of the undeformed solids. Using subscripts 1 and 2 to distinguish the bodies, the normal displacement boundary condition takes the form

$$u_{z1} + u_{z2} + \tfrac{1}{2} r^2 (R_1^{-1} + R_2^{-1}) - \alpha = 0, \quad 0 \leq r \leq a . \tag{14}$$

* See [19], p. 487, equation 11.4.41.

† See [20], p. 3, equation 1.4.

Here α, a constant, is the relative approach of the two solids. Outside the contact zone the left-hand side of (14) must be positive. It appears from equations (10) that if stresses are to match in the contact zone (thus satisfying Newton's third law) the functions ϕ_n must be the same for each body. Equation (7b) then implies that, provided $h_1 = h_2$,

$$\frac{\mu_1}{1-\nu_1}\, u_{z1}(r,\theta,0) = \frac{\mu_2}{-\nu_2}\, u_{z2}(r,\theta,0). \tag{15}$$

Equations (14) and (15) make

$$u_{z1}(r,\theta,0) = \frac{1-\nu_1}{\mu_1}\,\kappa\left(\alpha - \frac{r^2}{2R}\right), \qquad 0 \le r \le a \tag{16a}$$

$$\kappa = \left(\frac{1-\nu_1}{\mu_1} + \frac{1-\nu_2}{\mu_2}\right)^{-1}, \tag{17}$$

$$R^{-1} = R_1^{-1} + R_2^{-1}. \tag{18}$$

Clearly only the term $n = 0$ corresponding to ϕ_0 in the general solution (12) is needed in order to satisfy (16). In the remainder of this paper we consider

Table 1. Normal Loading: Relative approach (α), contact radius (a), peak stress $(\tau_{zz})_{\max}$: $\alpha = C_1(a^2/R)$, $a^3 = C_2(P_z R/\pi\kappa)$, $(\tau_{zz})_{\max} = -C_3(P_z/\pi a^2)$

$\nu = 0$			a/h	$\nu = 0.15$		
C_1	C_2	C_3		C_1	C_2	C_3
1.00000	1.17810	1.500	0	1.00000	1.17810	1.500
0.9088	1.17453	1.500	0.2	0.89682	1.17411	1.500
0.81789	1.15321	1.501	0.4	0.81071	1.15040	1.501
0.75416	1.10912	1.507	0.6	0.74689	1.10224	1.508
0.70773	1.04893	1.521	0.8	0.70070	1.03743	1.524
0.67388	0.98199	1.542	1.0	0.66757	0.96660	1.548
0.59263	0.69561	1.681	2.0	0.58995	0.67525	1.693
0.56233	0.52367	1.772	3.0	0.56116	0.50647	1.781
0.54683	0.41743	1.825	4.0	0.54636	0.40343	1.830
0.53734	0.34639	1.853	5.0	0.53711	0.33475	1.855
$\nu = 0.30$				$\nu = 0.45$		
1.00000	1.17810	1.500	0	1.00000	1.17810	1.500
0.88805	1.17336	1.500	0.2	0.86871	1.17181	1.500
0.79593	1.14535	1.502	0.4	0.76253	1.13503	1.503
0.72861	1.08916	1.510	0.6	0.68571	1.06242	1.514
0.68143	1.01517	1.529	0.8	0.63267	0.96917	1.540
0.64830	0.93608	1.558	1.0	0.99556	0.87222	1.578
0.57359	0.62736	1.721	2.0	0.51062	0.51806	1.787
0.54774	0.54960	1.813	3.0	0.48161	0.34239	1.906
0.53494	0.36050	1.861	4.0	0.46945	0.24681	1.976
0.52727	0.29608	1.881	5.0	0.46551	0.18948	2.008

only one of the two bodies in contact, say body 1, and drop the numerical suffixes where they are not required. If the contact is smooth or if the two layers have the same value of Poisson's ratio* the boundary conditions for normal loading are equations (1) with $w_0(r)$ given by the right-hand side of (16a). Then (12) reduces to

$$\phi_0\ s) - \frac{2}{\pi}\int_0^a \phi_0(t)\,\mathrm{d}t \int_0^\infty L(\beta, \nu)\cos(\xi s)\cos(\xi t)\,\mathrm{d}\xi$$

$$= \sqrt{\frac{2}{\pi}}\frac{\kappa}{\mu_1}\left(\alpha - \frac{s^2}{R}\right) \tag{19}$$

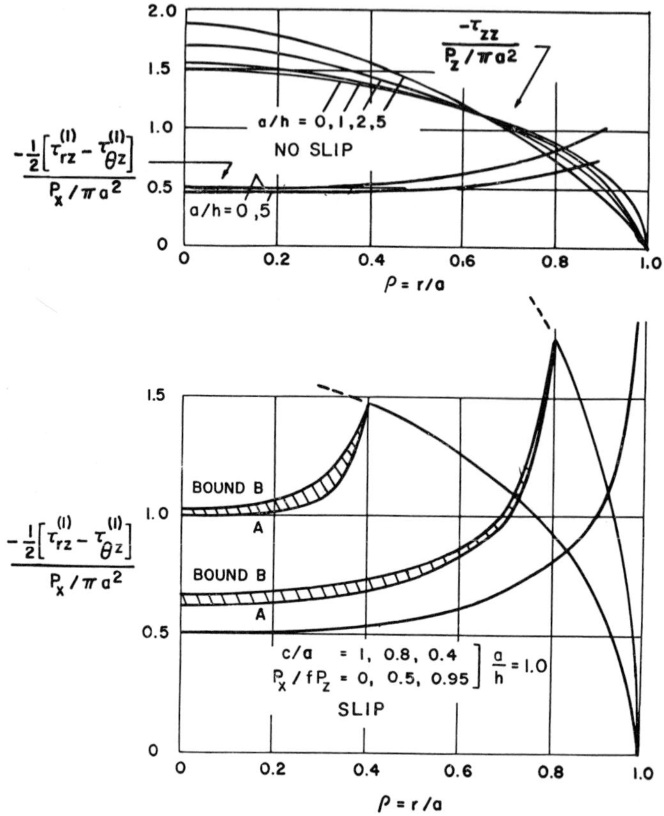

Fig. 1. Normal and shear stress distribution at surface of elastic layer
a/h = radius of contact region/thickness of elastic layer
c/a = radius of no-slip region/radius of contact region
Poisson's ratio = ν = 0.3

* If the bodies are rough, as we take them to be in the remainder of this paper, and if $\nu_1 \neq \nu_2$, normal loading will produce a radial shear stress in the contact region. This effect, sometimes ignored, is analyzed in [21] and [22].

The determination of ϕ_0 by means of the symmetric Fredholm integral equation (19) is straightforward. The magnitude of the relative approach, α, is determined by the condition (11).

If the surface layer is infinitely thick ($a/h = 0$), we recover the Hertz theory: $\alpha = a^2/R$. In general, however,

$$\alpha = C_1(a^2/R), \tag{20a}$$

where $0 < C_1 \leq 1$ is a dimensionless coefficient. The determination of the radius, a, of the contact area follows from (19) and (11). We write:

$$a/R = C_2(P_z/\pi a^2 \kappa). \tag{20b}$$

The normal pressure in the contact region is given by (10a) with $n = 0$ and $\phi_0(a) = 0$. For the present case in which the substrate is rigid this normal pressure is a maximum at the center of the contact circle and decreases monotonically to zero at the edge. We write

$$(\tau_{zz})_{\max} = -C_3(P_z/\pi a^2) \tag{20c}$$

so that C_3 is a type of stress concentration factor. The numerical values of C_1, C_2 and C_3 are given in Table 1 for $0 \leq a/h \leq 5$ and for representative values of Poisson's ratio. Representative profiles of the normal pressure are shown in Fig. 1.

3. TANGENTIAL LOADING

Suppose now that the two spheres described in the previous section, with bonded elastic surface layers of equal thickness and equal elastic constants, after being pressed together by normal forces P_z so as to produce a contact area of radius a, are subjected to forces P_x parallel to the contact surface. We suppose that the bodies are perfectly rough so that as long as P_x is less than a limiting value, fP_z, no slip occurs in the contact region. The state of stress in each elastic layer due to the addition of tangential loading is then governed by the boundary conditions

$$\text{On } z = 0, \quad \tau_{zz} = u_y = 0, \quad u_x = \Delta; \qquad 0 \leq r \leq a \tag{21a}$$

$$\tau_{zz} = \tau_{zx} = \tau_{zy} = 0; \qquad a < r < \infty \tag{21b}$$

$$\text{On } z = h, \quad u_r = u_\theta = u_z = 0; \qquad 0 < r \leq \infty \tag{21c}$$

Here the symbol Δ represents a uniform translational motion whose magnitude will depend on P_x. Little additional complexity is created by taking Δ to be an arbitrary function. We therefore write

$$\Delta = \tfrac{1}{2}\Delta_0(r) + \sum_{n=1}^{\infty} \Delta_n(r) \cos n\theta \tag{22}$$

for the present.

Proceeding along the lines of the previous section take

$$2\mu u_r = \frac{\partial F}{\partial r} + z\frac{\partial G}{\partial r} + \frac{2}{r}\frac{\partial H}{\partial \theta} \tag{23a}$$

$$2\mu u_\theta = \frac{1}{r}\frac{\partial F}{\partial \theta} + \frac{z}{r}\frac{\partial G}{\partial \theta} - 2\frac{\partial H}{\partial r} \tag{23b}$$

$$2\mu u_z = \frac{\partial F}{\partial z} - (3-4\nu)\, G + z\frac{\partial G}{\partial z} \tag{23c}$$

where F, G, H are harmonic functions.

$$\tau_{zr} = \frac{\partial^2 F}{\partial r \partial z} - (1-2\nu)\frac{\partial G}{\partial r} + z\frac{\partial^2 G}{\partial r \partial z} + \frac{1}{r}\frac{\partial^2 H}{\partial \theta \partial z} \tag{24a}$$

$$\tau_{z\theta} = \frac{1}{r}\frac{\partial^2 F}{\partial \theta \partial z} - (1-2\nu)\frac{1}{r}\frac{\partial G}{\partial \theta} + \frac{z}{r}\frac{\partial^2 G}{2\theta \partial z} - \frac{\partial^2 H}{\partial r \partial z} \tag{24b}$$

$$\tau_{zz} = \frac{\partial^2 F}{\partial z^2} - 2(1-\nu)\frac{\partial G}{\partial z} + z\frac{\partial^2 G}{\partial z^2} \tag{24c}$$

We take

$$F = \sum_{n=0}^{\infty} F_n(r,z)\cos n\theta, \qquad G = \sum_{n=0}^{\infty} G_n(r,z)\cos n\theta \tag{25a, b}$$

$$H = \sum_{n=0}^{\infty} H_n \sin n\theta \tag{25c}$$

with

$$F_n = \int_0^\infty [A_n \operatorname{sh}(\xi z) + B_n \operatorname{ch}(\xi z)]\, \xi^{-1} J_n(\xi r)\, d\xi \tag{26a}$$

$$G_n = \int_0^\infty [C_n \operatorname{ch}(\xi_z) + D_n \operatorname{sh}(\xi z)]\, J_n(\xi r)\, d\xi \tag{26b}$$

$$H_n = \int_0^\infty [K_n \operatorname{sh}(\xi z) + L_n \operatorname{ch}(\xi z)]\, \xi^{-1} J_n(\xi r)\, d\xi\,. \tag{26c}$$

It is convenient to re-write the boundary conditions (21) in cylindrical polar coordinates. With

$$u_r = \sum_{n=0}^{\infty} u_r^{(n)}(r,z)\cos n\theta, \qquad u_\theta = \sum_{n=0}^{\infty} u_\theta^{(n)}(r,z)\sin n\theta \tag{27a, b}$$

$$u_z = \sum_{n=0}^{\infty} u_z^{(n)}(r,z)\cos n\theta \tag{27c}$$

$$\tau_{zr} = \sum_{n=0}^{\infty} \tau_{zr}^{(n)}(r,z)\cos n\theta\,, \qquad \tau_{z\theta} = \sum_{n=0}^{\infty} \tau_{z\theta}^{(n)}(r,z)\sin n\theta \tag{27d, e}$$

the boundary conditions (21, 22) may be written:

$$\text{On } z=0, \quad \tau_{zz}(r,\theta)=0, \quad 0 \le r < \infty \tag{28a}$$

$$[u_r^{(n)} + u_\theta^{(n)}] = \Delta_{n+1}, \qquad n = 0,1,\ldots, \qquad 0 \le r \le a \tag{28b}$$

$$[u_r^{(n)} - u_\theta^{(n)}] = \Delta_{n-1}, \qquad n = 1,2,\ldots, \qquad 0 \le r \le a \tag{28c}$$

$$[\tau_{zr}^{(n)} + \tau_{z\theta}^{(n)}] = 0, \qquad n = 0,1,2, \qquad a < r < \infty \tag{28d}$$

$$[\tau_{zr}^{(n)} - \tau_{z\theta}^{(n)}] = 0, \qquad n = 0,1,2,\ldots, \qquad a < r < \infty \tag{28e}$$

$$\text{On } z=h, \quad u_z^{(n)} = 0, \qquad n = 0,1,2,\ldots, \qquad 0 \le r \le \infty \tag{28f}$$

$$[u_r^{(n)} + u_\theta^{(n)}] = 0, \qquad n = 0,1,2,\ldots, \qquad 0 \le r < \infty \tag{28g}$$

$$[u_r^{(n)} - u_\theta^{(n)}] = 0, \qquad n = 0,1,2,\ldots, \qquad 0 \le r \le \infty \tag{28h}$$

The boundary conditions (28a,f,g,h), each of which holds over an entire plane, are satisfied by a procedure analogous to that used in the previous section to satisfy boundary conditions (1a,d). It is easy to verify that (28a,f,g,h) are satisfied when the six coefficients $A_n, \ldots, L_n$ are expressed in terms of two new coefficients $S_n(\xi)$ and $T_n(\xi)$ by the equations

$$qA_n = \tfrac{1}{2}(S_n - T_n)\,[\beta^2 + 2(1-\nu)(3-4\nu)\,\mathrm{ch}^2\beta] \tag{29a}$$

$$qB_n = (S_n - T_n)(1-\nu)\,[\beta + (3-4\nu)\,\mathrm{sh}\,\beta\,\mathrm{ch}\,\beta] \tag{29b}$$

$$qC_n = \tfrac{1}{2}(S_n - T_n)\,[(1-2\nu) - (3-4\nu)\,\mathrm{ch}^2\beta] \tag{29c}$$

$$qD_n = \tfrac{1}{2}(S_n - T_n)\,[\beta + (3-4\nu)\,\mathrm{sh}\,\beta\,\mathrm{ch}\,\beta] \tag{29d}$$

$$K_n = \tfrac{1}{2}(S_n + T_n), \quad L_n = -\tfrac{1}{2}(S_n + T_n)\tanh\beta \tag{29e, f}$$

Here $\beta = \xi h$, as before.

The mixed boundary conditions (28b,c,d,e) require that in $a < r < \infty$

$$[\tau_{zr}^{(n)}(r,0) + \tau_{z\theta}^{(n)}(r,0)] = \int_0^\infty S_n \xi\, J_{n+1}(\xi r)\,\mathrm{d}\xi = 0, \tag{30a}$$

$$[\tau_{zr}^{(n)}(r,0) - \tau_{z\theta}^{(n)}(r,0)] = \int_0^\infty T_n \xi\, J_{n-1}(\xi r)\,\mathrm{d}\xi = 0 \tag{30b}$$

and in $0 \le r \le a$

$$2\mu[u_r^{(n)}(r,0) + u_\theta^{(n)}(r,0)] = \int_0^\infty [-\nu T_n - (2-\nu)\,S_n]\,J_{n+1}(\xi r)\,\mathrm{d}\xi \,+$$

$$+\frac{1}{2}\int_0^\infty [(M-N)\,T_n - (M+N)\,S_n]\,J_{n+1}(\xi r)\,\mathrm{d}\xi = 2\mu\Delta_{n+1}(r)$$

$$2\mu[u_r^{(n)}(r,0) - u_\theta^{(n)}(r,0)] = \int_0^\infty [-(2-\nu)\,T_n - \nu S_n]\,J_{n-1}(\xi r)\,\mathrm{d}\xi \,-$$

$$-\frac{1}{2}\int_0^\infty [-(M+N)\,T_n + (M-N)\,S_n]\,J_{n-1}(\xi r)\,\mathrm{d}\xi = 2\mu\Delta_{n-1}(r)$$ (30d)

where

$$qM(\beta,\nu) = 2(1-\nu)\,[\beta-\beta^2 - 4(1-\nu)^2 + (3-4\nu)\,\mathrm{sh}\,\beta\,\mathrm{e}^{-\beta}] \qquad (31)$$

and

$$N(\beta) = -2(1-\tanh\beta)\,. \qquad (32)$$

These mixed boundary conditions are satisfied by taking

$$S_n = \xi^{1/2}\int_0^a t^{1/2}\,\psi_n(t)\,J_{n+1/2}(\xi t)\,\mathrm{d}t \qquad (33a)$$

$$T_n = \xi^{1/2}\int_0^a t^{1/2}\,\chi_n(t)\,J_{n-3/2}\ \xi t)\,\mathrm{d}t - \frac{\nu}{2-\nu}\times$$

$$\times\,\xi^{1/2}\int_0^a t^{1/2}\,\psi_n(t)\,J_{n+1/2}(\xi t)\,\mathrm{d}t\,. \qquad (33b)$$

When we substitute (33) in (30) we find, after some manipulation,

$$[\tau_{zr}^{(n)}(r,0) + \tau_{z\theta}^{(n)}(r,0)] = \sqrt{\frac{2}{\pi}}\left\{\frac{r^{n+1}\,a^{-n-1}}{(a^2-r^2)^{1/2}}\,\psi_n(a) - \right.$$

$$\left. - r^{n+1}\int_r^a \frac{[t^{-n-1}\,\psi_n(t)]'}{(t^2-r^2)^{1/2}}\,\mathrm{d}t\right\} \quad 0 \le r \le a \quad (43a)$$

$$= 0, \qquad a < r < \infty \quad (34b)$$

and

$$[\tau_{zr}^{n}(r,0) - \tau_{z\theta}^{(n)}(r,0)] = \sqrt{\frac{2}{\pi}}\,r^{n-1}\left\{\frac{a^{-n+1}}{(a^2-r^2)^{1/2}}\left[\chi(a) + \frac{\nu}{2-\nu}\,\psi_n(a)\right] - \right.$$

$$-\int_r^a \frac{[t^{-n+1}\,\chi(t)]'}{(t^2-r^2)^{1/2}}\,\mathrm{d}t -$$

$$\left. -\frac{\nu}{2-\nu}\int_r^a \frac{t^{-2n+1}\,[t^n\,\psi_n(t)]'}{(t^2-r^2)^{1/2}}\,\mathrm{d}t\right\}, \qquad 0 \le r \le a \qquad (34c)$$

$$= 0, \qquad a < r < \infty\,. \qquad (34d)$$

The boundary conditions (28 d,e) are therefore automatically satisfied. Furthermore, the resultant tangential loading is given by the expression

$$P_x = -\int_0^{2\pi}\int_0^a \tau_{zx}(r,\theta,0)\,r\,\mathrm{d}r\mathrm{d}\theta = -\pi\int_0^a [\tau_{zr}^{(1)}(r,0) - \tau_{z\theta}^{(1)}(r,0)]\,r\,\mathrm{d}r\,.$$ (35a)

And on substituting from (34c) we find that

$$P_x = -\sqrt{2\pi} \int_0^a \chi_1(t)\, dt. \tag{35b}$$

It follows from this that the resultant tangential force depends only on the function χ_1.

It remains to consider boundary conditions (28b,c). When we substitute (33a,b) in (30) we find

$$\begin{aligned} 2\mu[u_r^{(n)}(r,0) - u_\theta^{(n)}(r,0)] &= -(2-\nu) \int_0^\infty \xi^{1/2} J_{n-1}(\xi r)\, d\xi \times \\ &\times \int_0^a t^{1/2} \chi_n(t) J_{n-3/2}(\xi t)\, dt + \\ &\frac{1}{2}\int_0^\infty [-(M+N)\, T_n + (M-N)\, S_n]\, J_{n-1}(\xi r)\, d\xi \\ &= 2\mu \Delta_{n-1}(r) \end{aligned} \tag{36a}$$

or

$$\begin{aligned} &-(2-\nu)\sqrt{\frac{2}{\pi}}\, r^{-n+1} \int_0^r \frac{t^{n-1} \chi_n(t)\, dt}{(r^2-t^2)^{1/2}} + \\ &+ \frac{1}{2}\int_0^\infty [-(M+N)\, T_n + (M-N)\, S_n]\, J_{n-1}(\xi r)\, d\xi = 2\mu\Delta_{n-1}(r). \\ &0 \le r \le a \end{aligned} \tag{36b}$$

Now multiply both sides of (36b) by $r^n(s^2-r^2)^{-1/2} dr$ and integrate over the region $0 \le r \le s$. After interchanging the order of integration we can carry out the integration with respect to r. Then after differentiating with respect to s (36b) becomes

$$\begin{aligned} &(2-\nu)\, \chi_n(s) + \tfrac{1}{2} s^{1/2} \int_0^a t^{1/2} \chi_n(t)\, dt \int_0^\infty (M+N)\, \xi J_{n-3/2}(\xi s) \times \\ &\quad \times J_{n-3/2}(\xi t)\, d\xi + s^{1/2} \int_0^a t^{1/2} \psi_n(t)\, dt \times \\ &\quad \times \int_0^\infty \left[\frac{-1}{2-\nu} M + \frac{1-\nu}{2-\nu} N\right] \xi J_{n-3/2}(\xi s)\, J_{n+1/2}(\xi t)\, d\xi \\ &\quad = -2\mu \sqrt{\frac{2}{\pi}}\, s^{-n+1} \frac{d}{ds} \int_0^s \frac{r^n \Delta_{n-1}(r)}{(s^2-r^2)^{1/2}}\, dr. \end{aligned} \tag{37}$$

Boundary condition (30c) is manipulated in the same way to yield

$$\nu\chi_n(s) - \frac{4(1-\nu)}{2-\nu}\, \psi_n(s) - \nu(2n-1)\, s^{-n} \int_0^s t^{n-1} \chi_n(t)\, dt +$$

$$+\tfrac{1}{2}s^{1/2}\int_0^a t^{1/2}\,\chi_n(t)\,\mathrm{d}t\int_0^\infty (M-N)\,\xi J_{n+1/2}(\xi s)\,J_{n-3/2}(\xi t)\,\mathrm{d}\xi\,-$$

$$-\frac{s^{1/2}}{2-\nu}\int_0^a t^{1/2}\,\psi_n(t)\,\mathrm{d}t\int_0^\infty [M+(1-\nu)N]\,\xi J_{n+1/2}(\xi t)\,\times$$

$$\times\, J_{n+1/2}(\xi s)\,\mathrm{d}\xi = 2\mu\sqrt{\frac{2}{\pi}}\,s^{-n-1}\frac{\mathrm{d}}{\mathrm{d}s}\int_0^s \frac{r^{n+2}\,\Delta_{n+1}(r)}{(s^2-r^2)^{1/2}}\,\mathrm{d}r\,. \tag{38}$$

The functions M, N are defined by (31,32).

The simultaneous Fredholm integral equations (37) and (38) determine the functions $\chi_n(s)$ and $\psi_n(s)$. In order to facilitate this determination we put the results in dimensionless form, by setting

$$s=\sigma a,\quad t=\tau a,\quad r=\rho a,\quad \beta=\xi h \tag{39a = 14}$$

$$\chi_n(s)=\mu a X_n(s),\quad \psi_n(s)=\mu a\Psi_n(s)\,; \tag{39b}$$

also

$$\Phi_0(s)=\mu a\phi_0(s) \tag{39c}$$

will be needed later. Then (37) and (38) become

$$(2-\nu)\,X_n(\sigma)+\frac{1}{2}\left(\frac{a}{h}\right)^2\sigma^{1/2}\int_0^1 \tau^{1/2}\,X_n(\tau)\,I_1(\sigma,\tau)\,\mathrm{d}\tau\,-$$

$$-\frac{1}{2-\nu}\left(\frac{a}{h}\right)\sigma^{1/2}\int_0^1 \tau^{1/2}\,\Psi_n(\tau)\,I_2(\sigma,\tau)\,\mathrm{d}\tau$$

$$=-2\sqrt{\frac{2}{\pi}}\,\sigma^{-n+1}\frac{\mathrm{d}}{\mathrm{d}\sigma}\int_0^\sigma \frac{\rho^n\,\Delta_{n-1}(\rho)\,\mathrm{d}\rho}{a(\sigma^2-\rho^2)^{1/2}} \tag{40a}$$

and

$$\nu X_n(\sigma)-\nu(2n-1)\int_0^\sigma X_n(\tau)\,\mathrm{d}\tau+\frac{1}{2}\left(\frac{a}{h}\right)^2\sigma^{1/2}\,\times$$

$$\times\int_0^1 \tau^{1/2}\,X_n(\tau)\,I_3(\sigma,\tau)\,\mathrm{d}\tau-\frac{4(1-\nu)}{2-\nu}\,\Psi_n(\tau)-\frac{1}{2-\nu}\left(\frac{a}{h}\right)^2\times$$

$$\times\int_0^1 \tau^{1/2}\,\Psi_n(\tau)\,I_4(\sigma,\tau)\,\mathrm{d}\tau=2\sqrt{\frac{2}{\pi}}\,\sigma^{-n-1}\,\times$$

$$\times\frac{\mathrm{d}}{\mathrm{d}\sigma}\int_0^\sigma \frac{\rho^{n+2}\,\Delta_{n+1}(\rho)}{a(\sigma^2-\rho^2)^{1/2}}\,\mathrm{d}\rho\,, \tag{40b}$$

where

$$I_1(\sigma,\tau) = \int_0^\infty (M+N)\,\beta J_{n-3/2}\left(\beta\sigma\,\frac{a}{h}\right) J_{n-3/2}\left(\beta\tau\,\frac{a}{h}\right) \mathrm{d}\beta \tag{40c}$$

$$I_2(\sigma,\tau) = \int_0^\infty [M-(1-\nu)\,N]\,\beta J_{n-3/2}\left(\beta\sigma\,\frac{a}{h}\right) J_{n+1/2}\left(\beta\tau\,\frac{a}{h}\right) \mathrm{d}\beta \tag{40d}$$

$$I_3(\sigma,\tau) = \int_0^\infty (M-N)\,\beta J_{n+1/2}\left(\beta\sigma\,\frac{a}{h}\right) J_{n-3/2}\left(\beta\tau\,\frac{a}{h}\right) \mathrm{d}\beta \tag{40e}$$

$$I_4(\sigma,\tau) = \int_0^\infty [M+(1-\nu)\,N]\,\beta J_{n+1/2}\left(\beta\sigma\,\frac{a}{h}\right) J_{n+1/2}\left(\beta\tau\,\frac{a}{h}\right) \mathrm{d}\beta\,. \tag{40f}$$

For the semi-infinite solid $a/h=0$ the equations provide explicit solutions for the functions X_n and ψ_n. These explicit solutions are in fact used as the initial step in the iterative procedure by which X_n and ψ_n are obtained when $a/h \neq 0$.

There being no further need for complete generality in this section, we return to the problem of tangential loading of plated spheres in contact. If there is no slip,

$$\Delta_0 = 2\Delta,\ \text{a constant,} \quad \text{and} \quad \Delta_n = 0, \quad n \neq 0\,. \tag{41}$$

Correspondingly, we take

$$\Psi_n = X_n = 0\,, \qquad n \neq 1\,. \tag{42}$$

Then equations (40), for $n=1$, reduce to

$$(2-\nu)\,X_1(\sigma) + \frac{1}{\pi}\left(\frac{a}{h}\right)\int_0^1 X_1(\tau)\,I_5(\sigma,\tau)\,\mathrm{d}\tau - \frac{2\pi^{-1}}{2-\nu}\left(\frac{a}{h}\right) \times$$

$$\times \int_0^1 \Psi_1(\tau)\,I_6(\sigma,\tau)\,\mathrm{d}\tau = -4\sqrt{\frac{2}{\pi}}\,\frac{\Delta}{a} \tag{43a}$$

$$\nu X_1(\sigma) - \nu\sigma^{-1}\int_0^\sigma X_1(\tau)\,\mathrm{d}\tau + \frac{1}{\pi}\left(\frac{a}{h}\right)\int_0^1 X_1(\tau)\,I_7(\sigma,\tau)\,\mathrm{d}\tau -$$

$$-\frac{4(1-\nu)}{2-\nu}\,\Psi_1(\sigma) - \frac{2\pi^{-1}}{2-\nu}\left(\frac{a}{h}\right)\sigma^{-1/2} \times$$

$$\times \int_0^1 \Psi_1(\tau)\,I_8(\sigma,\tau)\,\mathrm{d}\tau = 0 \tag{43b}$$

with

$$I_5(\sigma,\tau) = \int_0^\infty (M+N)\cos\left(\beta\sigma\,\frac{a}{h}\right)\cos\left(\beta\tau\,\frac{a}{h}\right)\mathrm{d}\beta \tag{43c}$$

$$I_6(\sigma,\tau) = \int_0^\infty [M-(1-\nu)\,N] \cos\left(\beta\sigma\frac{a}{h}\right) \times$$

$$\times \left[-\cos\left(\beta\tau\frac{a}{h}\right) + \left(\beta\tau\frac{a}{h}\right)^{-1} \sin\left(\beta\tau\frac{a}{h}\right)\right] \mathrm{d}\beta \tag{43d}$$

$$I_7(\sigma,\tau) = \int_0^\infty (M-N)\left[-\cos\left(\beta\sigma\frac{a}{h}\right) + \left(\beta\sigma\frac{a}{h}\right)^{-1} \sin\left(\beta\sigma\frac{a}{h}\right)\right] \times$$

$$\times \cos\left(\beta\tau\frac{a}{h}\right) \mathrm{d}\beta \tag{43e}$$

$$I_8(\sigma,\tau) = \int_0^\infty [M+(1-\nu)\,N]\left[-\cos\left(\beta\sigma\frac{a}{h}\right) + \left(\beta\sigma\frac{a}{h}\right)^{-1} \times\right.$$

$$\left.\times \sin\left(\beta\sigma\frac{a}{h}\right)\right]\left[-\cos\left(\beta\tau\frac{a}{h}\right) + \left(\beta\tau\frac{a}{h}\right)^{-1} \sin\left(\beta\tau\frac{a}{h}\right)\right] \mathrm{d}\beta \tag{43f}$$

Expressions for the resultant tangential loading and for the distribution of shearing traction on the contact surface are also wanted. These are easily obtained from (35b) and (34):

$$\frac{P_x}{\mu a^2} = -\sqrt{2\pi} \int_0^1 X_1(\tau)\,\mathrm{d}\tau \tag{44}$$

$$\mu^{-1}[\tau_{zr}^{(1)} + \tau_{z\theta}^{(1)}] = \sqrt{\frac{2}{\pi}} \left[\frac{\rho^2\,\Psi_1(1)}{(1-\rho^2)^{1/2}} - \rho^2 \int_\rho^1 \frac{[\tau^{-2}\,\Psi_1(\tau)]'}{(\tau^2-\rho^2)^{1/2}}\,\mathrm{d}\tau\right] \tag{45a}$$

$$\mu^{-1}[\tau_{zr}^{(1)} - \tau_{z\theta}^{(1)}] = \sqrt{\frac{2}{\pi}} \left\{(1-\rho^2)^{-1/2}\left[X_1(1) + \frac{\nu}{2-\nu}\,\Psi_1(1)\right] - \right.$$

$$\left. - \int_\rho^1 \frac{X_1'(\tau)}{(\tau^2-\rho^2)^{1/2}}\,\mathrm{d}\tau - \frac{\nu}{2-\nu} \int_\rho^1 \frac{\tau^{-1}[\tau\Psi_1(\tau)]'}{(\tau^2-\rho^2)^{1/2}}\,\mathrm{d}\tau\right\}. \tag{45b}$$

The last two equations, of course, hold only for the region $z=0$, $r \leq a$. Since

$$\tau_{zx}(r,\theta,0) = \tfrac{1}{2}[\tau_{zr}^{(1)} - \tau_{z\theta}^{(1)}] + \tfrac{1}{2}[\tau_{zr}^{(1)} + \tau_{z\theta}^{(1)}] \cos 2\theta \tag{46}$$

and

$$\tau_{zy}(r,\theta,0) = \tfrac{1}{2}[\tau_{zr}^{(1)} + \tau_{z\theta}^{(1)}] \sin 2\theta \tag{47}$$

equations (45a,b) serve to give the Cartesian components of the shearing traction. It should be noted that there is a singularity in both $[\tau_{zr}^{(1)} - \tau_{z\theta}^{(1)}]$ and $[\tau_{zr}^{(1)} + \tau_{z\theta}^{(1)}]$ at $\rho = 1$ $(r = a)$. The former singularity is radially symmetric but the latter is not. For the semi-infinite solid $a/h = 0$. It follows from

(43 a, b) that when $a/h = 0$

$$X_1 = \frac{-4}{2-\nu}\sqrt{\frac{2}{\pi}\frac{\Delta}{a}}, \qquad \Psi_1 = 0 \tag{48}$$

so that in the contact of semi-infinite solids only the radially symmetric singularity appears. This is true even if the semi-infinite bodies are anisotropic. The presence of this second, non-symmetric, singularity is a *bona fide* layer effect. The effect is due to the transverse component of traction, τ_{zy}, that is created by the change in material stiffness with depth. For the semi-infinite body τ_{zy} vanishes on the contact surface, which greatly simplifies the analysis.

The physical quantity of greatest interest to emerge from the analysis made on the assumption of no slip is the ratio $2\Delta/P_x$, the initial tangential compliance of the contact. The variation of this quantity with layer thickness is given in Table 2. The distribution of shear traction on the contact plane is shown in Fig. 1 for $\nu = 0.3$ and representative values of a/h.

Table 2. Initial tangential compliance

a/h	$(2\Delta/P_x)\mu a = C_x\mu a$			
	$\nu = 0$	$\nu = 0.15$	$\nu = 0.30$	$\nu = 0.45$
0.0	0.50000	0.4625	0.42500	0.3875
0.2	0.45192	0.42010	0.38708	0.35374
0.4	0.40726	0.38004	0.35198	0.32272
0.6	0.36788	0.34508	0.32114	0.29662
0.8	0.33402	0.31494	0.29458	0.27252
1.0	0.30512	0.28912	0.27176	0.25262
2.0	0.21018	0.20266	0.19406	0.18392
3.0	0.15904	0.15482	0.14982	0.14366
4.0	0.12764	0.12694	0.12170	0.11762
5.0	0.10650	0.10464	0.10236	0.09944

The function Ψ_1 determines τ_{zy} and the θ-dependent component of τ_{zx}, both of which are small compared with the radially symmetric component of τ_{zx}. Furthermore Ψ_1 does not enter the compliance equation (44); it affects the compliance only through its influence on X_1. These facts suggest that we may set $\Psi_1 = 0$, so effecting a simplification in calculation at little cost in accuracy. Such an approximation is equivalent to replacing the boundary condition u_y $(r < a, 0) = 0$ by the condition τ_{zy} $(r < a, 0) = 0$. Although, in principle, this approximation yields a value for the compliance that is slightly too large, comparison with the results of the rigorous calculation over the range $0 \leq \nu \leq 0.45$ and $0 \leq a/h \leq 5$ show complete agreement, to six decimal digits. In order to make an estimate of the error entailed by the approximation it is necessary to use a point variable. The most sensitive of these is the value of $\frac{1}{2}[\tau_{rz}-\tau_{z\theta}]$ at $r=0$. Over the range $0 \le a/h \le 5$ it has been found that the error in $\frac{1}{2}[\tau_{rz}-\tau_{z\theta}]$ due to taking $\Psi_1 = 0$ never exceeds 0.505 per cent.

4. THE EFFECT OF MICRO-SLIP

The singularity that would arise, in the absence of slip, at the perimeter of the tangentially loaded contact region requires consideration. In reality, as Mindlin notes, slip must occur in an area near the outer boundary of the contact region. If it is assumed that the slip region is a circular annulus of inner radius c, this inner radius can be chosen so as to eliminate the radially symmetric singularity in τ_{zx} produced by the first term on the right-hand side of equation (45b). That is essentially what has been done by Mindlin [3] in his treatment of the contact of homogeneous semi-infinite bodies. In the present case of the plated body we have also to remove the singularity due to the previously mentioned layer effect. This is the singularity which appears in the first term on the right-hand side of equation (45a) and which therefore appears in a non-radially symmetric manner in both τ_{zx} (46) and τ_{zy} (47). Physically, the layer effect produces a slip region that is not bounded internally by a circle. Although this circumstance presents an obstacle to the rigorous analysis of the effect of micro-slip on the tangentially loaded layered contact, it is possible to obtain upper and lower bounds to all the quantities that are of technological interest.

When the two plated spheres under discussion are brought into contact by normal loading the state of stress is given by the stress function ϕ_0 whose determination is described in section 2. After tangential load is added the additional stress state is given by equations (24,25,26,29). Instead of equations (33), however, we now write

$$S_n = \xi^{1/2}\left[\int_0^c t^{1/2}\,\Psi_n(t)\,J_{n+1/2}(\xi t)\,\mathrm{d}t + \frac{\delta}{a^2} \times \right.$$

$$\left. \times \int_c^a t^{5/2}\,\phi_0(t)\,J_{n+1/2}(\xi t)\,\mathrm{d}t\right] \tag{49a}$$

$$T_n = \xi^{1/2}\left[\int_0^c t^{1/2}\,\chi_n(t)\,J_{n-3/2}(\xi t)\,\mathrm{d}t - \frac{\nu}{2-\nu} \times \right.$$

$$\times \int_c^0 t^{1/2}\,\psi_n(t)\,J_{n+1/2}(\xi t)\,\mathrm{d}t - 2f\int_c^a t^{1/2}\phi_0(t)\,J_{n-3/2}(\xi t)\,\mathrm{d}t -$$

$$\left. - \frac{\nu}{2-\nu}\,\frac{\delta}{a^2}\int_c^a t^{5/2}\,\phi_0(t)\,J_{n+1/2}(\xi t)\,\mathrm{d}t\right] \tag{49b}$$

Here f denotes the coefficient of limiting friction and δ is a dimensionless constant whose magnitude is, for the time being, at our disposal. The contact region is divided by (49) into two parts. There is an inner circle, $0 \le r \le c$, in which there is effectively no slip and an outer annulus, $c \le r \le a$, where the shearing traction effectively follows the Amontons-Coulomb friction law. In order that there be no singularity in stress at the junction of these two regions it is necessary that the functions S_n and T_n be continuous. This is accomplished by imposing the conditions

$$\chi_n(c) = -2f\phi_0(c) \qquad \text{and} \qquad \psi_n(c) = \delta \frac{c^2}{a^2} \phi_0(c). \tag{50}$$

The determination of χ_n and ψ_n is the main object of the present section. We proceed as in section 3, substituting (49) in (30) and making use of (50). After some manipulation it appears that in terms of the dimensionless forms given in (39), on $z=0$

$$\frac{1}{2\mu}\left[\tau_{zr}^{(n)} + \tau_{z\theta}^{(n)}\right] = \frac{-\rho^{n+1}}{(2\pi)^{1/2}}\left[\int_\rho^{c/a} \frac{[\tau^{-n-1}\,\Psi_n(\tau)]'}{(\tau^2-\rho^2)^{1/2}}\,d\tau + \right.$$

$$\left. + \delta \int_{c/a}^{1} \frac{[\tau^{-n+1}\,\Phi_0(\tau)]'}{(\tau^2-\rho^2)^{1/2}}\,d\tau\right],$$

$$0 \le \rho \le c/a \tag{51a}$$

$$= \frac{-\rho^{n+1}}{(2\pi)^{1/2}}\,\delta \int_\rho^{1} \frac{[\tau^{-n+1}\,\Phi_0(t)]'}{(\tau^2-r^2)^{1/2}}\,d\tau,$$

$$c/a \le \rho > \infty. \tag{51b}$$

$$\frac{1}{2\mu}\left[\tau_{zr}^{(n)} - \tau_{z\theta}^{(n)}\right] = \frac{-\rho^{n-1}}{(2\pi)^{1/2}}\left[\int_\rho^{c/a} \frac{[\tau^{-n+1}\,X_n(\tau)]'}{(\tau^2-\rho^2)^{1/2}}\,d\tau - \right.$$

$$-2f \int_{c/a}^{1} \frac{[\tau^{-n+1}\,\Phi_0(\tau)]'}{(\tau^2-\rho^2)^{1/2}}\,d\tau +$$

$$+\frac{\nu}{2-\nu} \int_\rho^{c/a} \frac{\tau^{-2n+1}[\tau^{n}\,\Psi_n(\tau)]'}{(\tau^2-\rho^2)^{1/2}}\,d\tau +$$

$$\left. + \frac{\nu}{2-\nu}\,\delta \int_{c/a}^{1} \frac{\tau^{-2n+1}[\tau^{n+2}\,\Phi_0(\tau)]'}{(\tau^2-\rho^2)^{1/2}}\,d\tau\right],$$

$$0 \le \rho \le c/a \tag{51c}$$

$$= \frac{\rho^{n-1}}{(2\pi)^{1/2}}\left[2f \int_\rho^{1} \frac{[\tau^{-n+1}\,\Phi_0(\tau)]'}{(\tau^2-\rho^2)^{1/2}}\,d\tau - \right.$$

$$- \frac{\nu}{2-\nu} \delta \int_{\rho}^{1} \frac{\tau^{-2n+1} [\tau^{n+2} \Phi_0(\tau)]'}{(\tau^2 - \rho^2)^{1/2}} \, d\tau \Big]_{,\xi}$$

$$c/a \leq \rho < \infty . \tag{51d}$$

In the above expressions primes denote differentiation with respect to τ. An expression for the resultant tangential loading is derivable directly from (35a) and (51c):

$$\frac{P_x}{\mu a^2} = -(2\pi)^{1/2} \left[\int_0^{c/a} X_1(\tau) \, d\tau - 2f \int_{c/a}^{1} \Phi_0(\tau) \, d\tau \right]. \tag{52}$$

In what follows only the case $n=1$ is required; Ψ_n and X_n are therefore taken to vanish for $n \neq 1$. Displacements in the contact plane are given by the expressions

$$u_x(r, \theta, 0) = \tfrac{1}{2}[u_r^{(1)} - u_\theta^{(1)}] + \tfrac{1}{2}[u_r^{(1)} + u_\theta^{(1)}] \cos 2\theta \tag{53a}$$

$$u_y(r, \theta, 0) = \tfrac{1}{2}[u_r^{(1)} + u_\theta^{(1)}] \sin 2\theta \tag{53b}$$

which are analogous to expressions (46,47) for $\tau_{zx}(r, \theta, 0)$ and $\tau_{zy}(r, \theta, 0)$. Proceeding as in section 3, the requirement that the central portion of the contact region be free of slip and experience a rigid-body translation of magnitude Δ in the x-direction finally yields, the integral equations

$$\pi(2-\nu) X_1(\sigma) + \left(\frac{a}{h}\right) \int_0^{c/a} X_1(\tau) I_5(\sigma, \tau) \, d\tau - \frac{2}{2-\nu} \left(\frac{a}{h}\right) \int_0^{c/a} \times$$

$$\times \Psi_1(\tau) I_6(\sigma, \tau) \, d\tau = 2f \left(\frac{a}{h}\right) \int_{c/a}^{1} \Phi_0(\tau) I_5(\sigma, \tau) \, d\tau - 2\delta \times$$

$$\times \left(\frac{a}{h}\right) \int_{c/a}^{1} \Phi_0(\tau) \tau^2 I_6(\sigma, \tau) \, d\tau - 4(2\pi)^{1/2} \frac{\Delta}{a} \tag{54a}$$

and

$$\pi\nu X_1(\sigma) - \pi\nu\sigma^{-1} \int_0^{\sigma} X_1(\tau) \, d\tau + \left(\frac{a}{h}\right) \int_0^{c/a} X_1(\tau) I_7(\sigma, \tau) \, d\tau -$$

$$- \frac{4(1-\nu)}{2-\nu} \pi \Psi_1(\sigma) - \frac{2}{2-\nu} \left(\frac{a}{h}\right) \int_0^{c/a} \Psi_1(\tau) I_8(\sigma, \tau) \, d\tau$$

$$= 2f \left(\frac{a}{h}\right) \int_{c/a}^{1} \Phi_0(\tau) I_7(\sigma, \tau) \, d\tau + 2\delta \times$$

$$\times \left(\frac{a}{h}\right) \int_{c/a}^{1} \Phi_0(\tau) \tau^2 I_8(\sigma, \tau) \, d\tau . \tag{54b}$$

The functions $I_5(\sigma, \tau)$, ..., $I_8(\sigma, \tau)$ are defined by equations (43c–43f). The auxiliary conditions derived from (50) are

$$X_1\left(\frac{c}{a}\right) = -2f\Phi_0\left(\frac{c}{a}\right) \quad \text{and} \quad \Psi_1\left(\frac{c}{a}\right) = \delta\left(\frac{c^2}{a^2}\right)\Phi_0\left(\frac{c}{a}\right). \tag{54c, d}$$

The shearing components of stress given by (51 b) and (51 d) are to be compared with the product of the normal traction and the coefficient of friction. It follows from (10 a, b), with $n=0$, that

$$f\tau_{zz}(r,\theta,0) = \frac{2\mu f}{(2\pi)^{1/2}}\int_\rho^1 \frac{\Phi_0'(\tau)\,\mathrm{d}\tau}{(\tau^2-\rho^2)^{1/2}}. \tag{55}$$

A. 'Stiff' approximation

In this bound the requirement of zero micro-slip in $0 \le \rho \le c/a$ is maintained. Equations (54 a, b) are satisfied (for a pre-assigned value of c/a) and then δ and Δ are determined so as to satisfy equations (54 c, d). The corresponding tangential load may then be calculated through (52). In the no-slip region, $0 \le \rho \le c/a$, the shearing traction is given by (51 a, c) with $n=1$. In the slip region, $c/a \le \rho \le 1$, the shearing components of stress, in view of (51 b, d), are

$$\tau_{zx} = f\tau_{zz} - \tfrac{1}{2}\delta\tau_{zz}(e_1 + \rho^2\cos 2\theta) \tag{56a}$$

$$\tau_{zy} = -\tfrac{1}{2}\delta\tau_{zz}(\rho^2\sin 2\theta) \tag{56b}$$

where

$$e_1 = \frac{\nu}{2-\nu}\int_\rho^1 \frac{\tau^{-1}[\tau^3\,\Phi_0(\tau)]'}{(\tau^2-\rho^2)^{1/2}}\,\mathrm{d}\tau \Bigg/ \int_\rho^1 \frac{\Phi_0'(\tau)\,\mathrm{d}\tau}{(\tau^2-\rho^2)^{1/2}} < 1. \tag{56c}$$

The quantities in brackets in equations (56 a, b) represent 'error' terms in the sense that if they were zero the Amontons-Coulomb friction law would be satisfied at every point in the slip region.* The error in this approximation will be small provided (i) Poisson's ratio is small, or (ii) a/h is small, or (iii) $\delta < f$. As a/h becomes vanishingly small the approximation reduces to the exact solution (48) with $\delta = 0$.

In this approximation the ability of the outer region to muster resistance to slip is overestimated. It follows that the value of c/a corresponding to any given value of P_x/fP_z will be larger than it would be if the Amontons-Coulomb friction law were strictly satisfied. Since the relationship between these two quantities is monotonic the procedure provides a lower bound to P_x/fP_z for given c/a. On the other hand the compliance, for given c/a, is slightly overestimated.

* Satisfied in the sense that $(\tau_{xz}^2+\tau_{yz}^2)^{\frac{1}{2}} = -f\tau_{zz}$. The Amontons-Coulomb friction law also requires that $\tau_{xz}/\tau_{yz} = u_x/u_y$; that is, the shearing traction must have the same direction as the relative slip. This second requirement is usually ignored. It is not satisfied in the case of the contact of semi-infinite bodies treated in [3].

B. 'Soft' approximation

The computations described at the close of section 3 suggest that a very approximate lower bound to the compliance will be obtained by setting $\Psi_1 = \delta = 0$. Then

$$\tau_{zx} = -\frac{\mu}{(2\pi)^{1/2}}\left[\int_\rho^{c/a} \frac{X_1'(\tau)\,d\tau}{(\tau^2-\rho^2)^{1/2}} - 2f\int_{c/a}^{1} \frac{\Phi_0'(\tau)\,d\tau}{(\tau^2-\rho^2)^{1/2}}\right], \qquad 0 \le \rho \le c/a \tag{57a}$$

$$= \frac{2\mu f}{(2\pi)^{1/2}}\int_\rho^{1} \frac{\Phi_0'(\tau)\,d\tau}{(\tau^2-\rho^2)^{1/2}} = f\tau_{zz}, \qquad c/a \le \rho \le 1 \tag{57b}$$

$$\tau_{zy} = 0, \qquad 0 \le \rho < \infty \tag{57c}$$

The Amontons-Coulomb friction law is satisfied at every point in $c/a \le \rho \le 1$. In order to have the mean displacements in the region $0 \le \rho \le c/a$ correspond to an x-displacement of magnitude Δ it is necessary to satisfy (54a) with $\Psi_1 = \delta = 0$. This determines the function X_1. The relationship between P_x and Δ is then given by (52).

Physically, the 'soft' bound corresponds to neglect of the constraint provided by the layer effect. The ability of the inner region to provide resistance to slip is therefore underestimated. Since the contact area is constant, the size of the slip region needed to make up the total resistance is overestimated. For given P_x/fP_z the approximation provides a lower bound to c/a and, conversely, for given c/a it provides an upper bound to P_x/fP_z. The compliance corresponding to a given c/a is slightly underestimated. As a/h approaches zero this approximation approaches the exact solution[3]. When a/h is not equal to zero the error in this approximation may be measured either by the fractional change in shear traction as described in section 3 or by the ratio of the maximum θ-dependent component of displacement in the region $0 \le \rho \le c/a$ to Δ. If this error estimate is denoted e_2,

$$e_2 = \frac{1}{2\Delta}[u_r(c,0,0) - u_\theta(c,0,0)] = \frac{a/\Delta}{2(2\pi)^{1/2}} \times$$

$$\times \left[\nu\left(\frac{c}{a}\right)^{-2}\int_0^{c/a} X_1'(\tau)\,\tau\left(\frac{c^2}{a^2}-\tau^2\right)^{1/2} d\tau + \right.$$

$$+\frac{1}{2}\left(\frac{a}{h}\right)\int_0^{c/a} X_1(\tau)\,d\tau\int_0^\infty (M-N)\,J_2\left(\beta\frac{c}{a}\frac{a}{h}\right)\cos\left(\beta\tau\frac{a}{h}\right)d\beta -$$

$$\left. -f\left(\frac{a}{h}\right)\int_{c/a}^{1} \Phi_0(\tau)\,d\tau\int_0^\infty (M-N)\,J_2\left(\beta\frac{c}{a}\frac{a}{h}\right)\cos\left(\beta\tau\frac{a}{h}\right)d\beta\right]. \tag{58}$$

Of course comparison of upper and lower bounds also provides an estimate of the accuracy of each of them, at least when the bounds are close.

5. NUMERICAL RESULTS

Computations of compliance, slip radius and stress distribution have been carried out for $\nu = 0$, 0.15, 0.30, 0.45 and for $a/h = 0$, 0.2, 0.4, 0.6, 0.8, 1.0, 2.0, 3.0, 4.0, 5.0 over the complete range $0 \leq c/a \leq 1$. The thinnest layer for which calculations were carried out therefore had a thickness one-tenth the diameter of the contact region. The calculations could have been extended to thinner layers had that been desired.

For the above range of variables the dependence of c/a on P_x/fP_z is shown in Fig. 2. This figure locates the boundary that separates the slipped and unslipped regions. The ratio of the initial tangential compliance (given in

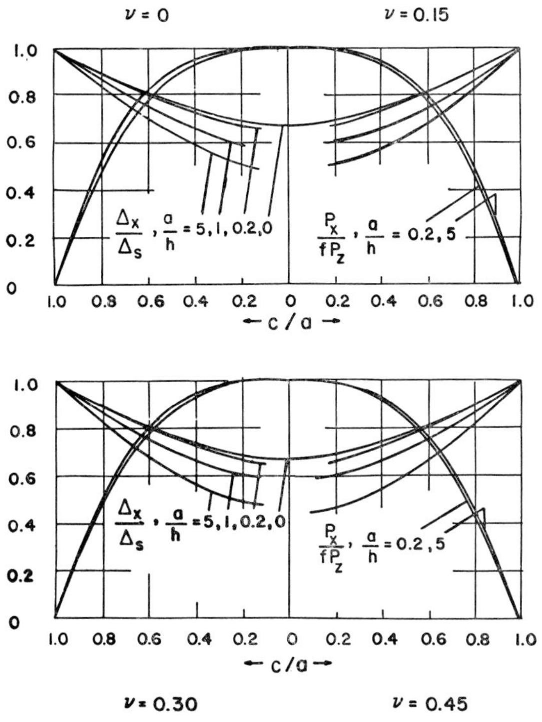

Fig. 2. Interrelation between slip zone dimension, lateral motion and tangential load

a/h = radius of contact region/thickness of elastic layer
c/a = radius of no-slip region/radius of contact region
Δ_x/Δ_s = lateral motion in absence of slip (Table 2)/lateral motion with slip

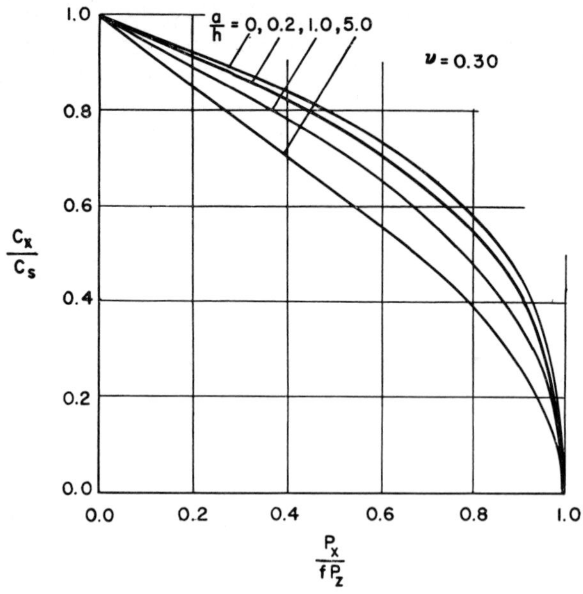

Fig. 3. Ratio of initial tangential compliance, C_x, to compliance with slip, C_s
$C_s = d\Delta/dP_x$

Table 2) to the actual compliance when slip occurs is shown in Fig. 3 as a function of P_x/fP_z. The actual compliance when slip occurs is relatively insensitive to variations in Poisson's ratio when the surface layer is thin. In all cases the results converge to the known solution for $a/h = 0$.

It is found that for the complete range of variables investigated the upper and lower bounds on compliance described in section 3 are identical in the first four decimal digits. Differences appear only in the fourth decimal place, and then only for large a/h. The curves given in the charts may therefore be regarded as correct for engineering purposes. The same remarks are true for the relation between P_x/fP_z and c/a. As has been shown in the previous section, the requirement $\delta < f$ is a rational one for assessing the adequacy of the 'hard' approximation. The ratio δ/f, which is independent of f, has been calculated for each case investigated. It is largest when c/a is small and a/h and ν are large. The maximum value encountered was $\delta/f = 0.23668$ for $a/h = 5$, $c/a = 0.2$, $\nu = 0.45$. As long as $a/h \leq 4$ and $\nu \leq 0.3$ the ratio $\delta/f < 0.1$. The way in which these bounds vary for $\nu = 0.3$ is shown in Table 3. Of course point-variable quantities are more sensitive to the approximations made than are quantities that represent integrated effects. Typical and most sensitive of these pointvariable quantities is stress. We recall that

$$\tau_{zx} = \tfrac{1}{2}[\tau_{zr}^{(1)} - \tau_{z\theta}^{(1)}] + \tfrac{1}{2}[\tau_{zr}^{(1)} + \tau_{z\theta}^{(1)}] \cos 2\theta \tag{46}$$

Table 3. Bounds for tangential load, compliance and stress (Poisson's ratio 0.30; stress components evaluated at the center of contact)

a/h		Bound	$\frac{c}{a} = \frac{\text{radius of non-slip region}}{\text{radius of contact region}}$ 0.2	0.4	0.6	0.8
0.2	$\frac{P_x}{fP_z}$	A	0.99201	0.93607	0.78420	0.48829
		B	0.99201	0.93607	0.77420	0.48829
	$\frac{\Delta}{P_x}\mu a$	A	0.28934	0.26696	0.24115	0.21613
		B	0.28934	0.26696	0.24115	0.21613
	$\frac{\frac{1}{2}[\tau_{zr}^{(1)}-\tau_{z\theta}^{(1)}]}{P_x/\pi a^2}$	A	−1.20988	−0.96117	−0.76550	−0.61488
		B	−1.21011	−0.96226	−0.76627	−0.61510
	$\frac{\frac{1}{2}[\tau_{zr}^{(1)}+\tau_{z\theta}^{(1)}]}{P_x/\pi a^2}$	A	0.00051	0.00043	0.00028	0.00033
	δ/f	A	0.00084	0.00090	0.00093	0.00106
1.0	$\frac{P_x}{fP_z}$	A	0.99203	0.93655	0.78576	0.49051
		B	0.99206	0.93658	0.78577	0.49051
	$\frac{\Delta}{P_x}\mu a$	A	0.22532	0.20372	0.17936	0.15632
		B	0.22531	0.20370	0.17933	0.15625
	$\frac{\frac{1}{2}[\tau_{zr}^{(1)}-\tau_{z\theta}^{(1)}]}{P_x/\pi a^2}$	A	−1.25216	−0.98855	−0.78046	−0.62127
		B	−1.26959	−1.02475	−0.83099	−0.68088
	$\frac{\frac{1}{2}[\tau_{zr}^{(1)}+\tau_{z\theta}^{(1)}]}{P_x/\pi a^2}$	A	0.023094	0.01756	0.01165	0.00705
	δ/f	A	0.03857	0.03616	0.03075	0.02372
5.0	$\frac{P_x}{fP_z}$	A	0.99366	0.94505	0.79271	0.47408
		B	0.99399	0.94552	0.79292	0.47409
	$\frac{\Delta}{P_x}\mu a$	A	0.10282	0.08994	0.07540	0.06201
		B	0.10276	0.08984	0.07533	0.06198
	$\frac{\frac{1}{2}[\tau_{zr}^{(1)}-\tau_{z\theta}^{(1)}]}{P_x/\pi a^2}$	A	−1.50330	−1.15685	−0.86914	−0.64554
		B	−1.67379	−1.43067	−1.18893	−0.97332
	$\frac{\frac{1}{2}[\tau_{zr}^{(1)}+\tau_{z\theta}^{(1)}]}{P_x/\pi a^2}$	A	−0.09246	−0.06618	−0.04198	−0.02479
	δ/f	A	−0.12898	−0.12794	−0.11669	−0.10150

$$\tau_{zy} = \tfrac{1}{2}\left[\tau_{zr}^{(1)}+\tau_{z\theta}^{(1)}\right]\sin 2\theta . \tag{47}$$

The quantity $\frac{1}{2}[\tau_{zr}^{(1)}-\tau_{z\theta}^{(1)}]$ evaluated at $r=0$ and normalized by division by $P_x/\pi a^2$ is given in Table 3 for both approximations. The agreement is excellent for small a/h, deteriorating somewhat as layer thickness decreases.

The stress components on the contact surface have also been calculated for the cases investigated. It is found that $\frac{1}{2}(\tau_{zr}+\tau_{z\theta})$, which is the peak magnitude of τ_{zy}, and which is also the peak magnitude of the θ-dependent part of τ_{zx}, does not exceed one tenth of $\frac{1}{2}(\tau_{zr}-\tau_{z\theta})$, which is the magnitude of the radially symmetric part of τ_{zx}. This suggests that it is reasonable to neglect these θ-dependent terms in many technological applications of the theory. Typical distributions of shearing stress in the presence of slip are shown in Figure 1 and values of $\frac{1}{2}[\tau_{zr}^{(1)}+\tau_{z\theta}^{(1)}]$, evaluated at $r=0$ are given in Table 3.

ACKNOWLEDGMENT

We wish to thank the authorities of Northwestern University who allowed one of us (L.E.G.) the academic privileges of a Visiting Professor during the preparation of this work and we thank the authorities of the University of Minnesota for leave of absence granted during this period. We wish to thank the National Science Foundation, Grant GK 37270, for the support of one of us (L.M.K.) during the course of this research. The Center for Numerical Computation at the University of Minnesota provided a grant to cover the cost of using their high-speed digital computer. Our special thanks are due to Dr. Sudarej Chuntranuluck, whose assistance in the programming of the computations has been most valuable.

REFERENCES

1. H. Hertz, 'Über die Berührung fester elastischer Körper', *J. f. d. Reine u. Angewandte Math.*, **92**, 1882, pp. 156–171.
2. C. Cattaneo, 'Sul Contatto di Due Corpi Elastici', *Acc. dei Lincei, Rend. Ser.* **6**, **27**, Pt. I, pp. 342–348, pt. II, pp. 434–436, pt. III, pp. 474–478.
3. R.D. Mindlin, 'Compliance of Elastic Bodies in Contact', *Trans. ASME ser. E, J. Applied Mech.*, **16**, 1949, pp. 259–268.
4. K.L. Johnson, 'Surface Interaction between Elastically Loaded Bodies under Tangential Forces', *Proc. Roy. Soc.* (London), **A230**, 1955, pp. 531–548.
5. J.J. O'Connor and K.L. Johnson, 'The Role of Surface Asperities in Transmitting Tangential Forces between Metals', *Wear*, **6**, 1963, pp. 118–139.
6. R.D. Mindlin, W.P. Mason, T.F. Osmer and H. Deresiewicz, 'Effects of an Oscillating. Tangential Force on the Contact Surfaces of Elastic Spheres', *Proc. First U.S Nat'l Cong. Applied Mech.*, 1951, pp. 203–208.
7. J. Duffy and R.D. Mindlin, 'Stress-Strain Relations and Vibrations of a Granular Medium', *Trans. ASME*, ser. **E79**, *J. Applied Mech.*, **24**, 1957, pp. 585–593.
8. L.E. Goodman and G.E. Bowie, 'Experiments on Damping at the Contacts of a Sphere with Flat Plates', *Experimental Mech.*, **1**, 1961, pp. 48–54.
9. L.E. Goodman and C.B. Brown, 'Energy Dissipation in Contact Friction: Constant Normal and Cyclic Tangential Loading', *Trans. ASME*, ser. E, *J. Applied Mech.*, 29, 1962, pp. 17–22.

10. K.L. Johnson, 'Energy Dissipation at Spherical Surfaces in Contact Transmitting Oscillating Forces', *J. Mech. Eng. Sci.*, **3**, 1961, pp. 362–368.
11. Y. Tu and D.C. Gazis, 'The Contact Problem of a Plate Pressed between Two Spheres', *Trans. ASME*, ser. E, *J. Applied Mech.*, **31**, 1964, pp. 659–666.
12. J. Lysmer and J.M. Duncan, (eds), *Stresses and Deflections in Foundations and Pavements*, 5th ed., Dept. of Civil Eng., Univ. of California, Berkeley (Cal.), 1972.
13. W.C. Hayes, L.M. Keer, G. Herrmann and L.F. Mockros, 'A Mathematical Analysis for Indentation Tests of Articular Cartilage', *J. Biomechanics*, **5**, 1972, pp. 541–551.
14. A.E. Green and W. Zerna, *Theoretical Elasticity*, 1st ed., Oxford Univ. Press, 1954, pp. 169–170.
15. L.M. Keer, 'Coupled Pairs of Dual Integral Equations', *Q. Applied Math.*, 25, 4, 1968, pp. 453–457.
16. R.A. Westmann, 'Simultaneous Pairs of Dual Integral Equations', *SIAM Review*, **7**, **3**, 1963, pp. 341–348.
17. I.N. Sneddon, 'The Elementary Solution of Dual Integral Equations', *Proc. Glasgow Math. Assoc.*, **4**, 1960, pp. 108.
18. E.T. Copson, 'On Certain Dual Integral Equations', *Proc. Glasgow Math. Assoc.*, **5**, 1961, pp. 19–24.
19. M. Abramowitz and I.A. Stegun, (eds.), *Handbook of Mathematical Functions*, U.S. Gov't Printing Office, Washington (D.C.), 1965.
20. I.N. Sneddon, *Fractional Integration and Dual Integral Equations*, North Carolina State College, Raleigh (N.C.), 1962.
21. L.E. Goodman, 'Contact Stress Analysis of Normally Loaded Rough Spheres', *Trans. ASME*, ser. E, **84**, *J. Applied Mech.*, **29**, 1962, pp. 515–522.
22. D.A. Spence, 'Self Similar Solutions to Adhesive Contact Problems with Incremental Loading', *Proc. Roy. Soc.*, A, London, 305, 1968, pp. 55–80.
23. H. Deresiewicz, '*Bodies in Contact witn Applications to Granular Media*', Chapter IV, (pp. 105–147) of G. Herrmann (ed.,) *R. D. Mindlin and Applied Mechanics*, Pergamon Press, Elmsford (N.Y.).

Small scale plastic flow associated with rolling

J. Christoffersen

1. INTRODUCTION

A heavy, rigid roller is assumed to move steadily over the surface of an elastic-plastic half-space deforming in plane strain.

The elastic properties of the half-space are homogeneous and isotropic, and the plastic deformation is assumed to be in-plane shear on planes normal to the direction of rolling. The shear is taken to be of constant magnitude γ within a surface layer of thickness h, extending backwards through the overrolled zone from the leading point of contact between the roller and the surface.

The distortion of the surface due to plastic strains is calculated in section 2. The theory of Muskhelishvili[1] on contact between elastic bodies, briefly reviewed in section 3, is applied in section 4 to the, essentially elastic, problem of contact between the roller and the distorted surface. In particular, the horizontal driving force on the roller is calculated.

An energy balance is set up in section 5, equating the work of the driving force and the plastic work, or dissipation. It is shown that the assumed steady state is only possible if the contact pressure exceeds a certain critical value. Assuming, in the sprit of thermodynamics, that the system dissipates as much energy as possible within the given constraints, we finally arrive at a set of equations, from which the parameters γ and h may be inferred.

It is finally demonstrated that the case in which the roller is elastic may be given an equivalent treatment.

The problem has been treated by Merwin and Johnson[2]. They assume, in contrast with the present work, that the surface is smooth, i.e. wedge-free. Their findings are, consequently, that plastic flow is confined to a sub-surface layer. They find, moreover, forward flow as the net result of a cycle of plastic shear, while we assume forward flow through the entire process. Our result, that a steady state of plastic shearing is impossible at nominal Hertzian pressures lower than 4.77 times the yield point in simple shear, seems, however, not to be inconsistent with Merwin and Johnson's shake-down limit of 4.00 times this yield point.

Collins[3] treats the case of a rigid-plastic roller. His slip-line fields for the case of a force-driven roller resemble our deformation fields in the formulation of a wedge.

An experimental investigation, supporting the views of Merwin and Johnson, are to be found in a paper by Hamilton[4]. Wedge-formation at high loads (higher than those contemplated by Merwin and Johnson) is illustrated in a paper by Johnson[5]. The present model should probably be regarded as a competitor to Collin's high-load model, rather than to that of Merwin and Johnson. Still, it may be argued that a wedge, once formed, is probably hard to get rid of, no matter the magnitude of Hertzian pressure.

2. PLASTIC DISTORTION OF THE SURFACE

Let, in a Cartesian frame (x_1, x_2, x_3), the half-space $x_2 \leq 0$ be occupied by an elastic-plastic body, isotropic and homogeneously elastic and subject to conditions of plane strain normal to the x_3-axis.

The surface of the half-space is loaded by a rigid roller assumed to move steadily in the positive direction of the x_1-axis, leaving behind it a plastically deformed surface layer with thickness h. Due to the assumed steadiness of the movement, the plastic deformation, when checked behind the roller, will be constant at a given depth under the surface.

In front of the surface layer there will be a region in which plastic deformation takes place. We assume, tentatively, that this region is a narrow slip-band emerging from the leading point of contact perpendicularly to the surface.

Removal of the material in the slip-band would leave the unloaded body with displacements

$$v_1 = v_1^s(x_2); \quad v_2 = 0 \quad \text{for} \quad -\infty < x_1 < b \quad \text{and} \quad -h \leq x_2 \leq 0, \tag{2.01}$$

and $v_1 = v_2 = 0$ elsewhere. Here, b is the x_1-coordinate of the slip-band. In particular, $v_1^s(-h) = 0$, and so the body is left stress-free.

To calculate the deformation of the surface in the actual body due to plastic straining, let, with Greek letters as subscripts ranging over the values 1 and 2, $g_{\lambda\alpha}(x_1 - t, x_2)$ be the x_α-direction displacement of the point (x_1, x_2, x_3) of the body, when deforming purely elastic subject to a unit concentrated line-load acting in the x_λ-direction at $(x_1, x_2) = (t, 0)$. Also let

$$G_{\lambda\alpha\beta}(x_1 - t, x_2) = -\frac{2}{\pi r} l_\lambda l_\alpha l_\beta \tag{2.02}$$

be the in-plane stress of the body thus loaded. Here,

$$r = \sqrt{(x_1 - t)^2 + x_2^2}\,; \qquad l_\alpha = \frac{1}{r}(x_1 - t, x_2) = (\cos\theta, \, -\sin\theta) \tag{2.03}$$

are, respectively, the distance between the points (x_1, x_2) and $(t, 0)$, and the unit director pointing from $(t, 0)$ to (x_1, x_2). The stress satisfies the following equilibrium conditions, with 'comma β' denoting the partial derivative with respect to x_β,

$$G_{\lambda\alpha\beta,\beta} = 0 \quad \text{for} \quad x_2 \leq 0; \quad G_{\lambda\alpha 2} = \delta_{\lambda\alpha}\,\delta(x_1 - t) \quad \text{for} \quad x_2 = 0 \tag{2.04}$$

Here $\delta_{\lambda\alpha}$ is Kronecker's delta, and $\delta(x_1 - t)$ is Dirac's delta function. The summation convention is adopted.

To restore the body to its state previous to removal of the slip-band, it is necessary to subject the flanks at $x_1 = b^-$ and $x_1 = b^+$ to tractions $T_\alpha^-(x_2)$ and $T_\alpha^+(x_2)$, respectively, with

$$T_\alpha^-(x_2) + T_\alpha^+(x_2) = 0 \tag{2.05}$$

In this process, the flanks will experience displacements $v_\alpha^-(x_2)$ and $v_\alpha^+(x_2)$, respectively. The tractions are determined by the conditions

$$v_1^-(x_2) + v_1^s(x_2) = v_1^+(x_2); \qquad v_2^-(x_2) = v_2^+(x_2) \tag{2.06}$$

This loading will give rise to fields of displacements $v_\alpha(x_1, x_2)$, and residual stresses $\sigma_{\alpha\beta}(x_1, x_2)$ satisfying

$$\sigma_{\alpha\beta,\beta} = 0 \quad \text{for} \quad x_2 \leq 0; \quad \sigma_{\alpha 2} = 0 \quad \text{for} \quad x_2 = 0 \tag{2.07}$$

From the reciprocity relations of Maxwell-Betti,

$$\int_{S^- - C^s} G_{\lambda\alpha\beta}(x_1 - t, x_2)\, v_{\alpha,\beta}(x_1, x_2)\, dS = \int_{S^- - C^s} g_{\lambda\alpha,\beta}(x_1 - t, x_2)\, \sigma_{\alpha\beta}(x_1, x_2)\, dS$$

Here, S^- is the region $x_2 \leq 0$, taken up by the body, and C^s denotes the slip-band. Application of the divergence theorem is admissible for the region $S^- - C^s$. With (2.04) through (2.08),

$$v_\lambda(t) = \int_{-h}^{0} G_{\lambda 11}(b - t, x_2)\, v_1^s(x_2)\, dx_2 \tag{2.09}$$

Here, $v_\lambda(t) = v_\lambda(t, 0)$ are the surface displacements.

For future purpose we need the derivative $v'_\lambda(t)$, rather than v_λ itself. Using (2.04) and partial integration

$$v'_\lambda(t) = -\int_{-h}^{0} G_{\lambda 11,1}(b - t, x_2)\, v_1^s(x_2)\, dx_2$$

$$= \int_{-h}^{0} G_{\lambda 12,2}(b-t, x_2)\, v_1^s(x_2)\, dx_2$$

$$= -\int_{-h}^{0} G_{\lambda 12}(b-t, x_2)\, \gamma(x_2)\, dx_2 ,$$

where

$$\gamma(x_2) = dv_1^s(x_2)\, dx_2 \tag{2.10}$$

is the magnitude of slip. Finally, cf. (2.03), as $l_1\, dx_2 = -r\, d\theta$,

$$\begin{cases} v_1'(t) = -\dfrac{2}{\pi}\displaystyle\int_0^{\theta_0} \gamma \cos\theta \sin\theta \, d\theta; \quad v_2'(t) = \dfrac{2}{\pi}\displaystyle\int_0^{\theta_0} \gamma \sin^2\theta \, d\theta; \\ \theta = \arctan \dfrac{-x_2}{b-t}; \quad \theta_0 = \arctan \dfrac{h}{b-t} \quad \text{for} \quad t < b. \\ v_1'(t) = \dfrac{2}{\pi}\displaystyle\int_{\theta_0}^{\pi} \gamma \cos\theta \sin\theta \, d\theta; \quad v_2'(t) = -\dfrac{2}{\pi}\displaystyle\int_{\theta_0}^{\pi} \gamma \sin^2\theta \, d\theta; \\ \theta = \pi - \arctan \dfrac{-x_2}{t-b}; \quad \theta_0 = \pi - \arctan \dfrac{h}{t-b} \quad \text{for} \quad t > b. \end{cases} \tag{2.11}$$

Assume for simplicity that γ is constant. In this case

$$v_1'(t) = -\frac{\gamma}{\pi} \frac{h^2}{h \;\; + (b-t)^2}; \quad v_2'(t) = \frac{\gamma}{\pi}\left\{\arctan \frac{h}{b-t} - \frac{h(b-t)}{h^2 + (b-t)^2}\right\} \tag{2.12}$$

It is noted that for $\gamma > 0$ the surface has a wedge at $t = b$ with a jump in inclination of magnitude γ. We assume the wedge to be situated at the leading point of contact. There are no shear stresses available to drive the plastic deformation in front of he contact zone, and the wedge would be flattened under the roller.

3. THE ELASTIC CONTACT PROBLEM

A full account of the problem of contact between an elastic half-plane and a rigid stamp is given by Muskhelishvili[1]. We recall the features of his theory, pertinent to our problem.

The equations of equilibrium and compatibility are satisfied for zero body force with stress components expressed in terms of two analytic functions $\Phi(z)$ and $\Psi(z)$ of the complex argument $z = x_1 + ix_2$ as

$$\frac{\sigma_{11}+\sigma_{22}}{2} = \Phi(z) + \overline{\Phi(z)}; \quad \frac{\sigma_{11}-\sigma_{22}}{2} - i\sigma_{12} = -\bar{z}\Phi'(z) - \Psi(z), \tag{3.01}$$

where an overbar denotes complex conjugation. The functions, assumed to vanish at infinity, are, so far, only defined in the lower half-plane S^-, $x_2 < 0$. Extending the definition of $\Phi(z)$ to the upper half-plane S^+, $x_2 > 0$ by putting

$$\Phi(z) = -\overline{\Phi(\bar{z})} - \overline{z\Phi'(\bar{z})} - \overline{\Psi(\bar{z})} \quad \text{for} \quad z \in S^+ \tag{3.02}$$

we may express $\Psi(z)$ in terms of $\Phi(z)$. The extension of $\Phi(z)$ to the upper half-plane does, moreover, represent the analytic continuation of $\Phi(z)$ across the unloaded part of the x_1-axis.

We contemplate a problem of frictionless contact between a roller and a half-plane. In this case

$$\overline{\Phi(\bar{z})} = -\Phi(z), \tag{3.03}$$

and

$$\frac{\sigma_{11}+\sigma_{22}}{2} = \Phi(z) - \Phi(\bar{z}); \quad \frac{\sigma_{11}-\sigma_{22}}{2} - i\sigma_{12} = (z-\bar{z})\,\Phi'(z). \tag{3.04}$$

Let the shape of the stamp be given by the function $f(t)$. In general, stress singularities will be present at the ends of the line of contact, $t = a$ and $t = b$. If, however, the condition

$$\int_a^b f'(t) \frac{\mathrm{d}t}{\sqrt{(t-a)(b-t)}} = 0 \tag{3.05}$$

is satisfied, then stresses will be bounded everywhere. It is, moreover, required that

$$\int_a^b tf'(t) \frac{\mathrm{d}t}{\sqrt{(t-a)(b-t)}} = \frac{1-\nu}{G}\,Q, \tag{3.06}$$

where Q is the contact force per unit length of the roller, G is the shear modulus, and ν is Poisson's ratio. The conditions (3.05) and (3.06) determine the unknowns a and b.

With a and b thus determined, $\Phi(z)$ may be calculated from

$$\Phi(z) = \frac{G}{2\pi(1-\nu)} \sqrt{(z-a)(b-z)} \int_a^b \frac{f'(t)}{\sqrt{(t-a)(b-t)}} \frac{\mathrm{d}t}{t-z}, \tag{3.07}$$

with $\sqrt{(z-a)(b-z)}$ approaching $\sqrt{(t-a)(b-t)}$ as z approaches t from S^+. Finally, the contact pressure $q(t)$ may be determined from

$$q(t) = \Phi^+(t) - \Phi^-(t), \tag{3.08}$$

where $\Phi^+(t)$ and $\Phi^-(t)$ are the boundary values of $\Phi(z)$ as z approaches t from S^+ and S^-, respectively.

4. THE CONTACT PROBLEM IN PRESENCE OF PLASTIC DISTORTION

We intend to apply the theory of Muskhelishvili[1], outlined in the previous section, to our problem of a rigid roller on an elastic-plastic foundation. The shape of the roller is given by the function

$$f_0(t) = \frac{(t-c)^2}{2R}, \tag{4.01}$$

where c is the x_1-coordinate of the centre of the roller, and R is its radius. The line of contact is ab, where b is situated at the plastically deforming slip-band (cf. section 2), and a at a distance $2l$ behind it,

$$b-a = 2l. \tag{4.02}$$

The shape of the unloaded surface is given by the function $v_2(t)$, and so the 'effective' shape of the stamp is

$$f(t) = f_0(t) - v_2(t), \tag{4.03}$$

whence, cf. (2.12) and (4.01),

$$f'(t) = \frac{t-c}{R} - \frac{\gamma}{\pi}\left\{\arctan\frac{h}{b-t} - \frac{h(b-t)}{h^2+(b-t)^2}\right\} \tag{4.04}$$

It is convenient for the evaluation of the integrals of the previous section to introduce the transformation

$$t = \frac{b\tau^2+a}{\tau^2+1}; \quad \tau = \sqrt{\frac{t-a}{b-t}}, \tag{4.05}$$

whence

$$f'\{t(\tau)\} = \frac{l}{R}\frac{\tau^2-1}{\tau^2+1} + \frac{e}{R} - \frac{\gamma}{2\pi i}\left\{\log\left(-\frac{\tau^2-\alpha^2}{\tau^2-\bar{\alpha}^2}\right) - \frac{\tau^2+1}{\tau^2-\alpha^2} + \frac{\tau^2+1}{\tau^2-\bar{\alpha}^2}\right\}. \tag{4.06}$$

Here,

$$e = \frac{a+b}{2} - c \tag{4.07}$$

is the eccentricity of the line of contact relative to the roller, and

$$\alpha = \sqrt{-1+2il/h} \tag{4.08}$$

is a square root with a positive real part. Thus

$$\int_a^b f'(t)\frac{dt}{\sqrt{(t-a)(b-t)}} = \frac{l}{R}\int_{-\infty}^{\infty}\frac{\tau^2-1}{(\tau^2+1)^2}d\tau + \frac{e}{R}\int_{-\infty}^{\infty}\frac{d\tau}{\tau^2+1} -$$

$$-\frac{\gamma}{2\pi i}\left\{\int_{-\infty}^{\infty}\log\left(-\frac{\tau^2-\alpha^2}{\tau^2-\bar{\alpha}^2}\right)\frac{d\tau}{\tau^2+1}-\int_{-\infty}^{\infty}\left(\frac{1}{\tau^2-\alpha^2}-\frac{1}{\tau^2-\bar{\alpha}^2}\right)d\tau\right\},$$

and

$$\int_a^b\left(t-\frac{a+b}{2}\right)f'(t)\frac{dt}{\sqrt{(t-a)(b-t)}}=\frac{l^2}{R}\int_{-\infty}^{\infty}\frac{(\tau^2-1)^2}{(\tau^2+1)^3}d\tau+$$

$$+\frac{le}{R}\int_{-\infty}^{\infty}\frac{\tau^2-1}{(\tau^2-1)^2}d\tau-\frac{\gamma l}{2\pi i}\left\{\int_{-\infty}^{\infty}\log\left(-\frac{\tau^2-\alpha^2}{\tau^2-\bar{\alpha}^2}\right)\frac{\tau^2-1}{(\tau^2+1)^2}d\tau-\right.$$

$$\left.-\int_{-\infty}^{\infty}\left(\frac{\tau^2-1}{\tau^2-\alpha^2}-\frac{\tau^2-1}{\tau^2-\bar{\alpha}^2}\right)\frac{d\tau}{\tau^2+1}\right\}.$$

These integrals may be evaluated as Cauchy-integrals. Only the logarithms require special attention, due to their lines of discontinuity. In this way we obtain the conditions (3.05) and (3.06) as

$$\frac{\pi e}{R}-\gamma\,\mathrm{Re}\left\{\frac{1}{i}\log\frac{\alpha+i}{\alpha-i}-\frac{1}{\alpha}\right\}=0, \tag{4.09}$$

$$\frac{\pi l^2}{2R}-\gamma l\,\mathrm{Re}\left(\frac{1}{\alpha}\right)=\frac{1-\nu}{G}Q. \tag{4.10}$$

Introducing the parameter ω by

$$\omega=\mathrm{Re}\left\{\frac{1}{i}\log\frac{\alpha+i}{\alpha-i}\right\};\qquad \alpha=\cot\omega+\mathrm{i}\,\mathrm{cosec}\,\omega \tag{4.11}$$

we obtain

$$\frac{\pi e}{R}=\gamma\left(\omega-\frac{\sin\omega\cos\omega}{1+\cos^2\omega}\right), \tag{4.12}$$

$$\frac{\pi l_0}{2R}(\lambda^2-1)=\gamma\lambda\frac{\sin\omega\cos\omega}{1+\cos^2\omega}, \tag{4.13}$$

where l_0 is the semi-length of the line of contact on an undistorted surface,

$$l_0=\sqrt{\frac{2(1-\nu)QR}{\pi G}};\quad \lambda=l/l_0 \tag{4.14}$$

With the mapping

$$z=\frac{b\xi^2+a}{\xi^2+1};\quad \xi=\sqrt{\frac{z-a}{b-z}} \tag{4.15}$$

the expression (3.07) for $\Phi(z)$ transforms into

$$\Phi\{z(\xi)\} = \frac{G\xi}{2\pi(1-\nu)} \int_{-\infty}^{\infty} \frac{f'\{t(\tau)\}}{\tau^2-\xi^2}\, \mathrm{d}\tau \tag{4.16}$$

By Cauchy-integration

$$\Phi\{z(\xi)\} = \frac{Gi}{2\pi(1-\nu)} \left\{\frac{\pi l}{R}\frac{\xi-i}{\xi+i} + \frac{\pi e}{R} - \right.$$

$$\left. - \frac{\gamma}{2i}\left\{\log\left(i\,\frac{\xi+\alpha}{\xi-\bar{\alpha}}\right) - \frac{\xi-\alpha^{-1}}{\xi+\alpha} + \frac{\xi+\bar{\alpha}^{-1}}{\xi-\bar{\alpha}}\right\}\right\} \quad \text{for} \quad z \in S^+ \tag{4.17}$$

For $z \in S^-$, ξ should everywhere be substituted by $-\xi$. The stresses calculated from (4.17) are not the total stresses of the half-plane. The residual stresses from plastic strains remain still to be added.

The distribution of contact pressure is, however, not affected by residual stresses. From (3.08),

$$q\{t(\tau)\} = \frac{G}{\pi(1-\nu)} \left\{\frac{\pi l}{R}\frac{2\tau}{\tau^2+1} - \gamma \operatorname{Re}\left\{\log\frac{\tau+\alpha}{\tau-\alpha} + \frac{\alpha^2+1}{\alpha}\frac{\tau}{\tau^2-\alpha^2}\right\}\right\} \tag{4.18}$$

We still need the moment of the pressure distribution with respect to the centre of the line of contact. The moment is

$$M = \int_a^b q(t)\left(t - \frac{a+b}{2}\right) \mathrm{d}t = 2l^2 \int_{-\infty}^{\infty} q\{t(\tau)\} \frac{\tau(\tau^2-1)}{(\tau^2+1)^3}\, \mathrm{d}\tau, \tag{4.19}$$

which may be evaluated as

$$M = -\frac{Gl^2\gamma}{2(1-\nu)} \operatorname{Re}\left\{\frac{1}{\alpha}\left(1 + \frac{2}{\alpha^2+1} - \frac{4}{(\alpha^2+1)^2}\right)\right\}, \tag{4.20}$$

or, in terms of ω,

$$M = -\frac{Gl^2\gamma}{2(1-\nu)} \frac{\sin\omega\cos^3\omega}{1+\cos^2\omega}. \tag{4.21}$$

The horizontal driving force on the roller is

$$P = \frac{Qe+M}{R}. \tag{4.22}$$

The work of P is consumed in the formation of plastic strains. Note that, with the assumed plastic deformation, the elastic energy of the system remains constant.

From (4.22) we obtain, with (4.12), (4.13),.(4.14), and (4.21),

$$P = \frac{Q\gamma}{2\pi}\left\{\frac{2\omega - \sin 2\omega}{\sin 2\omega} - (\lambda^2 - 1)\frac{\cos^2\omega}{1+\cos^2\omega}\right\}\sin 2\omega \tag{5.01}$$

as an expression for the driving force in steady rolling. Assuming the material of the elastic-plastic body to be non-hardening, the plastic work per unit of overrolled area is

$$P^* = \tau^*\gamma h, \tag{5.02}$$

where $\tau^* > 0$ for $\gamma > 0$ is the yield stress in shear. From (4.08) and (4.14),

$$h = l_0 \lambda \frac{\sin^2\omega}{\cos\omega}. \tag{5.03}$$

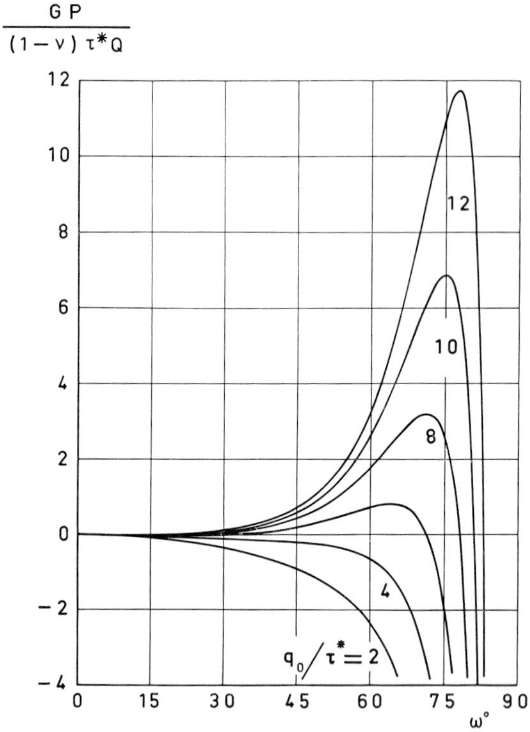

Fig. 1. Rolling resistance P as function of the parameter ω for various values of nominal pressure q_0

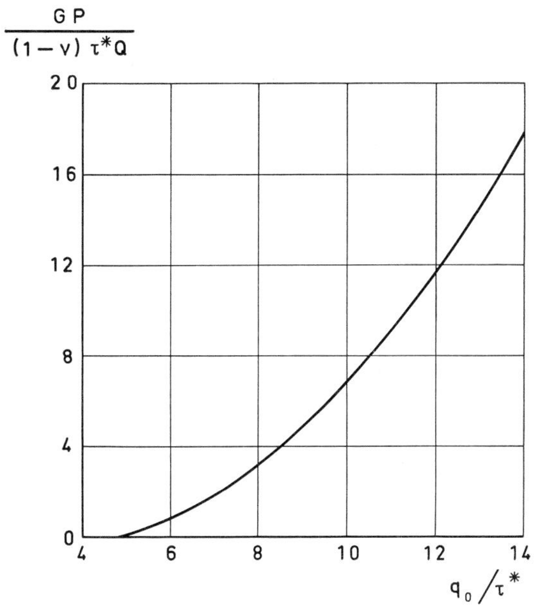

Fig. 2a. Rolling resistance P corresponding to maximum plastic work, as function of nominal pressure q_0

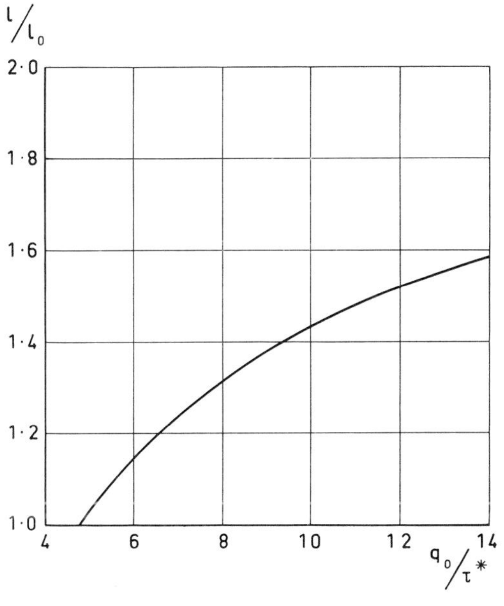

Fig. 2b. Semi-contact length 1 as function of nominal pressure q_0

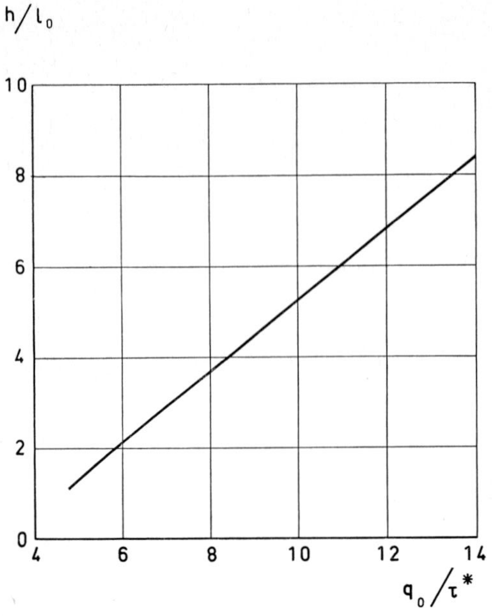

Fig. 2c. Depth of plastic zone h as function of nominal pressure q_0

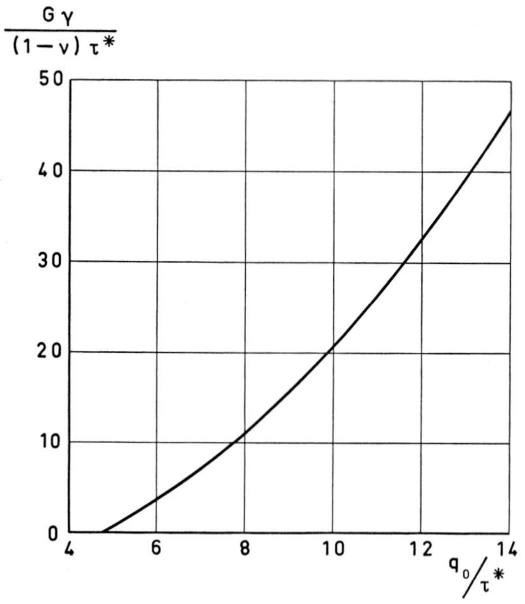

Fig. 2d. Magnitude of shear γ as function of nominal pressure q_0

Equating the work of P and the plastic work we obtain

$$\frac{2\omega - \sin 2\omega}{\sin 2\omega} - (\lambda^2 - 1)\frac{\cos^2\omega}{1+\cos^2\omega} = 2\frac{\tau^*}{q_0}\lambda\frac{\sin\omega}{\cos^2\omega}, \tag{5.04}$$

where

$$q_0 = \frac{2Q}{\pi l_0} = \sqrt{\frac{2GQ}{\pi(1-\nu)R}} \tag{5.05}$$

is a nominal pressure, viz. the maximum pressure on an undistorted surface.

Once the parameter λ is determined in terms of ω and q_0 from (5.04), we may calculate P, using (4.13), as

$$P = \frac{(1-\nu)\tau^* Q}{G}(\lambda^2 - 1)\frac{1+\cos^2\omega}{\cos^2\omega}\sin\omega. \tag{5.06}$$

Fig. 1 shows P as a function of ω for several values of q_0. Negative values of P correspond to negative values of γ, which are inconsistent with the shape of the roller (also then τ^* should be negative). The assumed type of plastic deformation cannot take place under steady rolling, unless the nominal pressure is sufficiently high to produce a range of ω with positive values of P. The critical value is

$$q_0^* = 4.77\tau^*. \tag{5.07}$$

This is, incidentally, pretty close to the value of q_0 at which out-of-plane plastic strains may be expected for values of ν in the vicinity of 0.3.

For pressures above q_0^* there is a range of values of ω for which a steady state is possible. There is one distinct value, at which the driving force attains a maximum. We propose an assumption of maximum plastic work (or dissipation), which ties the steady state to this particular value of ω. Values of P, l, h, and γ, consistent with this assumption, are plotted versus q_0 in Figs. 2a–d.

We mention, finally, that the problem, in which the elastic-plastic body is circular cylindrical, while the roller is elastic rather than rigid, may be solved similarly. It is, actually, only necessary to substitute $1/R$ by

$$\frac{1}{R_e} = \frac{1}{R_1} + \frac{1}{R_2}, \tag{5.08}$$

where R_1 and R_2 are the respective radii, and $(1-\nu)/G$ by

$$\left(\frac{1-\nu}{G}\right)_e = \frac{1-\nu_1}{G_1} + \frac{1-\nu_2}{G_2}. \tag{5.09}$$

Here, G_1 and G_2 are the respective shear moduli, while ν_1 and ν_2 are the respective Poisson's ratios.

REFERENCES

1. N.I. Muskhelishvili, *Singular Integral Equations* (English translation by J.R.M. Radok), Noordhoff, Groningen, 1953, ch. 13.
2. J.E. Merwin and K.L. Johnson, 'An Analysis of Plastic Deformation in Rolling Contact', *Proc. Instn. Mech. Engrs.*, **177**, 1963. pp. 676–685.
3. I.F. Collins, 'A Simplified Analysis of the Rolling of a Cylinder on a Rigid/Perfectly Plastic Half-Space', *Int. J. Mech. Sci.*, **14**. 1972, pp. 1–14.
4. G.M. Hamilton, 'Plastic Flow in Rollers Loaded above the Yield Point', *Proc. Inst. Mech. Engrs*, Lond., **177**, 25, 1963.
5. K.L. Johnson, 'Rolling Resistance of a Rigid Cylinder on an Elastic-Plastic Surface', *Int. J. Mech. Sci.*, 1972, pp. 145–148.

Heat effects in rolling contact

F. F. Ling

1. INTRODUCTION

By thermal effects, it is meant the effects on the mechanical behavior of rolling bodies within the framework of linear elasticity. Moreover, the material is assumed to be homogeneous and isotropic. For many technically important rolling problems fundamental solutions, or their approximations, dealing with thermoelastic response of the half-space are useful. In this paper two such solutions are shown. In addition, examples are shown of their applications.

2. SURFACE DISPLACEMENT OF THE HALF-SPACE

A solution of the normal displacement of the elastic half-space under an arbitrarily distributed fast moving heat source of constant velocity within the two-dimensional quasi-static, uncoupled thermoelasticity theory is given here. The surface of the half-space is allowed to dissipate heat by convection.

Mechanically, the equations of equilibrium, the stress-displacement relations and the boundary conditions are, respectively [1, 2],

$$(r+2)\frac{\partial^2 u}{\partial \xi^2}+\frac{\partial^2 u}{\partial \eta^2}+(r+1)\frac{\partial^2 v}{\partial \xi \partial \eta}=(3r+2)\,\beta\,\frac{\partial \varphi}{\partial \xi}$$

$$\frac{\partial^2 v}{\partial \xi^2}+(r+2)\frac{\partial^2 v}{\partial \eta^2}+(\gamma+1)\frac{\partial^2 u}{\partial \xi \partial \eta}=(3r+2)\,\beta\,\frac{\partial \varphi}{\partial \eta}, \tag{1}$$

$$\left.\begin{aligned}\sigma_\xi &= (r+2)\frac{\partial u}{\partial \xi}+r\frac{\partial v}{\partial \eta}-(3r+2)\,\beta\varphi\\ \sigma_\eta &= r\frac{\partial u}{\partial \xi}+(r+2)\frac{\partial v}{\partial \eta}-(3r+2)\,\beta\varphi\\ \sigma_{\xi\eta} &= \frac{\partial u}{\partial \eta}+\frac{\partial v}{\partial \xi},\end{aligned}\right\} \tag{2}$$

$$\sigma_{\xi\eta} = \sigma_\eta = 0 \qquad (\xi = \xi,\ \eta = 0), \tag{3}$$

$$\sigma_\xi,\ \sigma_\eta,\ \sigma_{\xi\eta},\ \sigma,\ v \to 0 \quad [(\xi^2+\eta^2)^{1/2} \to \infty]. \tag{4}$$

In the above, and referring to Fig. 1, $\xi \equiv x_1/l$; $\eta \equiv x_2/l$; $r \equiv \lambda/\mu$ with λ and μ being the Lamé constants; $u \equiv u_1/l$ and $v \equiv u_2/l$ with u_1 and u_2 being the

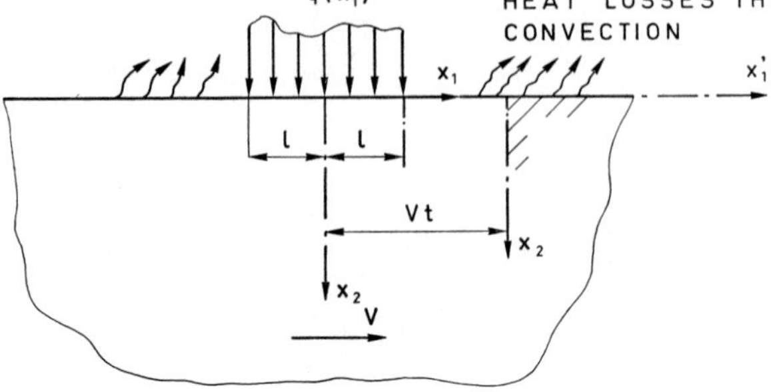

Fig. 1. Schematic diagram of the mechanical model

displacements in the x_1 and x_2 directions, respectively; $\beta \equiv \alpha q_0 l/K$ with α being the linear coefficient of thermal expansion, q_0 a measure of heat flux per unit area, and K the thermal conductivity; $\varphi \equiv TK/q_0 l$ with T the temperature above ambient; $\sigma_\xi \equiv \sigma_{xx}/\mu$, $\sigma_\eta \equiv \sigma_{xy}/\mu$ and $\sigma_\eta \equiv \sigma_{yy}/\mu$ with σ's being the stress components.

Thermally, the heat equation referred to the coordinates (x_1, x_2) which move with a constant velocity V with respect to the half-space is

$$\frac{\partial^2 \varphi}{\partial \xi^2} + \frac{\partial^2 \varphi}{\partial \eta^2} = R\,\frac{\partial \varphi}{\partial \xi}. \tag{5}$$

The boundary conditions are

$$\frac{\partial \varphi}{\partial \eta} = \begin{cases} -P(\xi) & (0 \le |\xi| < 1,\ \eta = 0) \\ H\varphi & (1 < |\xi| < \infty,\ \eta = 0), \end{cases} \tag{6}$$

$$\varphi \to 0 \quad \text{as} \quad (\xi^2+\eta^2)^{1/2} \to \infty. \tag{7}$$

In the above $R \equiv Vl/\kappa$ with κ being the thermal diffusivity; $H \equiv hl/K$ with h being the film coefficient of convection heat transfer; $P \equiv q/q_0$ with q being the arbitrarily distributed heat input at the surface.

For $R \equiv Vl/\kappa \gg 1$, $\partial^2 \varphi / \partial \xi^2$ may be neglected and (5) becomes

$$\frac{\partial^2 \varphi}{\partial \eta^2} = R \frac{\partial \varphi}{\partial \xi}, \tag{8}$$

subject to the boundary conditions (6) and (7).

Taking the Fourier transform and denoting the transformed quantities by a superposed $\sim$, (1) through (8) except (5) become, with some grouping

$$\begin{aligned} &\frac{\partial^2 \tilde{u}}{\partial \eta^2} - s^2(r+2)\,\tilde{u} - is(r+1)\frac{\partial \tilde{v}}{\partial \eta} = -\,is(3r+2)\,\beta\tilde{\varphi} \\ &(r+2)\frac{\partial^2 \tilde{v}}{\partial \eta^2} - s^2\tilde{v} - is(r+1)\frac{\partial \tilde{u}}{\partial \eta} = (3r+2)\frac{\partial \tilde{\varphi}}{\partial \eta}, \end{aligned} \tag{9}$$

$$\left.\begin{aligned} \tilde{\sigma}_\xi &= -\,is(r+2)\,\tilde{u} + r\frac{\partial \tilde{v}}{\partial \eta} - (3r+2)\,\beta\tilde{\varphi} \\ \tilde{\sigma}_\eta &= -\,isr\tilde{u} + (r+2)\frac{\partial \tilde{v}}{\partial \eta} - (3r\Psi+2)\,\beta\tilde{\varphi} \\ \tilde{\sigma}_{\xi\eta} &= \frac{\partial \tilde{u}}{\partial \eta} - is\tilde{v}, \end{aligned}\right\} \tag{10}$$

$$\frac{\partial^2 \tilde{\varphi}}{\partial \eta^2} = -\,isR\tilde{\varphi}, \tag{11}$$

$$\tilde{\sigma}_{\xi\eta} = \tilde{\sigma}_\eta = 0 \qquad (\eta = 0), \tag{12}$$

$$-\frac{\partial \varphi}{\partial \eta} = \tilde{P} \equiv \pi^{-1/2}\int_{-\infty}^{\infty} P^*(\xi)\,\mathrm{e}^{is\xi}\,\mathrm{d}\xi \quad (\eta = 0), \tag{13}$$

$$\tilde{\sigma}_\xi,\ \tilde{\sigma}_\eta,\ \tilde{\sigma}_{\eta\xi},\ \tilde{u},\ \tilde{v},\ \tilde{\varphi} \to 0 \qquad (\eta \to \infty), \tag{14}$$

where

$$P^*(\xi) \equiv \begin{cases} P(\xi) & (0 \le |\xi| \le 1) \\ -H\varphi(\xi, 0) & (1 < |\xi| < \infty). \end{cases} \tag{15}$$

First, the heat equation is solved. Let $\tilde{\varphi} = \tilde{\varphi}_1 + i\tilde{\varphi}_2$ and $\tilde{P} = \tilde{P}_1 + i\tilde{P}_2$, where $\tilde{\varphi}_1$, $\tilde{\varphi}_2$, $\tilde{P}_1$, and $\tilde{P}_2$ are real. Equations (11), (13) and (14) lead to, for $s > 0$,

$$\begin{aligned} \tilde{\varphi}_1 &= (2sR)^{-1/2}\,\mathrm{e}^{-\sqrt{sR/2}\eta} \times \\ &\quad \times \left[(\tilde{P}_1 - \tilde{P}_2)\cos\sqrt{\frac{sR}{2}}\eta - (\tilde{P}_1 + \tilde{P}_2)\sin\sqrt{\frac{sR}{2}}\eta\right] \\ \tilde{\varphi}_2 &= (2sR)^{-1/2}\,\mathrm{e}^{-\sqrt{sR/2}\eta} \times \end{aligned}$$

$$\times\left[(\tilde{P}_1+\tilde{P}_2)\cos\sqrt{\frac{sR}{2\eta}}+(\tilde{P}_1-\tilde{P}_2)\sin\sqrt{\frac{sR}{2\eta}}\right]. \tag{16}$$

For $s<0$, (16) may be used, provided s is replaced by $n\equiv -s$ and a negative sign is added to $\tilde{\varphi}_2$.

Proceed now to the solution of the displacement equations. From (9)

$$is\tilde{u} = s^{-2}(r+1)^{-1}\frac{\partial^3\tilde{v}}{\partial\eta^3}+r(r+1)^{-1}\frac{\partial\tilde{v}}{\partial\eta}-s^{-2}(r+1)\,m\frac{\partial^2\tilde{\varphi}}{\partial\eta^2}-m\tilde{\varphi}, \tag{17}$$

and

$$\frac{\partial^4\tilde{v}}{\partial\eta^4}-2s^2\frac{\partial^2\tilde{v}}{\partial\eta^2}+s^4\,\tilde{v}=m\frac{\partial^3\tilde{\varphi}}{\partial\eta^3}-s^2 m\tilde{\varphi}, \tag{18}$$

where

$$m=\frac{(3r+2)\,\beta}{r+z}.$$

With (16) in (18), the latter may be solved and the complimentary solution is

$$\tilde{v}_c=(A_1+iA_2)\,e^{-s\eta}+(B_1+iB_2)\,\eta\,e^{-s\eta}, \tag{19}$$

where A_1, A_2, B_1, and B_2 are constants of integration, and condition (14) has been taken into consideration without the benefit of the details concerning the coefficients.

For the particular integral,

$$\tilde{v}_p=m(s^2+R^2)^{-1}\,e^{-\sqrt{sR/2\eta}}\left\{\left[P_1+\left(\frac{R}{s}\right)P_2\right]+\right.$$

$$\left.+i\left[-\left(\frac{R}{s}\right)\tilde{P}_1+\tilde{P}_2\right]\right\}\cos\sqrt{\frac{sR}{2\eta}}+$$

$$+\left\{\left[\left(\frac{R}{s}\right)\tilde{P}_1-\tilde{P}_2\right]+i\left[\tilde{P}_1+\left(\frac{R}{s}\right)\tilde{P}_2\right]\right\}\sin\sqrt{\frac{sR}{2\eta}}. \tag{20}$$

Enforcement of (12) leads to expressions of A_1, A_2, B_1, and B_2.

For $s<0$, with $n\equiv -s$, A_1 and B_1 take on the same form but with n replacing s; A_2 and B_2 take on the same form but with n replacing s and multiplied by a negative sign.

Inversion of (16) yields

$$\varphi_0=\pi^{-1}\sqrt{\frac{2}{R}}\int_{-\infty}^{\infty}P^*(\xi')\,d\xi'\int_0^{\infty}[\cos(\xi-\xi')\,s+\sin(\xi-\xi')\,s]\,\frac{ds}{\sqrt{s}}, \tag{21}$$

where φ_0 denoted φ for $\eta \to 0$. Equation (21) may be shown to be

$$\varphi_0(\xi) = 2(\pi R)^{-1/2} \int_{-\infty}^{\infty} P^*(\xi')\, K_\varphi^*(\xi, \xi')\, d\xi', \tag{22}$$

where

$$K_\varphi^* = \begin{cases} (\xi - \xi')^{1/2} & [(\xi - \xi') \geq 0] \\ 0 & [(\xi - \xi') < 0]. \end{cases} \tag{23}$$

Recalling the definition of $P^*(\xi)$ in (15),

$$\varphi_0(\xi) = \begin{cases} 0 & (q\infty \leq \xi \leq -1) \\ 2(\pi R)^{-1/2} \displaystyle\int_{-1}^{\xi} P(\xi')\,(\xi - \xi')^{-1/2}\, d\xi' & (-1 \leq \xi \leq 1) \\ 2(\pi R)^{-1/2} \displaystyle\int_{-1}^{1} P(\xi')\,(\xi - \xi')^{-1/2}\, d\xi' - H \int_{1}^{\xi} \times & \\ \quad \times\, \varphi_0(\xi')\,(\xi - \xi')^{1/2}\, d\xi' & (1 < \xi \leq \infty). \end{cases} \tag{24}$$

Inverse transform of the displacement component ($v = v_1 + iv_2$), similar treatment as φ_0, $v_0(\xi)$ which is $v(\xi, \eta)$ for $\eta \to 0$ may be shown to be [2]

$$v_0(\xi) = mR^{-1}(r+2)\,(r+1)^{-1} \int_{-1}^{\infty} P^*(\xi')\, K^*(\xi, \xi')\, d\xi', \tag{25}$$

where

$$K^* = \begin{cases} -2 & [(\xi - \xi') \geq 0] \\ -2\, e^{-R(\xi - \xi')} \operatorname{erfc} \sqrt{R(\xi - \xi)} & [(\xi - \xi') \leq 0]. \end{cases} \tag{26}$$

Though an approximate solution in that the (5) has been replaced by (8), the solution has been studied numerically by several individuals. It was found that, for a variety of problems, (8) gives the same results as (5) provided $R > 1.3$.

3. TEMPERATURE DISTRIBUTION IN THE HALF-SPACE

For this section, refer to Fig. 1 again, ignoring the convection losses and, for the ease of representation for the solution, replacing x_1 by $x_1 + l$ (i.e. shifting the origin by l to the left).

The governing heat equation for the high speed approximation, (8) which is now written somewhat differently:

$$\frac{\partial^2 \theta}{\partial \eta^2} = R \frac{\partial \theta}{\partial \xi}, \tag{27}$$

with the associated boundary conditions

$$-\frac{\partial\theta}{\partial\eta} = \begin{cases} Q(\xi) & (0 \le \xi \le 2A,\ \eta = 0) \\ 0 & (-\infty < \xi < 0,\ 2A < \xi < \infty,\ \eta = 0), \end{cases} \tag{28}$$

$$\theta \to 0 \quad [(\xi^2 + \eta^2)^{1/2} \to \infty], \tag{29}$$

where the following dimensionless quantities have been introduced: $\theta \equiv TK/q_0 l_0$, $R \equiv Vl_0/\kappa$, $A \equiv l/l_0$, $\xi \equiv x_1/l_0$, $\eta \equiv x_2/l_0$, $Q \equiv q/q_0$ and l_0 is reference length.

Analysis

The fundamental solution of the above system, i.e. the Green's function, is the classical Laplace solution

$$\theta = \begin{cases} 0 & (\xi < 0) \\ (\pi R)^{-1/2} \int_0^{\xi} Q(\xi')\,(\xi - \xi')^{-1/2} \exp[-R\eta^2/4(\xi - \xi')]\,\mathrm{d}\xi' & (0 \le \xi \le 2A) \\ (\pi R)^{-1/2} \int_0^{2A} Q(\xi')\,(\xi - \xi')^{-1/2} \exp[-R\eta^2/4(\xi - \xi')]\,\mathrm{d}\xi' & (\xi \ge 2A). \end{cases} \tag{30}$$

Since $Q(\xi)$ is arbitrary, it may be represented by a Fourier series

$$Q(\xi) = \sum_{n=1}^{\infty} Q_n \sin\frac{n\pi\xi}{2A} \qquad (0 \le \xi \le 2A). \tag{31}$$

Letting θ_n be defined in such a way that

$$\theta = \sum_{n=1}^{\infty} \theta_n, \tag{32}$$

then θ_n may be obtained from (30) for the nth term of the series representation (31). It should be noted that $Q(\xi)$ is capable of representation by a cosine series as an alternative.

Equation (30) for a general term of (31) is

$$\theta_n = \begin{cases} 0 & (\xi < 0) \\ Q_n(\pi R)^{-1/2} I_1 & (0 \le \xi \le 2A) \\ Q_n(\pi R)^{-1/2} I_2 & (\xi > 2A), \end{cases} \tag{33}$$

where

$$I_1 = \int_0^{\xi} \sin\frac{n\pi\xi'}{2A}\,(\xi - \xi')^{-1/2} \exp[-R\eta^2/4(\xi - \xi')]\,\mathrm{d}\xi'$$

$$I_2 = \int_0^{2A} \sin\frac{n\pi\xi'}{2A}\,(\xi-\xi')^{-1/2}\exp[-R^2\eta/4(\xi-\xi')]\,\mathrm{d}\xi'.$$

Now I_1 may be written as

$$I_1 = \tfrac{1}{2}\sin\frac{n\pi\xi}{2A}[I_{11}+I_{12}] + \frac{i}{2}\cos\frac{n\pi\xi}{2A}[I_{11}-I_{12}], \tag{34}$$

where

$$I_{11} = \int_0^{\xi} \xi'^{-1/2}\exp\left(i\frac{n\pi\xi'}{2A} - \frac{R\eta'}{4\xi'}\right)\mathrm{d}\xi$$

$$I_{12} = \int_0^{\xi} \xi'^{-1/2}\exp\left(-i\frac{n\pi\xi'}{2A} - \frac{R\eta^2}{4\xi'}\right)\mathrm{d}\xi'.$$

I_{11} may be expressed as

$$I_{11} = 2\int_0^{\sqrt{\xi}} \exp\left[-c_{11}^2\left(x^2+\frac{a_1^2}{x^2}\right)\right]\mathrm{d}x, \tag{35}$$

where

$$\begin{cases} c_{11} = (n\pi/4A)^{1/2}(1-i) \\ a_1 = (AR/4n\pi)^{1/2}(1+i)\,\eta. \end{cases}$$

Differentiating (35) twice with respect to a_1, the following linear second order, ordinary differential equation may be obtained:

$$\frac{\mathrm{d}^2 I_{11}}{\mathrm{d}a_1^2} - 4c_{11}^4 I_{11} = 4c_{11}^2\frac{1}{\sqrt{\xi}}\exp\left[-c_{11}^2\left(\frac{a^2}{\xi}+\xi\right)\right].$$

This equation possesses the solution[3] under the restriction that I_{11} remains finite and it takes on the limiting value at $a_1 = 0$. Similarly, a differential equation exists for I_{22}.

Equation (34) may now be put into the form

$$I_1 = I_1^0 + I_1^1 + I_1^2, \tag{36}$$

where

$$I_1^0 = \sqrt{\left(\frac{A}{n}\right)}\,\mathrm{e}^{-Y}[\sin\Psi\, c(z_0) - \cos\Psi\, s(z_0)]$$

$$I_1^1 = \frac{1}{2}\sqrt{\left(\frac{A}{n}\right)}\,e^{-Y}\{\sin\Psi[c(\bar{z})+c(z)] - \cos\Psi[s(\bar{z})+s(z)]\}$$

$$I_1^2 = \frac{i}{2}\sqrt{\left(\frac{A}{n}\right)}\,\mathrm{e}^{-Y}\{\cos\Psi[c(\bar{z})-c(z)] + \sin\Psi[s(\bar{z})-s(z)]\}.$$

Or

$$I_1 = \sqrt{\left(\frac{A}{n}\right)}\, e^{-Y}\{\sin \Psi[c(z_0) + \text{Re}\, c(z) + \text{Im}\, s(z)] - $$
$$- \cos \Psi[s(z_0) + \text{Re}\, s(z) - \text{Im}\, c(z)\},$$

where

$$Y = \left(\frac{n\pi R}{A}\right)^{1/2} \eta, \qquad \Psi = \frac{n\pi\xi}{2A} - Y$$

$$z_0 = \left(\frac{n\xi}{A}\right)^{1/2}$$

$$z = \left(\frac{n}{A\xi}\right)^{1/2} \left\{\left[\xi - \left(\frac{AR}{4n\pi}\right)^{1/2} \eta\right] + i\left(\frac{AR}{4n\pi}\right)^{1/2} \eta\right\}.$$

In the same manner

$$I_2 = \sqrt{\left(\frac{A}{n}\right)}\, e^{-Y}\{\sin \Psi[c(z_0) + \text{Re}\, c(z) + \text{Im}\, s(z) - c(z_{02}) - $$
$$- \text{Re}\, c(z_2) - \text{Im}\, s(z_2)] - \cos \Psi[s(z_0) + \text{Re}\, s(z) - \text{Im}\, c(z) - $$
$$- s(z_{02}) - \text{Re}\, s(z_2) + \text{Im}\, c(z_2)]\},$$

where

$$z_{02} = \left[n\left(\frac{\xi}{A} - 2\right)\right]^{1/2},$$

$$z_2 = \left[\frac{n}{A(\xi - 2A)}\right]^{1/2} \left\{\left[\xi - 2A - \left(\frac{AR}{4\pi n}\right)^{1/2} \eta\right] + i\left(\frac{AR}{4n\pi}\right)^{1/2} \eta\right\}.$$

Therefore,

$$\theta_n = \begin{cases} 0 & (\xi < 0) \\ Q_n\left(\frac{A}{n\pi R}\right)^{1/2} e^{-Y}\{\sin \Psi[c(z_0) + \text{Re}\, c(z) + \text{Im}\, s(z)] - \\ \quad - \cos \Psi[s(z_0) + \text{Re}\, s(z) - \text{Im}\, c(z)]\} & (0 \le \xi \le 2A) \\ Q_n\left(\frac{A}{n\pi R}\right)^{1/2} e^{-Y}\{\sin \Psi[c(z_0) + \text{Re}\, c(z) + \text{Im}\, s(z) - c(z_{02}) - \\ \quad - \text{Re}\, c(z_2) - \text{Im}\, s(z_2)] - \cos \Psi[s(z_0) + \\ \quad + \text{Re}\, s(z) - \text{Im}\, c(z) - s(z_{02}) - \\ \quad - \text{Re}\, s(z_2) + \text{Im}\, c(z_2)]\}, & (\xi < 2A). \end{cases} \quad (37)$$

4. EXAMPLES

As an example of the application of (25), it is used in an elastohydrodynamic calculation. Here heat generated within a fluid film is being considered in this rolling situation.

Table 1 gives a typical set of data of $q(x)$ encountered in elastohydrodynamic studies. In the table let $l = 0.0125$ in.; $\alpha = 0.5 \times 10^{-6}$in./in. deg F;

Table 1. q(x) data encountered in elastohydrodynamics

x (in.)	$q(x)$ (Btu/in.^{2} hr) $\times 10^{-4}$
−0.012500	0.2015
−0.087500	0.3139
−0.005000	2.3780
−0.002500	6.0700
0	12.5300
0.002500	10.9700
0.005000	16.2000
0.007500	5.5130
0.008125	2.3890
0.008750	1.2430
0.009375	0.6064
0.010000	0.2714
0.011250	0.0565
0.012500	0.1069

$V = 150$ in./sec.; Young's modulus $E = 30 \times 10^6$ lb/in.^{2} and Poisson's ratio $\nu = 0.33$ for which $\lambda = E\nu(1-2\nu)^{-1}(1+\nu)^{-1} = 21.9 \times 10^6$ lb/in.^{2} and $\mu = E2^{-1}(1+\nu)^{-1} = 11.3 \times 10^6$ lb/in.^{2}; $\kappa = 0.0196$ in.^{2}/sec; $\rho = 0.283$ lb/in.^{3}; $C = 0.11$ Btu/lb deg F. Using the data in Table 1, $\varphi_0(\xi)$ is found by iteration for the region $1 \le \xi \le \infty$. With $\varphi(\xi)$ thus found $P^*(\xi)$ can now be found:

$$P^*(\xi) = \begin{cases} P(\xi) & (-1 \le \xi \le 1) \\ -H\varphi_0(\xi) & (-1 < \xi \le \infty). \end{cases}$$

With $P^*(\xi)$ in hand, (25) is used to find $v_0(\xi)$ for $-1 \le \xi \le 1$ numerically The result is shown in Fig. 2, with the displacement expressed in physical units. The solid curve represents the zeroth approximation; i.e., the case of $H = 0$; the correction of the higher approximation (as closely as desired) is shown by the dotted portion. For fast-moving body, i.e., $R \gg 1$, the correction is not large and not far reaching. Therefore, for all practical purposes, as long as $R \gg 1$, the simpler zeroth approximation should suffice for $-1 \le \xi \le 1$.

In the first example, rolling is associated with no sliding within the so-called elastohydrodynamic theory[4]. In many situations, rolling is associated with a good deal of sliding. Moreover, surface roughness plays a dominant role in many cases. In such cases plastic collapse of asperities

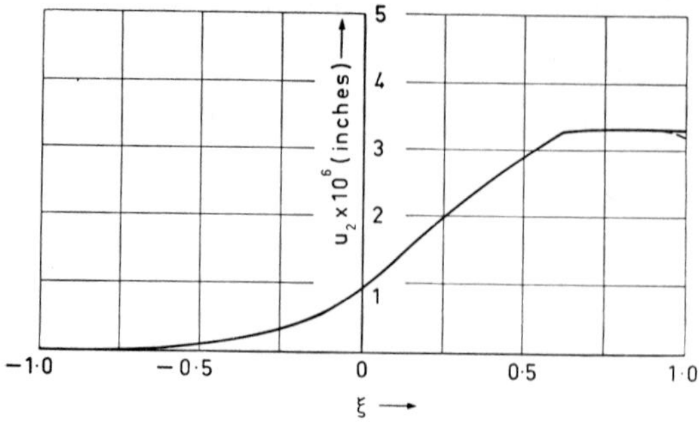

Fig. 2. Vertical displacement $U_2(\xi_1 0)$ *versus* ξ *for* $1 < \xi$ *for* $< \xi \leq 1$ *and data given in Table 1*

on surfaces in sliding contact, which are separated by a fluid, causes the nature of the sliding process to change from the state wherein the asperities are deformed only elastically. The outward manifestation may be a relatively high frictional resistance in the elastic case and a decreasing frictional resistance subsequent to such a collapse. Hence of interest is the influence of the thermo-processes upon the collapse of these asperities.

The thermo-mechanical model of an asperity is that of an ideal elasto-plastic semi-infinite solid with temperature dependent yield stress. The material is taken to be isotropic and in plane strain. All other moduli are considered temperature independent. Although the speed is high based on thermal consideration, i.e., the Péclét number $R \equiv Vl_0/\kappa \gg 1$, it is still negligible when compared with the speeds of propagation of mechanical disturbances, for example, the speed of propagation of the shear waves. Thus the mechanical inertia can still be neglected.

In this model, the coefficient of friction in the small, λ, is assumed to be load and temperature independent. The body is under the mechanical loading of $p(x_1)$ and $\lambda p(x_1)$ in the region $0 \leq x_1 \leq 2l$. It is assumed that the heat flux into the body is given by the expression $V\lambda p(x_1)$. An analysis of the model described above will establish the incipient plastic yielding on the surface. It should be noted that it is the physical spreading of the material under yielding which changes certain interface phenomenon. This in turn causes a decrease in the frictional coefficient. More elaborate models which take these and other contact phenomena into consideration are available but would be outside of the province of this paper[5].

Inasmuch as the appropriate temperature field is to be used in the uncoupled thermoplasticity theory with temperature dependent yield stress, it stands to reason the relevant temperature would be such that the thermal properties are also temperature dependent. For the case where thermal conductivity $K = K_0 f(\theta)$ and the product of density and specific heat $\rho C = \rho_0 C_0 g(\theta)$, where the subscript refers to a reference state, it has been shown[6] that such temperature dependence, however strong, may be ignored so long as $(fg)^{-1/2} \sim 1$. With the above provisal, the use of the heat equation with thermal properties which are temperature independent in a thermo-elastoplastic analysis where the yield stress is temperature dependent may not be inconsistent.

A typical contour map of the temperature field is given in Fig. 3. This is for the case $R = 444$ and $q(\xi)$ is parabolically distributed in $0 \leq \xi \leq 2A$, $A = 1$.

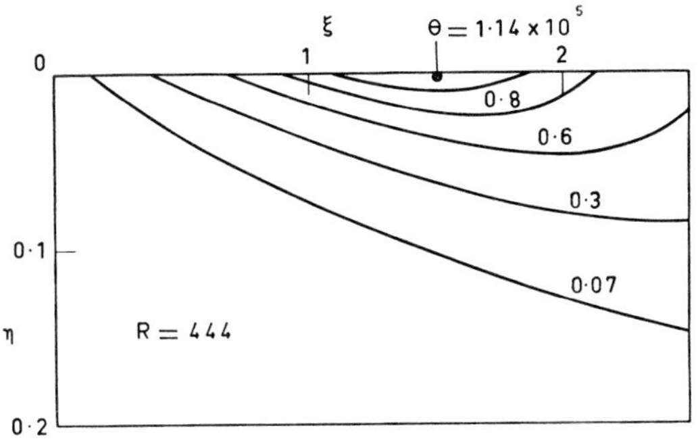

Fig. 3. Contour map of a typical temperature field in dimensionless measure $\theta(\xi, \eta)$

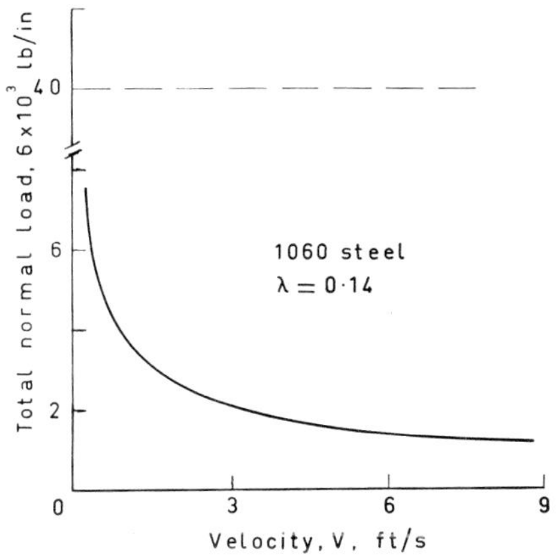

Fig. 4. Total load-velocity curve as demarcation for incipient thermoelastic plastic yielding
Coefficient of friction = 0.14
Dotted line for incipient elastoplastic yielding

A thermo-elastoplastic analysis has been made in which the yield stress is a function of temperature. For this example the data for 1060 steel[7] were used. Young's modulus is 30×10^6 psi and Poisson's ratio is 0.3. Fig. 4 shows the total load vs. velocity curve which is the demarcation for incipient thermo-elastoplastic yielding. This is for the case where the coefficient of friction in the small is 0.14. The dotted line is for incipient yielding based on isothermal theory.

Note that Fig. 4 has a break near the ordinate value of 8. This is done to give some detail of the load-velocity curve while providing a measure relative to the case of isothermal yielding, i.e., elastoplastic. It is clear, within magnitudes of loads and velocities encountered in practice, that the thermal effects reported here, if neglected, could lead to drastically different conclusions.

REFERENCES

1. F.F. Ling, *Surface Mechanics*, Wiley Interscience, John Wiley & Sons, New York, 1973, pp. 125–131.
2. F.F. Ling and V.C. Mow, 'Surface Displacement of a Convective Elastic Half-Space Under an Arbitrarily-Distributed Fast-Moving Heat Source', *Transactions of the American Society of Mechanical Engineers*, **87**, 1965, pp. 729–734.
3. F.F. Ling and C.C Yang, 'Temperature Distribution in a Semi-Infinite Solid Under. a Fast-Moving Arbitrary Heat Source', *International Journal of Heat and Mass Transfer* **14**, 1971, pp. 199–206.
4. D. Dowson, *Elastohydrodynamic Lubrication*, Institute of Mechanical Engineers, London, 1967.
5. C.C. Yang, 'Incipient Plastic Yielding of an Elastoplastic Half-Space Under an Arbitrarily Distributed Moving Heat Source', *International Journal of Engineering Science*, **9**, 1971, pp. 507–520.
6. F.F. Ling and J.S. Rice, 'Surface Temperature with Temperature-Dependent Thermal Properties', *Transactions of the American Society of Lubrication Engineers*, **9**, 1966, pp. 195–201.
7. B.A. Boley and J.H. Weiner, *Theory of Thermal Stresses*, John Wiley & Sons, New York, 1960.

Thermoelastic contact problems

J. R. Barber

1. INTRODUCTION

This paper will be concerned with problems of elastic contact in the presence of heat flow. A number of practical applications of this type have recently been discussed, in which the heat flow is due to a difference in temperature between the contacting solids[1, 2], or to frictional heat generation at a sliding interface[3, 4]. In the former case, thermoelastic distortion will generally cause the contact geometry to change with the magnitude and direction of heat flow, giving rise to the phenomenon of 'thermal rectification' if the two solids have different material properties. Frictional heat generation also produces a change in contact geometry and in suitable circumstances can cause a cyclic variation in contact pressure known as 'thermoelastic instability'[5, 6, 7].

The thermoelastic contact problem is very much simplified if the temperature field can be assumed to have reached a steady state, since in this case it is possible to express the stress and displacement field in terms of one or more harmonic potential functions. We shall restrict our attention to problems of this type, but it is worth noting that time dependent solutions can be obtained in integral form provided that the thermal boundary conditions are unmixed, making use of the properties of the instantaneous point source[8].

Suitable solutions of the steady state thermoelastic equations are developed in section 2 and applied to the contact problem in section 3. Section 4 is devoted to some particular examples, with special reference to the case where the contact area is circular, but the boundary conditions are not axisymmetric. Finally, some consideration is given in section 5 to situations in which the extent of the contact area has to be found as a part of the solution.

Solution A

A suitable solution to the equations of thermoelastic equilibrium in terms of two harmonic functions (ϕ, ψ) is

$$\mathbf{u} = z\nabla \frac{\partial \phi}{\partial z} - \nabla\phi + \mathbf{k}\frac{\partial \phi}{\partial z} - (1-2\nu)\nabla\psi -$$

$$- z\nabla\frac{\partial\psi}{\partial z} + (3-4\nu)\,\mathbf{k}\frac{\partial\psi}{\partial z}, \qquad (1)$$

$$T = \frac{2(1-\nu)}{\alpha(1+\nu)}\frac{\partial^2\phi}{\partial z^2}, \qquad (2)$$

where $\mathbf{u}$ is the displacement vector, ν, α are respectively Poisson's ratio and the coefficient of thermal expansion for the material, and T is the temperature rise above a suitable datum[9]. It is emphasised that this solution is restricted to cases in which T is a harmonic function.

The component of stress acting on the z plane ($\mathbf{s}_z$) is given by

$$\mathbf{s}_z/2G = z\nabla\frac{\partial^2}{\partial z^2}(\phi-\psi) - \mathbf{k}\frac{\partial^2}{\partial z^2}(\phi-\psi), \qquad (3)$$

where G is the modulus of rigidity.

On the surface plane, $z=0$, this component reduces to a purely normal stress

$$\sigma_{zz} = 2G\frac{\partial^2}{\partial z^2}(\psi-\phi), \qquad (4)$$

whilst the normal displacement at this surface is

$$u_z = 2(1-\nu)\frac{\partial\psi}{\partial z}. \qquad (5)$$

As a particular case, we note that, if ψ is identically zero throughout the solid, the normal surface displacement will be everywhere zero, whilst the surface traction becomes

$$\sigma_{zz} = -2G\frac{\partial^2\phi}{\partial z^2} \qquad (6)$$

$$= -\frac{G\alpha(1+\nu)\,T}{(1-\nu)}, \qquad (7)$$

from equation (2). In other words, a normal surface traction which is everywhere proportional to the surface temperature, is sufficient to keep the surface plane. This result greatly facilitates the solution of a certain class of heated punch problems (see section 3 below).

Solution B

For certain problems, it is convenient to represent the solution of equations (1)–(3) in terms of a new harmonic function, ω, such that

$$\psi = \phi - \omega \tag{8}$$

Substituting for ψ in equations (1)–(3) we obtain

$$\mathbf{u} = 4(1-\nu)\,\mathbf{k}\,\frac{\partial\phi}{\partial z} - 2(1-\nu)\,\nabla\phi + (1-2\nu)\,\nabla\omega + {}$$

$$+ z\nabla\,\frac{\partial\omega}{\partial z} - (3-4\nu)\,\mathbf{k}\,\frac{\partial\omega}{\partial z}, \tag{9}$$

$$\mathbf{s}_z/2G = z\nabla\,\frac{\partial^2\omega}{\partial z^2} - \mathbf{k}\,\frac{\partial^2\omega}{\partial z^2}, \tag{10}$$

whilst the temperature is still given by equation (2).

The normal components at the surface are

$$u_z = 2(1-\nu)\,\frac{\partial}{\partial z}\,(\phi-\omega), \tag{11}$$

$$\sigma_{zz} = -2G\,\frac{\partial^2\omega}{\partial z^2}, \tag{12}$$

If ω is identically zero throughout the solid, it follows from equation (10) that all z planes will be stress-free and hence we have the solution appropriate to the thick plate or the semi-infinite solid with traction-free surfaces [9].

We also note that the heat flow per unit area in the z direction

$$q_z = -K\,\frac{\partial T}{\partial z} = -\frac{2K(1-\nu)}{\alpha(1+\nu)}\,\frac{\partial^3\phi}{\partial z^3}, \tag{13}$$

from equation (2), where K is the thermal conductivity of the material. Since $\partial\phi/\partial z$ is harmonic, we have

$$q_z = \frac{2K(1-\nu)}{\alpha(1-\nu)}\left\{\frac{\partial^2}{\partial x^2} + \frac{\partial^2}{\partial y^2}\right\}\frac{\partial\phi}{\partial z} \tag{14}$$

$$= \frac{K}{\alpha(1+\nu)}\left\{\frac{\partial^2 u_z}{\partial x^2} + \frac{\partial^2 u_z}{\partial y^2}\right\}, \tag{15}$$

from equation (11) with $\omega = 0$.

Thus, the normal displacement, u_z, can be regarded as the logarithmic potential due to a source distribution proportional to q_z. Various results concerning the normal surface displacement due to surface heating can therefore be deduced from corresponding theorems of two-dimensional potential theory. For example, the greatest negative value of normal surface displacement must occur in a heated region or at infinity[2]. We note that equations (11) and (13)–(15) apply for all values of z and can therefore be applied to both surfaces of the traction-free thick plate.

3. THE CONTACT PROBLEM

The solution developed in the preceding section will now be applied to the problem of the frictionless heated punch, pressed into the surface of a semi-infinite elastic solid. Either of the two forms (A, B) may be used for any problem, but the statement of boundary conditions can be simplified by a judicious choice.

Following George and Sneddon[10], we can identify four idealised types of thermal boundary condition at the surface plane as follows:

Type a
$T = T(x, y)$ on A,
$T = 0$ on $\bar{A}$;

Type b
$\partial T/\partial z = -q(x, y)/K$ on A,
$T = 0$ on $\bar{A}$;

Type c
$T = T(x, y)$ on A,
$\partial T/\partial z = 0$ on $\bar{A}$;

Type d
$\partial T/\partial z = -q(x, y)/K$ on A,
$\partial T/\partial z = 0$ on $\bar{A}$;

where A denotes the contact area between the punch and the half-space, $\bar{A}$ denotes the rest of the surface plane, and $T(x, y)$, $q(x, y)$ are prescribed functions. (Type d is not considered by George and Sneddon, but is a logical extension of their classification.)

The mechanical boundary conditions for all problems will be of the form

$$u_z = u(x, y) \quad \text{on} \quad A; \qquad \sigma_{zz} = 0 \quad \text{on} \quad \bar{A};$$

$$\sigma_{xz} = \sigma_{yz} = 0 \quad \text{on} \quad A \text{ and } \bar{A};$$

where $u(x, y)$ is a prescribed function representing the profile and penetration of the punch.

Types a and b

For problems of type a, b we use solution A of section 2. Expressing the boundary conditions in terms of the potential functions ϕ, ψ, we obtain from equations (2), (4), (5) and (13):

$$\text{On } A, \quad \frac{\partial \psi}{\partial z} = \frac{u(x, y)}{2(1-\nu)}, \tag{16i}$$

$$\frac{\partial^2 \phi}{\partial z^2} = \frac{\alpha(1+\nu)\, T(x, y)}{2(1-\nu)} \quad \text{or} \tag{16iia}$$

$$\frac{\partial^3 \phi}{\partial z^3} = -\frac{\alpha(1+\nu)\, q(x, y)}{2K(1-\nu)}. \tag{16iib}$$

On $\bar{A}$,

$$\frac{\partial^2 \psi}{\partial z^2} - \frac{\partial^2 \phi}{\partial z^2} = 0, \tag{16iii}$$

$$\frac{\partial^2 \phi}{\partial z^2} = 0 \quad \text{and hence} \tag{16iv}$$

$$\frac{\partial^2 \psi}{\partial z^2} = 0. \tag{16v}$$

It is clear that the two functions can be found separately and the solution for ψ is in fact identical to that of the equivalent isothermal indentation problem.

The type *a* boundary conditions on ϕ are unmixed and hence the additional stress in the body due to heat flow can be found by integration. In particular, we note from equation (7) that the contact stress is increased by an amount

$$\sigma_{zz} = -\frac{G\alpha(1+\nu)\, T(x, y)}{(1-\nu)} \tag{17}$$

as a result of heat flow, corresponding to a 'thermal load'

$$P_1 = \frac{G\alpha(1+\nu)}{(1-\nu)} \iint_A T(x, y)\, \mathrm{d}A. \tag{18}$$

For the axisymmetric problem, this reduces to equation (1.31) of George and Sneddon[10].

The type *b* boundary conditions on the function $\partial\phi/\partial z$ are identical in form to those on ψ in the related isothermal problem of a flat crack extending over the area A and opened by a prescribed distribution of normal pressure (see for example Green[11], equation (2.13); Green and Sneddon[12], equation (3.5)). Various solutions are known to this latter problem and can be applied to equivalent thermoelastic contact problems.

Types c and d

For problems in which the thermal boundary conditions are of type *c* or *d*, it is more convenient to use solution B of section 2. In physical terms, this amounts to finding the thermal distortion of the solid in the absence of surface traction and then solving the equivalent isothermal problem in which the distorted solid is indented by the rigid punch.

Stating the boundary conditions in terms of ϕ and ω we obtain, from equations (2) and (11)–(13):

On A, $$\frac{\partial \phi}{\partial z} - \frac{\partial \omega}{\partial z} = \frac{u(x, y)}{2(1-\nu)}, \tag{19i}$$

$$\frac{\partial^2 \phi}{\partial z^2} = \frac{\alpha(1+\nu)\, T(x, y)}{z(1-\nu)} \quad \text{or} \tag{19iic}$$

$$\frac{\partial^3 \phi}{\partial z^3} = -\frac{\alpha(1+\nu)\, q(x, y)}{zK(1-\nu)}. \tag{19iid}$$

On $\bar{A}$,

$$\frac{\partial^2 \omega}{\partial z^2} = 0, \tag{19iii}$$

$$\frac{\partial^3 \phi}{\partial z^3} = 0. \tag{19iv}$$

We first find the function ϕ from the thermal boundary conditions (ii, iv) and then substitute for $\partial\phi/\partial z$ in the displacement condition (i) to set up the boundary value problem for ω. If the contact pressure only is required, it is sufficient to solve the thermal problem for surface values only. In problems of type d, surface values of $\partial\phi/\partial z$ can be found directly from the solution of equation (14).

Continuity conditions

If we put $u(x, y) = 0$ in the boundary condition (19(i)), we can obtain the additional stress field due to thermal effects alone.

We should then have on A,

$$\frac{\partial^3 \phi}{\partial z^3} - \frac{\partial^3 \omega}{\partial z^3} = -\left\{\frac{\partial^2}{\partial x^2} + \frac{\partial^2}{\partial y^2}\right\} \frac{u(x, y)}{2(1-\nu)} \tag{20}$$

(since $\partial\phi/\partial z$, $\partial w/\partial z$ are harmonic)

$$= 0 \tag{21}$$

and hence for problems of type d

$$\frac{\partial^2 \omega}{\partial z^3} = -\frac{\alpha(1+\nu)\, q(x, y)}{2K(1-\nu)}, \tag{22}$$

from equations (19iid) and (21).

The boundary conditions on ω (equations (19iii) and (22)) are thus analogous to those on ϕ in problems of type b (equations (16iib, v)) for the same distribution of heat input $q(x, y)$ and it is tempting to deduce that the contact pressure distribution is unaffected by a change from a constant temperature boundary condition on $\bar{A}$ to one of zero heat flux. However, such a deduction would be invalid, since the analogy does not extend to the requirements of continuity at the edge of the contact area. Temperature and displacement must be continuous in both cases, but this imposes continuity on $\partial w/\partial z$ for type d and $\partial^2 \phi/\partial z^2$ for type b. The effect of this difference is illustrated by an example in section 4.

4. SOME PARTICULAR EXAMPLES

The method described in the last section can be applied to any heated punch problem, but closed form solutions to the corresponding potential theory problem can only be obtained for a very restricted range of configurations. The axisymmetric heated punch has been discussed extensively by George and Sneddon[10], whilst the equivalent problem for a thick plate has been treated by Keer and Fu[13]. More recently, Fu[14] has analysed the case of simultaneous indentation of a half-space by two separately axisymmetric heated punches.

In this section, we shall briefly consider some particular examples in which the contact area A is circular or elliptical, making use of existing solutions to the appropriate boundary value problems. In addition to extending the range of existing solutions, this will serve to illustrate some of the points discussed in general terms in the preceding section.

i. *The elliptical punch*

Suppose that the contact area A is the ellipse $\frac{x^2}{a^2}+\frac{y^2}{b^2} \leq 1$ and the boundary conditions are of type b (equation (16)) with $q(x, y)$, $u(x, y)$ constant and equal to q_0, u_0 respectively. In other words, the end of the punch is flat and uniformly heated.

The isothermal indentation problem corresponding to the function ψ is considered by Green and Sneddon[12] who also give a solution for the flat elliptical crack opened by a constant pressure which can be adapted to give the solution for $\partial\phi/\partial z$. We find that

$$\frac{\partial \psi}{\partial z} = \frac{u_0 a}{4(1-\nu)\,\mathbf{K}(k)} \int_{\xi}^{\infty} \frac{ds}{\sqrt{Q(s)}}, \tag{23}$$

$$\frac{\partial \phi}{\partial z} = \frac{ab^2\,\alpha(1+\nu)\,q_0}{8K(1-\nu)\,\mathbf{E}(k)} \int_{\xi}^{\infty} \frac{F(s)\,ds}{\sqrt{Q(s)}}, \tag{24}$$

where

$$F(s) = x^2/(a^2+s^2) + y^2/(b^2+s^2) + z^2/s^2 - 1, \tag{25}$$

$$Q(s) = s(a^2+s)\,(b^2+s), \tag{26}$$

$\mathbf{K}(k)$, $\mathbf{E}(k)$ are the complete elliptic integrals of the first and second kind respectively with modulus $k = \sqrt{1-b^2/a^2}$ and ξ is the largest root of the equation $F(s)Q(s) = 0$.

The stress component $\mathbf{s}_z$ and the displacement component u_z can be written down in terms of these functions using equations (1) and (3). In particular, we find that the contact stress under the punch is

$$\begin{aligned}\sigma_{zz} &= 2G(\partial^2\psi/\partial z^2 - \partial^2\phi/\partial z^2)\\ &= \frac{-Gu_0}{(1-\nu)\,b\mathbf{K}(k)}\left\{1-\frac{x^2}{a^2}-\frac{y^2}{b^2}\right\}^{-1/2} - \frac{G\alpha(1+\nu)\,q_0\,b}{K(1-\nu)\mathbf{E}(k)}\left\{1-\frac{x^2}{a^2}-\frac{y^2}{b^2}\right\}^{1/2},\end{aligned} \tag{27}$$

corresponding to a total load

$$P = \frac{2\pi aG}{(1-\nu)}\left\{\frac{b^2\,\alpha(1+\nu)\,q_0}{K\,\mathbf{E}(k)} + \frac{u_0}{\mathbf{K}(k)}\right\}. \tag{28}$$

Alternative expressions for the potential functions in terms of Jacobian elliptic functions can be derived from the results of Green and Sneddon[12] (see also Kassir and Sih[15] who give a solution for the elliptical crack with linearly varying pressure, from which can be deduced that for the elliptical punch with linearly varying heat input).

ii. *The non-axisymmetric cylindrical punch*

We now consider the case of a cylindrical punch making contact with the half-space over a circle of radius a, for which the thermal boundary conditions are not axisymmetric. A general solution of the appropriate mixed boundary value problem is given by Green[11] in connection with the corresponding isothermal punch and crack problems. The prescribed potential (or derivative) in the area A is decomposed into a series of functions of the form $f_n(r)\cos n(\theta+\theta_0)$ in cylindrical polar co-ordinates whose origin is at the centre of the circular contact area. The mixed boundary value problem is then formulated and solved for a general function of this type in terms of surface values only. Finally, the internal values of the potential function are found by integration of the surface values or by other methods. The details of the solution are adequately discussed by Green[11] and will not be given here.

To illustrate the application of this method to thermoelastic contact, we consider the rigid, flat-ended cylindrical punch, pressed into the surface by a normal force P acting through the centre of the contact area, over which the thermal boundary conditions are of type b with

$$q(x, y) = q_0 x = q_0 r \cos\theta, \tag{29}$$

where q_0 is a constant.

From considerations of geometry and symmetry, it must follow that

$$u(x, y) = c + c_2 r \cos\theta, \tag{30}$$

where the constants c_1, c_2 will be determined by considerations of force and moment equilibrium on the punch.

From equations (16) and (29) we have

$$\frac{\partial^3 \phi}{\partial z^3}(r, \theta, 0) = -\frac{\alpha(1+\nu)\, q_0 r \cos\theta}{2K(1-\nu)}, \quad a \geqslant r \geqslant 0; \tag{31}$$

$$\frac{\partial^2 \phi}{\partial z^2}(r, \theta, 0) = 0, \qquad r > a \tag{32}$$

and hence

$$\frac{\partial \phi}{\partial z}(r, \theta, 0) = \frac{\alpha(1+\nu)\, q_0 r^3 \cos\theta}{16K(1-\nu)} + Ar\cos\theta, \quad a \geqslant r \geqslant 0, \tag{33}$$

where A is an arbitrary constant, to be determined from the requirement that temperature, and hence $\partial^2\phi/\partial z^2$ should be continuous at $r = a$.

Using Green's solution ([11], equations (2.17); (2.18)) and applying this condition, we find

$$A = -\frac{\alpha(1+\nu)\, q_0 a^2}{12K(1-\nu)} \tag{34}$$

and

$$\frac{\partial^2 \phi}{\partial z^2}(r, \theta, 0) = \frac{2\alpha(1+\nu)\, q_0 r \sqrt{a^2 - r^2}\cos\theta}{3\pi K(1-\nu)}, \quad a \geqslant r \geqslant 0. \tag{35}$$

The solution for the corresponding isothermal stress function, ψ, is given explicitly by Green ([11], § 8) and leads to

$$\frac{\partial^2 \psi}{\partial z^2}(r, \theta, 0) = -\frac{(c_1 + 2c_2 r\cos\theta)}{\pi(1-\nu)\sqrt{a^2 - r^2}}, \quad a \geqslant r \geqslant 0. \tag{36}$$

Thus, the contact stress distribution, from equations (4), (36) and (37) is

$$\sigma_{zz}(r,\theta,0) = -\frac{2Gc_1}{\pi(1-\nu)\sqrt{a^2-r^2}} - \frac{4Gr\cos\theta}{\pi(1-\nu)\sqrt{a^2-r^2}} \times$$

$$\times\left\{c_2 + \frac{\alpha(1+\nu)\,q_0(a^2-r^2)}{3K}\right\}, \qquad a \geqslant r \geqslant 0. \tag{37}$$

Finally, applying the condition that the net force on the punch is P whilst the net moment about the origin is zero, we find

$$c_1 = \frac{P(1-\nu)}{4Ga}, \tag{38}$$

$$c_2 = -\frac{\alpha(1+\nu)\,q_0 a^2}{15K} \tag{39}$$

and hence

$$\sigma_{zz}(r,\theta,0) = \frac{-P}{2\pi a\sqrt{a^2-r^2}} + \frac{4G\alpha(1+\nu)\,q_0 r(r^2-\frac{4}{5}a^2)\cos\theta}{3\pi K(1-\nu)\sqrt{a^2-r^2}},$$

$$a \geqslant r \geqslant 0. \tag{40}$$

As we would expect, the mean penetration of the punch is unaffected by the heat flow, which however causes a tilt through an angle $(-\alpha q_0(1+\nu)a^2/15K)$ as well as a redistribution of contact stress.

To illustrate the similarities and differences between problems of types b and d, we now consider the equivalent problem in which the external boundary condition is $q=0$; i.e., equation (32) is replaced by

$$\frac{\partial^2\phi}{\partial z^3}(r,\theta,0) = 0, \qquad r > a. \tag{41}$$

The derivation of equation (33) follows as in the previous example, but we should expect the arbitrary constant A to take a different value in view of the different external boundary condition. We also have

$$\frac{\partial\phi}{\partial z}(r,\theta,0) = \frac{B\cos\theta}{r}, \qquad r > a, \tag{42}$$

from equation (41), where B is another arbitrary constant.

One equation for determining these two constants can be obtained from the condition that the 'thermoelastic displacement', $\partial\phi/\partial z$ (see equation (11)) is continuous at $r=a$, whilst the other follows from the fact that $\partial^3\phi/\partial z^3$ is everywhere finite and hence $\partial^2\phi/\partial r\,\partial z$ must be continuous at $r=a$.

From these two conditions, we find

$$A = -\frac{\alpha(1+v)\,q_0 a^2}{8K(1-\nu)} \tag{43}$$

differing from equation (34) as expected, and hence

$$\frac{\partial \omega}{\partial t} = -\frac{(c_1 + c_2 r \cos\theta)}{2(1-\nu)} + \frac{\alpha(1+\nu)\, q_0 r(r - 2a^2)\cos\theta}{16K(1-\nu)}, \quad a \geqslant r \geqslant 0, \tag{44}$$

from equations (19i), (33) and (43).

We also have

$$\partial^2 \omega / \partial z^2 = 0, \quad r > a, \tag{19iii}$$

and we can therefore use Green's solution to find

$$\sigma_{zz}(r,\theta,0) = -2G\frac{\partial^2 \omega}{\partial z^2}(r,\theta,0) = -\frac{2G(c_1 + 2c_2 r\cos\theta)}{\pi(1-\nu)\sqrt{a^2 - r^2}} - $$

$$ - \frac{G\alpha(1+\nu)\, q_0 r(5a^2 - 4r^2)\cos\theta}{3\pi K(1-\nu)\sqrt{a^2-r^2}}, \quad a \geqslant r \geqslant 0. \tag{45}$$

Applying the punch equilibrium conditions, we find

$$c_2 = -\frac{3\alpha(1+\nu)\, q_0 a^2}{20K}, \tag{46}$$

whilst c_1 and the contact stress are unchanged from equations (38) and (40) respectively. Thus, the change in external boundary conditions from type b to type d causes an additional tilt of the punch (through an angle $(-\alpha(1+\nu) q_0 a^2/12K)\lambda$ without changing the contact stress distribution. There will, however, be a change in the stress field *within* the solid.

A comparable result in the case of *axisymmetric* problems is that a change in external thermal boundary conditions merely produces a uniform normal displacement of the contact area with no consequent change in contact pressure distribution, provided the total load on the punch rather than its displacement is specified[16]. We note that if the net heat inflow to the solid is not zero, this rigid body displacement will be unbounded. Hence, the usual assumption that the half-space solution approximates the local behaviour of a corresponding practical solid needs to be applied with some caution. We also note that, if the heat input over the contact area involves terms in $\cos n\,\theta$ where $n > 1$, the external boundary condition will generally influence the contact stress distribution as well as the rigid body displacement of the punch.

The method illustrated in these examples constitutes a general solution to the problem in which the contact area is circular and the thermal boundary conditions are of one of the four types (a–d) listed in section 3. No new principles are involved in the solution of problems of type a or c, although the latter case involves the successive solution of two mixed boundary value problems and hence leads to less tractable integrals. An axisymmetric example of this type is treated in reference [17]. For type a problems the additional contact stress due to the temperature field can be written down immediately by virtue of equation (17).

5. THE CONTINUOUS PUNCH

In all the preceding analysis, it has been assumed that thermal and mechanical boundary conditions are given at all points on the surface of the half-space, but there is an important class of problems in which this is not so – i.e., those in which the surface of the punch is a continuous curve. In such cases, the extent of the contact area has to be found as part of the solution, usually from considerations of symmetry and continuity. This method is practically restricted to axisymmetric problems, for which relationships between contact radius and load can be derived without working through a complete solution to the problem.

For the corresponding isothermal axisymmetric contact problem, Shield[18] has shown that the load on the punch is given by

$$P(a) = \frac{4G}{(1-\nu)} \int_0^a \frac{r u(r)\, \mathrm{d}r}{\sqrt{a^2-r^2}}, \tag{47}$$

where $u(r)$ represents the profile of the punch, and the depth of penetration $u(0)$ is determined by the condition

$$u(0) = -a \int_0^a \frac{u'(r)\, \mathrm{d}r}{\sqrt{a^2-r^2}}. \tag{48}$$

Integrating equation (47) by parts and substituting for $u(0)$ we find

$$P(a) = -\frac{4G}{(1-\nu)} \int_0^a \frac{r^2\, u'(r)\, \mathrm{d}r}{\sqrt{a^2-r^2}} \tag{49}$$

$$= -\frac{4G}{(1-\nu)} \int_0^a \sqrt{a^2-r^2}\, \frac{\mathrm{d}}{\mathrm{d}r}(r u'(r))\, \mathrm{d}r, \tag{50}$$

which relates contact radius to load for the continuously curved punch, since $u'(r)$ is independent of the depth of penetration.

Now suppose there is an arbitrary axisymmetric heat input, $q(r)$ per unit area over the contact area, tending to produce an additional thermal surface displacement, u_z, where

$$\frac{1}{r}\frac{\mathrm{d}}{\mathrm{d}r}\left(r\,\frac{\mathrm{d}u_z}{\mathrm{d}r}\right) = \frac{\alpha(1+\nu)\, q(r)}{K}, \tag{51}$$

from equation (15).

It follows from equations (50 and (51) that the additional 'thermal' load needed to suppress these displacements and maintain the contact radius,

a, is

$$P_1(a) = -\frac{4G\alpha(1+\nu)}{K(1-\nu)} \int_0^a r\sqrt{a^2-r^2}\, q(r)\, \mathrm{d}r\,. \tag{52}$$

By taking the sum of P, P_1 from equations (49 and (52), we obtain a relationship between load and contact radius for the thermoelastic contact problem which enables the latter to be determined without working through a complete solution. In view of the arguments of the last section, this result applies equally to boundary conditions of types b or d. We can derive a corresponding result for type c boundary conditions by applying the reciprocal theorem to the integral in equation (52). Thus,

$$\int_0^a r\sqrt{a^2-r^2}\, q(r)\, \mathrm{d}r = \int_0^a r\, q_1(r)\, T(r)\, \mathrm{d}r\,, \tag{53}$$

where $q_1(r)$ is the heat input over the contact circle which is just sufficient to produce the surface temperature distribution

$$T_1(r) = \sqrt{a^2-r^2}\,, \qquad a \geqslant r \geqslant 0\,, \tag{54}$$

the rest of the surface being unheated. This latter problem can be solved to give

$$q_1(r) = \frac{2K}{\pi} \mathscr{F}(\sqrt{1-r^2/a^2}) \tag{55}$$

where

$$\mathscr{F}(x) = \frac{1+\log x}{x} + 2\left(\chi_2(0) - \chi_2\left(\frac{1-x}{1+x}\right)\right) \tag{56}$$

and χ_2 is Legendre's chi function[19].

In terms of these functions, we obtain

$$P_1(a) = \frac{8G\alpha(1+\nu)}{\pi(1-\nu)} \int_0^a r\mathscr{F}(\sqrt{1-r^2/a^2})\, T(r)\, \mathrm{d}r\,, \tag{57}$$

from equations (52), (53) and (55). However, the usefulness of this result is limited by the intractable nature of the integral.

A corresponding result for type a boundary conditions follows immediately from equation (18) and is

$$P_1(a) = \frac{2\pi G\alpha(1+\nu)}{(1-\nu)} \int_0^a rT(r)\, \mathrm{d}r\,. \tag{58}$$

REFERENCES

1. A.M. Clausing, 'Heat Transfer at the Interface of Dissimilar Metals—the Influence of Thermal Strain', *Int. J. Heat Mass Transfer*, **9**, 1966, pp. 791–801.
2. J.R. Barber, 'The Effect of Thermal Distortion on Constriction Resistance', *Int. J. Heat Mass Transfer*, **14**, 1971, pp. 751–766.
3. M.V. Korovchinskii, 'Plane-Contact Problem of Thermoelasticity during Quasi-Stationary Heat Generation on the Contact Surfaces', *Trans. ASME,* **D87**, 1965, pp. 811–817.
4. R.A. Burton, S.R. Kilaparti and V. Nerlikar, 'A Limiting Stationary Configuration with Partially Contacting Surfaces, *Wear*, **24**, 1973, pp. 199–206.
5. J.R. Barber, 'Thermoelastic Instabilities in the Sliding of Conforming Solids', *Proc. Roy. Soc. London*, **A312**, 1969, pp. 381–394.
6. T.A. Dow and R.A. Burton, 'Thermoelastic Instability of Sliding Contact in the Absence of Wear', *Wear*, **19**, 1972, pp. 315–328.
7. F.E. Kennedy and F.F. Ling, 'A Thermal, Thermoelastic and Wear Simulation of a High-Energy Sliding Contact Problem', *Trans. ASME,* **F96**, 1974, pp. 497–507.
8. J.R. Barber, 'Distortion of the Semi-Infinite Solid Due to Transient Surface Heating', *Int. J. Mech. Sci.*, **14**, 1972, pp. 377–393.
9. W.E. Williams, 'A Solution of the Steady State Thermoelastic Equations', *Z. angew. Maths. Phys.*, **12**, 1961, pp. 452–455.
10. D.L. George and I.N. Sneddon, 'The Axisymmetric Boussinesq Problem for a Heated Punch', *J. Math. Mech.*, **11**, 1962, pp. 665–689.
11. A.E. Green. 'On Boussinesq's Problem and Penny-Shaped Cracks', *Proc. Cambridge Phil. Soc.*, **45**, 1949, pp. 251–257.
12. A.E. Green and I.N. Sneddon, 'The Distribution of Stress in the Neighbourhood of a Flat Elliptical Crack in an Elastic Solid', *Proc. Cambridge Phil. Soc.*, **46**, 1950, pp. 159–164.
13. L.M. Keer and W.S. Fu, 'Some Stress Distributions in an Elastic Plate Due to Rigid Heated Punches', *Int. J. Engng. Sci.*, **5**, 1967, pp. 555–570.
14. W.S. Fu, 'Indentation of an Elastic Half-Space Due to Two Coplanar Heated Punches', *Int. J. Engng. Sci.*, **8**, 1970, pp. 337–349.
15. M.K. Kassir and G.C. Sih, 'Geometric Discontinuities in Elastostatics', *J. Math. Mech.*, **16**, 1967, pp. 927–948.
16. J.R. Barber, 'The Solution of Heated Punch Problems by Point Source Methods', *Int. J Engng. Sci.*, **9**, 1971, pp. 1165–1170.
17. J.R. Barber, 'Indentation of the Semi-Infinite Elastic Solid by a Hot Sphere', *Int. J. Mech. Sci.*, **15**, 1973, pp. 813–819.
18. R.T. Shield, 'Load-Displacement Relations for Elastic Bodies', *Z. angew. Math. Phys.*, **18**, 1967, pp. 682–693.
19. L. Lewin, *Dilogarithms and Associated Functions*, MacDonald, London, 1958.

An axisymmetric contact patch configuration for two slabs in frictionally heated contact

R. A. Burton

LIST OF SYMBOLS

- a_i coefficient of pressure distribution equation
- E Young's modulus
- $\hat{G}$ dimensionless quantity shown relative thermal expansion influence
- $\hat{H}$ dimensionless measure of effect of initial surface curvature
- K thermal conductivity of the non-conductive body
- l radius of contact patch
- m exponent in pressure distribution equation
- $\hat{m}$ equivalent radius of surrounding field belonging to one contact patch
- n designates order of differentiation
- p pressure on interface
- $\hat{p}$ average pressure on interface
- p_0 pressure at center of contact patch
- q heat flow per unit of surface area
- r radius measured in interface, out from center of contact patch
- V sliding speed
- x coordinate axis located in interface
- y coordinate axis located in interface
- w displacement normal to interface

- α coefficient of thermal expansion
- ζ ratio of average pressure to p_0 in contact patch
- Λ ratio of contact area to gap-area + contact area, $(l/\hat{m})^2$
- μ friction coefficient
- ν Poisson's ratio
- ξ measure of radial position
- ρ distance from point of application of load.

1. INTRODUCTION

In the contact of real solid bodies two major effects produce local stresses and temperatures in excess of those predicted from the initial macroscopic geometry of the bodies. One such effect results from the fact that real bodies are far from smooth, and surface contact occurs only at the peaks of microscopic roughness asperities. The second effect results from overall distortion as a consequence of frictional heating, and would lead to contact at the peaks of thermal asperities even if the surfaces were nearly geometrically flat before sliding began. For a given pair of materials the transition from smooth contact to patches of contact, separated by gaps, may appear suddenly as sliding speed is raised, and may therefore be spoken of as the onset of an instability.

Much of the recent interest in such instabilities stems from the work of J.R. Barber[1,2,3] who has demonstrated the phenomenon experimentally and has provided analyses which partially explain his observations, these explanations drawing upon the idea of modification of asperity contact by wear. The present investigators were drawn to the problem after studies of thermoelastic instabilities in journal bearings[4], ball bearings[5] and a related instability in liquid films[6]. Subsequently we were able to show[7] that neither asperity contact nor wear were essential to explain the instability of sliding contact. In the analysis of a two-dimensional scraper it was shown that geometrically flat contact, once disturbed, would redistribute interface pressures, relieving some regions and loading others, above a critical sliding speed. One difficulty in this analysis arose from the fact that the predicted critical sliding speed was quite low. Reintroduction of wear into the analysis[8] led to higher critical speeds, which were felt to be realistic. Subsequent work[9] was devoted to a seal-like geometry where boundary lubrication was assumed to prevail and wear was assumed to be negligible. This geometry consisted to two thin walled, concentric tubes meeting end to end, with sliding in the tangential direction. The first results of analyses were perplexing, in that instability was predicted only at extremely high friction coefficients for a material sliding on its own kind. It was also shown[9] that even a modest difference in thermal conductivity for the two materials would call for instability at sliding speeds under 30 m/sec with friction coefficient well below unity.

Further investigation showed that thin surface fllms of thermally insulating material, such as metallic oxide, could produce the same effect by causing one body to act as an insulator relative to disturbances moving with the second body[10]. Barber[11] has suggested that constriction resistance would act similarly, and preliminary calculations bear this out, although the influence is not great. More important is the fact that the reduced true area of contact alters the resistance of the insulating film[12], increasing the apparent film thickness by the ratio $A_{\text{nominal}}/A_{\text{real}}$ which is typically near 10^3. This would call for strong participation by extremely thin films. For such films, however, the frictional heating is thought to occur in the substrate rather than at the interface, thus calling for modification of the analysis. Irrespective of such uncertainties, there are many instances where thin insulating films may cause two conductive bodies to interact as conductor on insulator.

When T.A. Dow began the investigations described in his dissertation[13] he was specifically motivated to explore the effects reported by Sibley and Allen[14] and shown vividly in a film which they prepared at Battelle Institute. These studies were concerned with a puck pressed against the side of a rotating disk, and showed red-hot streaks on the surface of the disk as it moved out from under the trailing edge of the puck. These streaks would sometimes move harmonically back and forth (radially on the disk) and would sometimes move erratically. The question of instability in an idealized version of such a system was treated in Refs. [7,8]; where a thin scraper moves normal to the line of contact on the face of a large slab. A computer simulation of this phenomenon[12] showed erratic movement of concentrated-load spots, with traverse velocity comparable to the values observed[14]. Recently Kennedy and Ling[15] have reported a computer simulation of brake behavior and have found an oscillatory radial movement of the highly loaded contact zone.

An important difference between scraper contact and end-contact of tubes is that for two-dimensional heat flow, uniform edge heating of a free-expanding scraper will cause bowing of the edge, while uniform end heating of the tube will cause no bowing in the axial direction. In the absence of wear, and with increase of sliding speed the scraper will show continuously increasing departures from the zero-speed pressure distribution. The tubular contact will show uniform loading tangentially, until the sliding speed V_c' is reached, where,

$$V_c' = 2\pi K/\mu E^* \alpha m$$

Above this speed weak zero-average perturbations in pressure will grow[7]. It can also be shown that a limiting stable condition, with patches of contact separated by gaps, can exist above the critical sliding speed[16,18].

If there is a small but finite wear rate, and the scraper is operated below V_c' for a long period of time, contact pressure will become uniform. Above the critical speed the contact-patch configuration will form, will wear off and form again, at an oscillation frequency determined by wear rate[8]. Increase the wear rate and the critical speed is raised, along with the oscillation frequency.

The thermoelastic and elastic behavior of both scrapers and tubes have been treated as examples of plane stress. The analogous plane-strain case of slab-contact may be found directly from this, by using well-known transformations between the two cases, giving

$$V_c = (1-\nu)\, V_c' = 2(1-\nu)\, \pi K/\mu E^* \alpha m \qquad (1)$$

and the plane strain solution analogous to the patch-like contact would be a sequence of contact bands on the face of the slab. This effect is complicated by the same factors at work on the scraper, however, and one would expect bowing of the unconstrained slab, with rectilinear heat flow through it. Below the critical speed wear will ultimately remove this bowing, as in the case of the scraper. Mechanical constraints at the boundaries may also reduce it in some instances.

The above arguments suggest that we already know something about the frictional contact of slabs. With regard to the two-dimensional instability, at least, the foregoing arguments about relative conductivities, and insulating film effects apply as well to face contact of slabs as to edge contact of plates. An obvious question remains however, concerning the stability of the predicted contact bands themselves, especially since we observe contact spots rather than bands in our experimental studies. To answer this question we shall investigate the conditions under which a circular contact patch may exist.

Re-stating the problem: the sliding of a conductive slab on a semi-infinite thermal insulator will be investigated. Both bodies will show linear, Hookean elastic behavior, and linear thermal expansion. Heating will result from friction at the interface. It will be postulated that contact is not uniform but takes place at circular patches (resulting from initial

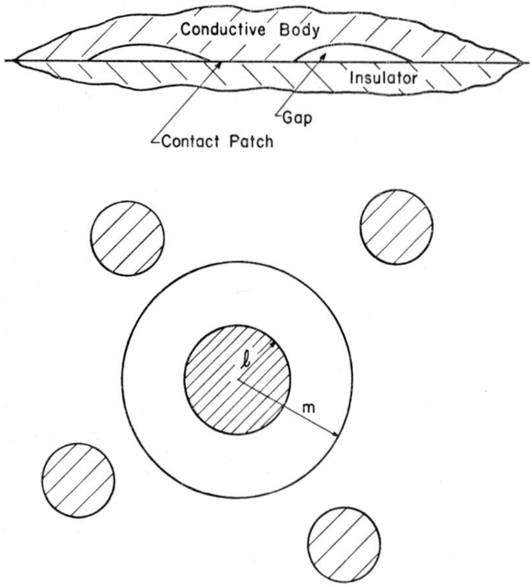

Fig. 1. Side view and top view of patchlike contact, showing patch dimension, l, and dimension of pro-rated gap area, m, for a patch and its nearest neighbors

paraboloidal peaks, or from thermally-produced asperities). Conditions will be determined where equilibrium can exist involving frictional heating, thermal expansion effects and elastic deformations. The height of the disturbances will be assumed to be small when compared with spacing; and radii of the peaks will consequently be assumed to be large.

Fig. 1 shows a possible patch distribution, where m is a measure of the unloaded region attributed to a given contact patch, and l is the radius of the patch. The analysis would just as correctly apply to a single contact patch on the face of a slab of face area πm^2.

2. THERMOELASTIC DISPLACEMENT UNDER FRICTIONAL HEATING

The following equation relates the displacement of the flat surface of a semi-infinite body to the rate of heat input per unit of area, and serves to define Γ

$$\Gamma = \frac{\partial^2 w}{\partial x^2} + \frac{\partial^2 w}{\partial y^2} = +q \frac{\alpha(1+\nu)}{K} \tag{2}$$

This has been derived by Barber[11] on the basis of his earlier work[19]. Some boundary constraint is implied, since this underestimates Γ by the ratio $(1+\nu)/(2)$ for rectilinear heat flow through the entire face of an unconstrained slab. The error reduces to ten percent for heat input on only half the face area, as in the cases of most interest here. When the heat input q is of frictional origin:

$$q = \mu p V \tag{3}$$

where μ is friction coefficient, V is sliding speed, and p is local pressure.

In the manner of Refs. [16, 17, 18] an origin will be taken at the center of a contact patch and the pressure distribution will be found where elastic and thermal displacements balance one another so as to lead to uniform contact out to some radius l, which is itself a function of operating variables and materials properties.

Returning to equations (2) and (3). and letting $p = p_0\left[1+\Sigma a_m \left(\frac{r}{l}\right)^m\right]$

$$\Gamma = \alpha \frac{(1+\nu)}{K} \mu V p_0 \left[1+\Sigma a_m \left(\frac{r}{l}\right)^m\right] \tag{4}$$

To avoid discontinuities of slope at $x = 0$, the series of equation (4) has been written so as to restrict the exponents to even, positive integers.

Upon differentiation, as follows, terms with $m < n$ disappear; and when the derivative is evaluated at $r = 0$ terms with $m > n$ also disappear, and only the term where $n = m$ remains. Hence

$$\left(\frac{d^n \Gamma}{dr^n}\right)_{r=0} = \frac{\alpha(1+\nu)}{K} \frac{\mu V p_0 a_n n_\sigma}{l^n} \tag{5}$$

An additional result will be of interest in further derivations, and this can be obtained from equation (4) for $r \to 0$.

$$(\Gamma)_{r=0} = (\Gamma)_0 = \frac{\alpha(1+\nu)}{K} \mu V p_0 \tag{6}$$

3. ELASTIC DISPLACEMENTS OF A SURFACE UNDER AN AXISYMMETRIC PRESSURE DISTRIBUTION

For a single concentrated load P_a, the displacement w_i at a distance ρ_{ia} from the point of application of load is [20,21]:

$$w_i = \frac{(1-\nu^2)}{\pi E} \frac{P_a}{\rho_{ia}} \tag{7}$$

This assumes an isotropic, Hookean material; and for distributed loading the effect of each component load is linearly additive, thus

$$w(\xi) = \frac{(1-\nu^2)}{\pi E} \int_0^l \int_0^{2\pi} \frac{pr}{\rho} \, dr d\theta \tag{8}$$

For the ring load shown in Fig. 2, and for $\xi < r$, letting $e_1 = \xi/r$:

$$\rho = \sqrt{(r+\xi \cos\theta)^2 + (\xi \sin\theta)^2} = r\sqrt{1+e_1^2+2e_1 \cos\theta} \tag{9}$$

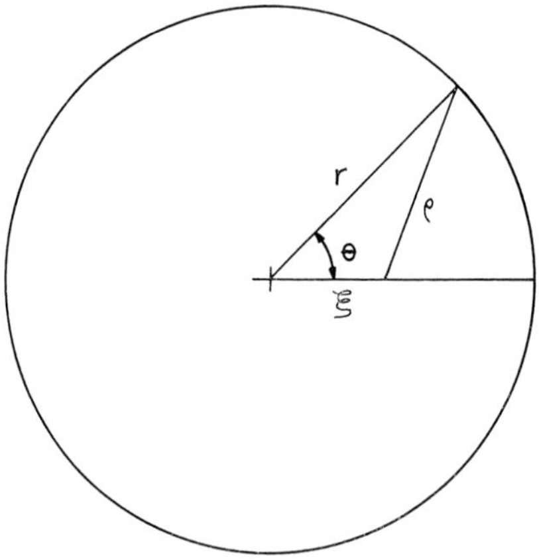

Fig. 2. Illustration of nomenclature for calculation of displacement at distance ξ from center of application of ring load

For $\xi > r$, letting $e_2 = -\xi/r$

$$\rho = \sqrt{(\xi - r\cos\theta)^2 + (r\sin\theta)^2} = r\sqrt{1 + e_2^2 + 2e_2\cos\theta} \tag{10}$$

Letting

$$\begin{aligned} \varphi_1 &= e_1^2 + 2e_1\cos\theta \\ \varphi_2 &= e_2^2 + 2e_2\cos\theta \end{aligned} \tag{11}$$

it is possible to rewrite equation (8), where $A = (1-v^2)/\pi E$.

$$w = A\int_\xi^l \int_0^{2\pi} \frac{p}{(1+\varphi_1)^{1/2}}\, drd\theta + \int_0^\xi \int_0^{2\pi} \frac{p}{(1+\varphi_2)^{1/2}}\, drd\theta \tag{12}$$

Using the binomial expansion

$$\begin{aligned} (1+\varphi)^{-1/2} &= [1 - 0.5\,\varphi + 0.375\,\varphi^2 - 0.3125\,\varphi_3 + 0.2734\,\varphi_4 \\ &\quad - 0.2460\,\varphi^5 + 0.2256\,\varphi^6 - 0.2095\,\varphi^7 + 0.1964\,\varphi^8 - \\ &\quad - 0.1855\,\varphi^9 + 0.1762\,\varphi^{10} - 0.1682\,\varphi^{11} + 0.1611\,\varphi^{12} - \\ &= \sum_0^\infty b_i\varphi^i \end{aligned} \tag{13}$$

and

$$\begin{aligned} \varphi &= e^2 + 2e\cos\theta \\ \varphi^2 &= e^4 + 4e^3\cos\theta + 4e^2\cos^2\theta \\ \varphi^3 &= e^6 + 6e^5\cos\theta + 12e^4\cos^2\theta + 8e^3\cos^3\theta \\ \varphi^4 &= e^8 + 8e^7\cos\theta + 24e^6\,\mathrm{cps}^2\,\theta + 32e^5\cos^3\theta + 16e^4\cos^4\theta \\ \varphi^5 &= e^{10} + 10e^9\cos\theta + 40e^8\cos^2\theta + 60e^7\cos^3\theta + 80e^6\cos^4\theta + \\ &\quad + 32e^5\cos^5\theta \end{aligned} \tag{14}$$

Writing

$$\int_0^{2\pi} (1+\varphi)^{-1/2}\, d\theta = \int_0^{2\pi} (\Sigma b_i\varphi^i)\, d\theta \tag{15}$$

Recall that

$$\int_0^{2\pi} \cos^2\theta = \pi; \quad \int_0^{2\pi} \cos^4\theta = 4\pi/3; \quad \int_0^{2\pi} \cos^6\theta = 5\pi/8 \tag{16}$$

and write the integral of equation (15) term by term for the quantities in equation (14), truncating above e^6.

Table 1

φ_s	$\frac{b_i}{\pi}\int_0^{2\pi}\varphi^i \mathrm{d}\theta$			
$\varphi^0 =$	2			
$\varphi^1 =$		$-e^2$		
$\varphi^2 =$		$1.5e^2$	$+0.75e^4$	
$\varphi^3 =$			$-3.75e^4$	$-0.625e^6$
$\varphi^4 =$			$3.28e^4$	$+6.56e^6$
$\varphi^5 =$				$-14.76e^6$
$\varphi^6 =$				$+9.024e^6$

Combining the quantities in Table 1

$$\frac{1}{\pi}\int_0^{2\pi}(1+\varphi)^{-1/2}\,\mathrm{d}\theta = 2+0.5e^2+0.28e^4+0.195e^6+0.148e^8+\ldots$$

$$= \sum_0^\infty c_j e^j, \qquad \text{for} \qquad j=0,2,4,6,\ldots \tag{17}$$

Returning to equation (12) with reference to equations (9), (10), (11)

$$w = A\pi\sum_{j=0}^{\infty}\left[\int_\xi^l pc_j\left(\frac{\xi}{r}\right)^j \mathrm{d}r + \int_0^\xi pc_j\left(\frac{-\xi}{r}\right)^j \mathrm{d}r\right] \tag{18}$$

When $p=p_0+\Sigma a_m(r/l)^m$ for m even and $a_0=1$, equation (18) becomes

$$w = A\pi p_0\sum_{j,m}\frac{\pi c_j a_m}{l^m}\left[\int_\xi^l r^{m-j}\xi^j\,\mathrm{d}r + \int_0^\xi r^{m-j}\xi^j\,\mathrm{d}r\right] \tag{19}$$

Integrating, and defining $B_{j,m}\equiv A\pi p_0 c_j a_m$

$$w = \sum_{j,m}B_{j,m}\left[\frac{l^{-j+1}}{m-j+1}\xi^j\right] \tag{20}$$

It is of passing interest that for $p=p_0$ or $m=0$

$$w_{\text{const.}\,p} = A\pi p_0 l\left[1-\left(\frac{\xi}{l}\right)^2+\frac{1}{3}\left(\frac{\xi}{l}\right)^4-\ldots\right] \tag{21}$$

Returning to equation (20) and differentiating

$$\frac{\mathrm{d}w}{\mathrm{d}\xi} = \sum B_{j,m}(j)\,(l^{-j+1})\,\xi^{j-1}/(m-j+1) \tag{22}$$

$$\frac{\mathrm{d}^2w}{\mathrm{d}\xi^2} = \sum B_{j,m}(j)\,(j-1)\,(l^{-j+1})\,\xi^{-2}/(m-j+1) \tag{23}$$

Since, for an axisymmetric displacement distribution:

$$\Gamma = \frac{\partial^2 w}{\partial \xi^2} + \frac{1}{\xi}\frac{\partial w}{\partial \xi} \tag{24}$$

$$\Gamma = \sum B_{j,m}\, l^{-j+1}\, \xi^{j-2}(j^2/m-j+1) \tag{25}$$

For this to exist at $\xi = 0$, the exponent of ξ must be zero or: $j-2=0$.

$$\Gamma_0 = \sum B_{2,m}\, l^{-1}(4/m-1) \tag{26}$$

Returning to equation (25) it is possible to generate higher derivatives of Γ by repeated differentiation. In that which follows we shall write $\mathrm{d}^n\Gamma/\mathrm{d}\xi^n$ as $\mathrm{d}^n\Gamma/\mathrm{d}r^n$ since there is no further need of distinguishing between these two coordinates.

$$\left(\frac{\mathrm{d}^n\Gamma}{\mathrm{d}r^n}\right)_0 = \sum B_{j,m}\, n!\, n^2/l(m-n+1) \tag{27}$$

and $j = n+2$.

4. INTRODUCTION OF BOUNDARY CONDITIONS

Let us require that relative displacements of the surfaces produced thermally, w_t, or by elastic deformation under pressure, w_e, or by small initial departures from flatness, w_h, sum to zero throughout the contact zone. It follows that the relative curvature near the origin must also vanish or

$$(\Gamma_t + \Gamma_e^* + \Gamma_h) = 0 \tag{28}$$

to assure contact of adjacent points. Indeed, the same must also be true for higher derivatives

$$(\mathrm{d}^n\Gamma_t/\mathrm{d}r^n + \mathrm{d}^n\Gamma_e^*/\mathrm{d}r^n + \mathrm{d}^n\Gamma_h/\mathrm{d}r^n)_{r=0} = 0 \tag{29}$$

These statements would be true for displacements expressible in McLauren's series near the origin of r. The Γ indicates that the relative elastic curvature components are due to deformation of both bodies. This is accounted for by replacing E by E^* in equation (12), where $E^* \equiv E_1 E_2/(E_1+E_2)$, and the subscripts refer to the two bodies [9], for the case of $\nu_1 = \nu_2$.

Letting Γ_h be a constant quantity, using equation (26) for $(\Gamma_e)_0$, and using equation (6) for $(\Gamma_t)_0$, then equation (28) becomes

$$\hat{G} + \hat{H} = -\, c_2\, \Sigma 4 a_m/(m-1) \tag{30}$$

where $m = 2, 4, 6, \ldots$, and

$$\hat{G} \equiv \frac{\alpha\mu V E^* l}{(1-\nu) K} \tag{31}$$

and

$$\hat{H} \equiv \frac{\Gamma_h E^* l}{(1-\nu^2)\, p_0} \tag{32}$$

Recall that for pressure loading on the contact zone equation (27) calls for

$$\left(\frac{\mathrm{d}^n \Gamma_e}{\mathrm{d}r^n}\right)_0 = \frac{(1-\nu^2)\, p_0}{E^* l^{1+m}} \left[\frac{(n+2)^2}{m-n-1}\right] n!\, c_{n+2}\, a_m \tag{33}$$

For thermal loading equation (5) becomes

$$\left(\frac{\mathrm{d}^n \Gamma_t}{\mathrm{d}r^n}\right) = \frac{\alpha(1+\nu)}{K} \frac{\mu V p_0 a_n n!}{l^n} \tag{34}$$

If Γ_h is constant, then for $n > 0$

$$\left(\frac{\mathrm{d}^n \Gamma_h}{\mathrm{d}r^n}\right)_0 = 0 \tag{35}$$

Equation (29) becomes upon substitution of (33), (34), (35)

$$\frac{a_n}{(n+2)^2} = -\sum \frac{c_{n+2}\, a_m}{m-n-1} \tag{36}$$

Returning now to equation (30) and expanding the summation

$$0.5[\hat{G}+\hat{H}] = 1 - a_2 - 0.333\, a_4 - 0.20\, a_6 \ldots \tag{37}$$

Similarly expanding equation (36) for $n = 2, 4, 6$

$$\begin{aligned} 0.6695\, \hat{G} a_2 &= 1 + 3 a_2 - 3 a_4 - 1 a_6 \\ 0.7110\, \hat{G} a_4 &= 1 + 1.6 a_2 + 5 a_4 - 5 a_6 \\ 0.7365\, \hat{G} a_6 &= 1 + 1.4 a_2 + 2.333\, a_4 + 7 a_6 \end{aligned} \tag{38}$$

Table 2. Contact parameters

	$\hat{G}$	$\hat{H}$	ζ	Λ	a_2	a_4	a_6	V/V_c
Herzian	0	2.901	0.75	—	−0.426	−0.0857	−0.0286	—
Punch	0	0	1.75	—	−0.600	0.600	1.00	—
Thermal	3.73	0	0.584	—	−0.94	0.18	−0.026	—
Thermal	3.73	—	0.584	0	−0.94	0.18	−0.026	∞
with $\hat{p}$								
removed	4.03	—	0.563	0.083	−1.01	0.24	−0.035	2.22
	4.33	—	0.539	0.156	−1.10	0.31	−0.043	1.74
	4.48	—	0.527	0.189	−1.15	0.35	−0.056	1.62
	4.63	—	0.513	0.221	−1.20	0.39	−0.066	1.557
	4.78	—	0.498	0.252	−1.25	0.94	−0.077	1.514
	4.93	—	0.482	0.281	−1.31	0.49	−0.091	1.47

and by integration of the pressure profile

$$\zeta = 1+0.5\,a_2 + 0.331\,a_4 + 0.25\,a_6 \tag{39}$$

These equations may be extended if desired to higher values of m and n.

For any selected $\hat{G}$, equations (38) may be solved for a_i quantities and these may then be inserted into equation (37). If $\hat{H}$ is taken as zero (surfaces are initially flat), then the value of G selected for use in equation (38) must equal the value computed in (39). This precise value may be found by successive trials.

5. CALCULATED RESULTS

There are two cases of special interest because they have well known exact solutions[21,22], these being

a. Hertzian contact of spheres and

b. the frictionless rigid punch.

Calculated contact parameters for these two cases are included in Table 2 where $\Lambda \equiv (l/\hat{m})^2$. In both instances, where $\hat{G}=0$, equations (38) become identical with their plane strain or stress counterparts[16,18], and pressure distributions in neither case change in going from the plane problem to axial symmetry. This is in accord with exact analyses. Table 3 shows good agreement between predictions of the truncated set of equations and exact analysis.

Table 3

Case	$\hat{H}$	ζ
Hertzian, calculated	2.90	0.75
Hertzian, exact	3.06	0.67
Punch, calculated	—	1.75
Punch, exact	—	2.00

Returning now to the thermal-asperity problem, where $H=0$, equations (37) and (38) are satisfied when $\hat{G}=1.674$. Drawing upon the definitions of $\hat{G}$, V_c and Λ,

$$\hat{G} = 0.629\,\Lambda^{1/2}\,V/V_c \tag{40}$$

Inserting the magnitude of $\hat{G}$, it follows that

$$\Lambda = 0.353\,(V_c/V)^2 \tag{41}$$

This result is plotted in Fig. 3, showing that $\Lambda=1$ when $V/V_c=0.594$;

hence, the solution with separated contact between circular patches does not exist below this dimensionless sliding speed.

Let us now examine the case where wear or edge restraints cause the surface to remain flat under uniform heat input, and do not affect the development of zero-average perturbations. To remove the effect of uniform heating from the curvature replace p in equation (31) by $(p-\hat{p})$,

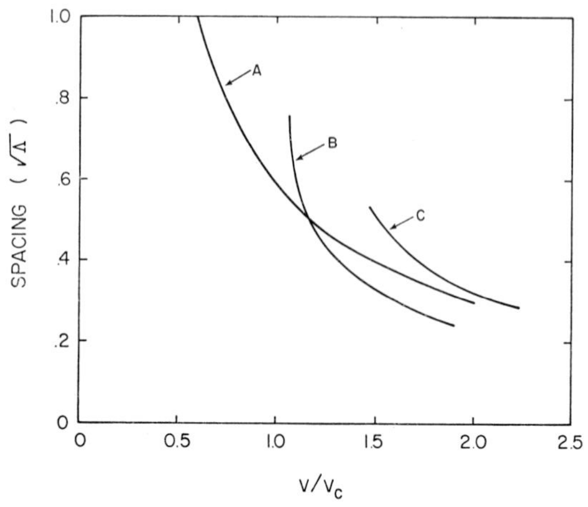

Fig. 3. Dimensionless contact spacing for three cases, as a function of dimensionless sliding speed. A. Circular patches for slabs with unconstrained bending. B. Banded contact from plane-strain analysis of slab contact. C. Circular patches for slabs where wear or edge constraints remove surface bowing under uniform heat input

where $\hat{p}$ is the average pressure on the entire interface. Drawing upon definitions of the quantities involved

$$\hat{p} = p_0 \zeta \Lambda \tag{42}$$

Carrying this through the above derivations, one arrives at the equivalent of equation (37):

$$0.747 \left[\hat{G}(1-\zeta\Lambda) + H\right] = 1 - a_2 - 0.333\, a_4 - \ldots \tag{43}$$

Equation (38) remains unchanged. Solving as before, $\hat{G}$ is assumed, inserted into (38) and a_i quantities are found. These are inserted into equation (43), and for $H=0$, $G(1-\zeta\Lambda)$ is found. Using a_i quantities in equation (39), ζ is found and Λ may then be determined. Typical results are shown in Table 2 under the designation 'Thermal with $\hat{p}$ removed'. The calculations were not carried to higher values of $\hat{G}$ since forbidden negative pressures in the contact zone would have appeared.

These results are plotted in Fig. 3, where it is seen that with small wear, slow advance of sliding speed would permit operation above V_c without axisymmetric contact patches. Results for the plane strain case of banded contacts have been adapted from [17], showing existence of this state for $V \geq V_c$. This would suggest that as speed is increased bands of contact form and these then degenerate into the contact spots which have been observed.

6. CONCLUSIONS

We conclude that with absence of wear, when a thermally conductive material slides on a thermal insulator there is a critical speed above which flat surface contact may change to patch-like contact in the form of nearly axisymmetric contact patches surrounded by areas where the surfaces pull away from one another. Although the series solution was truncated, its predictions are sufficiently close to earlier work to provide reassurance that the calculated quantities are meaningful.

In this analysis the effects of adjacent contact patches on the elastic deformation in a given patch are omitted. In [19] this influence was studied for the plane-stress case and found to be small. Fig. 4 illustrates the small difference between the exact analysis which includes these and the approximate analysis which does not. Similar considerations bear upon the validity of equation (18) when the patch extends too near to the edge

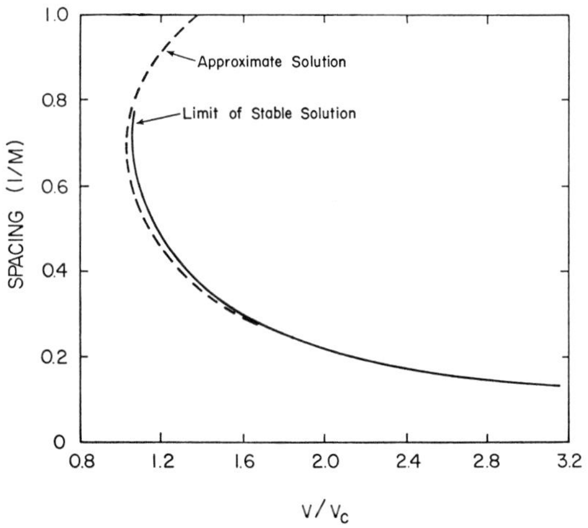

Fig. 4. Solutions for slabs with banded contact, with neglect of elastic influence of neighboring contacts (broken line), and inclusion of their influence (solid line)

of a slab, and restrict the usefulness of this analysis to $m/l \geq 2$ as a rough guide.

It is of significance that adjacent patches only affect the thermal curvature through their contribution of $\hat{p}$. It follows that there is no restriction on their distribution for equation (43) to be valid.

Throughout the analysis a question has been evaded, concerning the magnitude of m. For the case of a single patch m is a measure of face area of the conductive slab. For multiple contacts the spacing may be determined by an initial waviness. Because the longest wave lengths (maximum spacing) become unstable at the lowest speeds, one would expect the smallest permissible number of patches to prevail. This may be limited to three in some cases, to provide for static stability. Further study of this matter will require more careful consideration of the distant constraints on the slabs. Indeed further study is called for on other questions which were not treated here such as the effects of large initial curvature as discussed by Korovchinsky[22], and for cases where the contact patch is moving, for which Ling and Mow[23] have provided Green's functions for temperature and thermal displacement.

ACKNOWLEDGMENT

The research reported here was supported by the U.S. Office of Naval Research Contract N00014–67–A–0356–0022, Mr. Keith Ellingsworth and Lt. R.J. Miller, monitors.

I wish to thank: Mr. B. Banerjee, Mr. S. Heckmann and Mr. R. Kilaparti for their kind help.

REFERENCES

1. J.R. Barber, 'Thermoelastic Instabilities in the Sliding of Conforming Solids', *Proc. Roy. Soc.*, **A312**, 1969, pp. 381–391.
2. J.R. Barber, 'The Influence of Thermal Expansion on the Friction and Wear Process', *Wear*, **10**, 1967, p. 155.
3. J.R. Barber, *Thermal Effects in Friction and Wear*, Dissertation St. John's College, Cambridge, England, 1968.
4. R.A. Burton and H.E. Staph, 'Thermally Activated Seizure of Angular Contact Bearings', *ASLE Trans.*, **10**, 4, 1967, pp. 381–391.
5. R.A. Burton, 'Thermal Aspects of Bearing Seizures', *Wear*, 8, 9, 1965. pp. 157–172.
6. R.A. Burton, 'Effect of Two-Dimensional Sinusoidal Roughness on the Load Support Characteristics of a Lubricant Film', *Trans. ASME*, Series 7, *JOLT*, **85**, 1963, pp. 258–262.
7. T.A. Dow and R.A. Burton, 'Investigation of Thermoelastic Instabilities of Sliding Contact in the Absence of Wear', *Wear*, **19**, 1972, pp. 315–328.
8. T.A. Dow and R.A. Burton, 'The Role of Wear in the Initiation of Thermoelastic Instabilities of Rubbing Contact', *Trans. ASME*, Series 7, *JOLT*, **95**, 1, 1973, pp. 71–75.

9. R.A. Burton, V. Nerlikar and S.R. Kilaparti, 'Thermoelastic Instability in a Seal-like Configuration', *Wear*, **24**, 2, 1973, pp. 169–198.
10. R.A. Burton, 'The Role of Insulating Surface Films on Frictionally Excited Thermoelastic Instability', *Wear*, **24**, 2, 1973, pp. 189–198.
11. J.R. Barber, 'Letter to the Editor', *Wear*, **26**, 1973, pp. 423–428.
12. R.A. Burton, 'Letter to the Editor', *Wear*, **26**, 1973, pp. 425–427.
13. T.A. Dow, *Thermoelastic Instabilities in Sliding Contact*, Dissertation Northwestern University, 1972.
14. L.B. Sibley and C.M. Allen, 'Friction and Wear Behavior of Refractory Materials at High Sliding Velocities and Temperatures', *A*SME Paper 61-Lubs-15.
15. F.E. Kennedy and F.F. Ling, 'A Thermal, Thermoelastic and Wear Simulation of a High-Energy Sliding Contact Problem', *Trans. ASME*, Series 7, *JOLT* **96**, 1974, pp. 496–507.
16. R.A. Burton, S.R. Kilaparti and V. Nerlikar, 'A Limiting Stationary Configuration with Partially Contracting Surfaces', *Wear*, **24**, 2, 1973, pp. 199–206.
17. K.A. Burton and V. Nerlikar, 'Effect of Initial Surface Curvature on Frictionally Excited Thermoelastic Phenomena', *Wear*, **27**, 2, 1974, pp. 195–207.
18. R.A. Burton and V. Nerlikar, 'Large Disturbance Solutions for Initially Flat, Frictionally Heated, Thermoelastically Deformed Surfaces', to be printed in *ASME*, Series 7, *JOLT*.
19. J.R. Barber, 'Distortion of the Semi-Infinite Solid Due to Transient Surface Heating', *Int. J. Mech. Sci.*, **14**, 1972, pp. 377–393.
20. S. Timoshenko and J. Goodier, *Theory of Elasticity*, 3rd ed., McGraw-Hill, New York, 1970.
21. F.F. Ling, *Surface Mechanics*, John Wiley and Sons (New York, 1973).
22. M.V. Korovchinsky, 'The Plane Contact Problem of Thermoelasticity During Quasi Stationary Heat Generation on Contact Surfaces', *ASME Trans.*, Series D, *Journal of Basic*, **87**, 1965, pp. 811–817.
23. F.F. Ling and V.C. Mow, 'Surface Displacement of a Convective Elastic Half Space', *ASME Trans.*, Series D, *Journal of Basic Engng.*, **87**, 1966, pp. 814–816.

Dynamic contact stresses produced by impact in elastic plates of finite thickness

Y. M. Tsai

1. INTRODUCTION

The stresses set up when two smooth elastic bodies are pressed together and the force between them is normal to the surface of contact were first analyzed by Hertz[1]. Analytical expressions for the stress distribution within the two bodies were obtained in terms of the elastic constants of the bodies, the radii of curvature in the underformed bodies in the region of contact, and the total force between them. A more detailed account of the stress distribution for the contact between two elastic spheres or between a sphere and a half-space was later given by Huber[2]. On the basis of his static solutions, Hertz's theory of impact was established under the assumption that near the point of contact the stresses and strains may be computed at any instant as though the contact were static[1, 3]. By the solution of three-dimensional equations of motion, the above assumption was recently studied as a function of contact time and contact radius[4].

The static and dynamic Hertz theory has been found useful in many experimental investigations[5, 6]. In practice, however, the material body studied is not of infinite thickness, as is required in the Hertz solution, but of finite thickness. The effect that plate thickness has on the stress distribution around the contact area was recently studied both theoretically and experimentally[7–10]. For the case of a plate subjected to symmetrical rigid indentations on its upper and lower surfaces, the normal contact stress and the radial surface stress were shown to approach the Hertz solution for a small area of contact[7]. When the circle of the contact area became compatible with the thickness of the plate, however, the above stresses were shown to be much higher than the corresponding values in a half-space. This magnification of the tensile radial stress was demonstrated experimentally; it appreciably reduced the magnitude of the load required to produce conical fractures around the contact region in glass plates by pressing steel balls onto the glass surfaces[8]. The static solutions obtained[7] were extended to the dynamic theory and applications[9]. It was shown both theoretically and experimentally that for sufficiently large steel balls the contact time decreases with decreasing plate thickness as a result of the above stress magnification near the impact area.

The three-dimensional problem of a plate under transverse static contact pressure was studied by varying the indenter radius and the ratio of plate thickness to support radius[10]. Under the combined action of the symmetrical and antisymmetrical components at small spans of support, the loading capacity of a plate was shown to increase significantly, as compared to the corresponding symmetrical indentations[8,10]. This is a significant gain in loading capacity resulting from proper arrangements of support. The above three-dimensional combined action also exists in a finitely thick plate under transverse loading.

At present, the solutions of three-dimensional equations of motion are used to investigate the implication of the assumption of locally static response that was introduced in the earlier plate impact problem[9]. By a similar approach, a dynamic theory associated with the earlier static solution[10] is developed to account for the effect that the plate thickness has on the contact time and the contact stresses in a plate subject to transverse impact.

2. SYMMETRICAL CONTACT STRESS

The problem considered in the first place is the impact of a spherical body on the surface of an elastic plate overlying a smooth rigid foundation. The spherical projectile is assumed to be a rigid body as was done in earlier works[4,7–10]. The plate has a finite thickness H which is comparable to the radius 'a' of the contact area between the plate and the projectile. It is assumed that during impact an axis of symmetry of the projectile is perpendicular to the impact surface and that the shear stress vanishes inside the contact area. The contact area generally varies with respect to time. Thus, the problem is an axisymmetric one with moving boundary conditions, the loading area being finite. The three-dimensional equations of motion are to be solved using the cylindrical coordinates (r, φ, z). The problem considered is mathematically equivalent to that of a plate of thickness $2H$ subject to the same impact on both upper and lower surfaces. On the surfaces of impact, the shear stresses σ_{zr} and $\sigma_{z\varphi}$ are assumed to be vanishing, and the normal contact stresses $\sigma(r, t)$ are formally known functions of the radial distance r and the time t. In fact, the function $\sigma(r, t)$ can be determined as the solution of an integral equation. The dynamic boundary conditions concerned can be prescribed as on the surfaces $z = \pm H$ for $t > 0$

$$\sigma_{zr} = \sigma_{z\varphi} = 0 \tag{1}$$

and

$$\sigma_{zz} = \begin{cases} \sigma(r, t), & r \leq a(t) \\ 0 & r > a(t) \end{cases} \tag{2}$$

where σ_{zz} is the normal z-component of the stresses. The equations of motion to be satisfied for a homogeneous, isotropic solid are

$$(\lambda+2\mu)\,\nabla\,(\nabla\cdot\underline{U}) - \mu\nabla\times(\nabla\times\underline{U}) = \rho'\,\frac{\partial^2\underline{U}}{\partial t^2} \tag{3}$$

where $\underline{U}$ is the displacement vector, ρ' is the density of the medium, and λ and μ are the Lamé constants. The displacement $\underline{U}$, in general, has three components, i.e., $\underline{U} = (U_r, U_\varphi, U_z)$. Since the problem considered is an axially symmetric one, U_φ and all the derivatives with respect to φ must vanish everywhere. Under these conditions, the dilatation is

$$\Delta = \nabla\cdot\underline{U} = \frac{1}{r}\frac{\partial}{\partial r}(rU_r) + \frac{\partial U_z}{\partial z} \tag{4}$$

and the rotation has only one non-vanishing circumferential component

$$\Omega = \frac{\partial U_r}{\partial z} - \frac{\partial U_z}{\partial r} \tag{5}$$

The equations of motion (3) are satisfied if the following two wave equations are satisfied:

$$\frac{1}{r}\frac{\partial}{\partial r}\left(r\frac{\partial\Delta}{\partial r}\right) + \frac{\partial^2\Delta}{\partial z^2} = \frac{1}{C_1^2}\frac{\partial^2\Delta}{\partial t^2} \tag{6}$$

and

$$\frac{\partial}{\partial r}\left(\frac{1}{r}\frac{\partial}{\partial r}(r\Omega)\right) + \frac{\partial^2\Omega}{\partial z^2} = \frac{1}{C_2^2}\frac{\partial^2\Omega}{\partial t^2} \tag{7}$$

where the wave speeds are

$$C_1^2 = \frac{\lambda+2\mu}{\rho'}, \quad C_2^2 = \frac{\mu}{\rho'} \tag{8}$$

To solve the equations, Laplace transforms $f^*(p)$ are operated over the time t. Furthermore, Hankel transforms are applied over r and are defined as

$$\hat{U}^m = \int_0^\infty U(r)\,rJ_m(sr)\,\mathrm{d}r \tag{9}$$

Proper transformations reduce (6) and (7) into simple, ordinary equations of z[4, 11, 12, 13]. The transformed solutions consist of symmetrical and asymmetrical parts[13, 14]. The symmetrical components of the solutions which can satisfy the prescribed boundary conditions are

$$(\hat{\Delta}^0)_{\mathrm{I}}^* = A_1\cosh z\alpha' \quad\text{and}\quad (\hat{\Omega}^1)_{\mathrm{I}}^* = B_1\sinh z\beta' \tag{10}$$

where

$$\alpha' = (s^2+k_1^2)^{1/2}, \quad \beta' = (s^2+k_2^2)^{1/2}$$
$$k_1 = p/C_1 \quad \text{and} \quad k_2 = p/C_2 \tag{11}$$

If the transformed expressions for stresses obtained in the earlier work[11] are used, the boundary conditions (1) and (2), respectively, give

$$0 = (k_2^2+2s^2)\sinh H\beta' B_1 - 2k^2 s\alpha' \sinh H\alpha' A_1 \tag{12}$$

and

$$\rho p^2 \bar{\sigma}^*(s,p)/\mu^2 = k^2(k_2^2+2s^2)\cosh H\alpha' A_1 - 2s\beta' \cosh H\beta' B_1 \tag{13}$$

where the zero-order Hankel transform of the normal contact stress is

$$\bar{\sigma} = \int_0^a \sigma(r,t)\, r\, J_0(rs)\, \mathrm{d}r \tag{14}$$

and

$$k^2 = k_2^2/k_1^2 = 2(1-\nu)(1-2\nu) \tag{15}$$

The boundary condition $\sigma_{z\varphi}=0$ is automatically satisfied when equations (4) and (5) are written. The solutions of (12) and (13) are

$$A_1 = k_2^2(2s^2+k_2^2)\,\bar{\sigma}^* \sinh H\beta/\mu k^2 G_{\mathrm{I}} \tag{16}$$

and

$$B_1 = 2s\alpha' \rho p^2\, \bar{\sigma}^* \sinh H\alpha'/\mu^2 G_{\mathrm{I}} \tag{17}$$

where the symmetrical characteristic equation is

$$G_{\mathrm{I}}(s,p) = (2s^2+k_2^2)^2 \cosh H\alpha' \sinh H\beta' -$$
$$- 4s^2 \alpha'\beta' \sinh H\alpha' \cosh H\beta' \tag{18}$$

In terms of (10), (16), and (17), the transformed vertical displacement on the surfaces can be determined[11] as

$$U_z^* = \int_0^\infty k_2 F_{\mathrm{I}} \bar{\sigma}^* J_0(sr)\, \mathrm{d}s \tag{19}$$

where

$$F_{\mathrm{I}} = sk_2 \alpha' \sinh H\alpha' \sinh H\beta'/G_{\mathrm{I}} \tag{20}$$

The Laplace inversion of equation (19) can be obtained by considering $F_{\mathrm{I}}(s,p)$ as the transform of $f_{\mathrm{I}}(s,t)$. To obtain the inversion (20), let $p=sC_2\zeta$ and cut the complex ζ-plane as shown in the earlier work[4]. The inversion of (20) can be written as

$$f_{\mathrm{I}}(s,t) = \frac{C_2}{2\pi i}\int_{\delta-i\infty}^{\delta+i\infty} h_{\mathrm{I}}/g_{\mathrm{I}}\, \mathrm{d}\zeta \tag{21}$$

where

$$h_{\mathrm{I}} = \zeta\alpha \sinh Hs\alpha \sinh Hs\beta\, e^{sC_2\zeta t} \tag{22}$$

$$g_{\mathrm{I}} = (2+\zeta^2)^2 \cosh Hs\alpha \sinh Hs\beta - $$
$$- 4\alpha\beta \sinh Hs\alpha \cosh Hs\beta \tag{23}$$

$$\alpha = (1+\zeta^2/k^2)^{1/2} \quad \text{and} \quad \beta = (1+\zeta^2)^{1/2} \tag{24}$$

Along the imaginary axis for $|\eta| < 1$, g_{I} has no root except for a double root at the origin $\zeta = 0$, around which equation (23) has the expansion

$$g_{\mathrm{I}}(s, \zeta) = (2Hs + \sinh 2Hs)\, \zeta^2/2(1-\nu) + 0(\zeta^3) \tag{25}$$

For $1 < |\eta| < k$, the root of g_{I}, γ_2, is determined in terms of $u = 2Hs$ as shown in Fig. 1. The root occurs periodically as a period of 2π. When H

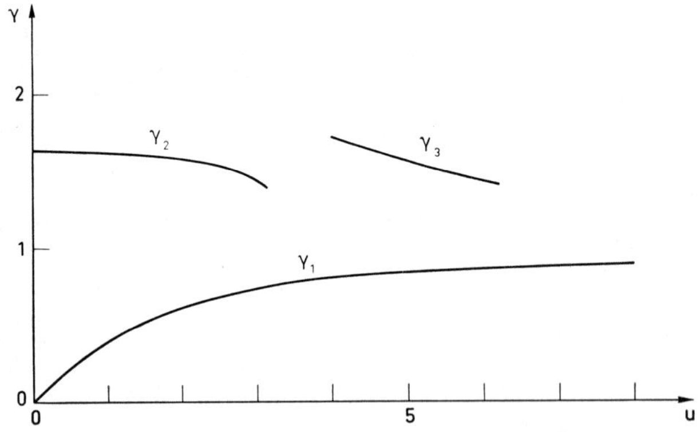

Fig. 1. Symmetrical root γ_2 and antisymmetrical roots γ_1 and γ_3 for $\nu = 0.25$

approaches zero, γ_2^2 is equal to $2/(1-\nu)$, which gives the thin plate wave speed $C_L = \gamma_2 C_2$. For $k < |\eta| < \infty$, g_{I} has infinitely many roots γ_{2m}, $m = 2, 3, \ldots,$. The integrand in equation (21) has the same expression on both sides of the branch cut. Therefore, the integration of equation (21) along the contour shown in the earlier work[4] gives

$$f_{\mathrm{I}} = C_2[2(1-\nu)\, H(t) \sinh^2 Hs/(2Hs + \sinh 2Hs) - $$
$$- 2 \sum_{m=1}^{\infty} Q_1(s, \gamma_{2m}) \cos sC_2\gamma_{2m} t] \tag{26}$$

where $H(t)$ is the Heaveside function and

$$Q_{\mathrm{I}}(s, \gamma_{2m}) = 2h_{\mathrm{I}}(s, \gamma_{2m})/g_{\mathrm{I}}'(s, \gamma_{2m}) \tag{27}$$

The term $g_{\mathrm{I}}'(s, \gamma_{2m})$ is the derivative of g_{I} evaluated at $\eta = \gamma_{2m}$. In terms of equation (26), the inversion of equation (19) can be written as

$$U_z(r,t) = \frac{2(1-\nu)}{\mu} \int_0^\infty \frac{\sinh^2 Hs}{2Hs+\sinh 2Hs}\, \bar{\sigma}\, J_0(sr)\, \mathrm{d}s -$$

$$-\frac{1}{\mu}\frac{\partial}{\partial t}\int_0^\infty \sum_{m=1}^{\infty} Q_1(s,\gamma_{2m})\, J_0(sr) \int_0^t \times$$

$$\times \cos[sC_2\, \gamma_{2m}(t-\tau)]\, \bar{\sigma}\, \mathrm{d}\tau \mathrm{d}s \tag{28}$$

The first term on the right-hand side of equation (28) is the associated static normal displacement and is the same as that obtained in an earlier work[7]. The second term apparently accounts for the effect of stress waves. Equation (28) will result in an integral equation for determining the unknown $\sigma(r,t)$. In order to find $\sigma(r,t)$, it is assumed that the vertical displacement projectile inside the contact area can be described by $g(r,t)$, i.e., $U_z(r,t)=g(r,t)$ for $r \leq a(t)$ at $z=H$. For projectiles of simple geometry, $g(r,t)$ can generally be determined. When equation (28) is inverted, the associated static equation is split into half-space and thickness-effect terms as before[7]. If inversion procedures similar to those used in earlier works[4,7] are applied, equation (28) results in the following two equations:

$$\sigma(r,t) = \sigma_{st} - \frac{2}{\pi(1-\nu)}\int_r^a \frac{\partial^2 W_1(\xi,t)/\partial t \partial \xi}{(\xi^2-r^2)^{1/2}}\, \mathrm{d}\xi \tag{29}$$

and

$$g(0,t) + a\int_0^a \frac{g'(m,t)\, \mathrm{d}m}{(a^2-m^2)^{1/2}} + \frac{C}{2H}\int_0^\infty B_1(u) \cos\left(\frac{au}{2H}\right) \bar{\sigma}\, \mathrm{d}\lambda \mathrm{d}u +$$

$$+\frac{1}{\mu}\frac{\partial W_1(a,t)}{\partial t} = 0 \tag{30}$$

where the associated static plate contact stress is

$$\sigma_{st} = -\frac{2}{\pi C}\int_r^a \frac{\mathrm{d}\xi}{(\xi^2-r^2)^{1/2}} \frac{\partial}{\partial \xi}\int_0^\xi \frac{\xi g'(m,t)\, \mathrm{d}m}{(\xi^2-m^2)^{1/2}} +$$

$$+\frac{1}{2\pi H^2}\int_0^a \sigma(\lambda,t)\, \lambda \int_0^\infty B_1(u)\, J_0\left(\frac{ru}{2H}\right)\int_r^a \frac{\sin(\xi u/2H)\, \mathrm{d}\xi}{(\xi^2-r^2)^{1/2}}\, \mathrm{d}u \mathrm{d}\lambda \tag{31}$$

$$B_1(u) = (1+u-\mathrm{e}^{-u})/(u+\sinh u) \tag{32}$$

$$u = 2Hs \quad \text{and} \qquad C = (1-\nu)/\mu \tag{33}$$

Furthermore, the wave effect integral is

$$W_{\mathrm{I}}(r,t) = \frac{1}{2H}\int_0^{\infty}\sum_{m=1}^{\infty} Q_{\mathrm{I}}(u,\gamma_{2m})\cos\left(\frac{ru}{2H}\right)\times$$

$$\times\int_0^t \cos\left[\frac{uC_2\,\gamma_{2m}(t-\tau)}{2H}\right]\bar{\sigma}\,d\tau du \tag{34}$$

Equations (29) and (30), together with Newton's equation of motion, form a system of three integral-differential equations which are sufficient to determine the three dynamic unknowns $\sigma(r, t)$, $a(t)$, and $g(0, t)$. Thus, the problem is well defined and the procedures similar to those used in the earlier work [4] can be used to solve for the unknowns. In the evaluation of the wave-effect integral (34), the integration has values only over the intervals of u in which the symmetrical roots γ_{2m} have definite values. The formulation of a simplified theory will be discussed in Section 4.

3. FLEXURE CONTACT STRESS

The problem here concerns an elastic plate of finite thickness $2H$ which is subjected to the impact of an axisymmetric projectile on its upper surface $z = H$. On its lower surface $z = -H$ it is axisymmetrically supported on a ring foundation with radius ρ. The size of ρ is, in general, comparable to the size of H. Therefore, this is a three-dimensional impact problem. In order to solve the problem, it is assumed that the shear stress vanishes on the surfaces, that the normal contact stress can be described by the formally known function $\sigma(r, t)$, and that the reaction along the support can be regarded as a line load [10]. Indeed, the function $\sigma(r, t)$ can be solved from an integral equation. Thus the boundary conditions with $\sigma_{r\phi} = \sigma_{rz} = 0$ for all r at $z = \pm H$ can be prescribed as

$$\text{at } z = H:\ \sigma_{zz} = \begin{cases}\sigma(r,t) & r \le a(t)\\ 0 & r > a(t)\end{cases} \tag{35}$$

but

$$\text{at } z = -H:\ \sigma_{zz} = \frac{P}{2\pi r}\,\delta(\rho - r),\ P = 2\pi\int_0^a \sigma r\,dr \tag{36}$$

where $\delta(r)$ is the delta function. The above normal stresses may be considered as the sum of the following symmetrical and antisymmetrical stresses:

$$\text{symmetrical: } \sigma_{zz}^{\mathrm{I}} = \frac{1}{2}\left[\sigma(r,t) + \frac{P}{2\pi r}\,\delta(\rho - r)\right] \quad \text{at } z = \pm H \tag{37}$$

$$\text{antisymmetrical: } \sigma_{zz}^{\mathrm{II}} = \pm\frac{1}{2}\left[\sigma(r,t) - \frac{P}{2\pi r}\,\delta(\rho - r)\right] \quad \text{at } z = \pm H \tag{38}$$

The superscripts or subscripts I and II will hereafter indicate quantities in symmetrical and antisymmetrical problems, respectively. Consequently, the quantities in the problem considered will be the sum of the corresponding symmetrical (I) and antisymmetrical (II) quantities. The procedures for solving the problem are similar to those used in the preceding section. If σ in the preceding section is replaced here by σ^{I}_{zz}, the results obtained above can be regarded as the symmetrical quantities. In a form similar to equation (10), the solutions which can satisfy the antisymmetrical boundary conditions are

$$(\hat{\Delta}^{0})^{*}_{\mathrm{II}} = A_2 \sinh z\alpha' \quad \text{and} \quad (\hat{\Omega}^{1})^{*}_{\mathrm{II}} = B_2 \cosh z\beta' \tag{39}$$

In satisfying the boundary conditions, the antisymmetrical characteristic equation in the form similar to equation (23) is

$$g_{\mathrm{II}} = (2+\zeta^2)^2 \sinh Hs\alpha \cosh Hs\beta - 4\alpha\beta \sinh Hs\beta \cosh Hs\alpha \tag{40}$$

The term g_{II} has a double root at $\zeta = 0$ and single roots γ_1 for $|\eta| < 1$ and γ_3 for $1 < \eta < k$. The value of γ_1 ranges from zero to that corresponding to Rayleigh wave speed. Both γ_1 and γ_3 are determined and shown in Fig. 1. For $k < |\eta| < \infty$, g_{II} has infinitely many roots γ_{2m+1}, $m = 2, 3, \ldots,$. The equation similar to equation (28) is evaluated for the vertical displacement as

$$U^{\mathrm{II}}_z(r,t) = \frac{2(1-\nu)}{\mu} \int_0^\infty \frac{\cosh^2 Hs}{\sinh 2Hs - 2Hs}\, \bar{\sigma}^{\mathrm{II}}_{zz}\, J_0(sr)\, \mathrm{d}s - $$
$$ - \frac{1}{\mu}\frac{\partial}{\partial t} \int_0^\infty \sum_{m=1}^{\infty} Q_{\mathrm{II}}(s, \gamma_{2m-1})\, J_0(sr) \times $$
$$ \times \int_0^t \cos\left[sC_2\, \gamma_{2m-1}(t-\tau)\right] \bar{\sigma}^{\mathrm{II}}_{zz}\, \mathrm{d}\tau \mathrm{d}s \tag{41}$$

where

$$Q_{\mathrm{II}} = 2\, h_{\mathrm{II}}(s, \gamma_{2m+1}) / g'_{\mathrm{II}}(s, \gamma_{2m+1}) \tag{42}$$

and

$$h_{\mathrm{II}} = \zeta\alpha \cosh Hs\alpha \cosh Hs\beta\, \mathrm{e}^{sC_2\zeta t} \tag{43}$$

The term $\bar{\sigma}^{\mathrm{II}}_{zz}$ is the zero-order Hankel transform of $\sigma^{\mathrm{II}}_{zz}$ in equation (38). The first term on the right-hand side of equation (41) is the associated static normal displacement which is the same as that obtained in earlier work[10]. The second term apparently accounts for the effect of stress waves. Equations similar to (29) and (30) can be established, following the preceding and those procedures used in the earlier work[10]. The equations obtained after calculations are

$$\sigma(r,t) = \sigma_{st} - \frac{2}{\pi(1-\nu)} \int_r^a \frac{\partial^2 W_{\mathrm{II}}(\xi, t)/\partial t \partial \xi}{(\xi^2 - r^2)^{1/2}}\, \mathrm{d}\xi \tag{44}$$

and

$$g(0,t) + a\int_0^a \frac{g'(m,t)\,\mathrm{d}m}{(a^2-m^2)^{1/2}} + \frac{C}{2H}\int_0^\infty \{[\bar{\sigma}^{\mathrm{I}}_{zz}\, B_{\mathrm{I}}(u) - \bar{\sigma}^{\mathrm{II}}_{zz}\, B_{\mathrm{II}}(u)] \times$$

$$\times \cos\left(\frac{au}{2H}\right) + B_{\mathrm{III}}(u)\,\bar{\sigma}^{\mathrm{II}}_{zz}\}\,\mathrm{d}u + \frac{1}{\mu}\frac{\partial W_{\mathrm{II}}(a,t)}{\partial t} = 0 \qquad (45)$$

where the associated static plate contact stress is

$$\sigma_{st} = -\frac{2}{\pi C}\int_r^a \frac{\mathrm{d}\xi}{(\xi^2-r^2)^{1/2}}\frac{\partial}{\partial \xi}\int_0^\xi \frac{\xi g'(m,t)\,\mathrm{d}m}{(\xi^2-m^2)^{1/2}} +$$

$$+\frac{1}{2\pi H^2}\int_0^\infty [\bar{\sigma}^{\mathrm{I}}_{zz} B_{\mathrm{I}} - \bar{\sigma}^{\mathrm{II}}_{zz} B_{\mathrm{II}}]\, u \int_r^a \frac{\sin(\xi u/2H)\,\mathrm{d}\xi}{(\xi^2-r^2)^{1/2}}\,\mathrm{d}u \qquad (46)$$

$$B_{\mathrm{II}} = (1+u+\mathrm{e}^{-u})/(\sinh u - u) \qquad (47)$$

$$B_{\mathrm{III}} = [u\sinh u/2 + 4(1-\nu)\cosh u/2]/[(\sinh u - u)\,2(1-\nu)] \qquad (48)$$

and

$$W_{\mathrm{II}}(r,t) = \frac{1}{2H}\int_0^\infty \int_0^t \sum_{m=1}^\infty \left\{\bar{\sigma}^{\mathrm{I}}_{zz}\, Q_{\mathrm{I}}(u,\gamma_{2m})\cos\left[\frac{uC_2\,\gamma_{2m}(t-\tau)}{2H}\right] - \right.$$

$$\left. - \bar{\sigma}^{\mathrm{II}}_{zz}\, Q_{\mathrm{II}}(u,\gamma_{2m-1})\cos\left[\frac{uC_2\,\gamma_{2m-1}(t-\tau)}{2H}\right]\right\}\cos\left(\frac{ru}{2H}\right)\mathrm{d}\tau\mathrm{d}u \qquad (49$$

The term $\bar{\sigma}^{\mathrm{I}}_{zz}$ is the zero-order Hankel transform of σ^{I}_{zz} in equation (37). The methods for determining the three dynamic unknowns $\sigma(r,t)$, $a(t)$, and $g(0,t)$ in terms of equations (44) and (45) are similar to those mentioned previously in regard to equation (29) and (30).

4. PRACTICAL THEORY

It was assumed in the Hertz impact theory[1,3] that near the point of contact the stresses and strains may be computed at any instant as though the contact were static. The same assumption was introduced in earlier work[9] concerning the impact problem of a finitely thick plate overlying a rigid foundation. It can now be seen from equation (28) that the assumption is equivalent to keeping the first order term, the first term, and dropping the higher order terms, the second term on the right-hand side of equation (28). For the flexure impact problem with the ρ comparable to H described above, it is assumed that in the neighborhood of the contact region the stresses and strains may be computed at any instant as though the contact were static. Under this assumption, the second terms on the right-hand sides of equations (28) and (41) are both dropped. When only the first order terms are retained, the field equations obtained reduce to the corresponding equations in the earlier work[10].

To obtain a complete set of equations, an additional dynamic equation in terms of those reduced field equations will be set up here on the basis of Newton's equation of motion. A practical, general theory is to be established following the procedures used before[9]. For a rigid sphere of radius R impinging on the surface of the plate, the shape of the penetration can be described[4, 10] as

$$g(r,t) = r^2/2R - \alpha(t) \tag{50}$$

where α is the maximum depth of penetration. In terms of equation (50), the associated half-space normal stress is

$$p_H(x,t) = -4a(1-x^2)^{1/2}/C\pi R \tag{51}$$

where $x = r/a$. Equation (51) is the integration of the first term on the right-hand side of (46). In terms of equations (50) and (51), the normal contact stress obtained from equation (46) was written[10] as

$$\sigma_{st}(x,t) = p_H(0,t)\, k_p(x, a/H, \rho/H) \tag{52}$$

The nondimensional function k_p was determined by the process of successive approximation[10]. The total force acting on the plate may be obtained from equation (52) by integrating over r, and the result was written[10] as

$$P = P_H k_P(a/H,\ \rho/H) \tag{53}$$

where the half-space load corresponding to a is

$$P_H = -8a^3/3CR \tag{54}$$

and the nondimensional function computed[10] is

$$k_P = 3\int_0^1 k_p(x,\ a/H,\ \rho/H)\ x\ \mathrm{d}x \tag{55}$$

To establish a dynamic relationship between the mass of a projectile and the reaction of a plate, equation (45), excluding the last term, needs to be evaluated. In evaluations, the denominators in equations (47) and (48) vanish at the order of u^3 as u tends to zero[10, 15]. This singularity will be removed later with proper arrangement of the terms involved. In terms of equations (50) and (52), the transforms of (37) and (38) are, respectively,

$$\bar{\sigma}^{\mathrm{I}}_{zz} = -\frac{2a^3}{C\pi R}\,\bar{\sigma}^{\mathrm{I}} \quad \text{and} \quad \bar{\sigma}^{\mathrm{II}}_{zz} = -\frac{2a^3}{C\pi R}\,\bar{\sigma}^{\mathrm{II}} \tag{56}$$

where

$$\bar{\sigma}^{\mathrm{I}} = \int_0^1 k_p\left(x, \frac{a}{H}, \frac{\rho}{H}\right) x \left[J_0\left(\frac{axu}{2H}\right) + J_0\left(\frac{u\rho}{2H}\right)\right] \mathrm{d}x \tag{57}$$

and

$$\bar{\sigma}^{\mathrm{II}} = \int_0^1 k_p\left(x, \frac{a}{H}, \frac{\rho}{H}\right) x \left[J_0\left(\frac{axu}{2H}\right) - J_0\left(\frac{u\rho}{2H}\right)\right] \mathrm{d}x \tag{58}$$

Equation (45) can now be written in terms of (50) and (56) as

$$\alpha = \alpha_H k_\alpha(a/H,\ \rho/H) \tag{59}$$

where

$$\alpha_H = a^2/R \tag{60}$$

and the nondimensional function is

$$k_\alpha = 1 - \frac{2}{\pi}\left(\frac{a}{2H}\right)\int_0^\infty \left\{\bar{\sigma}^{\mathrm{I}} B_{\mathrm{I}} \cos\left(\frac{ua}{2H}\right) - \left[B_{\mathrm{II}} \cos\left(\frac{ua}{2H}\right) - B_{\mathrm{III}}\right]\bar{\sigma}^{\mathrm{II}}\right\} du \tag{61}$$

For convenience later, the first derivative of α is obtained from equation (59) as

$$\frac{\partial \alpha}{\partial a} = \alpha'_H k_{\alpha'}(a/H,\ \rho/H) \tag{62}$$

where

$$\alpha'_H = 2a/R \tag{63}$$

and

$$\begin{aligned} k_{\alpha'} = {} & 1 + \frac{1}{\pi}\left(\frac{a}{2H}\right)^2 \int_0^\infty (B_{\mathrm{I}}\bar{\sigma}^{\mathrm{I}} - B_{\mathrm{II}}\bar{\sigma}^{\mathrm{II}})\, u \sin\left(\frac{ua}{2H}\right) du - \\ & - \frac{H}{2\pi a}\int_0^\infty \left\{B_{\mathrm{I}} \cos\left(\frac{ua}{2H}\right) \partial\left[\left(\frac{a}{H}\right)^3 \bar{\sigma}^{\mathrm{I}}\right] \Big/ \partial\left(\frac{a}{H}\right) - \right. \\ & \left. - \left[B_{\mathrm{II}} \cos\left(\frac{ua}{2H}\right) - B_{\mathrm{III}}\right] \partial\left[\left(\frac{a}{H}\right)^3 \bar{\sigma}^{\mathrm{II}}\right] \Big/ \partial\left(\frac{a}{H}\right)\right\} du \end{aligned} \tag{64}$$

As u tends to zero, the singularities of the denominators mentioned above are removed from the expressions. In other words, the integrands in equations (61) and (64) assume definite values as u vanishes. Thus, the nondimensional functions k_α and $k_{\alpha'}$ are now amenable to numerical computation for various values of a, H and ρ.

If a spherical projectile with mass 'm' impinges on the upper surface of the plate, Newton's equation of motion can be written

$$m\ddot{\alpha} = -P \tag{65}$$

where the dot means differentiation with respect to time. If equation (65) is multiplied by $\dot{\alpha}$ and if the values of P and α' are substituted from equations (53) and (62), respectively, the integration of (65) over t gives

$$\frac{m}{2}(\dot{\alpha}^2 - V^2) = -\int_0^a P_H \alpha'_H \, da - \int_0^a P_H \alpha'_H [k_P k_{\alpha'} - 1] \, da \tag{66}$$

where V is the initial velocity of approach of the sphere. If the parameter a/H of k_p and $k_{\alpha'}$ is replaced by y, equation (66) can be written as

$$\frac{m}{2}(\dot{\alpha}^2 - V^2) = -\frac{16a^5}{15CR^2} k_V^2 \left(\frac{a}{H}, \frac{\rho}{H}\right) \tag{67}$$

where the nondimensional function is

$$k_V^2 = 1 + 5\left(\frac{H}{a}\right)^5 \int_0^{a/H} y^4 \left[k_P\left(y, \frac{\rho}{H}\right) k_{\alpha'}\left(y, \frac{\rho}{H}\right) - 1\right] dy \tag{68}$$

When the impact reaches its maximum distance of approach with maximum contact radius a_1, the velocity of the projectile vanishes, i.e., $\dot{\alpha} = 0$. Under this condition equation (67) becomes

$$V = V_H k_V(a_1/H, \rho/H) \tag{69}$$

where the associated impact velocity is

$$V_H = (32a_1^5/15\,CR^2 m)^{1/2} \tag{70}$$

For a given impact velocity, equation (69) can be used to determine the maximum contact radius a_1 [9], as will be explained later. By the procedures used in the earlier work[9], equation (67) can be integrated over the time t and written as

$$t = \frac{2a_1^2}{RV_H} \int_0^{a/a_1} \frac{x k_{\alpha'}(a_1 x/H, \rho/H) \, dx}{[k_V^2\,(a_1/H, \rho/H) - x^5\, k_V(a_1 x/H, \rho/H)]^{1/2}} \tag{71}$$

This equation can be used to determine the dynamic loading curve and the value of $a(t)$ during the process of impact. When a reaches its maximum a_1, t in equation (71) equals half the total contact time T. From (71), the total contact time can be written as

$$T = T_H k_T(a_1/H, \rho/H) \tag{72}$$

where the associated half-space contact time is

$$T_H = k' a_1^2/RV_H, \quad k' = 2.94324 \tag{73}$$

and

$$k_T = \frac{4}{k'} \int_0^1 \frac{x k_{\alpha'}(a_1 x/H, \rho/H) \, dx}{[k_V^2(a_1/H, \rho/H) - x^5\, k_V^2(a_1 x/H, \rho/H)]^{1/2}} \tag{74}$$

If a_1 is determined from equation (69), eq. (72) can be used to determine the contact time T.

All the above complicated nondimensional functions are calculated by using an electronic computer as a function of the nondimensional parameters a/H and ρ/H. Only the curves for $k_V(a/H, \rho/H)$ in equation (69) and $k_T(a/H, \rho/H)$ in (72) are shown in Fig. 2 and Fig. 3, respectively,

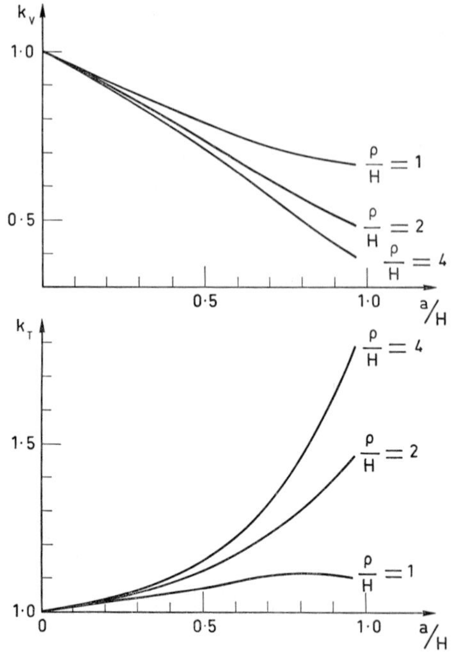

Fig. 2. Velocity correction factor $k_V(a/H, \rho/H) = V/V_H$

Fig. 3. Contact time correction factor $k_T(a/H, \rho/H) = T/T_H$

a_1 being replaced by a for convenience in notation and application. These two curves are of practical interest for determining the maximum contact time and the critical tensile stresses [9]. The calculation of k_T from equation (74) involves singular integrands when x approaches unity. This difficulty was overcome using procedures similar to those used in earlier work [9]. The calculated value of k_T (Fig. 3) indicates that the contact time increases with increasing a/H for $a/H < 0.8$. In other words, for the same impact velocity, indenter, and support span, the contact time increases with decreasing plate thickness. The variation of contact time respect to plate thickness obtained here is opposite to that obtained before for a different type of impact [9].

The results obtained can be applied to the impact problem in which a sphere of radius R impinges on the upper surface of an elastic plate of thickness $2H$ resting on a ring foundation. The associated static problem was investigated in the earlier work[10]. For determining dynamic quantities associated with the impact problem, a trial-and-error method similar to that used in earlier works[8,9,10] is to be described here. For a given V, R, H, and ρ, the maximum contact radius a_1 may first be estimated from equation (70) by choosing V_H larger than V. For this a_1, a_1/H is calculated to determine the value of k_V in Fig. 2. These k_V and V_H values are then substituted in equation (69) to see of their product equals the given value V. Usually, two or three such trials will give a sufficiently accurate result. Once a_1 has been calculated, T and P can be determined from equations (72) and (53), respectively. Furthermore, the stresses concerned can be calculated as described in the earlier work[10]. The stresses which are of importance in practical problems are the following two possible maximum tensile stresses. The first maximum tensile stress is the radial stress along the contact circle:

$$\sigma_{rr}(a, H) = {}_H\sigma_{rr}(a)\, k_\sigma^+(\nu, a/H, \rho/H) \tag{75}$$

where the associated half-space tensile stress is

$$_H\sigma_{rr}(a) = -(1-2\nu)\, P_H/2\pi a^2 \tag{76}$$

The second possible maximum tensile stress occurs at the center of the lower surface as follows:

$$\sigma_{\phi\phi}(0, -H) = \sigma_{rr}(0, -H) = {}_H\sigma_{rr}(a)\, k_\sigma^-(\nu, a/H, \rho/H) \tag{77}$$

The curves for the nondimensional functions k_P, k_σ^+ and k_σ^- were all determined and given in the earlier work[10].

5. CONCLUSIONS

The dynamic contact stresses between an axisymmetric projectile and elastic plates of finite thickness are solved by the method of integral transforms in terms of three-dimensional equations of motion. The stresses obtained are written as the sums of the associated static plate stresses and wave-effect integrals. It is shown that the assumption of locally static response during impact is equivalent to keeping only the first order term, but dropping all higher order terms obtained from contour integrations.

Under the above assumption, a practical theory is developed for the impact problem in which a spherical body impinges on the upper surface of an elastic plate of finite thickness resting on a ring foundation. A general method of calculation is described for determining the maximum dynamic tensile stresses and the maximum contact time for the impact of a sphere at arbitrary speeds on the plate surface.

ACKNOWLEDGMENT

The present research was supported by the Engineering Research Institute, Iowa State University, Ames, Iowa 50010.

REFERENCES

1. H. Hertz, *Collected Works*, **1**, Barth, Leipzig, 1882, p. 174.
2. M.T. Huber, 'Zur Theorie der Berührung fester elastischer Körper', *Annln. Phys.*, **14**, 1904, pp. 153–163.
3. A.E.H. Love, *A Treatise on the Mathematical Theory of Elasticity*, Cambrigde University Press, 1927, pp. 198–203.
4. Y.M. Tsai, 'Dynamic Contact Stresses Produced by the Impact of an Axisymmetrical Projectile on an Elastic Half-Space', *Int. J. Solids Struc.*, **7**, 1971, pp. 543–558.
5. R.M. Davies, 'The Determination of Static and Dynamic Yield Stresses Using a Steel Ball', *Proc. Roy. Soc.*, **A197**, 1949, pp. 416–432.
6. Y.M. Tsai and H. Kolsky, 'A Theoretical and Experimental Investigation of the Flaw Distribution on Glass Surfaces', *J. Mech. Phys. Solids.*, **15**, 1967, pp. 29–46.
7. Y.M. Tsai, 'Stress Distributions in Elastic and Viscoelastic Plates Subjected to Symmetrical Rigid Indentations', *Quar. Appl. Math.*, **27**, 1969, pp. 371–380.
8. Y.M. Tsai, 'Thickness Dependence of the Indentation Hardness of Glass Plates', *Int. Frac. Mech.*, **5**, 3, 1969, pp. 157–165.
9. Y.M. Tsai and K. Dilmanian, 'Impact of Spheres on Elastic Plates of Finite Thickness', *Devel. in Mech.*, **6**, Proc. 12th Midwest. Mech. Conf., 1971, pp. 1009–1022.
10. Y.M. Tsai, 'Stress Distribution and Fracture Produced by a Spherical Indenter in a Glass Plate Resting on a Ring Foundation', *Devel. in Mech.*, **5**, 11th Midwest. Mech. Conf., 1969, pp. 795–809.
11. Y.M. Tsai, 'Stress Waves Produced by Impact on the Surface of a Plastic Medium', *J. Franklin Institute*, **285**, 3, 1968, pp. 204–221.
12. Y.M. Tsai, 'Exact Stress Distribution, Crack Shape and Energy for a Running Penny-Shaped Crack in an Infinite Elastic Solid', *Int. J. Fracture.*, **9**, 1973, pp. 157–169.
13. L.R. Maier, Jr. and Y.M. Tsai, 'Wave Propagation in Linear Viscoelastic Plates of Various Thickness', Iowa State University Engineering Research Institute, Preprint ERI-72266, 1972, (To appear in *J. Mech. Phys. Solids*).
14. H. Pursey, 'The Launching and Propagation of Elastic Waves in Plates', *Quart. J. Mech. and Appl. Math.*, **10**, 1, 1957, pp. 45–62.
15. A.I. Luré, *Three Dimensional Problems of the Theory of Elasticity*, Interscience Publishers, New York, 1964, p. 146.

Transition of collision contact force between a viscoelastic half-space and a flat-headed rigid body

K. Kawatate

1. INTRODUCTION

Most of the mathematical analyses of collision have been done along the fundamental line suggested by Hertz, where a local compressive pressure is regarded as a static effect[1]. According to his theory, a total pressure which is initially zero increases in the early stage following the first contact. Such a result is effective for explaining the result of an experiment made on a sphere, where the greater part of the total pressure after collision is produced by the restitutive force caused by a deformation of contact surface.* One of the reasons why Hertz's theory of collision has been widely accepted is found in this respect. Problems such as those arising when a flat-headed body collides against a flat plane cannot, however, be solved rationally by applying Hertz's theory of collision. The pressure caused by a rapid change of velocity has to be accounted for. It seems that few systematic studies have so far been made concerning this problem, except for the well-known result with respect to a one-dimensional bar.

In the present paper both the impinging and the impinged body are considered to be flat at the parts where contact occurs. Though the impinging body is limited to a rigid one, the impinged body is assumed to be a semi-infinite linear viscoelastic space, including Hooke's elastic one as a special case, in view of the possibility of attenuation due to the viscosity of the body in dealing with dynamic problems. An attempt is made here to clarify the transient behavior of force induced by collision: the distributed pressure, the friction and the total pressure on the contact surface.

The process for evaluating the contact force is as follows. In order to facilitate treatment by separation of time, the Laplace transformation is applied in formulation. Incidentally, the term 'Laplace transform' will often be omitted in the following considerations, as long as there is no danger of confusion. First, all the components of displacement and stress

* Validity of Hertz's theory of collision has been discussed by Y. M. Tsai [2].

of the viscoelastic body are evaluated, assuming the force to be sought as given on its contact surface. Next, an equation of motion of the rigid body which receives the contact force as a reaction is written. From the conditions of the displacement on the contact surface, a set of Fredholm-type integral equations of the first kind in terms of the Laplace transforms of the components of contact force is induced. They are solved. The method of the numerical Laplace inversion developed by Bellman et al.[3] is applied to them. The result obtained can also be used to evaluate all the components of displacement and stress in the viscoelastic body.

2. BASIC EQUATIONS OF THE VISCOELASTIC BODY

The constitutive equations of a linear viscoelastic body are expressed in terms of the Lamé type relaxation moduli $\lambda(T)$ and $\mu(T)$. The mass density of the body is denoted by ρ. The Laplace transformation is performed with respect to the space-like time $T = c_0 t$, in which t is time and c_0 an equivoluminal wave velocity given by $(\mu_0/\rho)^{1/2}$. In the above and hereafter the subscript 0 is attached as in $\mu_0 = \mu(0)$ to represent a value at $T = 0$. The parameter S is used in the course of transformation. The Laplace transform of $f(T)$ is denoted by, for instance, $\hat{f} = \hat{f}(S)$. Defining S-varying moduli $\lambda*(S)$ and $\mu*(S)$ respectively by $S\hat{\lambda}(S)$ and $S\hat{\mu}(S)$[4], we shall use S-varying Poisson's ratio $\nu* = \lambda*[2(\lambda*+\mu*)]^{-1}$ together with the following notations such as $\beta* = (\lambda*+2\mu*)/\mu* = (1-\nu*)/(0.5-\nu*)$, $s = S(\mu_0/\mu*)^{1/2}$, and $s_1 = s/\beta*$.

An equation of slight motion of the body is written in terms of a displacement $\hat{u}(S)$,

$$s^2 \hat{u} = \beta_*^2 \text{ grad. div } \hat{u} - \text{curl. curl } \hat{u} . \tag{1}$$

By defining the vector Laplacian $\nabla^2 = \text{grad.div} - \text{curl.curl}$[5] and expressing the displacement through the Galerkin function $\hat{G}$[6]

$$2\mu_* \hat{u} = 2(1-\nu_*)(\nabla^2 - s_1^2)\hat{G} - \text{grad. div } \hat{G} . \tag{2}$$

the equation of motion becomes

$$(\nabla^2 - s_1^2)(\nabla^2 - s^2)\hat{G} = 0 . \tag{3}$$

This is a repeated wave equation in terms of $\hat{G}$. It has been shown that one of the three components of $\hat{G}$ can be taken as zero[7].

3. COLLISON OF A FLAT-HEADED RIGID PUNCH

3.1 *Formulation as a problem of plane strain*

Here we examine a case in which a rigid body of width $2a$ with mass per unit length m and velocity v is collided against the viscoelastic body. It is assumed that the rigid body has a plane surface and is infinitely long in one direction. The problem to be considered can therefore be approximated by a state of plane strain. All the components of strain and stress are independent of z $(\partial/\partial z = 0)$ and the z component of displacement u_z vanishes. Time is measured from the instant when the collision commences. Cartesian coordinates (x, y) are selected so that the viscoelastic half-space may occupy $x \geqq 0$ and the surface may be represented by $x = 0$. By placing the origin (0, 0) at the center of contact surface, the phenomenon becomes symmetrical with respect to the x axis. In this case the solution of the repeated wave equation (3), which tends to zero as x tends to infinity, will be expressed in the form

$$\hat{G}_x(x, y, S) = (2/\pi)^{1/2} \int_0^\infty [A(\eta, S) \exp(-h_1 x) + B(\eta, S) \exp(-hx)] \times$$
$$\times \cos y\eta \, \mathrm{d}\eta \,, \tag{4}$$

where $h = (\eta^2 + s^2)^{1/2}$ and $h_1 = (\eta^2 + s_1^2)^{1/2}$. Unknown functions A and B are to be determihed from the boundary conditions. By writing equation (2) in terms of the components

$$2\mu_* \hat{u}_x = 2(1 - \nu_*)\,(\partial^2/\partial x^2 + \partial^2/\partial y^2 - s_1^2)\, \hat{G}_x - \partial^2 \hat{G}_x/\partial x^2 \,, \tag{5}$$

$$2\mu_* \hat{u}_y = \qquad - \partial^2 \hat{G}_x/\partial x \partial y \,, \tag{6}$$

and furthermore by using the linear viscoelastic constitutive equation

$$\hat{\sigma}_{xx} = \lambda_* (\partial \hat{u}_x/\partial x + \partial \hat{u}_y/\partial y) + 2\mu_* \partial \hat{u}_x/\partial x \,, \ldots \tag{7}$$

$$\hat{\sigma}_{xy} = \mu_* (\partial \hat{u}_y/\partial x + \partial \hat{u}_x/\partial y) \,, \ldots \tag{8}$$

all the components of displacement and stress are represented in the integral forms, including A and B.

Let p denote the distributed pressure and q the friction on the contact surface $x = 0$, so that

$$\hat{\sigma}_{xx}(0, y, S) = -\hat{p}(y, S) \,, \quad (|y| < a) \,; \qquad 0(a < |y|) \,, \tag{9}$$

$$\hat{\sigma}_{xy}(0, y, S) = -\hat{q}(y, S) \,, \quad (|y| < a) \,; \qquad 0(a < |y|) \,, \tag{10}$$

in which p is even in terms of y, while q is odd. When the friction acts in the direction of widening the surface, $\hat{q}(y, S)$ is taken positive for $y > 0$. Using Fourier transforms of $\hat{p}$ and $\hat{q}$,

$$P(\eta, S) = (2/\pi)^{1/2} \int_0^a \hat{p}(y, S) \cos \eta y \, \mathrm{d}y \,, \tag{11}$$

$$Q(\eta, S) = (2/\pi)^{1/2} \int_0^a \hat{q}(y, S) \sin \eta y \, \mathrm{d}y \,, \tag{12}$$

we can express A and B as follows:

$$A = 2\Omega_A[-P(\eta^2+s^2/2) + Q\eta h]\, s^{-2} h_1^{-2}, \tag{13}$$

$$B = 2\Omega_A[P\eta h_1 - Q(\eta^2+s^2/2)]\, s^{-2} h_1^{-1} \eta^{-1}, \tag{14}$$

in which

$$\left.\begin{aligned} &\Omega_A(\eta, S) = 2^{-1} s^2 h_1/D, \quad \Omega_B = \eta[h_1 h - (\eta^2+s^2/2)]/D, \\ &\Omega_C = 2^{-1} s^2 h/D, \quad D = (\eta^2+s^2/2)^2 - \eta^2 h_1 h \end{aligned}\right\}. \tag{15}$$

An equation of motion of the flat-headed rigid punch is written as

$$\hat{u}(S) = S^{-2}[v c_0^{-1} - \hat{f}(S)\, m^{-1} c_0^{-2}], \tag{16}$$

where $\hat{u}(S)$ represents a displacement of the rigid punch measured from the instant of first contact and $\hat{f}(S)$ is the total pressure per unit length in z direction, given by

$$\hat{f}(S) = 2\int_0^a \hat{p}(y, S)\, \mathrm{d}y = (2\pi)^{1/2}\, P(0, S). \tag{17}$$

Let us consider an early stage of collision in which the impinging punch moves with the impinged half-space in the direction of the normal line on the contact surface. We have

$$\hat{u}_x(0, y, S) = \hat{u}(S), \quad (|y| < a). \tag{18}$$

When the head of the rigid punch is so rough that no slippage occurs on the contact surface, we obtain

$$\hat{u}_y(0, y, S) = 0, \quad (|y| < a). \tag{19}$$

By using these conditions (18) and (19) together with (5), (6) and (16), a set of Fredholm-type integral equations of the first kind, valid for $|y| < a$, is derived

$$(2/\pi)^{1/2} \int_0^\infty [\Omega_A(\eta, S)\, P(\eta, S) - \Omega_B(\eta, S)\, Q(\eta, S)] \cos y\eta\, \mathrm{d}y$$

$$= \frac{2\mu_*}{S^2}\left[\frac{v}{c_0} - (2\pi)^{1/2}\frac{P(0, S)}{mc_0^2}\right], \tag{20}$$

$$(2/\pi)^{1/2} \int_0^\infty [-\Omega_B(\eta, S)\, P(\eta, S) + \Omega_C(\eta, S)\, Q(\eta, S)] \sin y\eta\, \mathrm{d}\eta = 0, \tag{21}$$

from which the components of contact force p and q can be determined.

Incidentally, when the head is so smooth that no friction acts on the contact surface, there is no need to consider (19) and hence (21). The distributed pressure p is determined by putting $Q = 0$ in (20).

3.2 *Contact force immediately after collision*

Since no stress wave is yet propagated in the viscoelastic body at the instant immediately after collision, the contact force is assumed to be distributed uniformly by

$$p_0 = p(y,0^+), \qquad q_0 = q(y,0^+). \tag{22}$$

This proposition has been confirmed by the result of a numerical calculation to be presented later. We apply one of the limit theorems in Laplace transformation

$$\lim_{S\to\infty} S\hat{p}(y,S) = p(y,0^+) = p_0, \qquad \lim_{S\to\infty} S\hat{p}(y,S) = q(y,0^+) = q_0, \tag{23}$$

and similarly introduce relations such as

$$\left.\begin{array}{l} \mu_*(S) \to \mu_0, \quad \beta_*^2 \to (\lambda_0+2\mu_0)/\mu_0 \equiv \beta_0^2, \\ S\Omega_A \to 2/\beta_0, \quad S\Omega_B \to 0, \quad S\Omega_C \to 2 \end{array}\right\}, \tag{24}$$

which also hold when $S\to\infty$, to equations (20) and (21) together with (11) and (12). By virtue of the inversion theorem of the Fourier transformation, we obtain

$$p_0 = \mu_0 v c_0^{-1} \beta_0 = v\rho[(\lambda_0+2\mu_0)/\rho]^{1/2}, \tag{25}$$

$$q_0 = 0. \tag{26}$$

The total pressure is written as

$$f_0 = 2ap_0 = 2av\rho[(\lambda_0+2\mu_0)/\rho]^{1/2}. \tag{27}$$

Equation (25) is fundamentally identical with the compressive stress $\sigma = v\rho(E/\rho)^{1/2}$ [8], induced in a one-dimensional bar with a mass density ρ and Young's modulus E struck at one end with the velocity v, in which lateral inertia effects are neglected. Young's modulus E is replaced by the initial modulus $\lambda_0+2\mu_0$: the dilatational wave velocity takes the place of the longitudinal wave velocity of the bar. It is a reasonable and acceptable result. It is seen from equations (25) and (27) that the distributed pressure immediately after collision, and therefore the total pressure, are determined by the collision velocity v of the rigid punch and the characteristic impedance $\rho[(\lambda_0+2\mu_0)/\rho]^{1/2}$ of the viscoelastic body, and that it is independent of the mass of the impinging rigid punch. It is also known from equation (26) that no friction acts immediately after collision. Such a result is quite natural when we consider that the stress wave has not yet propagated in the viscoelastic body and no deformation has occurred in the whole range, including the free surface and the contact surface.

3.3 *Calculation of transition of contact force*

For evaluating a transition of the initial contact force obtained in the last section in (25) and (26) we have to seek the solution of a set of integral equations (20) and (21). An approximate solution is sought here by means of the function expansion method since an analytical solution cannot be obtained.

Since the components of the contact force are singular at $y = \pm a$ and, in addition, since the distributed pressure is even about y and the friction odd, we put

$$\hat{p}(y, S) = 2\mu_* v c_0^{-1} a \sum_{m=0}^{\infty} \hat{\phi}_m(\Sigma)\, T_{2m}(y/a)\, [1 - (y/a)^2]^{-1/2}, \tag{28}$$

$$\hat{q}(y, S) = 2\mu_* v c_0^{-1} a \sum_{m=0}^{\infty} \hat{\psi}_m(\Sigma)\, T_{2m+1}(y/a)\, [1 - (y/a)^2]^{-1/2}, \tag{29}$$

in which $T_m(z)$ is the Chebyshev polynomial and $\Sigma = aS$. We substitute these into equations (11), (12), (17), again substitute the results into equations (20), (21) and expand the trigonometrical function in terms of the Chebyshev polynomial. In this way, we can obtain a set of equations which is to determine the unknown coefficients $\hat{\phi}_m$ and $\hat{\psi}_m$ as follows:

$$\sum_{m=0}^{\infty} [a_{km}(-)^m \hat{\phi}_m(\Sigma) - b_{km}(-)^m \hat{\psi}_m(\Sigma)] = \frac{\delta_{k0}}{\Sigma^2} \times$$

$$\times \left[1 - \frac{4a^2 \rho}{m} \frac{\mu_*}{\mu_0} \frac{\pi}{2} \hat{\phi}_0(\Sigma)\right], \tag{30}$$

$$\sum_{m=0}^{m} [-b_{mk}(-)^m \hat{\phi}_m(\Sigma) + c_{km}(-)^m \hat{\psi}_m(\Sigma)] = 0, \tag{31}$$

where $k = 0, 1, \ldots$ and δ_{k0} is Kronecker's delta, i.e., $\delta_{k0} = 1$ $(k = 0)$; $= 0$ $(k \neq 0)$. Coefficients a_{km} and others are also induced from the known quantities already given by equation (15). Expressing what has been assumed as $\eta a = x$ in $a^{-1}\Omega_A(\eta, S)$ as $\Gamma_A(x, \Sigma)$, we obtain the following integrals

$$a_{km} = \int_0^{\infty} \Gamma_A(x, \Sigma)\, J_{2m}(x)\, J_{2k}(x)\, \mathrm{d}x, \tag{32}$$

$$b_{km} = \int_0^{\infty} \Gamma_B(x, \Sigma)\, J_{2m+1}(x)\, J_{2k}(x)\, \mathrm{d}x, \tag{33}$$

$$c_{km} = \int_0^{\infty} \Gamma_C(x, \Sigma)\, J_{2m+1}(x)\, J_{2k+1}(x)\, \mathrm{d}x, \tag{34}$$

all of which are known. In an actual calculation only a finite number of coefficients $\hat{\phi}_m(\Sigma)$ and $\hat{\psi}_m(\Sigma)$ $(m = 0, 1, \ldots, M)$ can be determined from equations (30) and (31). We take M from 2 to 7 with respect to $\Sigma = 1, 2, \ldots, N$; N ranges 3 through 7.

By following the technique of the numerical Laplace inversion developed by Bellman et al., we evaluate $\phi_m(tc_0/a)$ and $\psi_m(tc_0/a)$. The distributed pressure p can be calculated without difficulty by replacing $\hat{\phi}_m$ in equation (28) by ϕ_m as can also the friction q by replacing $\hat{\psi}_m$ by ψ_m. It is also possible to calculate all the components of displacement and stress at arbitrary points (x, y) in a similar way by applying the numerical Laplace inversion since their Laplace transforms are to be evaluated in the form containing $\hat{\phi}_m$ and $\hat{\psi}_m$.

4. COLLISION OF A FLAT-HEADED RIGID CYLINDER

We shall now consider a case in which a rigid cylinder with flat surface of mass m and diameter $2a$ is collided against the viscoelastic body at velocity v. The cylindrical coordinate system (r, θ, z) is chosen so that the viscoelastic body may occupy $z \geqq 0$ and its surface is given by $z=0$. The origin of the coordinate is placed at the center of the contact surface. The problem becomes axially symmetrical $(\partial/\partial\theta = 0)$. Let us follow the same procedure as in the last section. From the conditions of $\hat{u}_z(r, 0, S) = \hat{u}(S)$, and $\hat{u}_r(r, 0, S) = 0$, we obtain a set of Fredholm-type integral equations of the first kind, valid for $r < a$, which determine the components of contact force. The result is expressed as follows:

$$\int_0^\infty [\Omega_A(\eta, S)\, P(\eta, S) - \Omega_B(\eta, S)\, Q(\eta, S)]\, \eta\, J_0(r\eta)\, \mathrm{d}\eta = \frac{2\mu_*}{S^2}\left[\frac{v}{c_0} - \frac{2\pi P(0, S)}{mc_0^2}\right], \tag{35}$$

$$\int_0^\infty [-\Omega_A(\eta, S)\, P(\eta, S) + \Omega_C(\eta, S)\, Q(\eta, S)]\, \eta\, J_1(r\eta)\, \mathrm{d}\eta = 0, \tag{36}$$

where P and Q respectively represent the Hankel transform of the distributed pressure p and the friction q

$$P(\eta, S) = \int_0^a \hat{p}(r, S)\, r\, J_0(\eta r)\, \mathrm{d}r, \tag{37}$$

$$Q(\eta, S) = \int_0^a \hat{q}(r, S)\, r\, J_1(\eta r)\, \mathrm{d}r, \tag{38}$$

$J_m(z)$ being the Bessel function of the first kind.

The contact force immediately after collision is given by equations (25) and (26) again as in the last section. The total pressure becomes

$$f_0 = \pi a^2 p_0 = \pi a^2 v \rho [(\lambda_0 + 2\mu_0)/\rho]^{1/2}. \tag{39}$$

Incidentally, in the case of $q = 0$ and hence $Q = 0$, the distributed pressure p is to be sought by equation (35).

In order to seek the contact force which changes with time, we expand for $r < a$ the distributed pressure and the friction as follows:

$$\hat{p}(r, S) = 2\mu_* v c_0^{-1} a \sum_{m=0}^{\infty} \hat{\phi}_m(\Sigma)\ T_{2m+1}(r/a)\,(r/a)^{-1}\,[1-(r/a)^2]^{-1/2}, \tag{40}$$

$$\hat{q}(r, S) = 2\mu_* v c_0^{-1} a \sum_{m=0}^{\infty} \hat{\psi}_m(\Sigma)\ r/a \cdot C_{2m}^{3/2}[\{1-(r/a)^2\}^{1/2}] \times$$
$$\times [1-(r/a)^2]^{-1/2}, \tag{41}$$

where $C_m^{\alpha}(z)$ is Gegenbauer's function[9]. Let us substitute these into equations (37) and (38) and expand the Bessel function in terms of the Chebyshev polynomial. We now obtain a set of linear equations which is to determine the unknown coefficients $\hat{\phi}_m$ and $\hat{\psi}_m$ as follows:

$$\sum_{m=0}^{\infty} [a_{km}\,\hat{\phi}_m(\Sigma) - b_{km}\,\hat{\psi}_m(\Sigma)] = \frac{\delta_{k0}}{\Sigma^2}\left[1 - \frac{4\pi a^3\rho}{m}\frac{\mu_*}{\mu_0}\sum_{m=0}^{\infty}\frac{(-)^m\,\hat{\phi}_m(\Sigma)}{2m+1}\right], \tag{42}$$

$$\sum_{m=0}^{\infty} [-b'_{km}\,\hat{\phi}_m(\Sigma) + c_{km}\,\hat{\psi}_m(\Sigma)] = 0, \tag{43}$$

where $k = 0, 1, \ldots$ and

$$\left.\begin{aligned}
a_{km} &= \int_0^{\infty} \Gamma_A(x,\Sigma)\frac{\pi}{2}\,J_{m+1/2}(x/2)\,J_{-m-1/2}(x/2)\,xJ_k^2(x/2)\,\mathrm{d}x \\
b_{km} &= \int_0^{\infty} \Gamma_B(x,\Sigma)\frac{(2m+1)!!}{(2m)!!}\,j_{2m+1}(x)\,xJ_k^2(x/2)\,\mathrm{d}x \\
b'_{km} &= \int_0^{\infty} \Gamma_B(x,\Sigma)\frac{\pi}{2}\,J_{m+1/2}(x/2)\,J_{-m-1/2}(x/2)\,xJ_k(x/2)\,J_{k+1}(x/2)\,\mathrm{d}x \\
c_{km} &= \int_0^{\infty} \Gamma_C(x,\Sigma)\frac{(2m+1)!!}{(2m)!!}\,j_{2m+1}(x)\,xJ_k(x/2)\,J_{k+1}(x/2)\,\mathrm{d}x
\end{aligned}\right\}, \tag{44}$$

$j_m(z)$ being the spherical Bessel function of the first kind, $(2m)!! = 2.4\ldots m$, $0!! = 1$, $(2m+1)!! = 1.3\ldots(2m+1)$, and $(-1)!! = 1$. To obtain the distributed pressure and the friction as well as all the components of displacement and stress, we follow the same procedure as described in the preceding section.

5. RESULTS OF NUMERICAL CALCULATION AND DISCUSSIONS

5.1 *Influence of number of terms in function expansion*

A smooth-headed punch was assumed to impinge upon an elastic body, so that we could apply the plane strain formulation with no friction $q = 0$. Poisson's ratio was taken as $\nu = 0.25$ and the mass ratio as $4a^2\rho/m = 1$.

By changing $M = 1, 2, \ldots, 7$ in equations (28) and (30), it was found that the absolute value of ϕ_m ($m = 0, 1, \ldots, M$) rapidly decreases with the increase of m and that convergency is satisfactory. In practice it is sufficient to adopt $M = 4$ for evaluating the contact force and all the components of displacement and stress[10]. Even in the case of a rough-headed punch with $u_y(0, y) = 0$, for $|y| < a$, or also in the axially symmetrical problems, it was confirmed that adoption of $M = 4$ was satisfactory[11, 12, 13].

5.2 *Experimental verification using urethane rubber*

A test was conducted with a view to ascertaining the validity of the series of numerical solutions stated so far. One side of a urethane rubber plate 220 mm wide, 160 mm high, and 6 mm thick, was impacted with a falling flat-headed steel punch of width $2a = 60$ mm and effective weight $W = 870$ g. The resultant transient deformation was detected by means of Moire fringe and taken on a cine film Eastman Kodak 4X Negative 7224 (100 ft.), using of a 16 mm framing camera HIMAC 16 HB running at 8000 pictures per second. The measured value was compared with the numerical one obtained from the plane strain condition with $q = 0$.

A uniaxial tensile behavior of the urethane rubber plate obtained at room temperature $\theta = 23\,^\circ$C could be approximately represented by that of

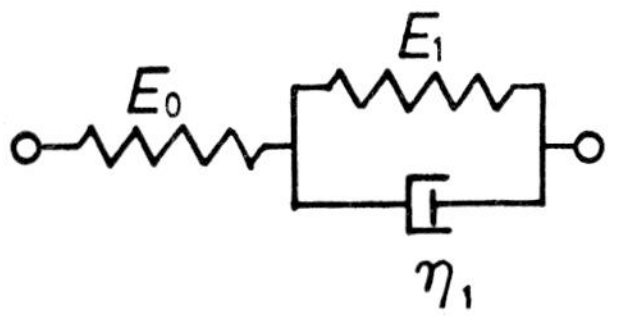

Fig. 1. Three-parameter solid model for urethane rubber with $E_0 = 0.202$ kg.mm^{-2}, $E_1 = 6.75$ kg.mm^{-2}, *and* $\eta_1 = 1.29 \times 10^3$ kg.s.mm^{-2}

the three parameter solid model shown in Fig. 1. By using Poisson's ratio $\nu = 0.45$[14], the relaxation modulus was expressed as

$$\mu(t) = \mu_0 [E_1 + E_0 \exp\{-(E_0 + E_1)\,\eta_1^{-1}\, t_1\}]\,(E_0 + E_1)^{-1}, \tag{45}$$

where $\mu_0 = E_0[2(1+\nu)]^{-1} = 0.0697$ kg.mm^{-2}. The weight density was $\rho g = 1.05 \times 10^{-6}$ kg.mm^{-3}. The equivoluminal wave velocity was therefore $c_0 = (\mu_0/\rho)^{1/2} = 25.5$ m.s^{-1}. Since the thickness was 6 mm, as mentioned above, the urethane rubber plate was considered to be subjected to an almost plane stress state when loaded in plane. The dilatational wave velocity then became $c_1 = [2(1-\nu)^{-1}]^{1/2} c_0 = 48.6$ m.s^{-1}, while the value corresponding to the plane strain condition was $c_1 = [2(1-\nu)(1-2\nu)^{-1}]^{1/2} c_0 = 84.6$ m.s^{-1}. In applying the formula of plane strain, an equivalent Poisson's ratio $\nu = 0.31$ was used. The line density of weight was $mg = W/h = 0.876/6 =$

0.145 kg.mm^{-1} and hence the mass ratio $4a^2\rho/m = 0.026$. The transition of the distributed pressure p calculated by using the above values is shown in Fig. 2. The pressure p which is uniformly distributed over the contact

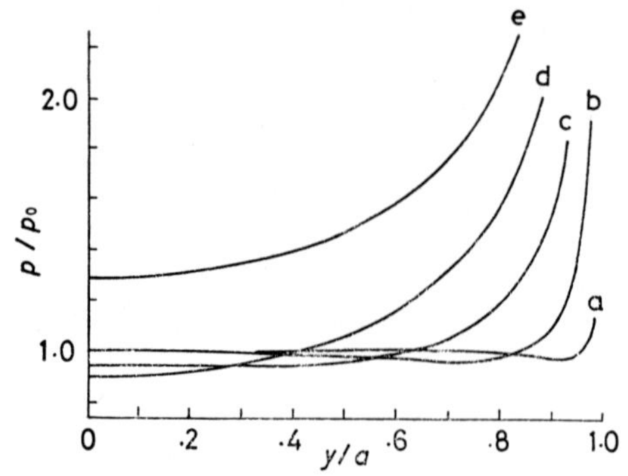

Fig. 2. Distributed pressure p, numerically obtained from a plane strain condition involving a smooth-headed punch, for urethane rubber with $\nu = 0.31$ *and* $4a^2\rho/m = 0.026$
Legend: $tc_0/a = 0.048$ *(a)*, 0.262 *(b)*, 0.693*(c)*, 1.466*(d) and* 3.060*(e)*

surface, as given by equation (25) immediately after collision, changes with the lapse of time as shown in the figure, causing conspicuous concentration at the corner region.

The surface of the specimen was undercoated by a vinyl acetate binding agent (Sebian A). Over this we put a photosensitive mixture composed of water 32 cc + PVA binding agent (Fueki) 8 g + black ink 10 cc + bichromate 2 g which was dried by using an infrared lamp. A model screen of 500 lines per inch in 100 × 100 mm area was then printed on it in a vacuum frame by using a carbon arc light (Toshi). The result obtained at collision velocity $v = 0.6$ m.s^{-1} is shown in Fig. 3.

Initial fringes were observed in a photograph taken 0.25 ms before the collision. The analysis of Moire fringe with original mismatch was used to correct the measured displacements[15]. The black part at the top was a shadow of the steel punch of width $2a = 60$ mm. Fringes appeared in succession. We can observe how the deformation generated at the impacted end progresses in the specimen.

⟶

Fig. 3. Sequence of Moire fringe in a 6 mm-*thick urethane rubber, resulting from the collision of a flat-headed steel punch,* 60 mm *in width and* 870 g *in weight, at a velocity of* 0.6 m.s^{-1}. *Photographs taken at* 8000 pps. *Patterns presented at* 0.25 ms *before first contact (a),* 0.0*(b),* 0.25*(c),* 0.50*(d),* 0.75*(e),* 1.00*(f),* 1.25*(g), and* 1.50*(h) each in* ms *after collision*

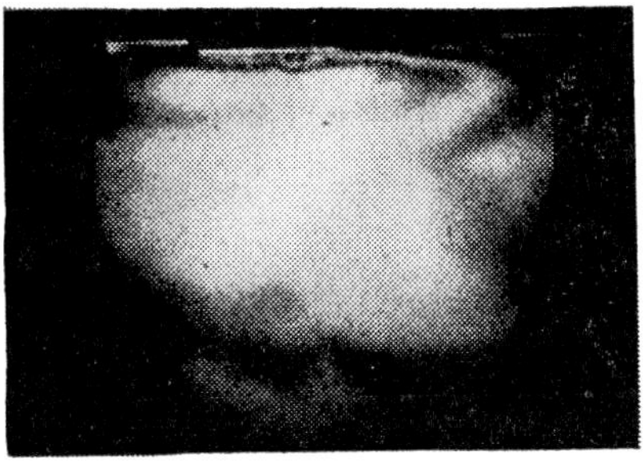

a

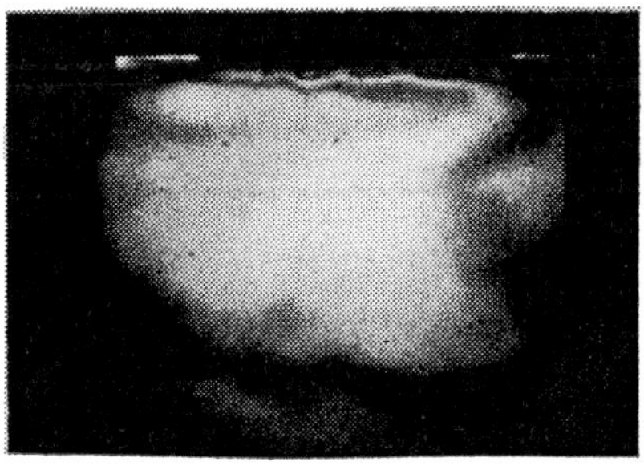

b

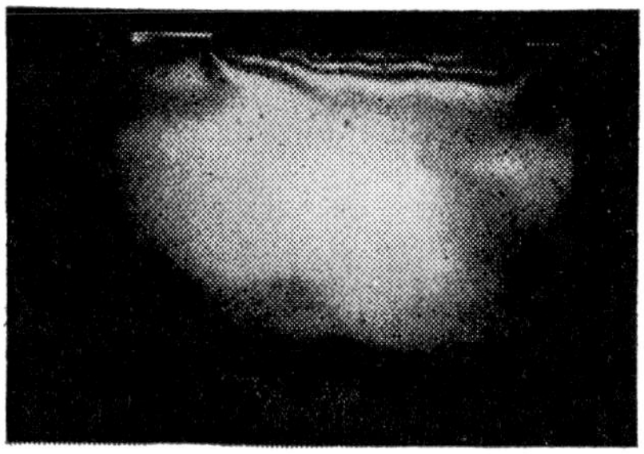

c

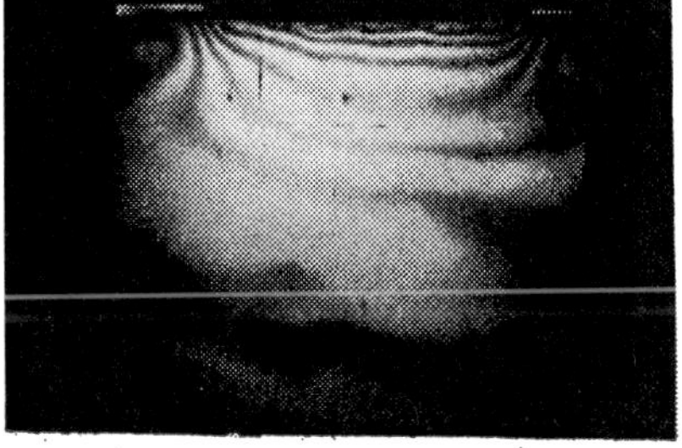

d

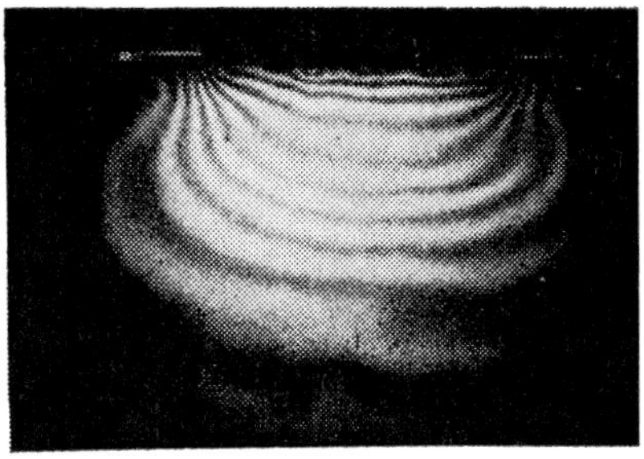

e

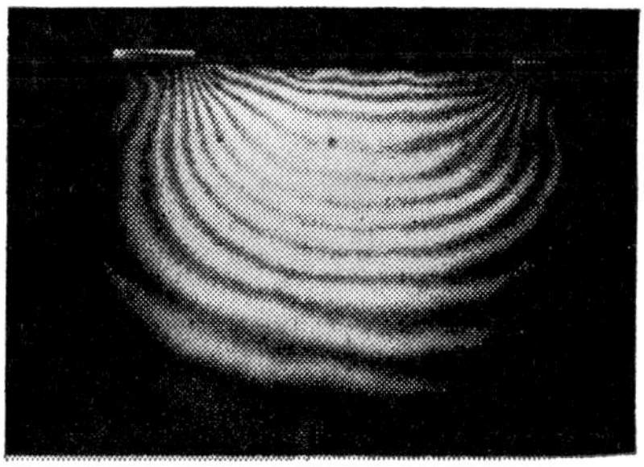

f

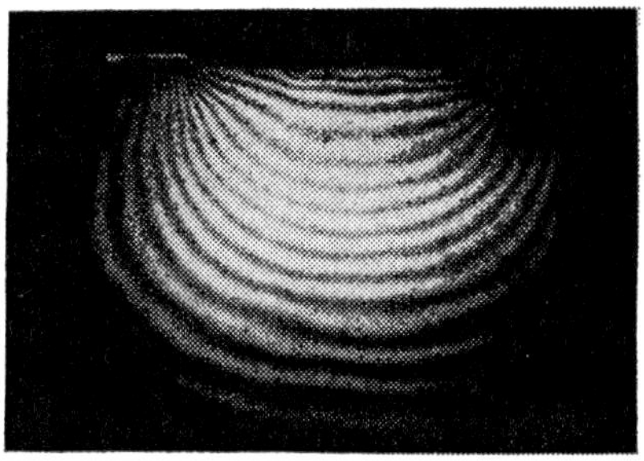

g

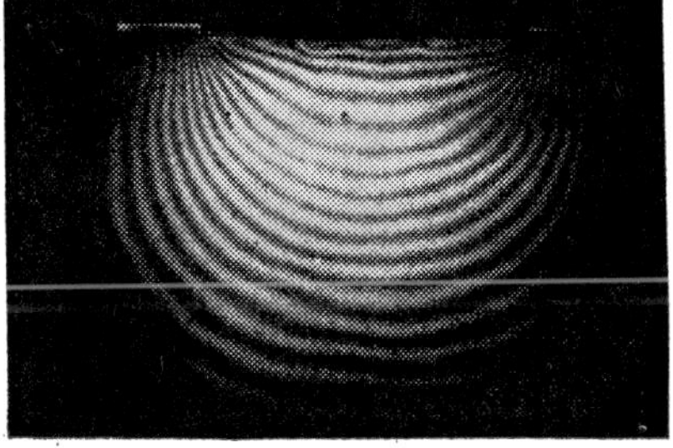

h

The component u_x, perpendicular to the contact surface of displacement along the center line $y=0$ is shown in Fig. 4 by denoting the measured value with ● and the calculated values with ○. Another result of an experiment carried out under the same conditions is also given in the figure

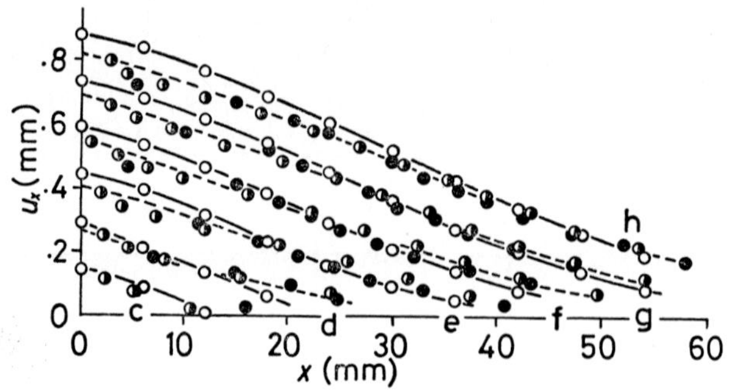

Fig. 4. Displacement u_x of urethane rubber at $y=0$ mm
Legend: Measured ●, ◑ *Calculated* ○: $t=0.25(c)$, $0.50(d)$, $0.75(e)$, $1.00(f)$, $1.25(g)$, *and* $1.50(h)$ *each in* ms

by the symbol ◑. Let us put $q=0$ and therefore $\psi_m=0$. By solving equation (30) we obtained $\phi_m(\Sigma)$. In accordance with equations (28), (11), (13), (14) ($Q=0$), (4), and (5), we calculated

$$\hat{u}_x(x, y, S) = vc_0^{-1} a^2 \sum_{m=0}^{M} (-)^m \hat{\phi}_m(\Sigma) \int_0^{\infty} [(\eta^2+s^2/2) \exp(-h_1 x) - $$

$$- \eta^2 \exp(-hx)] \, 2s^{-2} \, \Omega_A(\eta, s) \, J_{2m}(a\eta) \cos y\eta \, d\eta \,. \qquad (46)$$

The numerical inversion of Laplace transforms of Krylov and Skoblya[16] was applied to yield the calculated component u_x of displacement. Similarly, the component u_x near the corner region $y=22.5$ mm is shown in Fig. 5.

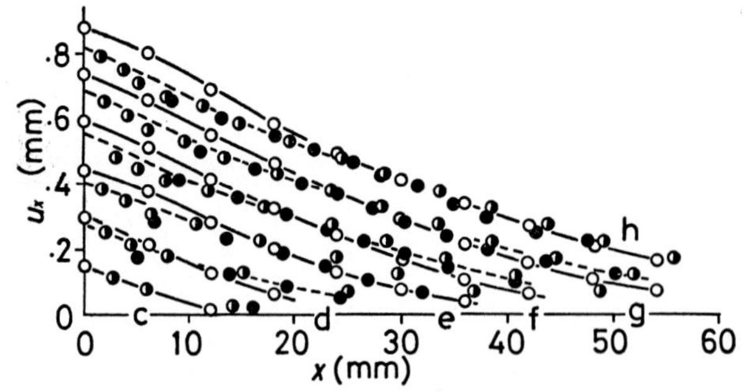

Fig. 5. Displacement u_x in connection with Fig. 4, at $y=22.5$ mm

In the neighborhood of the contact surface $x = 0$–20 mm, the measured value is smaller than the calculated one. The specimen was held between two iron plates on the bottom and lateral sides by means of bolts. A slight rigidity of the urethane rubber may presumably have caused the center of the free surface to fall in and so reduced the measured value. The farther we go from the contact surface, the better the agreement attained between the measured value and the calculated one. The propagation velocity read off for the displacement wave head was 50–70 m.s^{-1}. This is half way between the calculated value 48.6 m.s^{-1} of the dilatational wave velocity under plane stress condition and that of 84.6 m.s^{-1} under plane strain one.

The distribution of the component u_x of the displacement in y direction is given in Fig. 6 after Figs. 4 and 5. Since the stress wave does not propagate

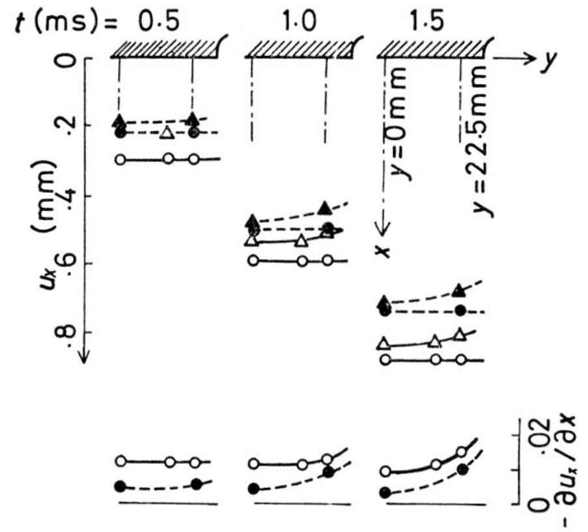

Fig. 6. Transient distribution of u_x in y direction at $x = 0$ and 6 mm *together with compressive strain $-\partial u_x/\partial x$ at $x = 0$ mm*

Legend:

x (mm)	0	6
calculated	○	△
measured	●	▲

deeply enough into the specimen in the early stage of collision $t = 0.5$ ms, the component u_x at $x = 6$ mm, for instance, is almost uniform in the y direction. As the stress wave propagates more deeply through the specimen, the component u_x at the corner part becomes smaller compared with that of the central region due to the inertia of the free surface and its neighbors. Thus the component u_x, for instance, at $x = 6$ mm, is no longer uniform in the y direction with the lapse of time, $t = 1.0$ and 1.5 ms. Since the component u_x at the contact surface $x = 0$ is uniform, we know immediately that the compressive strain in the x direction is concentrated in the corner part.

The distribution and propagation of the measured displacement were in good agreement with the calculated one as was also the compressive strain on the contact surface, which became concentrated at the corner region with the lapse of time.

5.3 *Influence of mass of an impinging rigid body*

We considered the axially symmetrical case in which the smooth-headed cylinder was involved so that $q = 0$. An elastic body and an asphalt were taken. For both Poisson's ratio was taken as $v = 0.4$. The uniaxial behavior of the latter [17] was assumed to be given by that of the four-parameter fluid, with $\rho g = 1.2 \times 10^{-6}$ kg.mm^{-3} and $a = 10^3$ mm, as shown in Fig. 7.

The total pressure f is given in Fig. 8, assuming the mass ratio to be $4\pi a^3 \rho/m = 0.1$ and 1. The stress relaxation appeared with the asphalt. The initial total pressure is constant and independent of an impinging mass in accordance with equation (39). In the initial stage of collision, when the

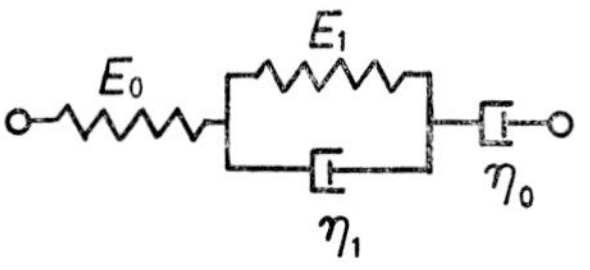

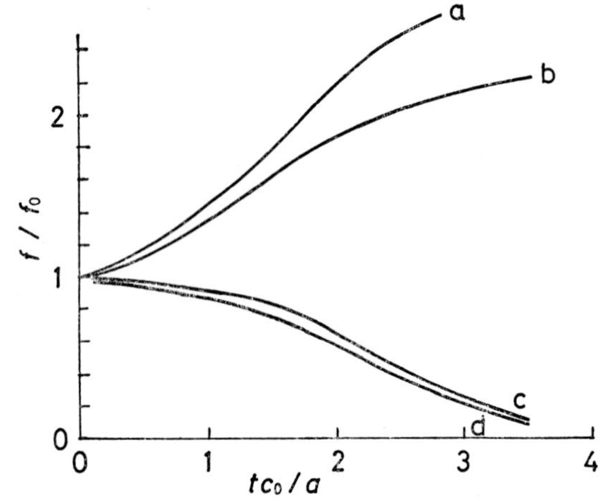

Fig. 7. Four-parameter fluid model for asphalt with $E_0 = 20.4$ kg.mm^{-2}, $E_1 = 1.02$ kg.mm^{-2}, $\eta_1 = 8.87$ kg.s.mm^{-2}, *and* $\eta_0 = 0.612$ kg.s.mm^{-2}

Fig. 8. Influence of an impinging mass on the total pressure, numerically obtained from an axially symmetrical condition involving a smooth-headed cylinder with $v = 0.4$

Legend:

$4\pi a^3 \rho/m$	0.1	1
elastic	*a*	*c*
asphalt	*b*	*d*

stress wave does not yet propagate widely, the influence of the mass of the rigid cylinder is not apparent. As the stress wave propagates more widely with the lapse of time, the total pressure increases in the case of a large impinging mass $4\pi a^3 \rho/m = 0.1$ but decreases in the case of a small one $4\pi a^3 \rho/m = 1$.

The larger the impinging mass, the more conspicuous the restitutive force caused by the deformation of the collided surface; while the mass is small, it is almost negligible. The impinging body decelerates. When its mass is large, the deceleration is almost negligible in the initial period of collision. According to the simplified one-dimensional theory[8], the total pressure is proportional to the impinging velocity. The total pressure caused by the velocity then remains nearly the same. While the mass is small, however, the deceleration is considerable and the total pressure decrease rapidly. Thus the transient behavior in Fig. 8 of the total pressure depending on the impinging mass can be qualitatively interpreted on the basis of the combination of the above-mentioned restitutive force and the deceleration.

5.4 *Friction*

The plane strain elastic problem was also considered. The rough-headed punch was involved so that $u_y(0, y) = 0$, $|y| < a$. The distributed pressure p and the friction q arrived at by putting Poisson's ratio $\nu = 0.25$ and the mass ratio $4a^2 \rho/m = 1$, are given in Figs. 9 and 10 respectively.

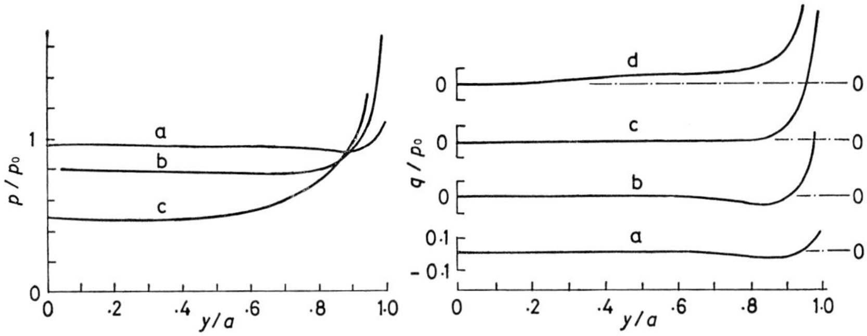

Fig. 9. The distributed pressure p numerically obtained from a plane strain elastic problem with $\nu = 0.25$ *and* $4a^2 \rho/m = 1$
Legend: $tc_0/a = 0.048(a)$, $0.262(b)$ *and* $0.603(c)$

Fig. 10. The friction q numerically resulting in connection with Fig. 9
Legend: $tc_0/a = 0.048(a)$, $0.262(b)$, $0.603(c)$ *and* $1.466(d)$

In the early stage of collision, when the stress wave has not yet propagated far enough, the loaded region is limited to a narrow area. Accordingly, the friction profile has, in rough approximation, become something like the profile obtained by compressing a block with finite width and height. It acts to prevent the contact surface from spreading, $q < 0$.

With the lapse of time, the stress wave propagates into the impinged body. The wider region is loaded. The friction profile is therefore considered to be expressed approximately by what is obtained through pressing a semi-infinite elastic plane with a rigid punch. The slippage at the contact surface where no friction is presented occurs in the direction of narrowing down the elastic surface[18]. Accordingly, when no slippage is allowed, the friction acts to prevent the contact surface from narrowing down, $q > 0$.

Incidentally, the similar conclusions can also be drawn with regard to the axially symmetrical problem[14].

5.5 *Comparison of total pressure on the basis of smoothness and roughness of surface*

An elastic body was taken in the axially symmetrical problem. Poisson's ratio was assumed to be $\nu = 0.25$ and the mass ratio was taken as $4\pi a^3 \rho/m = 1$. The total pressure is given in Fig. 11 for a smooth-headed

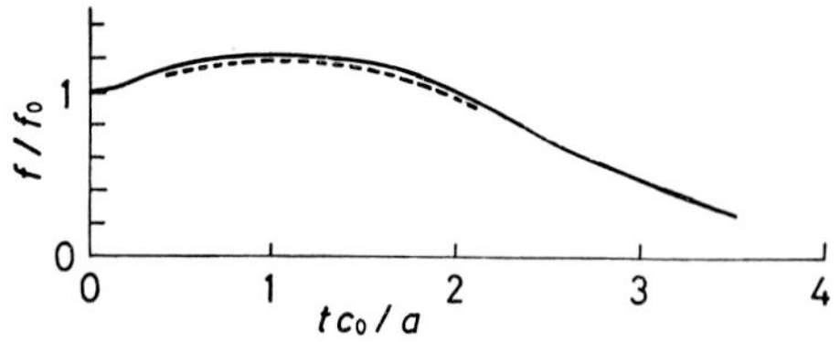

Fig. 11. Influence of the surface of an impinging head on the total pressure, numerically obtained from an axially symmetrical elastic problem with $\nu = 0.25$ *and* $4\pi a^3 \rho/m = 1$
Legend: Rough-headed —— *and smooth-headed* ----

cylinder and for a rough one. Within the range of $tc_0/a < 2.5$ the total pressure for the rough-headed cylinder is somewhat larger than that of the smooth one.

The friction keeps the contact surface from slipping. Thus with the rough-headed cylinder, the elastic body is constrained on the surface. The apparent elastic modulus is considered to increase. Hence, the restitutive force excited in the elastic body by the rough-headed projectile becomes predominant.

6. CONCLUSIONS AND ADDITIONS

1. The pressure and the friction immediately after collision are given by $v\rho[(\lambda_0+2\mu_0)/\rho]^{1/2}$ and 0 respectively.
2. The distributed pressure concentrates in the corner region as time passes. The total pressure for the rough-headed projectile assumes a somewhat larger value than the smooth one.
3. In the early stage of collision the friction serves to keep the contact surface from spreading, while in the later it prevents the surface from narrowing.
4. If the mass of the impinging projectile is large, the total pressure increases with time; while the mass is small, the total force decreases monotonically.
5. Fair agreement was established between the measured and the calculated displacement and strain of the urethane rubber.

There seems to be almost no difficulty in applying the present method of calculation to the collision problem of a viscoelastic body of finite dimension with a flat-headed rigid projectile. It is, however, impossible to apply this method in its present state, for instance, to the collision problem of an elastic sphere against an elastic body considered by Hertz. We have to determine a dimension of a contact surface in the course of analysis, that is, to solve the floating boundary value problem. It is a noteworthy achievement that Tsai [2] has succeeded in formulating the collision problem between a rigid sphere and an elastic half-space and in obtaining the solution using the successive approximation. Further development in the field of the collision contact theory is looked for.

ACKNOWLEDGMENTS

The author wishes to express his sincere gratitude to Professor M. Higuchi for stimulating and helpful discussions. He is also grateful to Mr. T. Shinozaki for assisting him in the experiments and drawing the figures. Thanks are due to Miss H. Oya for typing the manuscript. Numerical calculations were performed with the aid of a FACOM 230–60 at the Computer Center of Kyushu University.

REFERENCES

1. A. E. H. Love, *A Treatise on the Mathematical Theory of Elasticity*, Dover, New York, 1944, p. 198.
2. Y. M. Tsai, 'Dynamic Contact Stresses Produced by the Impact of an Axisymmetrical Projectile on an Elastic Half-Space', *Int. J. Solids Structures*, 7, 1971, pp. 543–5581
3. R.E. Bellman, R.E. Kalaba and J.A. Lockett, *Numerical Inversion of the Laplace Transform*, Elsevier, New York, 1966, Chapter 1 and 2.

4. D.R. Bland, *The Theory of Linear Viscoelasticity*, Pergamon, Oxford, 1960, p. 77.
5. P. Moon and D.E. Spencer, *Field Theory Handbook*, Springer, Berlin, 1961, p. 3.
6. E. Sternberg and R.A. Eubanks, 'On Stress Functions for Elastokinetics and the Integration of the Repeated Wave Equations', *Quart. Appl. Math.*, xv, 1957, pp. 149–153.
7. Y.C. Fung, *Foundations of Solid Mechanics*, Prentice-Hall, New Jersey. 1965, p. 194.
8. W. Johnson, *Impact Strength of Materials*, Arnold, London, 1972, p. 6 (1.7).
9. G.N. Watson, *A Treatise on the Theory of Bessel Functions*, Cambridge Univ. Press, Cambridge, 1966, p. 379.
10. K. Kawatate, 'Stress Distribution on the Colision Surface of a Viscoelastic Half-Plane with a Rigid Punch', *Proc. the 21st Japan Nat'l Congr. Appl. Mech. 1971, 21*, 1973, p. 150.
11. K. Kawatate, 'Stress Distribution on the Collision Surface of a Viscoelastic Body with a Smooth-Bottomed Rigid Cylinder', *Proc. the 22nd JNCAM 1972, 22*, 1974, p. 398.
12. K. Kawatate, 'Compression and Friction of the Colision Surface of a Viscoelastic Half-Plane with a Rough-Bottom-Rigid-Punch (Plane Strain Problem)', *Proc. the 23rd JNCAM 1973, 23*, 1975, p. 441.
13. K. Kawatate, 'Compression and Friction on the Collision Surface of a Viscoelastic Half-Space with a Rough-Bottom-Rigid-Cylinder (Axially Symmetrical Problem)', *Proc. the 23rd JNCAM 1973, 23*, 1975, pp. 451-452.
14. J.W. Dally, W.F. Riley and A.J. Durelli, 'A Photoelastic Approach to Transient Stress Problems Employing Low-Modulus Materials', *J. Appl. Mech.*, **26**, 1959, pp. 613–620.
15. K. Nagai, 'Strain Measurement by Moire Fringe Method' (in Japanese), *Data of the Structural Committee of West Japan, 66.34.1/1*, 1966, pp. 1–31.
16. V.I. Krylov and N.S. Skoblaya, *Handbook of Numerical Inversion of Laplace Transforms*, IPST, Jerusalem, 1969, pp. 33-34.
17. Soc. Polymer Sci., Japan (ed.), *Rheology Handbook* (in Japanese), Maruzen, Tokyo, 1965, p. 463.
18. I.N. Sneddon, *Fourier Transforms*, McGraw-Hill, New York, 1951, p. 433 (116).

Impact on a worn surface

P. A. Engel

NOMENCLATURE

a	=	contact radius
a_{ij}	=	influence coefficient for normal displacement
C	=	heat capacity
E	=	modulus of elasticity
E_r	=	reduced modulus of elasticity, $= \left(\dfrac{1-\nu_2^2}{\pi E_1} + \dfrac{1-\nu_2^2}{\pi E_2}\right)^{-1}$
$L(x)$	=	logic function, = 1 or 0 if x is true or false, respectively
$\underline{L}$	=	optimal wearpath
M	=	mass
p_i	=	pressure at i
P	=	contact force
q	=	Hertz pressure
r	=	radial dimension
R	=	radius of curvature
s	=	crater radius
t^*	=	contact time
T	=	temperature (rise)
U	=	volumetric strain energy
V	=	impact velocity
z	=	dimension of depth
α	=	elastic approach
β	=	$\dfrac{1}{2}\left(\dfrac{1}{R_1} + \dfrac{1}{R}\right)$
γ	=	density
ξ	=	non-dimensional crater radius, $= s/a$
ρ	=	non-dimensional radius, $= R_1/R$

ψ = non-dimensional contact radius, $= a/a_0$
σ = normal stress
τ = shear stress.

1. INTRODUCTION

The mechanical response of impulsively loaded contacts is an important consideration for machine designs. Modern design applications often require a large number of repeated load cycles to be sustained with minimum degradation of the contact. In order to determine rational criteria for material selection, surface preparation and lubrication, a quantitative theory for impact wear has been produced[1,2,3].

The local failure, while dependent on the materials, is frequently associated with a key parameter of macroscopic origin, such as the maximum subsurface shear stress. Thus, the Hertz theory enjoys great relevance whenever contact stresses are mild enough for the material to stay in the elastic range.

Since by wearing, the contact attempts to redistribute and reduce the stresses upon it, it is important to be able to estimate impact stresses at any stage of wear of the surface. For the given configuration, a critical stress-related failure parameter should be selected. The actual variation of the latter in terms of the wear geometry defines the optimal wearpath.

In the present paper the above program is pursued by simplified mechanical models, such as a parabolic crater approximating the shape of a wear scar; linearly elastic materials are stipulated. Section 2 deals with impact analysis on a cratered surface. Section 3 presents a few possible failure parameters, and in Section 4 two examples for the optimal wear-path are discussed.

The paper is motivated by the effects of the change of a contact, both during a single impact and in its history of wearing.

2. IMPACT ANALYSIS

In this section, a numerical method will be presented for quasi-static impact analysis on a worn surface[3]. Stress wave effects will be assumed negligible. For simplicity, a homogeneous, isotropic and elastic body (E_1, ν_1, M, R_1) will be considered impacting a cratered elastic half-space (E_2, ν_2); the parabolic crater has a radius s and meridional radius of curvature R, and the impact is axially symmetric (Fig. 1a).

The impact approach α is related to the impact force by:

$$M\ddot{\alpha} = -P(t) \tag{1}$$

with the initial conditions

$$t = 0: \alpha = 0, \quad \dot{\alpha} = V \tag{2}$$

The solution of this problem in the ordinary Hertz theory is well known. Here however the crater presents a discontinuity in the contact surface,

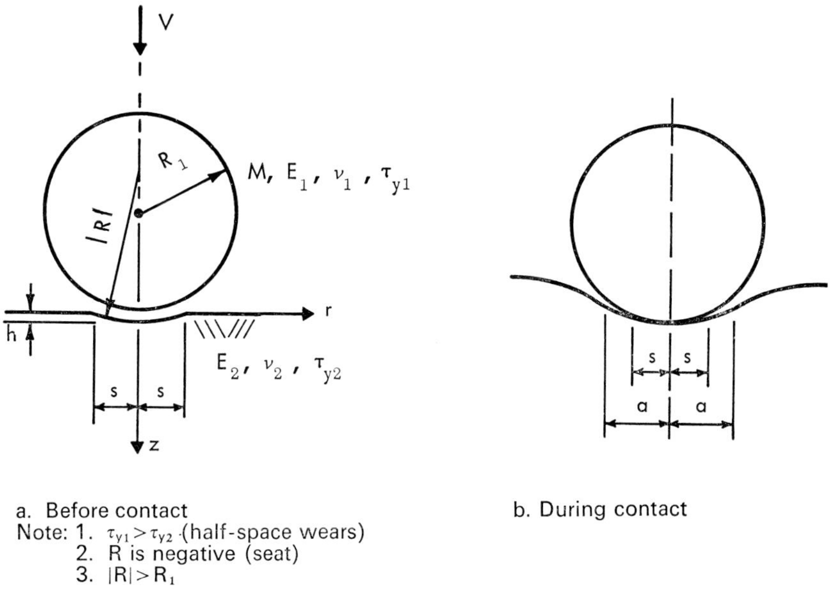

V
R_1
$|R|$
M, E_1, ν_1, τ_{y1}
h
r
s
s
E_2, ν_2, τ_{y2}
z
s
s
a
a
a. Before contact
Note: 1. $\tau_{y1} > \tau_{y2}$ (half-space wears)
2. R is negative (seat)
3. $|R| > R_1$
b. During contact

Fig. 1. Axisymmetric impact of a ball on a cratered elastic half-space

requiring formulation of the impact event in terms of the changing contact radius '*a*'. Recalling the Hertz solutions for $\alpha(a)$ and $P(a)$, we can generalize them for the crater contact:

$$\alpha = \left(\frac{1}{R} + \frac{1}{R_1}\right) a^2 \, \bar{F}_1(s/a, R, R_1) \tag{3}$$

$$P = \frac{4\left(\frac{1}{R} + \frac{1}{R_1}\right) E_r}{3\pi} a^3 \, \bar{F}_2(s/a, R, R_1) \tag{4}$$

where $\bar{F}_1$ and $\bar{F}_2$ are suitable 'modifying functions'.

Introducing the non-dimensional parameters

$$\rho \equiv R_1/R \tag{5}$$

$$\xi \equiv s/a \tag{6}$$

we then obtain from equations (3) and (4), respectively:

$$\alpha = \frac{s^2}{R_1} \frac{(1+\rho)}{\xi^2} F_1(\rho, \xi) \tag{7}$$

$$P = \frac{4E_r s^3}{3\pi R_1} \cdot \frac{(1+\rho)}{\xi^3} F_2(\rho, \xi) \tag{8}$$

When the contact does not extend to the shoulder of the crater, $\xi > 1$, and the modifying functions F_1 and F_2 take on the value of unity. However, for the quasi-static analysis of a contact reaching over and beyond the rim ($\xi < 1$), F_1 and F_2 must be evaluated.

The numerical method of point-matching[4] will be used to evaluate the displacement and force modifying functions, $F_1(\rho, \xi)$ and $F_2(\rho, \xi)$ respectively. The contact area is divided into n concentric annuli of equal thickness Δr. Out of the n rings, m cover the central crater, thus $s = m \cdot \Delta r$, $a = n\Delta r$, and $m \leq n$. If n is large enough, the pressure distribution over the contact circle ($0 < r < a$) may be approximated by n concentric steps p_i:

$$\sigma_z(r) = -p_i, \quad (i-1)\,\Delta r \leq r \leq i\Delta r \quad 1 \leq i \leq n \tag{9}$$

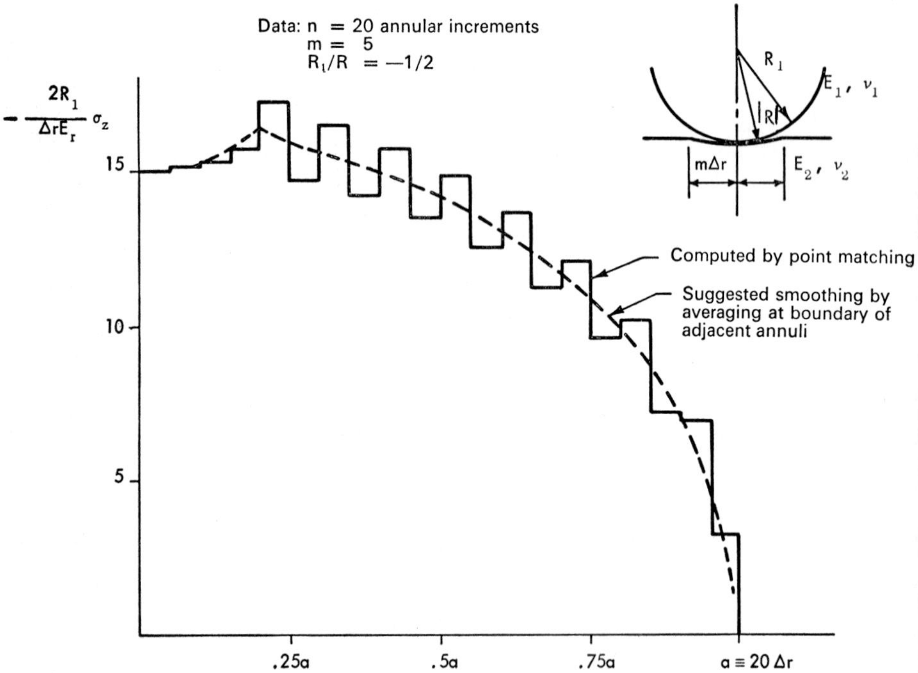

Fig. 2. Ball pressed against cratered half-space of same material (pressure distribution obtained by point-matching)

a. Normal approach modifying functions:

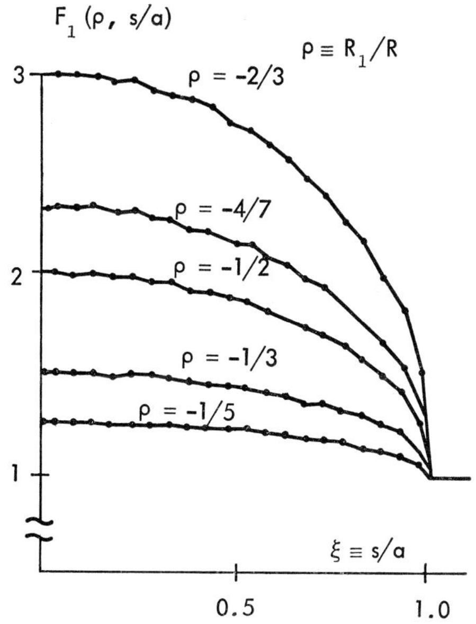

$$F_1(\rho, \xi) \approx 1 - \rho(1+\rho)^{-1}\sqrt{1-\xi^2}$$

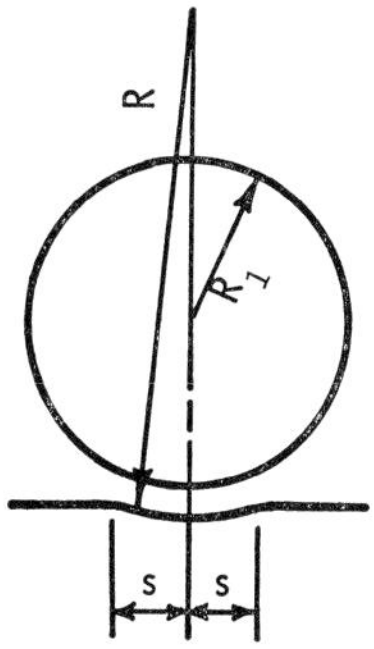

b. Contact force modifying functions:

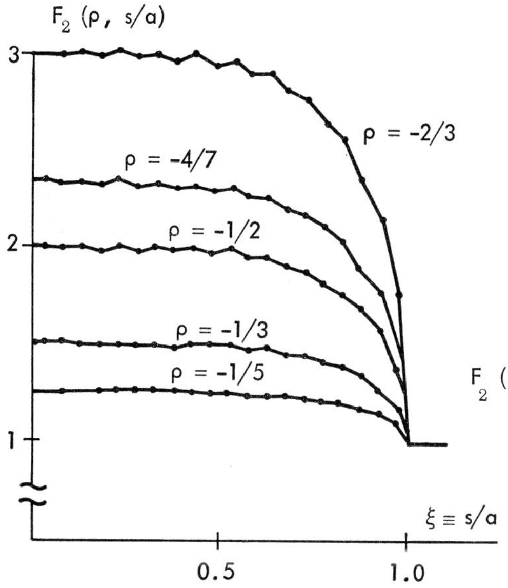

$$F_2(\rho, \xi) \approx 1 - \rho(1+\rho)^{-1}\sqrt{1-\xi^4}$$

Fig. 3. Two-segment modifying functions

The total elastic displacement at $r_i = i\Delta r$ on the contact surface may now be written as a linear combination of pressures,

$$w_i = \sum_{j=1}^{n} a_{ij} p_j, \qquad 0 \le i \le n \tag{10}$$

where the influence coefficients a_{ij} are displacements of the boundary circles i due to unit pressure j.

Now subtracting the central approach from w_i, the displacements of the two bodies are made compatible at the n boundary circles, yielding a system of simultaneous algebraic equations in the unknown pressures:

$$\sum_{j=1}^{n} (a_{ij} - a_{0j})\, p_j = \frac{r_i^2}{2R_1} + \frac{r_i^2}{2R} L(r_i \le r_m) + $$

$$+ \frac{r_m^2}{2R} L(r_i > r_m), \qquad (1 \le i \le n) \tag{11}$$

An example for the pressure distribution, for identical materials, $n = 20$ elements, $\rho \equiv R_1/R = -1/2$, and a crater radius ratio $s/a = m/n = 5/20$ is shown in Fig. 2. A smoothing of the curve appears justified. The total approach $\alpha_0 \equiv w_0$ and contact force P, however, are not influenced by this smoothing, since they are linear combinations of the p_j. Considering all integer values m between 1 and n, the modifying functions F_1 and F_2 can be calculated for discrete values of ρ then, as shown in Fig. 3.

It is remarked that for both modifying functions F_1 and F_2, the ordinates at $\xi = 0$ simply follow from Hertz theory, since they correspond to the indentation of a half-space, $R = \infty$. By the definitions implicit in equations (7) and (8), we get

$$F_1(\rho, 0) = F_2(\rho, 0) = (1+\rho)^{-1}$$

The modifying functions are well approximated by the function

$$F(\rho, \xi) = 1 - \rho(1+\rho)^{-1} \sqrt{1-\xi^n} \times L(\xi < 1) \tag{12}$$

with $n = 2$ for F_1 and $n = 4$ for F_2.

Having solved the static crater indentation problem, we now return to the analysis of impact, equation (1).

If the target area is cratered, there is a chance that as 'a' increases during the impact process, contact will extend to the flat portion of the half-space, on the periphery of the crater. This would affect direct solution of the equation of motion (equation (1)), because the modifying functions F_1 and F_2 must enter the analysis at $\xi \equiv s/a = 1$, and they do so with a discontinuity of their first and second derivatives.

The impact analysis is therefore performed in three stages in the general case of a cratered target area. The first small time interval at the beginning of impact is a simple Hertz analysis with time as the independent variable, justified for obtaining initial conditions for the second interval. The latter ends when contact is just completed up to s, the shoulder of the crater. The second interval is still conventional Hertz analysis between bodies of uninterrupted curvature, but the independent variable is $\xi = s/a$. If the kinetic energy of impact is still unexhausted, impact will continue into a third stage, in which contact extends to the shoulders. In this stage, the modifying functions enter into the equations of motion. This numerical integration of the equation of indenter motion is a further development of the approach used in solving impact problems on elastic layers[5].

First stage

We perform Hertz-analysis, based on equation (1) for a small fixed initial time interval Δt, so that

$$\xi > 1 \qquad (0 \leq t \leq \Delta t) \tag{13}$$

$$F_1(\rho, \xi) = F_2(\rho, \xi) = 1$$

Thus at $t = \Delta t$ we obtain:

$$a = a_{\Delta t}, \quad \alpha = \alpha_{\Delta t}, \quad \dot{\alpha} = \dot{\alpha}_{\Delta t}, \qquad \text{and} \qquad P = P_{\Delta t}$$

Second stage

We choose ξ as the independent variable for the next interval of the impact:

$$\xi_{\Delta t} \geq \xi \geq 1 \tag{14}$$

During this interval we still have $F_1(\rho, \xi) = 1$, $F_2(\rho, \xi) = 1$.
Transforming the equation of motion (1) into a set containing the new variables

$$z_1 \equiv t \tag{15}$$

$$z_2 \equiv \frac{dt}{d\xi} \tag{16}$$

the following system of first order ordinary differential equations arises:

$$\frac{d}{d\xi} z_1 = z_2 \tag{17}$$

$$\frac{d}{d\xi} z_2 = -\frac{2E_r z_r^3 s}{3\pi M} - \frac{3z_2}{\xi} \tag{18}$$

Subjected to the initial conditions:

$$\xi = \xi_{\Delta t}\colon z_1 = \Delta t \tag{19}$$

$$z_2 = -\frac{2(1+\rho)\, a_{\Delta t}^3}{s R_1 \dot{\alpha}_{\Delta t}} \tag{20}$$

These are readily solved by the Runge Kutta method.

For mesh size $\Delta\xi$, the quantity $(1-\xi_{\Delta t})/20$ can be conveniently used. At the end of the second stage ($\xi = 1$) the variables a, α, $\dot{\alpha}$ and P can be stored, for initial conditions to the third stage.

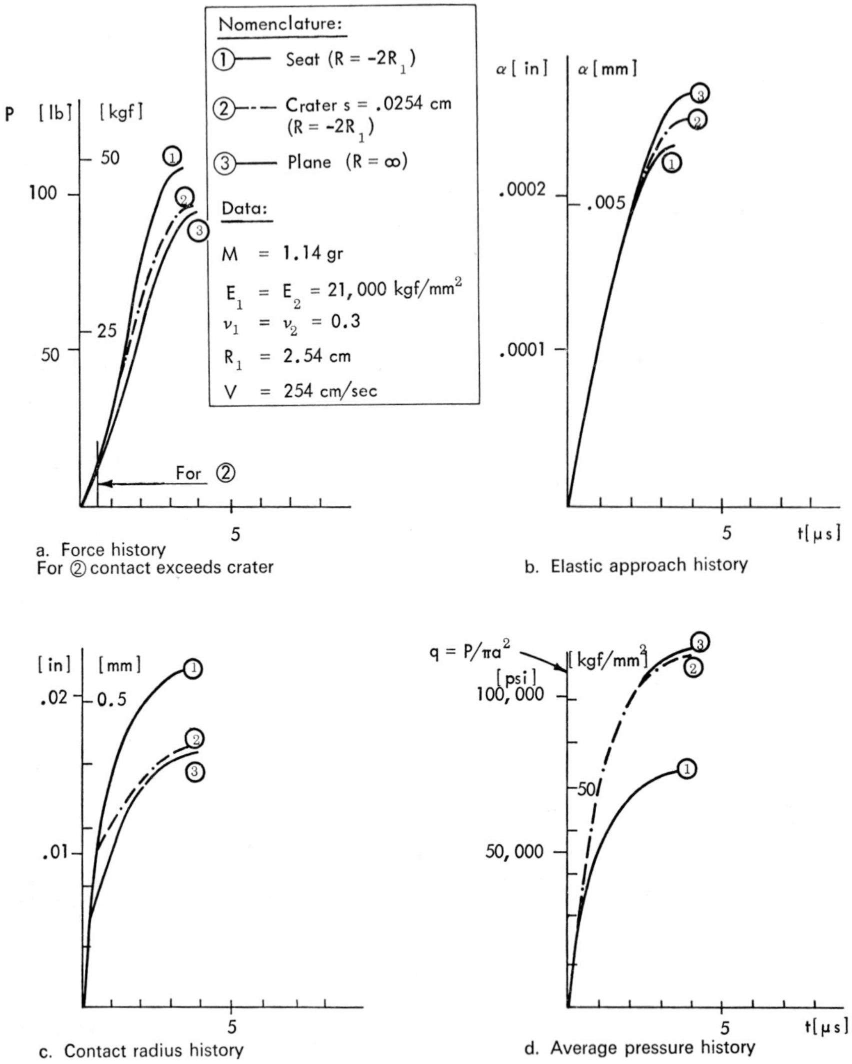

Fig. 4. Half-history of impact of sphere; ① on seat; ② in crater; ③ on plane

Third stage

In the third stage of impact, the independent variable is the time t; now $F_1(\rho, \xi)$ and $F_2(\rho, \xi)$ vary. Transformation of the equation of motion, with new variables,

$$y_1 \equiv a \qquad (21)$$

$$y_2 \equiv \frac{da}{dt}, \qquad (22)$$

yields the system of first-order, ordinary differential equations:

$$\frac{d}{dt} y_1 = y_2 \qquad (23)$$

$$\frac{d}{dt} y_2 = -\frac{\frac{4E_r y_1^3}{3\pi M} F_2 + y_2^2(2F_1 - 2\xi F_1' + \xi^2 F_1'')}{2y_1 F_1 - y_1 \xi F_1'} \qquad (24)$$

where the prime (′) denotes differentiation with respect to ξ. The initial conditions to equations (23) and (24) are obtained from compatibility of α and $\dot{\alpha}$ with their values calculated at the end of the second stage.

Example 1
For a demonstration of the effect of a cratered area on the impact, in comparison with a plane and also, a seated target, a numerical example is given in Fig. 4. Noteworthy is the rather sharp discontinuity in the $a(t)$ curve when at $t \simeq 0.4$ μs, 'a' reaches s.

3. SOME FAILURE PARAMETERS

Among one-shot failures we consider simple fracture and excessive plastic yielding. Repetitive impacting may cause various forms of fatigue and wear; the wear of surfaces impacted in the elastic stress range is also commonly fatigue-originated. In a broad sense, all these types of influences causing deterioration of the original surface are classifiable as failures[6].

3.1 *Maximum subsurface shear stress*

The maximum shear stress in a spherical Hertz contact arises at the axis at a depth of $z = 0.47a$, and its value is (from of the principal σ_z and σ_r stresses): $\tau_{max} = 0.31 q_0$. The impact of a mass M at impact velocity V gives rise to an average pressure of $q = P_0/\pi a^2$ where the peak contact force is calculated by the Hertz theory:

$$P_0 = 0.407 \left[\frac{V^6 M^3 R_1 E_r^2}{1+\rho}\right]^{1/5} \tag{25}$$

Fig. 4 shows that the non-Hertzian case of a partial crater formation reduces q with respect to that corresponding to a plane surface; the condition of a full seat offers maximum reduction to q.

3.2 *Maximum reversible radial stress*

The radial stress at the surface of a spherical contact is:

$$\sigma_r = \left[-\sqrt{1-r^2/a^2} - \frac{1-2\nu}{3}\left(\frac{a^2}{r^2}\left\{\left[1-\frac{r^2}{a^2}\right]^{3/2} - 1\right\}\right)\right] q_0 \tag{26}$$

This stress component is negative (compressive) up to $r = 0.991\,a$, for $\nu = 0.3$. Beyond this it is tensile, reaching a maximum tension at the edge $r = a$:

$$\sigma_r^+ = \frac{1-2\nu}{3}\, q_0 \tag{27}$$

During the impact process when the indentation radius increases, formerly tensile σ_r-zones become compressive. For the critical fatigue stress, the maximum reversible radial stress is sought at a distance $\bar{r} < a_0$, a_0 being the peak impact radius. Since by Hertz theory

$$q_0 = \frac{4\beta E_r}{\pi^2}\, a, \tag{28}$$

σ_r^+ at the position of the contact $a = \bar{r}$ becomes

$$\sigma_r^+ = \frac{4(1-2\nu)\,\beta E_r}{3\pi^2}\,\bar{r} \tag{29}$$

The (compressive) radial stress at $r = \bar{r}$, corresponding to $a = a_0$ is, by equation (26):

$$\sigma_r^- = \left[-\sqrt{1-\bar{r}^2/a_0^2} - \frac{1-2\nu}{3}\frac{a_0^2}{\bar{r}^2}\left\{\left[1-\frac{\bar{r}^2}{a_0^2}\right]^{3/2} - 1\right\}\right)\right] \times \frac{4\beta E_r}{\pi^2}\, a_0 \tag{30}$$

Table 1. The maximum reversible radial stress σ_R in terms of ν

ν =	0.0	0.1	0.2	0.3	0.4	0.5
$\bar{r}/a$ =	.933	.959	.979	.992	.998	1
σ_R/q_0 =	.310	.256	.196	.132	.066	0
σ^+/q_0 =	.333	.267	.200	.133	.067	0

Equating σ_r^+ and σ_r^- from (29) and (30), we obtain by the Newton-Raphson method the location $\bar{r}/a_0$ for the maximum reversible radial stress. Table 1 shows that the latter is very near the value of the maximum tensile stress at $r = a$.

3.3 *Peak strain energy*

Some workers associate impact fatigue and impact wear with the energies to be repeatedly absorbed by the contact[7]. Using the spherical contact solution[8] to the Hertz problem, a volumetric strain energy contour map $U(r, z)$ was constructed in Fig. 5*a*. The distortional part U_d of the strain energy is shown in Fig. 5b.

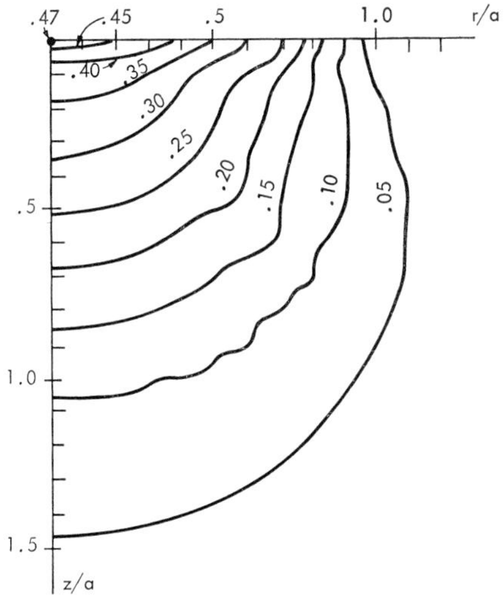

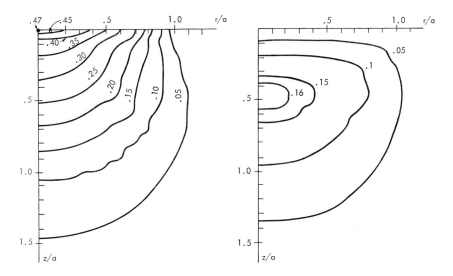

Fig. 5a. Contour map for the total volumetric strain energy EU/q_0^2 for spherical Hertz contacts

Fig. 5b. Contour map for the deviatoric volumetric strain energy EU_d/q_0^2 for Hertz contacts

These contour maps can also be used to obtain an estimate for the temperature rise in pure normal impact. The heat conduction process in a solid is orders of magnitude slower than the impact process[9, 10]. We may assume that heat is generated by an inelastic fraction of U (say $(1-e^2)U$), dissipated into the medium in a comparable time with t^*, very rapidly. The heat balance is now solved for an instantaneous temperature rise:

$$T = \frac{(1-e^2)\,U}{\gamma C} \tag{31}$$

The heat generated by the single impact of a smooth metal ball on like material stressed near its yield stress, is calculated on the basis of $q_{av} = 1.1\,\sigma_y$; thus, we obtain by equation (31),

$$T = \frac{1.278(1-e^2)\,\sigma_y^2}{\gamma CE} \tag{32}$$

For steel, considering $\sigma_y = 2.5 \times 10^9$ N/m^2, $C = 462$ J/kg°C and $e = 0.80$, $T \cong 4$°C results. This is comparable with some results reported by Bowden and Tabor[11].

For viscoelastic materials, the dilatational and distortional energies U_K and U_d, respectively, are associated with separate loss moduli. The right hand side of equation (32) will then have two terms,

$$\Delta T = \frac{1}{\gamma C}([1-e_K^2]\,U_K + [1-e_d^2]\,U_d) \tag{33}$$

where

$$U = U_K + U_d. \tag{34}$$

We remark in passing over Fig. 5 that Hertzian contact is a largely dilatational affair, shown by the relatively small distortional energy levels.

The temperature rise due to *repetitive* impact should also include consideration of the boundary conditions for heat conduction and convection, and the ratio of the interval between impact to the impact duration[12].

4. THE OPTIMAL WEARPATH

Let a contact stress-related failure parameter p be selected; the variation of p with the geometrical change of the wear crater dimensions s, ρ will be called the wearpath of p with respect to s and ρ. The actual wear path $p(s, \rho)$ is selected among an infinite variety of admissible surfaces.

We now postulate, from physical intuition, that the wear scar formation for a repetitively impacted surface follows the shape corresponding to the fastest relief of the critical failure parameter. This can be found as a steepest descent curve for p in (s, ρ)-space.

The correct descent is subjected however to some physical constraints from the nature of the repetitive impact problem (Fig. 1). It also depends on the initial conditions of crater geometry.

4.1 *Repetitive impact at constant momentum*

The constraints involve the size of the crater s with respect to the peak contact dimension a. On one hand s must at all times be smaller or equal to 'a' since wear cannot form in an unloaded area. Meanwhile s cannot be smaller than 'a' for a chosen failure parameter $p = q \propto \tau_{max}$, since the impact analysis of Section 2 (Fig. 4) shows that the average pressure q diminishes when a 'shouldered' contact is replaced by a full seat.

The above considerations[3] define the optimal wear path $\underline{L} = \underline{\rho} + \underline{s}$ for the failure parameter $p = q$:

$$\text{minimize} \quad p(\rho, s) \tag{35}$$

subject to

$$s = a \tag{36}$$

and the initial conditions of the contact geometry at the beginning of the wear process:

$$s = s_0, \qquad \rho = \rho_0, \qquad p = p_0 \tag{37}$$

The failure parameter is then by Hertz theory,

$$p = K_1(1+\rho)^{3/5} \tag{38}$$

where the constant $K_1 = 0.08387(E_r^4 \cdot V^2 \cdot M \cdot R_1^{-3})^{1/5}$.

The contact radius is also determined by Hertz theory,

$$a = K_2(1+\rho)^{-2/5} \tag{39}$$

where $K_2 = 1.2411(E_r^{-1} V^2 M R_1^{-2})^{1/5}$.

Since by the constraint (36) $s = a$, the optimal wear path is the sole admissible path, constituting a trivial solution,

$$s = K_2(1+\rho)^{-2/5} \tag{40}$$

4.2 *Repetitive impact on a previously worn small crater*

Assume that a small wear crater (s_0, R_0) has been made by previous activity at a target. Now repetitive impacting is started with a large momentum so that a_0 the resumed contact radius is greater than s_0. Thus at impact peak, we have contact on the shoulder of the crater (Fig. 1).

Selecting again $p = q = P/\pi a^2$ for failure parameter, and introducing the non-dimensional contact radius

$$\psi \equiv a/a_0 \tag{41}$$

we obtain from (8):

$$P = K(1+\rho)\, F_2(\rho, \xi) \cdot \psi \tag{42}$$

where $K = 4E_r a_0 / 3\pi^2 R_1$.

Now the modifying function $F_2(\rho, \xi)$ can be approximated:

$$F_2(\rho, \xi) \approx 1 - \rho(1+\rho)^{-1} \cdot \sqrt{1-\xi^4} \tag{43}$$

and a classical optimization problem for a non-dimensional pressure, $\bar{p} \equiv p/K$ results:

$$\text{minimize } \bar{p} = [1+\rho(1-\sqrt{1-\xi^4})]\cdot\psi \tag{44}$$

subject to the constraints:

$$0<\xi<1; \quad -1<\rho<0; \quad \psi>0 \tag{45}$$

and the initial conditions:

$$\xi=\xi_0, \quad \rho=\rho_0, \quad \psi=\psi_0. \tag{46}$$

This steepest descent problem can be numerically tackled by the gradient method to obtain the relation between ξ, ρ and ψ for the optimal curve $\underline{L}$ in (ξ, ρ, ψ) space. A step $\Delta\underline{L}$ on the optimal curve is always in the direction of the gradient of $\bar{p}$:

$$\Delta L_{\rightarrow} \propto \nabla\bar{p}(\xi,\rho,\psi) \tag{47}$$

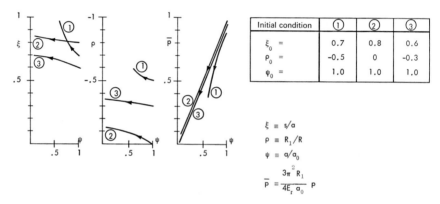

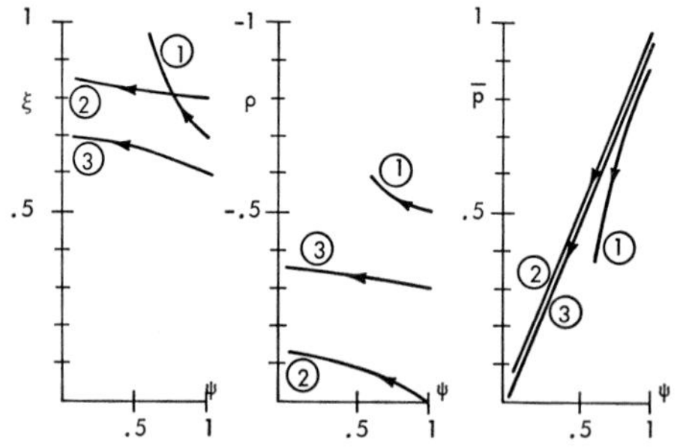

Initial condition	1	2	3
ξ_0 =	0.7	0.8	0.6
ρ_0 =	-0.5	0	-0.3
ψ_0 =	1.0	1.0	1.0

$\xi \equiv s/a$

$\rho \equiv R_1/R$

$\psi \equiv a/a_0$

$\bar{p} \equiv \dfrac{3\pi^2 R_1}{4E_r a_0}\, p$

Fig. 6. Optimal wear path for a preworn surface, with various initial conditions

Fig. 6 shows examples demonstrating that as s grows, and ρ approaches -1 (conformity), the contact radius 'a' diminishes with respect to the initial a_0.

5. CONCLUSIONS

We considered impact in its dynamic relation to the changing geometry of the repetitively clashing target surfaces. Failure parameters associated with Hertz contact stresses were discussed, among them the temperature rise due to impact. The optimal wearpath of repetitively impacted elastic bodies was defined and formulated as a classical optimization problem.

ACKNOWLEDGMENT

The author wishes to thank Prof. J. Geer for a discussion on the steepest descent problem.

ABSTRACT

This paper investigates the changes in the state of elastic impact stress arising in a surface worn by previous impacts. There are three basic problems discussed:

1. Given the worn surface geometry and the required geometric, material and impact parameters, the state of stress in the contact during impact is analyzed.
2. Failure criteria that correspond to further changes of the target surface are sought.
3. The geometric laws governing the change of the shape of the contact surface are investigated.

For an arbitrary cratered-surface geometry, quasi-static impact calculation is performed by a numerical method of analysis, using the Runge-Kutta technique.

Application of the appropriate yield theory leads to a stress quantity p, characterizing the gradual contact fatigue process.

To derive the relations for the change of the shape of the impact target area, the concept of the "optimal wearpath" is introduced as follows: the geometrical formation of the contact crater corresponds to the fastest relief (i.e., steepest descent) of the critical stress quantity, with respect to the selected crater parameters, among all permissible configurations.

REFERENCES

1. P.A. Engel, *Impact Wear*, Research Monograph, to be published by the Elsevier Scientific Publishing Co., Amsterdam.
2. P.A. Engel, T.H. Lyons and J.L. Sirico, 'Impact Wear Model for Steel Specimens', *Wear*, **23**, 1973, pp. 185–201.
3. P.A. Engel and R.G. Bayer, 'The Wear Process between Normally Impacting Elastic Bodies', *J. Lub. Techn.*, **96**, 1974, pp. 595–604.
4. H.D. Conway and P.A. Engel, 'Contact Stresses in Slabs due to Round Rough Indenters', *Int. Jl. Mech. Sci.*, **11**, 1969, pp. 709–722.
5. W.T. Chen and P.A. Engel, 'Impact and Contact Stress Analysis in Multilayer Media', *Int. Jl. Solids Structures*, **8**, 1972, pp. 1257–1281; XIIIth Intl. Congress of Theoretical and Applied Mechanics, Moscow, USSR, 1972.
6. S.V. Pinegin, *Kontaktnaya Proshnost'v Mashinach*, Mashinostroienie, Moscow, 1965.
7. K. Wellinger and H. Breckel, 'Kenngrößen und Verschleiß beim Stoß metallischer Werkstoffe', *Wear*, **13**, 1969, pp. 257–281.
8. G.M. Hamilton and L.E. Goodman, 'The Stress Field Created by a Circular Sliding Contact', *J. App. Mech.*, **33**, 1966, pp. 371–376.
9. J.F. Archard, 'The Temperature of Rubbing Surfaces', *Wear*, **2**, 1958/'59, pp. 438–455.
10. H. Blok, 'The Flash Temperature Concept', *Wear*, **6**, 1963, pp. 483–494.
11. F.P. Bowden and D. Tabor, *The Friction and Lubrication of Solids*, **1**, Oxford Univ. Press, 1950.
12. P.A. Engel and R.C. Lasky, 'Mechanical Response and Heat Buildup in Repetitively Impacted Elastomers', SESA Spring Conference, 1975, Paper No. 2397 A.

The normal contact of arbitrarily shaped multilayered elastic bodies

T. G. Johns and A. W. Leissa

1. INTRODUCTION

When two bodies are placed in contact, the bodies touch over a compressed or contact area. While the initial contact is determined by the geometric features of the bodies, the final contact area to a great extent is determined by the degree to which the bodies are deformed by the applied forces. The stresses developed during the contact of elastic bodies have long been of great concern in the design of mechanical systems. The mechanics of contacting elastic bodies has received considerable attention in the literature since 1831. Most of this attention has been directed toward homogeneous media. In recent years, however, a class of problems dealing with the contact of layered, elastic bodies has become increasingly important. In modern engineering practice, one frequently meets with stratified structures, laminated composites, or solid components coated with surface layers. Such layered contact problems stem from such areas as design of airport runways, roadways, foundations, plated or clad rollers, tracks, ball bearings, electrical contacts, as well as layered shells. In biomedical engineering, for example, problems arise in the study of joints in the skeletal system, and in designing equipment to measure the intraocular pressure within the eye.

The first results on the topic of contacting elastic bodies in the theory of elasticity were obtained in the last century. The first calculations directed toward determining the stresses during the contact of two bodies were obtained by Lame and Clapeyron[1] in 1831. The general theory of contact of elastic bodies was formulated later (1881) by Heinrich Hertz[2]. Hertz assumed the bodies to be 'flat' enough in the neighborhood of the area of contact to permit treatment by analytical methods of potential theory available for semi-infinite half-spaces. He assumed that the actual contact surfaces could be represented by general surfaces of the second degree. It has been shown that such problems, in fact those having surfaces with even higher order, are meaningful within the context of the linear theory of elasticity[3]. The solutions of Hertz, arrived at by means of the semi-inverse method, have long served as the basis of research into contact problems.

One of the most significant contributions to the solution of contact problems was made by Conry and Seireg[4] (1971). Realizing the intractable nature of problems involving bodies having more complex surface geometries than the ellipsoids of Hertz, Conry formulated a programming procedure utilizing a simplex-type algorithm. Somewhat later and independently, the method was also applied to contacting homogeneous bodies by Kalker[5] (1971). This procedure makes possible approximate solutions to the contact of homogeneous bodies of arbitrary geometry never before possible.

The first treatment of layered media also may be credited to Hertz[6] (1884). This Hertz problem can be classified as a force boundary-value problem as opposed to a true-contact problem. Naturally many force boundary-value problems have been solved since that time. One rather general treatment of boundary force loading of a layered system was performed by Kuo[7] (1969).

By using Hankel transform methods and a matrix method frequently used in elastodynamics, Kuo obtained a generalized solution to the problem of a multilayered medium under static loads. In this paper a method is described which incorporates the programming procedure utilizing a simplex-type algorithm like that of Conry and Kalker. Influence coefficients relating surface stresses to surface displacements of the bodies are determined by methods of analyses of problems in layered elastic systems by the Hankel transform technique like that presented by Kuo. The deformations are assumed small and the bodies are assumed to obey the laws of linear elasticity. Contact-stress distributions resulting from the contact of layered spheres are presented.

2. MATHEMATICAL FORMULATION

Formulation of general equations

Consider two bodies pressed together; let points (x, y, z_1) and (x, y, z_2) on the surfaces of body 1 and body 2, respectively (Fig. 1) come in contact within a compressed area. For points within the compressed area, it must be true that

$$w_1 + w_2 = \alpha - \varepsilon. \tag{1}$$

For points outside the contact area, it must be true that

$$w_1 + w_2 > \alpha - \varepsilon, \tag{2}$$

where

α = relative approach of the two bodies

w_1, w_2 = normal displacement of surface points of bodies 1 and 2, respectively, with respect to undeformed surfaces

ε = f_1+f_2

f_1, f_2 = equations of the surfaces of bodies 1 and 2, respectively, before deformation.

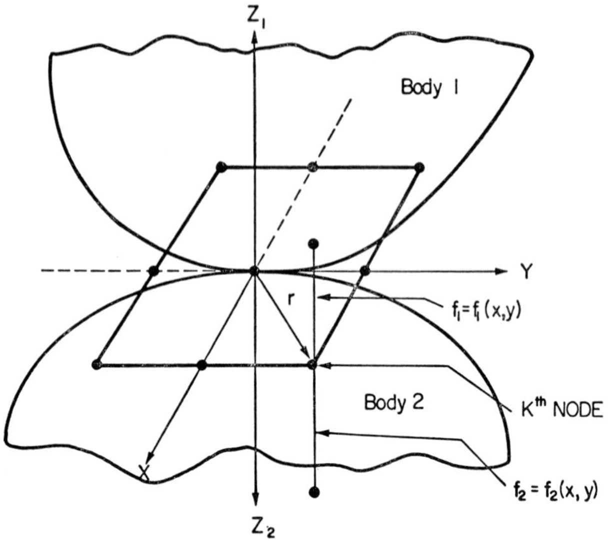

Fig. 1. Contact of two arbitrarily shaped elastic bodies

It is assumed that the contacting bodies are isotropic and linearly elastic and that the representative dimensions of the contact area are small compared to the radii of curvature of the undeformed surfaces in the vicinity of the contact interface.

Equations (1) and (2) represent displacement boundary conditions; in addition it is required that the tractions vanish on $Z_1 = Z_2 = 0$. Equilibrium is expressed in a simple way, namely

$$\int_{R^1} \sigma_{zz} \, \mathrm{d}A = \int_{K^2} \sigma_{zz} \, \mathrm{d}A = P, \tag{3}$$

where

R = the contact region

P = the total normal load forcing the bodies together.

Assume now that a proposed zone of contact is divided into a regular array of finite areas (Fig. 1). At the center of each area is placed a node. If a particular node lies within the area of contact, the sum of the elastic deformations and any initial separations must equal the relative approach of the bodies, or for the kth node equation (1) may be rewritten as

$$w_{1k}+w_{2k}+\varepsilon_k-\alpha = 0, \tag{4}$$

where

w_{1k}, w_{2k} = the normal deformations of bodies 1 and 2, respectively, at the kth node

$\varepsilon_k = f_{1k}+f_{2k}$ = the initial separation at the kth node.

Outside the contact area equation (2) becomes

$$w_{1k}+w_{2k}+\varepsilon_k-\alpha > 0. \tag{5}$$

Equations (4) and (5) can be combined into a single equation by introducing new variables, called 'slack variables', which are either positive or zero. Equation (5) can then be written as

$$w_{1k}+w_{2k}+\varepsilon_k-\alpha-Y_k = 0, \tag{6}$$

where Y_k are the slack variables defined such that

$$Y_k \geq 0. \tag{7}$$

The continuous pressure distribution is approximated by a set of uniform pressures acting over small areas about each of the N_0 nodes. The product of the pressure and the area at each individual node is F_k. For equilibrium, the sum of all F_j must equal the total normal force, P, forcing the bodies together; i.e.,

$$\sum_{j=1}^{N_0} F_j = P. \tag{8}$$

The contact problem is characterized by equations (6), (7) and (8) in addition to the criterion that inside the contact region, R,

$$Y_k = 0 \quad \text{and} \quad F_k \geq 0 \text{ (inside } R), \tag{9}$$

outside the contact region

$$Y_k > 0 \quad \text{and} \quad F_k = 0 \text{ (outside } R). \tag{10}$$

A solution to a discrete contact problem is, therefore, a set of stresses and displacements which satisfies the boundary and continuity conditions of equations (6), (7), (8), (9) and (10), and the field equations of elasticity of the two contacting bodies.

Because both bodies are assumed to follow laws of linear elasticity, the elastic deflection at node k is the superposition of the influence of all the forces F_j ($j = 1, 2, \ldots, N_0$). This is expressed by

$$w_{ik} = \sum_{j=1}^{N_0} a_{ikj} F_j, \tag{11}$$

where a_{ikj} is the deflection of the ith body at node k due to a unit force distributed over the area about the jth node; i.e., a_{ikj} is an influence coefficient describing at a node k due to a unit load applied at node j.

One method to solve equation (6) is that enumeration in which all the possible combinations of base vector sets of (F, α, Y) which satisfy equation (6) are chosen. Even with the aid of modern high-speed computers, this method is still economically impractical.

Conry and Seireg[4] demonstrated that an efficient method for sorting through the various combinations of base vectors to find a unique feasible solution is that of a modified simplex algorithm. The simplex method[8] of linear programming consists of finding values for a set of non-negative variables that satisfies a system of linear equations and minimizes some functional. The functional which is minimized is formed by the summation of artificial variables one each of which is added to equation (6) for each value from $k = 1$ to N_0, as well as to the equilibrium equation (8).

Formulation of influence coefficients

In formulating the influence coefficients we begin by replacing the array of uniformly loaded areas shown in Fig. 1 by an array of uniformly loaded circles of radius a, Fig. 2. The stress on each circular area (say at node k) is the static equivalent of the associated nodal force, (F_k). Or

$$F_k = \pi a^2 [\sigma_{zz}] \text{ evaluated at node } k. \qquad (12)$$

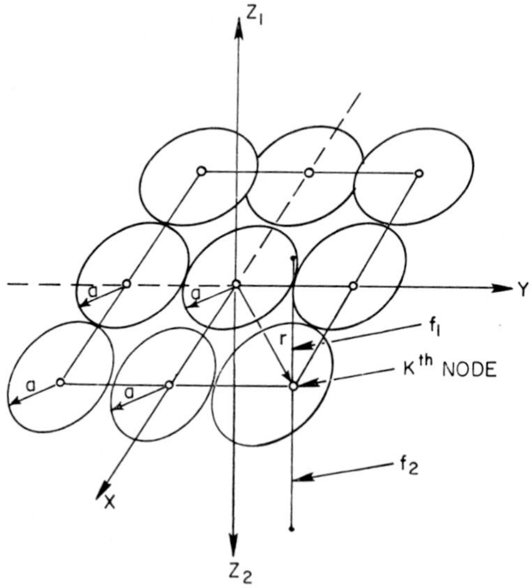

Fig. 2. Proposed zone of contact represented by an array of uniformly loaded circular regions

To determine the general influence coefficient (a_{ikj}), assume that each layer of the multilayered medium is elastic, homogeneous, and isotropic and that the interfaces of the layers are perfectly welded. Let the original of cylindrical coordinates coincide with the top of the lth layer of the multilayered medium and with the z axis directed into the layer below (Fig. 3).

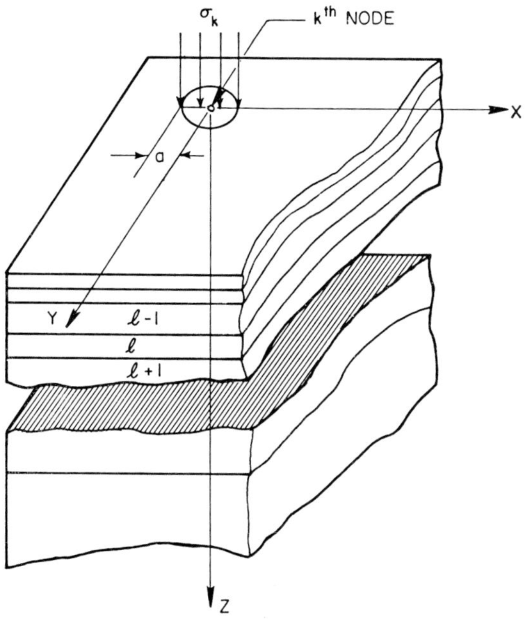

Fig. 3. Multilayered half-space

The axisymmetric equations of equilibrium for any layer, l, under no body forces in cylindrical coordinated (r, z) are,* in terms of the displacements U_{zk}, W_{zk} in the r and z directions, respectively, are

$$\nabla^2 U_k + \frac{1}{1-2\nu}\frac{\partial \Delta}{\partial r} - \frac{1}{r}\frac{U_k}{r} = 0$$

$$\nabla^2 W_k + \frac{1}{1-2\nu}\frac{\partial \Delta}{\partial z} = 0, \tag{13}$$

where ∇^2 and ν denote the harmonic operator and Poisson's ratio, respec-

* Here the coordinates r and z are taken at the kth node. For simplicity, no subscripts will be carried on the coordinate, but it should be kept in mind that we are considering a typical node.

tively, and Δ denotes the dilatation, which is given by

$$\Delta = \varepsilon_{rr} = \varepsilon_{\theta\theta} + \varepsilon_{zz} = \frac{\partial U_k}{\partial r} + \frac{U_k}{r} + \frac{\partial W_k}{\partial z}. \tag{14}$$

These equilibrium equations are trivially satisfied by using a potential function solution. However, for the axisymmetric problem, the Love's stress function, ϕ_k, possesses sufficient generality to allow the boundary conditions and continuity conditions for each layer to be satisfied. From the compatibility conditions of elasticity theory

$$\nabla^4 \phi_k = 0, \tag{15}$$

where ∇^4 is the biharmonic operator.

Applying the theory of Hankel transforms we reduce the biharmonic equation (15) in r and z to a fourth-order ordinary differential equation in z, i.e.,

$$\int_0^\infty r\nabla^4 \phi_k \, J_0(\eta r) \, \mathrm{d}r = \left(\frac{\mathrm{d}^2}{\mathrm{d}z^2} - \eta^2\right)^2 H_k(\eta, z), \tag{16}$$

where

$$H_k(\eta, z) = \int_0^\infty r\phi_k \, J_0(\eta r) \, \mathrm{d}r. \tag{17}$$

The stress function ϕ_k is then given in terms of $H_k(\eta, z)$ by the inverse transform

$$\phi_k = \int_0^\infty \eta \, H_k(\eta, z) \, J_0(\eta, r) \, \mathrm{d}\eta.$$

It follows that for ϕ_k to be a solution of equation (15), then the Hankel function $H_k(\eta, z)$ must be a solution to the ordinary differential equation

$$\left(\frac{\mathrm{d}^2}{\mathrm{d}z^2} - \eta^2\right)^2 H_k(\eta, z) = 0, \tag{18}$$

or

$$H_k(\eta, z) = [(A_k + C_k) + z(B_k + D_k)] \cosh(\eta z) + \\ + [(A_k - C_k) + z(B_k - D_k)] \sinh(\eta z). \tag{19}$$

The arbitrary constants $A_k \ldots D_k$ can be determined by

1. Satisfying the boundary conditions on the free surface

$$\left.\begin{aligned} \sigma_{zz}(r, 0) &= \sigma_k \\ \tau_{rz}(r, 0) &= 0 \end{aligned}\right\} r \leq a$$

$$\sigma_{zz}(r, 0) = \tau_{rz}(r, 0) = 0 \text{ elsewhere}.$$

2. Enforcing continuity of the U_k, W_k, σ_{zz} and τ_{rz} at the interface of each layer of the multilayered media.
3. Enforce conditions of finite deformations and stresses as z becomes large.

By enforcing conditions 1 through 3 above, it can be shown[9] that the normal displacement at the kth node of the free surface due to a force F_j uniformly distributed over a circular area of radius a about node j is

$$W_{ik} = a_{ikj} F_j . \tag{20}$$

Where the influence coefficients a_{ikj} are given by

$$a_{ijk} = \frac{1}{\pi a} \int_0^{\infty} \frac{(g_1 - g_2)}{F_k(\eta)} J_1(\eta a)\, J_0(r_{jk})\, \mathrm{d}\eta , \tag{21}$$

where r_{kj} is the distance between the kth and jth nodes and

$$\begin{aligned} F_k(\eta) &= (M_{11}+M_{41})\,(M_{22}+M_{32}) - (M_{12}+M_{42})\,(M_{21}+M_{31}) \\ g_1 &= (M_{11}+M_{41})\,(M_{23}+M_{33}) \\ g_2 &= (M_{13}+M_{43})\,(M_{21}+M_{31}) . \end{aligned}$$

The M_{ij} are coefficients of the matrix $\overline{M}_k$ for an N layer media where

$$\begin{aligned} \overline{M}_k &= [T_{kN}(0)]^{-1}\, [a_{k,N-1}]\, [a_{k,N-2}] \dots [a_{k1}] \\ a_{kl} &= T_{kl(d_l)}\, T_{kl}(0)^{-1} , \end{aligned}$$

and where d_l is the thickness of the lth layer, the matrix $T_{kl}(z)$ is given by

$$\begin{bmatrix} \eta^2 \cosh \eta z & \eta(\sinh \eta z + \eta z \cosh z) & \eta(\cosh \eta z + \eta z \sinh \eta z) & \eta^2 \sinh \eta z \\ \eta^2 \sinh \eta z & \eta[-2(1-2\nu_l) \sinh \eta z + \eta z \cosh \eta z] & \eta[-2(1-2\nu_l) \sinh \eta z + \eta z \cosh \eta z] & \eta^2 \cosh \eta z \\ -\mu_l \eta^3 \cosh \eta z & \mu_l\, \eta[(1-2\nu_l)\eta \sinh \eta z - \eta^2 z \cosh \eta z] & \mu_l\, \eta[(1-2\nu_l)\eta \cosh \eta z - \eta^2 z \sinh \eta z] & -\mu_l \eta^3 \sinh \eta z \\ \mu_l{}^3 \sinh \eta z & \mu_l\, \eta[2\nu_l \eta \cosh \eta z + \eta^2 z \sinh \eta z] & \mu_l\, \eta[2\nu_l \eta \sinh \eta z + \eta^2 z \cosh \eta z] & \mu_l \eta^3 \cosh \eta z \end{bmatrix}$$

3. NUMERICAL RESULTS AND APPLICATIONS

The simplex method was programmed to solve the set of equations (6) through (10). For an example problem, the geometrical shape of the bodies was kept simple in order to best present the effects of layering upon the contact stress phenomenon; more solutions to layered media problems arrived at using this method of analysis may be found in [9].

The program was used to solve for the normal contact stress distribution resulting from elastic contact of two-layered spheres. The spheres were identical in every way, and had a layer-thickness-to-radius ratio of $t/R = 0.01$. Fig. 4 shows the computed contact stress distribution for the

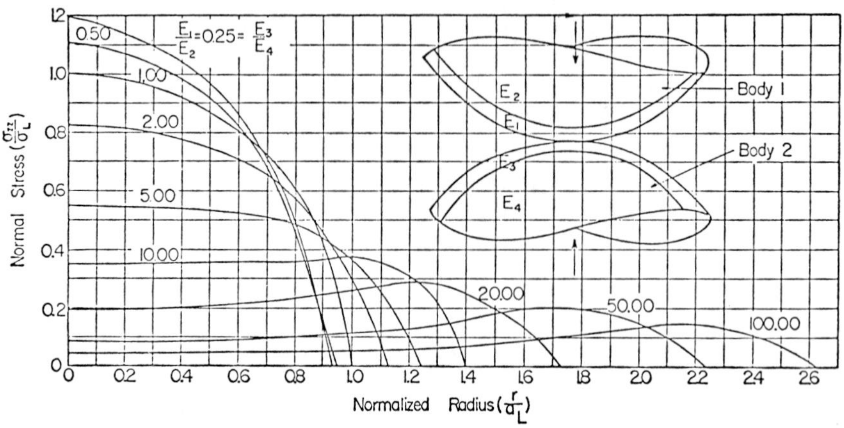

Fig. 4. Contact distribution for layered spheres and various layer modulus, $11 \times 11 = 121$ *node model*
$t/R = 0.01$, load $(P/E_1R^2) = 3.33 \times 10^{-6}$

two spheres for various ratios E_1/E_2. This problem was modeled using $11 \times 11 = 121$ nodes. The normal load $P/(E_1 R^2)$ was 3.33×10^6. The figure shows that for small E_1/E_2 the contact stress distribution (for this layer thickness) is governed strongly by the properties of the half-space. But as E_1/E_2 becomes larger, the location of maximum contact stress moves to, or concentrates, near the periphery of the contact region. As E_1/E_2 becomes large, the stress distribution approaches that of contacting spherical shells[10] and the foundation modulus plays a lesser role in supporting the contact force; a large amount of this force is supported through bending of the layer.

BIBLIOGRAPHY

1. G. Lame and P. B. E. Clapeyron, 'Mémoire sur l'équilibre Intérieur des Corps Solides Homogènes', *Crelle's Journal,* 7, 1831, pp. 400–404.
2. H. Hertz, 'Über die Berührung fester elastischer Körper', 1881, re-printed in *Gesammelte Werke von Heinrich Hertz,* 1, pp. 155–173.
3. J.J. Kalker, 'Minimum Principle for Frictionless Elastic Contact with Application to Non-Hertzian Half-Space Contact Problems', *Jour. Eng. Math.,* **6**, 2, April 1972, pp. 193–206.
4. T.F. Conry and A. Seireg, 'A Mathematical Programming Method for Design of Elastic Bodies on Contact', *Jour. Appl. Mech., Trans. ASME,* 70-WA/APM-52, 1971.

5. J.J. Kalker, 'A Minimum Principle for the Law of Dry Friction, With Applications to Elastic Cylinders in Rolling Contact', *Jour. Appl. Mech.*, 1971, pp. 875–887.
6. H. Hertz, 'Über das Gleichgewicht schwimmender elastischer Platten', *Wiedermann's Annalen der Physik und Chemie*, **22**, 1884, pp. 449–455.
7. J.T. Kuo, 'Static Response of a Multilayered Medium under Inclined Surface Loads', *Jour. Geophysics Res.*, **74**, 12, 1969, pp. 3195–3207.
8. P. Wolfe, 'Some Simplex-Like Nonlinear Programming', *Econometrica*, **27**, 1959, pp. 382–398.
9. T.G. Johns, *The Normal Contact of Arbitrarily Shaped Multilayered Elastic Contacts*, Ph.D. Dissertation, Ohio State University, 1973.
10. D.P. Updike and A. Kalnins, 'Contact Pressure between an Elastic Spherical Shell and a Rigid Plate', *Jour. Appl. Mech., ASME*, December 1972, pp. 1110–1114.

Contact stresses for multiply-connected Regions - the case of pitted spheres*

K. P. Singh, B. Paul and W. S. Woodward

NOMENCLATURE

A_{ij}	area of cell i, j
a_{ij}	coefficient in eq. (21)
B_{ij}	coefficient in eq. (23)
c_{ijl}	distance between field point l and a point in cell S_{ij}
$\bar{c}_{ijl}$	distance between field point l and centroid of cell S_{ij}
d	hypothetical interpenetration
d_0	initial interpenetration of spheres
e'	separation (gap) between corresponding surface points after load is applied
E	Young's modulus
$f(x, y)$ or $f(r)$	initial separation between surface points
f_i	$f(r)$ evaluated at field point i
f'_i	$f_{N+1} - f_i$
F	normal load
F^*	dimensionless normal load, $kF/(R)^2$
I_{ijl}	$\int_{S_{ij}} \mathrm{d}A_{ij}/c_{ijl}$
k	elastic parameter
m	number of equal sectors in Ω
n	number of cells in Ω
N	number of annular rings in Ω
N_{con}	number of cells within radius r_{con}
$p(x, y)$	interfacial contact pressure
p_i	piecewise constant pressure in annular ring i
p_i^*	dimensionless pressure, kp_i
r	radius defined from origin of Ω
r_b	radius of blending point of pit and sphere

* This research was partially supported by the Federal Railroad Administration of the U.S. Department of Transportation under Grant DOT-05-40093.

r_c	radius of curvature at 'edge' of pit
r_{con}	radius used in convergence studies
r_i	inner radius of contact region
r_0	outer radius of contact region
$r\Delta\phi$	width of cell
Δr	length of cell
r^*	r/R
r_b^*	r_b/R
r_c^*	r_c/R
R	effective radius $(2R_1R_2)/(R_1+R_2)$
R_1	radius of sphere with pit
R_2	radius of smooth sphere
S_{ij}	region included in ring i between rays j and $j+1$
SCF	stress concentration factor
v_j	vector in eq. (22)
x'	radius of point 0′ in Fig. 2
(x, y)	Cartesian coordinates in the fixed reference plane
α	$\Delta\phi/2$ half-angle of sector
β_{ij}	centroidal radius of sector S_{ij}
δ	relative approach
δ^*	δ/R
θ	semi-vertex angle (Fig. 1)
ν_1, ν_2	Poisson's ratio of bodies 1 and 2 respectively
ξ_i	inner radius of ring i
ρ	a boundary radius (r_i or r_0) of the contact region
ϕ	polar coordinate in fixed reference plane
$\Delta\phi$	$2\pi/m$
Ω	region of contact
Ω_i	region of cell i
Ω^*	candidate contact region.

1. INTRODUCTION

Contact problems involving multiply-connected contact regions have received little attention in the literature, possibly because of the non-Hertzian nature of such problems. Such problems arise, for example, whenever either of the contacting bodies have surface pits (e.g. casting defects, corrosion pits, machining faults, etc.). Barely perceptible surface flaws can cause high stress concentrations, and consequently, rapid fatigue failure. Experimental observations by Tallian[1], Martin and Eberhardt[2] and Littman and Widmer[3] indicate that such surface defects may be potential nuclei of microcrack propagation and can produce rapid destruction of rolling surfaces.

Based on the degree of difficulty associated with their solution, these problems may be divided into the following two categories:

i. Contact region known *a priori*:
   When the indentor contact surface is flat (or almost flat) it will be called a 'stamp', and the contact surface is defined *a priori* by the stamp boundary. When the indentor surface is not flat, but the indentor has a substantially higher elastic modulus than the indented body, the indentor can be treated as rigid, and the shape of the contact region becomes known for any given depth of penetration relative to the indentor tip.
ii. Elastic contact problems:
   When the indentor is not a stamp, and the two bodies have comparable elastic moduli, then the geometry of the contact region is unknown *a priori*, and it must be determined by solving the appropriate elasticity problem.

To the best of our knowledge, no solutions to problems of category (ii), for three-dimensional elastostatics with multiply-connected regions, have been reported in the literature. However, solution of a few special cases of rigid indentor problems (category (i)) have been found by Olesiak[4], Parlas and Michalopoulos[5] and Chiu[6].

Olesiak[4] solved the problem of an annular flat faced-stamp pressed on an elastic half space. Parlas, et al. proposed the solution for a 'bolt shaped' indentor pressed into an elastic half-space with a cylindrical hole. The cylindrical (bolt) section of the indentor was assumed to be rigidly bonded to the wall of the cylindrical hole while the bottom face of the bolt head presses against the half-space.

Chiu[6] solved the problem of an infinitely long rigid cylinder in contact with an elastic half-space, where the rigid cylinder has a groove running parallel to its axis.

In this paper, we give results which indicate that problems of both categories (i) and (ii) may be successfully solved by an extension of the method introduced by Singh and Paul[7].

A brief synopsis of the simply-discretized method of solution is given in Section 2, and some limitations and advantages of this method are discussed in Section 3. The example problem of a pitted sphere in contact with a complete sphere is described in Section 4. Techniques devised for an accurate numerical solution and rapid convergence are described in Section 5. Results for an example are given in Section 6, and conclusions are reviewed in Section 7.

2. THE SIMPLY DISCRETIZED METHOD OF SOLUTION

Singh and Paul[7] proposed a group of numerical methods for the solution of frictionless elastic contact problems where the surface profiles of the contacting bodies are allowed to have discontinuities in slope and curvature. One of their solution procedures, called the 'simply discretized method', is briefly recapitulated in this section. For a detailed description of the theoretical foundation and applicability of the method, the reader is referred to Singh[8] or Singh and Paul[9].

We will restrict our attention to 'nonconformal' contact problems where the dimensions of the contact region are small compared to appropriate radii of curvature of the undeformed bodies. Therefore, we may assume that the contact surfaces do not deviate significantly from a reference plane in which we imbed fixed Cartesian axes (x, y). Furthermore, we shall consider only those cases where the two bodies undergo a relative rigid body translation of amount δ, in a direction normal to the reference plane, plus an elastic deformation. The translation δ is called the 'relative approach' and is positive if it moves the bodies towards one another. We will also assume that the applied load consists of a force F, acting normal to the reference plane, and that the contacting surfaces have a sufficient degree of symmetry that the resultant of the contact pressures on each body is a force of magnitude F which acts through the origin 0 of the reference plane and equilibrates the applied force F.

It is well known (see for example Luré[10]) that the fundamental integral equation for nonconformal contact stress problems is

$$k \int_{\Omega} \frac{p(x', y')\, \mathrm{d}x' \mathrm{d}y'}{[(x-x')^2 + (y-y')^2]^{1/2}} - \delta + f(x, y) = e'(x, y) \tag{1}$$

where the 'elastic parameter' k is defined as

$$k = \frac{1-\nu_1^2}{\pi E_1} + \frac{1-\nu_2^2}{\pi E_2} \tag{2}$$

In the foregoing equations, ν_1, ν_2 and E_1, E_2 denote the Poisson's ratio and Young's modulus respectively for body 1 (indentor) and body 2 (indented); $p(x', y')$ is the normal pressure over the contact surface; Ω is the projection of the contact surface on the (x, y) reference plane; $f(x, y)$ represents the initial separation (or gap) between surface points on the two bodies, located at the same (x, y) coordinates, before the load F is applied; $e'(x, y)$ is the separation of the opposed surface points after the load is applied. Fig. 1 illustrates the initial separation f for a case of axial symmetry where f is a function $f(r)$ of the radial coordinate r.

The condition of impenetrability of matter requires that $e'(x, y)$ should vanish inside Ω and it should be positive outside of Ω. Conversely, the interfacial contact pressure $p(x, y)$ should be positive inside Ω, and it should vanish identically outside of it. In symbolic terms,

$$e' = 0 \qquad \text{for} \quad (x, y) \text{ inside } \Omega \tag{3a}$$

$$e' > 0 \qquad \text{for} \quad (x, y) \text{ outside of } \Omega \tag{3b}$$

$$p(x, y) = 0 \quad \text{for} \quad (x, y) \text{ outside of } \Omega \tag{4a}$$

$$p(x, y) \geq 0 \quad \text{for} \quad (x, y) \text{ inside } \Omega \tag{4b}$$

In short, a solution of the problem requires the determination of the boundaries of region Ω, a pressure field $p(x, y)$, and an approach δ which

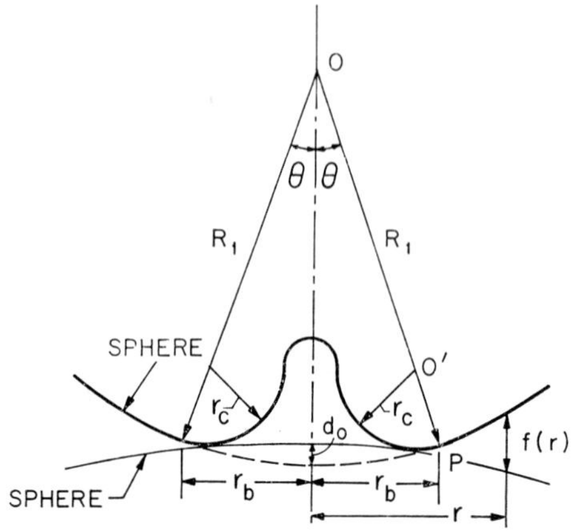

Fig. 1. Geometry of pitted surface; $f(r)$ is initial separation

satisfy relations (1)–(4). The associated load may be found from the expression

$$F = \int_{\Omega} p(x, y) \, dx dy$$

The absence of foreknowledge of the contact region Ω is a major impediment to a mathematical solution. This obstacle is overcome by postulating a tentative contact region Ω^*. Singh and Paul[7] proposed that the 'interpenetration curve' described by

$$f(x, y) = d \tag{5}$$

be used as a tentative contact region. Equation (5) defines the contour of the curve formed by interpenetration (without deformation) of the two surfaces through an arbitrary distance d. Picking a suitable value of d establishes the candidate contact region Ω^*. Using this as a preliminary estimate of Ω, equation (1) is readily recognized to be an integral equation of the first kind. A 'simply discretized' numerical solution of equation (1) is found by subdividing Ω into a large number of small cells. The pressure function $p(x, y)$ is replaced by a piecewise constant pressure field (pressure p_i

in cell i). Thus if Ω is subdivided into n cells, equation (1) becomes

$$\sum_{i=1}^{n} kp_i \int_{\Omega_i} \frac{\mathrm{d}x'\mathrm{d}y'}{[(x-x')^2+(y-y')^2]^{1/2}} - \delta + f(x,y) = e' \tag{6}$$

where Ω_i is the region of cell i. In equation (6), n values of p_i and the constant δ are unknowns to be determined. The centroids (x_i, y_i) of the cells are taken as field points (x, y) and equation (6) is written for each field point. The integrals in equation (6) are evaluated by numerical quadrature. Thus n linear algebraic equations are generated. An additional independent linear equation, essential for a unique solution, is generated by picking up a field point other than the cell centroids. The choice of this additional field point is otherwise arbitrary, however, it does affect the quality of the results, as discussed in Section 4.

Having thus generated a set of $n+1$ linear equations, the n unknown pressures, p_i, and the approach δ, are obtained through Gaussian elimination. The next step in the solution is to determine whether the tentatively selected region of integration Ω^* is indeed the true contact region. This is done by utilizing the inequalities (3) and (4), and systematically adjusting the boundaries of Ω until these inequalities are satisfied.

3. LIMITATIONS AND ADVANTAGES OF THE SIMPLY-DISCRETIZED METHOD

Singh and Paul[7] showed that the simply-discretized method just described is numerically unstable in the general case. This is due to the fact that the solution vector of the set of linear algebraic equations generated is very sensitive to small perturbations in elements of the coefficient matrix. Since such perturbations are unavoidable in the discretization process, the solution vector tends to be very erratic. Large oscillations in the solution vector correspond to small perturbations in the elements of the coefficient matrix. This behavior is similar to that observed in ill-posed problems of partial differential equations, as discussed by Hadamard[11]. For problems which do not show axisymmetry (or where axisymmetry is not utilized), Singh and Paul[7] found that the simply discretized method was incapable of predicting the proper stress distribution. For such problems they found it necessary to introduce stabilizing techniques known as the 'redundant field point method', and the 'functional regularization method' (see [7,9]).

The amount of numerical computation required for either of the two last named methods exceeds that of the simply discretized method. Accordingly, it is desirable to use the latter whenever circumstances permit.

In this paper we will focus on a problem with complete axisymmetry, and it will be shown that the simply discretized method provides an excellent solution, provided that the maximum possible use is made of the symmetry of the problem.

In other words, we recognize that all cells located at the same radius from the axis of symmetry have the same contact pressure at their centroids, and the number of unknown pressures p_i is reduced from n (the number of cells) to N (the number of annular rings formed by an axisymmetric distribution of cells). By using the simply discretized method, we are able to utilize inequality (4b) to iteratively refine the region of contact Ω. Upon satisfying inequality (4b), it was invariably found that inequality (3b) was satisfied.

The nature of the functional regularization method prohibits the use of inequality (4b) as a basis for refining Ω.

Numerical experiments have indicated that iteration procedures based up on inequality (4b) converge much faster than those based upon inequality (3b). Further details of the iteration procedures will be found in Sections 5 and 6.

4. PITTED SPHERE GEOMETRY

As a typical example, contact of a pitted elastic sphere of radius R_1 with an unpitted elastic sphere of radius R_2 is considered. A section of the pitted surface by a plane through the axis of symmetry is shown in Fig. 1. The local contour of the pitted surface is idealized as a torus smoothly blended into a sphere. The blending point P, where the pit joins the main surface, is located at a distance r_b from the load line. The center of curvature $0'$ of the pit blending arc lies on the conical surface of semi-vertex angle θ. The meridional radius of curvature of the torus is r_c.

Note that the discontinuity in curvature which occurs at P does not preclude the use of the method of solution described. A tentative contact

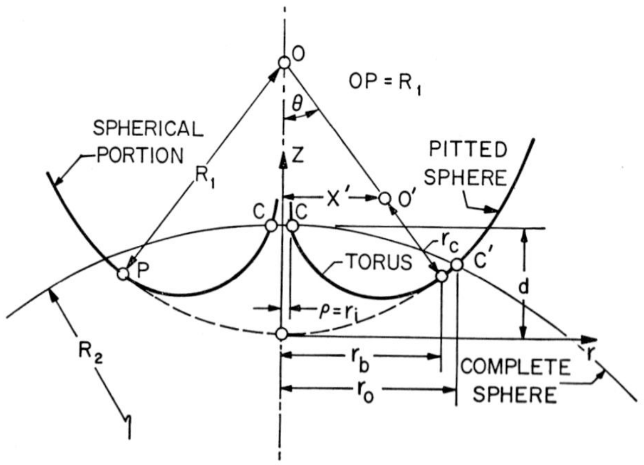

Fig. 2. Generation of annular interpenetration region

region, Ω, is established by a hypothetical interpenetration of the two spheres through a distance d. The annulus of contact so formed is bounded by an inner radius r_i and an outer radius r_0 as shown in Fig. 2, where suitable coordinate axes r, and z are indicated. The values of r_i and r_0 for a given problem are determined as follows. The z coordinate of a point $C(\rho, z_1)$ located at a distance ρ from the z-axis on the portion of body 1 (see Fig. 2), where

$$\rho < r_b, \tag{7a}$$

is:

$$z_1 = R_1 - (R_1 - r_c)\cos\theta - [r_c^2 - (x' - \rho)^2]^{1/2} \tag{7b}$$

where

$$x' = \frac{(R_1 - r_c)\, r_b}{R_1} \tag{7c}$$

$$\theta = \sin^{-1}\frac{r_b}{R_1} \tag{8}$$

The z-coordinate of a point on sphere 2, located at a distance ρ from the z-axis, is given by

$$z_2 = d - [R_2 - (R_2^2 - \rho^2)^{1/2}] \tag{9}$$

Since point C lies on both the torus and the lower sphere, $z_1 = z_2$; thus equations (7b) and (9) require that

$$d = R_1 - (R_1 - r_c)\cos\theta - [r_c^2 - (\rho - x')^2]^{1/2} + R_2 - (R_2^2 - \rho^2)^{1/2} \tag{10}$$

$$\rho < r_b \tag{10a}$$

Furthermore, the z-coordinate of a material point C' located on the spherical portion of body 1, at a distance ρ from the z-axis, is given by

$$z_1 = R_1 - (R_1^2 - \rho^2)^{1/2} \tag{11}$$

where

$$(\rho > r_b) \tag{11a}$$

Hence, for a given interpenetration d, the radius ρ of a point on the intersection of sphere 2 and spherical region of body 1 is given by

$$d = R_1 - (R_1^2 - \rho^2)^{1/2} + R_2 - (R_2^2 - \rho^2)^{1/2} \tag{12a}$$

$$(\rho > r_b) \tag{12b}$$

The geometry of the toroidal surface indicates that for $r_c < R_1$, equation (10) has two solutions for ρ. Let ρ_1 and ρ_2 ($\rho_1 < \rho_2$) be roots of equation (10). Two cases are readily identified.

Case i. When both inner and outer radii of the assumed contact region lie inside the blending radius, i.e.

$$\rho_2 < r_b \tag{13}$$

In this case the contact is assumed to be completely confined to the toroidal segment of body 1, in which case

$$\begin{aligned} r_i &= \rho_1 \\ r_0 &= \rho_2 \end{aligned} \tag{14}$$

Case ii. When the outer boundary of Ω lies beyond the blending radius (as shown in Fig. 2), i.e.

$$\rho_2 > r_b \tag{15}$$

In this case

$$r_i = \rho_1$$

and the outer radius r_0 is determined from the solution of equation (12). Note that equations (10) and (12) are transcendental in ρ, which can be found by an iterative procedure (e.g. Newton-Raphson).

In order to find the initial separation $f(r)$, shown in Fig. 1, it is only necessary to find

$$f(r) = z_1 - z_2 \tag{15a}$$

where z_2 is found from equation (9) with $\rho \equiv r$ and $d \equiv d_0$; d_0 is the value of d corresponding to initial contact as shown in Fig. 1. To find z_1, set $\rho \equiv r$ and use equation (7b) for points on the torus ($r < r_b$), or equation (11) for points on the upper sphere ($r > r_b$).

In order to find the initial separation d_0, it is necessary to note from Fig. 1, that when $d = d_0$, the slope of the torus matches that of the lower sphere at the contact point; i.e.

$$\frac{dz_1}{d\rho} = \frac{dz_2}{d\rho} \tag{15b}$$

where the derivatives are found from equation (7b) and equation (9). Equation (15b), together with equations (7b) and (9), suffice to find d_0, and the two coordinates (r, z) of the initial contact point.

Having found the boundaries (r_i and r_0) of the contact region Ω and the initial separation function $f(r)$, we may proceed to solve the governing integral equation (1).

5. NUMERICAL SOLUTION PROCEDURE

The contact region Ω is subdivided into N annular rings. Since a steep pressure gradient is expected near the pit, the annular rings near the inner boundary are very narrow in width. It was also learned from experience that the peak pressure always occurs at some radius r where $r < r_b$. Guided by this consideration, a majority of the rings are clustered in the region $r_i \leq r \leq r_b$. Exploiting the axisymmetry of the problem, we assume that the pressure is constant in each ring. The rings are numbered sequentially from 1 to N, from the inside out, and the pressure in the i-th ring is assumed to be an unknown constant p_i. Let ξ_i and ξ_{i+1} be the inner and outer bounding radii for cell i; thus $\xi_1 \equiv r_i$ and $\xi_{N+1} \equiv r_0$. Each ring is further subdivided circumferentially into m equal sectors by drawing (m) equispaced radial rays from the center of Ω; the angle $\Delta\phi$ between two adjacent rays is $2\pi/m$. The sector, bounded by radial rays 1 and 2, is shown in Fig. 3.

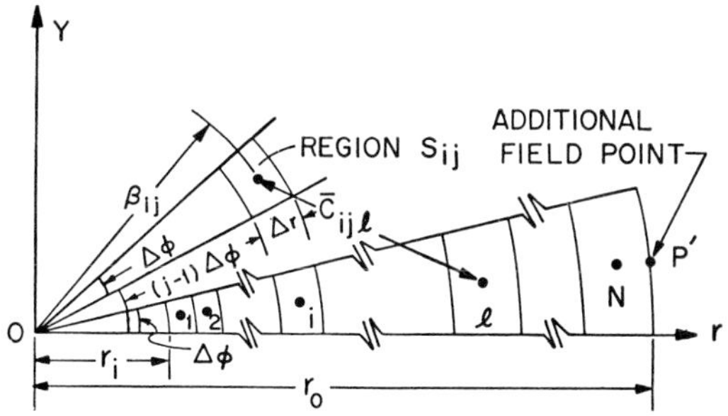

Fig. 3. Subdivided and labeled contact region (portion)

The region of the sector located in the i-th ring, between ray j and ray $(j+1)$, is identified as S_{ij}; and its centroidal radius by β_{ij}. Elementary calculations show that

$$\beta_{ij} = \frac{2 \sin\alpha}{3\alpha} \frac{(\xi_{i+1}^2 + \xi_i^2 + \xi_{i+1}\xi_i)}{(\xi_{i+1} + \xi_i)} \tag{16}$$

where

$$\alpha = \pi/m = \Delta\phi/2 \tag{17}$$

The centroids of the first sector shown in Fig. 3 (i.e. where $j = 1$) are selected as field points.

Thus for the field point l, equations (6) and (3a) reduce to

$$k \sum_{i=1}^{N} p_i \sum_{j=1}^{m} \int_{S_{ij}} \frac{dA_{ij}}{c_{ijl}} - \delta + f(\beta_{l1}) = 0 \tag{18}$$

where $f(r)$ is calculated for $r=\beta_{l1}$ from equation (15a). c_{ijl} is the radial distance from field point l to the elemental area dA_{ij} located in S_{ij}. For most cells, the integral in equation (18) may be replaced by the approximation

$$I_{ijl} = \int_{S_{ij}} \frac{dA_{ij}}{c_{ijl}} = \frac{A_{ij}}{\bar{c}_{ijl}} \tag{19}$$

where $\bar{c}_{ij}$ is the distance between field point l and the centroid of the region S_{ij}, whose area is denoted by A_{ij}. It was shown in Singh[8] that, in general, equation (19) is a very useful approximation which results in a significant reduction of computation time, without compromising the accuracy of results. However, for regions located in the immediate vicinity of the field point l, the errors due to the approximation (19) may be unacceptable. To avoid such errors, I_{ijl} is evaluated by numerical quadrature within cells located near the field point. The criterion which must be satisfied in order to use equation (19) is

$$\bar{c}_{ijl} > \max[r\Delta\phi, \Delta r] \tag{20}$$

In equation (20), $r\Delta\phi$ and Δr are the side lengths of a typical cell. Notice that when the field point l lies inside the region S_{ij} (i.e. $j=1$, $i=l$), $\bar{c}_{ijl}=0$, and hence the integrand in equation (14) has a singularity. However, for such cases, an approximate analytical solution for the integral is readily constructed.

In this manner, N linear equations corresponding to the N field points are generated. An additional linearly independent equation is generated by selecting point P' at the outermost boundary of the contact region as field point $(N+1)$. The location of this additional field point has a pronounced affect on the solution, which deteriorates as P' is moved inside the boundary. It is plausible to assume that this behavior is due to the gradual increase in cell width Δr with r (see Fig. 3), which was introduced to keep the aspect ratio of the cells from becoming excessive. With the cells so designed, the location of P' shown in Fig. 3 maximizes the distance between P' and its nearest neighboring field point. This in turn tends to maximize the amount of independent information supplied by the equation written for field point P', and should tend to minimize ill-conditioning effects on the coefficient matrix generated.

Thus $(N+1)$ equations in $(N+1)$ unknowns are generated, and equation (18) assumes the form

$$a_{ij}p_j = -f_i+\delta \tag{21}$$

and the equation constructed using P' as a field point becomes,

$$v_j p_j = -f_{N+1}+\delta \tag{22}$$

where f_i is the value of the 'initial separation' function $f(r)$ at the field point i. f_{N+1} is the value of $f(r)$ at P'; and summation from 1 to N is henceforth implied over repeated subscripts. From equations (21) and (22), δ may be eliminated to yield

$$B_{ij}p_j = f_i' \tag{23}$$

where

$$B_{ij} = A_{ij} - v_j \tag{24}$$

and

$$f_i' = f_{N+1} - f_i \tag{25}$$

When equation (23) is solved for p_i, using Gaussian elimination, the resulting pressure distribution is usually found to predict negative contact pressures in the immediate vicinity of the inside boundary, $r = r_i$. The axisymmetry of the problems enables us to maintain the outside boundary fixed, and iterate on the inside boundary where the predicted pressure is incorrect. The iteration scheme is best explained with the aid of the numerical example given in Section 6.

6. A NUMERICAL EXAMPLE

The following example was considered.

$R_1 = R_2 = 1$ in.

$\nu_1 = \nu_2 = 0.$

$E_1 = E_2 = 30 \times 10^6$ b/in^2

$r_c = .006$ in

$r_b = .00025$ in.

The results are presented in dimensionless form. Let

$$R = \frac{2R_1R_2}{R_1+R_2} \tag{26}$$

Then, we define

Dimensionless pressure in ring i, $p_i^* = kp_1$ (27)

Dimensionless load, $F^* = \frac{kF}{R^2}$ (28)

Dimensionless distance from origin of Ω, $r^* = r/R$ (29a)

Dimensionless approach, $\delta^* = \delta/R$ (29b)

$$r_b^* = r_b/R \tag{30a}$$

$$r_c^* = r_c/R \tag{30b}$$

Fig. 4 shows the pressure distribution near the inside boundary for the uniterated solution. The pressure distribution far from the pit agrees closely

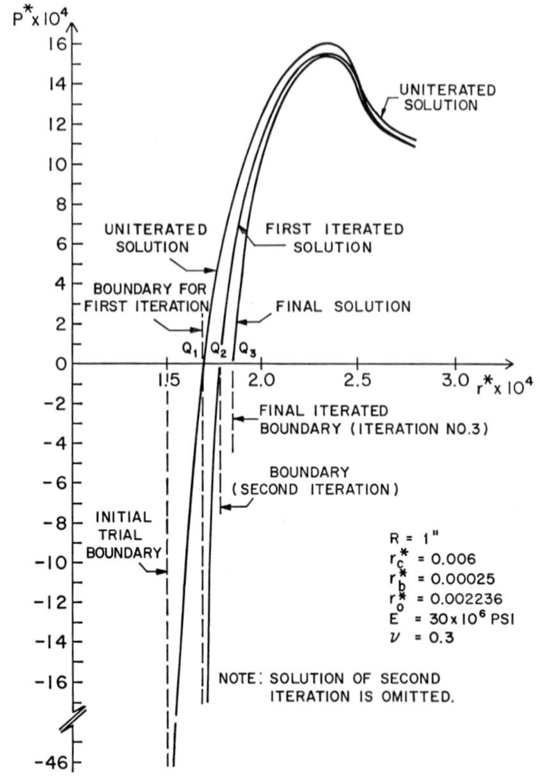

Fig. 4. Boundary iteration sequence

with the Hertzian solution for unpitted spheres (not shown in the figure). However, the pressure in cell # 1 is highly negative. The pressures in the successive cells are less and less negative, until at point Q_1 the pressure curve crosses the x-axis. The shape of the pressure curve readily suggests the iteration scheme. The new region of integration is assumed to have inner radius $r_i = 0Q_1$. The discretized equation set (23) is generated corresponding to this new region Ω, and thus a new pressure vector is generated (see first iteration, Fig. 4). This new curve also has a negative peak (weaker than that of the uniterated solution) at the innermost field point. The new point of intersection is Q_2, which defines the inner boundary of Ω for the next iteration. The process is thus continued until all pressures are positive. In Fig. 4, the third iteration yields the desired solution. It is found that this solution also satisfies inequality (3), thus qualifying as the 'true' solution of the contact problem. The complete pressure distribution is shown in Fig. 5. Notice the essentially Hertzian pressure distribution (corresponding to contact of unpitted spheres) at $r^* > 6 \times 10^{-4}$. Thus the

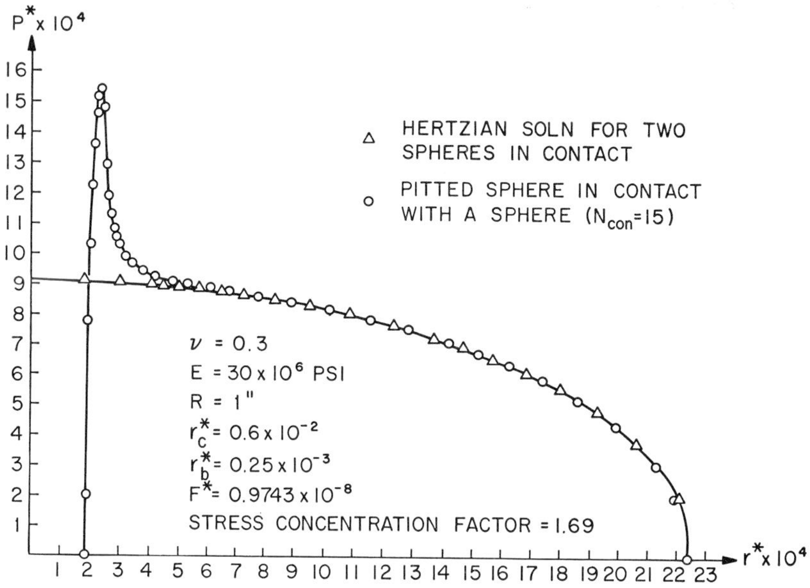

Fig. 5. Pressure distribution for pitted sphere pressed against a sphere

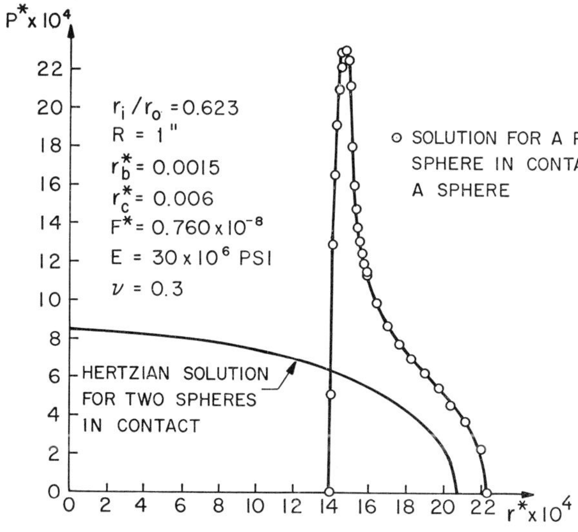

Fig. 6. Pressure distribution for large pit diameter

effect of the cavity is of a strictly localized nature. However, as the cavity is made larger (e.g. $r_i/r_0 \geq 0.3$) the pressure curve departs completely from the Hertzian case. For example, Fig. 6 shows a typical pressure distribution for $r_i/r_0 = 0.623$, along with the Hertzian solution for unpitted spheres corresponding to identical values of thrust F.

In order to establish confidence in the solution, it is necessary to study its convergence with change in the number of cells used. It must be recognized that it is necessary for the cells to be densely concentrated only in that region where a high pressure gradient exists. Therefore, for purposes of convergence studies, we have systematically varied the number of cells

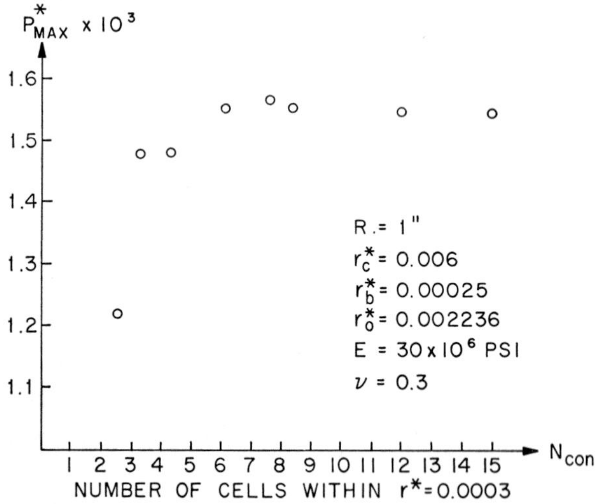

Fig. 7. Convergence of peak pressure with increasing number of cells

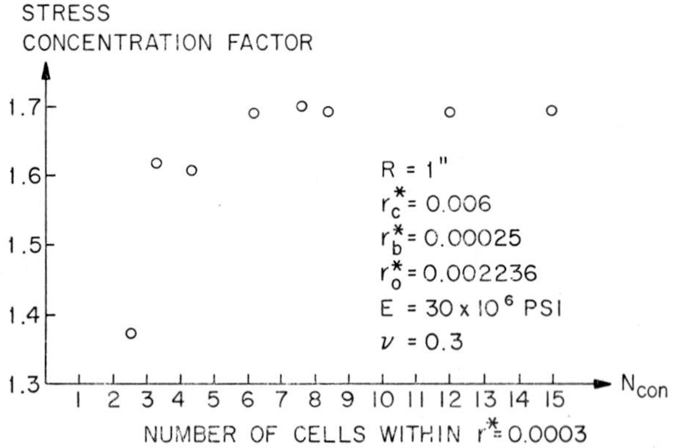

Fig. 8. Convergence of stress concentration with increasing number of cells

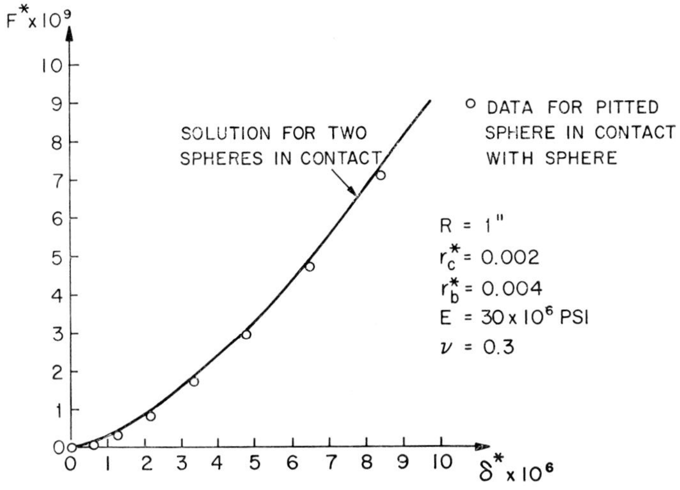

Fig. 9. Load approach relationship

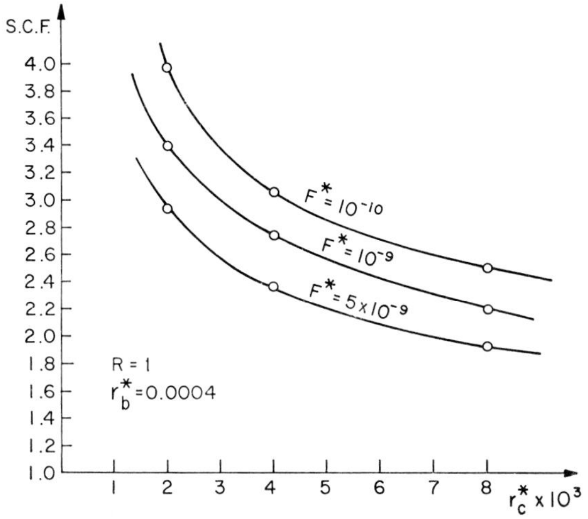

Fig. 10. The effect of load and r_c on stress concentration factor

within a fixed radius r_{con}. This radius is chosen arbitrarily for each problem in such a way that the major area of stress concentration lies inside the radius r_{con}. For the example problem considered, $r_{con} = .0003$. Let N_{con} be the number of rings located within radius r_{con}. Fig. 7 illustrates the convergence of the peak pressure, p^*_{max}. Fig. 8 shows the convergence of stress concentration factor with N_{con}. Stress concentration factor (SCF) is defined as the ratio of the peak computed pressure to the peak pressure for unpitted spheres under equal thrust. Notice both Figs. 7 and 8 exhibit convergence for $N_{con} > 8$.

The load-approach curve is shown in Fig. 9. It is obvious from Fig. 9 that the compliance characteristics of the balls (with small pits) remain essentially the same as that predicted by the Hertzian solution.

Fig. 10 shows SCF as a function of cavity edge radius r_c^*. Smaller values of r^* cause greater stress concentration. Due to the nonlinearity of the problem, the SCF is also a function of the applied load F_c^*. Table 1 shows

Table 1. Dependence of stress concentration factor on r_b

Case No.	$r_b^* \times 10^3$	*SCF*	$F^* \times 10^8$	$\delta \times 10^4$	$r_i^* \times 10^3$
1	0.25	1.692	0.9743	0.1023	0.1845
2	0.35	1.856	0.9737	0.1029	0.2753
3	0.50	2.049	0.9702	0.1041	0.4166

$R = 1''$, $r_c^* = 0.006$, $r_0^* = 0.002236$, $E = 30 \times 10^6$ *psi*, $\nu = 0.3$

the variation of SCF with the size of the pit (measured by blend point radius). Notice the SCF increases with increasing value of r_b^*. This variation of SCF with r_b^* may be related to the loss of load carrying area.

The computer program developed to solve this problem is moderately efficient. For example, the nine cases, needed to generate Fig. 10, required an average running time of 10 minutes each on the IBM/360/65 computer, corresponding to \$8.33 per case, with $N = 34$ nodes per case.

7. CONCLUSIONS

A non-Hertzian elastic contact problem involving an unknown multiply-connected contact region has been solved. The example problem considered is that of a pitted sphere in contact with an unpitted sphere. The axi-symmetry of the problem enabled us to use the 'simply-discretized method' with a polar coordinate grid. For problems with a lower degree of symmetry, it had been found in earlier work, that a more complicated (and less efficient) method of solution was necessary because of the numerical instability of the equations generated. It may be appropriate to describe the equation set (23) as 'quasi-stable' because it exhibits dependence on the location of the $(N+1)$th field point. Through experience and heuristic reasoning, it was established that locating the additional field point (P' in Fig. 3) at the outside boundary yields a well-conditioned matrix.

The variation of the SCF, contact region Ω and peak pressure p_{max}^* with changes in the pit blending radius r_b^*, and the pit edge radius r_c^*, was studied, and some numerical results were presented.

The numerical solution was shown to converge rapidly with a moderate cell density.

To the best of our knowledge, this is the first published solution of a multiply-connected contact region problem with *a priori* unknown contact boundary.

REFERENCES

1. T.E. Tallian, 'On Competing Failure Modes in Rolling Contact', *ASLE Trans.*, **10**, 4, 1967, pp. 418–435.
2. J.A. Martin and A.D. Eberhardt, 'Identification of Potential Failure Nuclei in Rolling Contact Fatigue', *J. of Basic Engrg., Trans. of the ASME,* D, **80**, 1967, pp. 932–942.
3. W.E. Littman and R.L. Widmer, 'Propagation of Contact Fatigue from Surface and Subsurface Origin', *J. of Basic Engrg., Trans. of the ASME,* D, **88**, 1966, pp. 624–636.
4. Z. Olesiak, 'Annular Punch on Elastic Semi-Space', *Archwm. Mech. Stosow,* **17**, 4, 1965, pp. 633–642.
5. S.C. Parlas and C.D. Michalopoulos, 'Axisymmetric Contact Problem for an Elastic Half-Space with a Cylindrical Hole', *Int. J. of Engrg. Sc.*, **10**, 1972, pp. 699–707.
6. Y.P. Chiu, 'On the Contact Problem of Cylinders Containing a Shallow Longitudinal Surface Depression', *J. of Appl. Mech.*, **36**, *Trans. of the ASME,* **91**, Series E, 4, 1969, pp. 852–858.
7. K.P. Singh and B. Paul, 'Numerical Solution of Non-Hertzian Elastic Contact Problems', *J. of Appl. Mech.*, **41**, *Trans. of the ASME,* **96**, Series E, 2, 1974, pp. 484–496.
8. K.P. Singh, *Contact Stresses in Elastic Bodies with Arbitrary Profiles*, Ph. D. thesis, University of Pennsylvania, 1972.
9. K.P. Singh and B. Paul, 'A Method for Solving Ill-Posed Integral Equations of the First Kind', *Computer Methods in Applied Mechanics and Engineering*, **2**, 1973, pp. 339–348.
10. A.I. Luré, *Three Dimensional Problems of the Theory of Elasticity*, Interscience, 1964.
11. J. Hadamard, *Lectures on Cauchy's Problem in Linear Partial Differential Equations*, Dover Publications, New York, 1952, pp. 33–34.

Stylus profilometry and the analysis of the contact of rough surfaces

J. F. Archard, R. T. Hunt and R. A. Onions

1. INTRODUCTION

All surfaces used in engineering practice are rough when judged by the standards of molecular dimensions. This fact has played a major role in the development of the science and technology of Tribology (both before and after the subject acquired that name). Ideas about the nature of surface contact are closely related to the techniques used for the measurement of surface topography and to the methods used to generate surfaces in engineering practice.

Therefore in the development of this subject three major interacting themes can be discerned. First, and foremost, there is the scientific problem of the nature of the contact between surfaces under load and its relationship to tribological problems such as friction, wear and the conduction of heat and electricity between bodies in contact. A major emphasis in this field has been the deduced relationships between the true area of contact and the applied load. A second important element in this subject has been the techniques used for the measurement of surface topography. Throughout these developments the stylus profilometer has played a significant role and, in recent years, this has been reinforced by the way in which the profilometer has been linked with the digital computer. The third element in the subject, which has assumed increasing importance in recent years, is the link between these theories and techniques and engineering practice. In production engineering a central practical problem is the adoption of appropriate and meaningful methods for the definition and specification of engineering surfaces.

We first present a review of some developments in this field. This review is not intended to be exhaustive; for the purposes of introducing what follows it concentrates upon a few elements of the English language literature and gives some emphasis to the models which have been used to represent surfaces in contact and to the role of the stylus profilometer. In recent years the profilometer has played an increasingly significant role through the use of analogue/digital conversion techniques and this has made possible the analysis of the data derived from surfaces using the powerful methods of the digital computer. In this way models for the representation of surfaces of increasing sophistication have been employed. Nevertheless these models usually represent the character of freshly prepared surfaces whereas, in practice, surfaces are modified, in important respects, by the rubbing process itself.

An alternative procedure is to use profilometer data derived directly from the surfaces used in the experiments and thus to present a computer simulation of the contacting surfaces. Whilst this procedure has obvious attractions, such a direct simulation could require an almost impossibly large store of data. Therefore, we shall describe methods which show that important and significant information about the nature of surface contact can be obtained from a much smaller sample of profilometer records taken from the surfaces. The application of these methods to the specific problem of the contact of surfaces through a lubricant film will be described.

2. EARLIER THEORIES OF CONTACT

Holm[1] emphasised the crucial fact that when bodies are placed in contact the true area of contact is small and that the contact pressures are high. Bowden and Tabor[2] further argued that, because surfaces contain asperities having a small radius of curvature, the true area of contact, A, arose from plastic deformation and that the major force of friction arose from a shearing of these junctions. Thus the relationship between the frictional force, F, the load, W, and the coefficient of friction μ became

$$W = AH, \quad F = sA, \quad \mu = s/H, \tag{1}$$

where H is the hardness of the softer of the contacting materials and s is the shear strength of the junctions.

Later Archard[3][4] using the ideas of Holm[1] and Burwell and Strang[5] applied these same concepts to the problem of wear. The worn volume, V, produced in a sliding distance, L, was related to the true area of contact, A, which was, again, assumed to be produced by plastic deformation of the contacting asperities. Thus,

$$\frac{V}{L} = \frac{1}{3}KA = \frac{1}{3}K\frac{W}{H}, \tag{2}$$

where K is a constant or wear coefficient, commonly assumed to represent the proportion of the contacts producing worn particles.

Equations (1) and (2) represent a view which persisted for approximately two decades. They show that the frictional force, F, and the wear rate, V/L, are proportional to the load, W, because the area A is proportional to the load. However the role of surface topography seems relatively minor since it plays no part in determining the severity of the contact conditions.

3. MODELS OF SURFACE CONTACT WITH ELASTIC DEFORMATION

Blok[6] and Halliday[7] considered the shape of asperities which could be pressed into the general level of the surface without plastic deformation. The criterion for the avoidance of plastic deformation is

$$\bar{m} = \frac{h}{d} = \frac{d}{2R} = \left(\frac{h}{2R}\right)^{1/2} < K\frac{H}{E'} \qquad (3)$$

where $\bar{m}$ is the mean upward or downward slope of the surface, H is the hardness, $E' = E/(1-\nu^2)$, E is Young's modulus and ν is Poisson's ratio. K is a numerical constant $(0.7 < K < 1.8)$ depending upon the assumed shape of the asperity. The central parts of equation (3) are based upon the familiar and common assumption that the asperity is a spherical cap of height, h, radius of curvature, R, and linear dimension, $2d$, as shown in Fig. 1. Halliday[7,8] used reflexion electron microscopy to show that worn

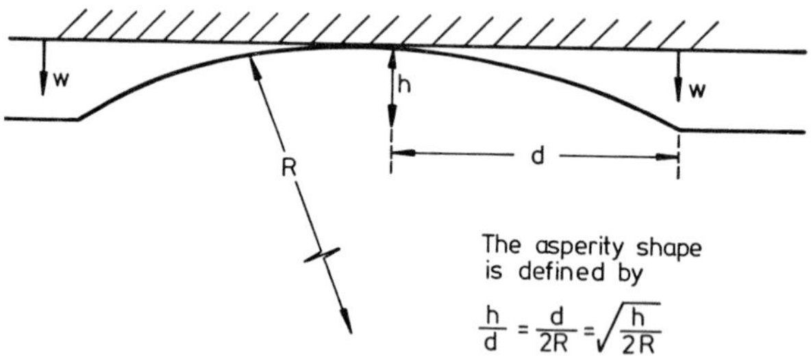

Fig. 1. Contact of a single asperity

For elastic deformation, the maximum pressure in the contact is

$$P_0 = \frac{2E'}{\pi}\left(\frac{w}{a}\right) = \frac{2E'}{\pi}\sqrt{\frac{w}{R}}$$

where w = deformation in depth, a = radius of contact area.

surfaces contained asperities which conformed to equation (3) and therefore could not be deformed plastically.

Equation (1) provides a simple and logical explanation of Amontons' laws of friction but it is based upon the assumption of plastic deformation of asperities. Archard[9,10] argued that, since values of K deduced from equation (2) and measured wear rates lie between 10^{-3} and 10^{-6}, many contacts in *continuous* rubbing must be entirely elastic; he showed that the area of contact could, for all practical purposes, be proportional to the load even if the deformation were entirely elastic. The models used to represent nominally flat surfaces in this theory are shown in Fig. 2; as the models become more complex the relationship between area and load moves more closely towards a relationship of direct proportionality. Two features of these models deserve comment. First, they involve the existence, upon the surface of features of a number of different scales of size. Second,

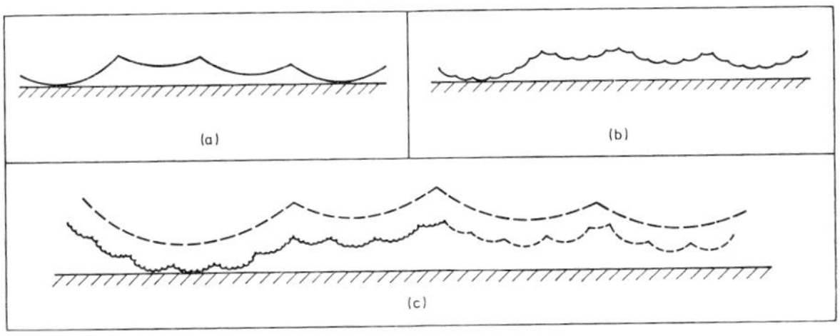

Fig. 2. Models of surfaces containing asperities of differing scales of sizes

For elastic deformations the relations between area of contact (A) and load (W) are: (a) $A \alpha W^{4/5}$, (b) $A \alpha W^{14/15}$, (c) $A \alpha W^{44/45}$.

it was recognised that they are artificial and it was pointed out that the feature which determines the relationship between area and load is the extent to which an increase in load is utilized in creating new areas rather than enlarging existing areas.

4. ANALYSIS OF SURFACE PROFILES AND THE GAUSSIAN MODEL

One significant consequence about the use of the stylus profilometer with analogue/digital conversion techniques is that it allows the development of models of surface contact which are more closely related to the shape of real surfaces. In engineering practice many, if not most, surfaces are finished by methods, such as grinding and abrasion, which involve random interactions between grits and the surface.

Using digital analysis of profiles Greenwood and Williamson[11] showed that for such surfaces the heights of the profile, and, more importantly, the heights of the peaks, were distributed in a form which was close to Gaussian. Therefore they used as a representation of surface contact a model, similar to that of Fig. 2(a), consisting of asperities, all having the same radius of curvature, whose heights had a Gaussian distribution. This model was therefore defined by three parameters; the standard deviation of the asperity heights, σ^*, the radius of curvature of the asperities, R, and the number of asperities per unit area, η.

The use of this model may be explained in a little more detail to emphasize the methods used and some significant features. The relationships between the area, δA, and the load, δW, at a single asperity contact and the deformation of the asperity in depth, w, are

$$\delta A = \pi R w, \quad \delta W = \tfrac{4}{3} E' R^{1/2} w^{3/2} \tag{4}$$

It will be noted that, for the sake of simplicity, the theory assumes the contact between a perfectly smooth rigid flat surface and a deformable model surface. The probability of finding an asperity at a height between y and $(y+\mathrm{d}y)$ is $\phi(y)\mathrm{d}y$. Then the total area of contact, A, and the total load, W (per unit apparent area) are

$$A = \Sigma \delta A = \eta \int_d^\infty \pi R(y-d)\, \phi(y)\, \mathrm{d}y \tag{5a}$$

$$W = \Sigma \delta W = \eta \int_d^\infty \tfrac{4}{3} E' R^{1/2} (y-d)^{3/2}\, \phi(y)\, \mathrm{d}y \tag{5b}$$

where d is the separation of the surfaces.

When the Gaussian distribution was used in this calculation it was shown that the relationship between A and W was close to direct proportionality. Following Greenwood and Williamson[11], the reasons for this can be seen more clearly by writing

$$\phi(y) = \frac{1}{\sigma^*} \exp(-y/\sigma^*), \tag{6}$$

which is a reasonable fit for the upper levels of a Gaussian distribution. Then, by changing the variable, equations (5) become

$$A = \frac{\pi R \eta}{\sigma^*} \exp(-d/\sigma^*) \int_0^\infty \exp(-w/\sigma^*)\, w\, \mathrm{d}w \tag{7a}$$

$$W = \frac{4 E' R^{1/2} \eta}{\sigma^*} \exp(-d/\sigma^*) \int_0^\infty \exp(-w/\sigma^*)\, w^{3/2}\, \mathrm{d}w \tag{7b}$$

and writing the equation for the number of contacts, n,

$$n = \int_d^\infty \frac{\eta}{\sigma^*} \exp(-y/\sigma^*)\, \mathrm{d}y = \eta \exp(-d/\sigma^*) \tag{7c}$$

Equations (7) show that both the load and the area of contact are proportional to the number of contacts and, because of this, the area is proportional to the load. This is a neat mathematical justification of the earlier assertion, based on physical reasoning, that the feature of importance is the way in which increased load is divided between creating new contacts and enlarging existing contacts. It will also be noted that this model shows that the area of contact is proportional to the load regardless of the mechanism of deformation. It is the geometry of the surfaces which primarily determines the area/load relation in this situation.

Stylus profilometers with A/D conversion have found increasing use in production engineering research and one consequence of this development has been attempts to characterize surfaces through analysis of their profiles. An important element of this approach is to examine the profile using the methods of random signal analysis. In this way Peklenik[12] has provided a classification of engineering surfaces and their methods of manufacture.

Since, as explained earlier, we are primarily concerned with surfaces prepared by random methods it is appropriate to regard the profile (Fig. 3)

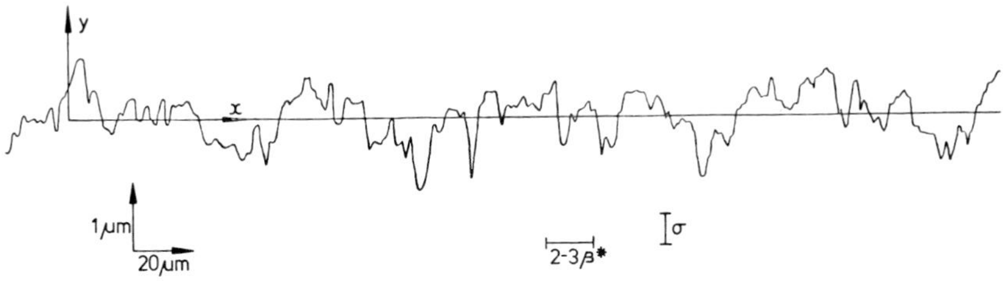

Fig. 3. A typical profile showing coordinate system

Digital analysis shows that this profile is a close approximation to the W & A model (see eq. (8)).

as a random signal. This can be defined in terms of its height distribution and its autocorrelation function. Based upon the examination and analysis of a large number of profiles drawn from engineering practice, Whitehouse and Archard[13] chose, as a suitable model, a profile with a Gaussian distribution of heights (with standard deviation, σ) and an exponential autocorrelation function, defined by

$$C(\beta) = \lim_{L\to\infty} \frac{1}{L} \int_{-1/2L}^{+1/2L} y(x)\cdot y(x+\beta)\,\mathrm{d}x = \exp(-\beta/\beta^*)\,, \tag{8}$$

where $y(x)$ is the ordinate height at a coordinate x and $y(x+\beta)$ is the height at an adjacent ordinate $(x+\beta)$ (Fig. 3). In this equation y is normalised by σ, the standard deviation of the height distribution. (The distinction between σ and σ^*, the standard deviation of the asperity height distribution, should be noted.) In the exponential model β^* is called the correlation distance and is a scaling factor indicating the extent of the wavelength distribution. Fig. 4 shows the relation between the autocorrelation function and the power spectral density. The exponential model can be regarded as white noise limited in the upper frequencies by a cut-off of 6 dB per octave. Thus it will be seen that the model consists of a major element of lower frequencies (longer wavelengths); higher frequencies (shorter wavelengths) also exist upon the surface but their magnitudes decline so that, in this range, amplitude is proportional to wavelength. Thus the model contains the random features of the Gaussian model of Greenwood and Williamson[11] but also includes the multiple scales of size of the models of Fig. 2.

There is now a major problem in the presentation of this model in a form suitable for a theory of elastic contact. Whitehouse and Archard[13] have

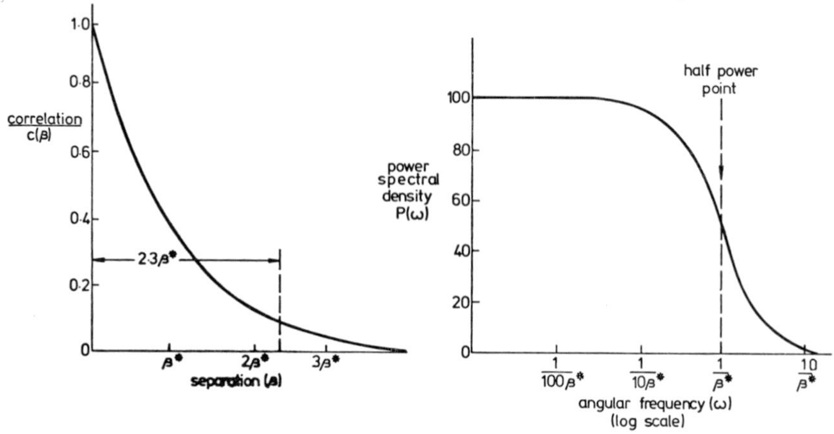

Fig. 4. The model. Autocorrelation function and power spectral density

shown that the information about profile peaks derived from the methods commonly used in digital analysis gives results which depend critically upon the sampling interval used in the digital presentation of the profile. Such methods reveal the structure of the profile associated with wavelengths comparable with the sampling interval. If we use asperity models, such as those of Fig. 2, how do we choose an appropriate scale of size (or scales of size)? Whitehouse and Archard argued that it was appropriate to start by considering the main scale of size associated with the major component of the power spectral distribution (Fig. 4); they suggested that, in the context of digital methods, this structure was revealed using a sampling interval of $2.3\beta^*$, i.e. by sampling events on the profile just sufficiently far apart for them to be regarded as independent.

Such a model has a distribution of asperity heights which is nearly Gaussian and, more importantly, has a distribution of asperity curvatures which is dependent upon the asperity height. Onions and Archard[14] have derived a theory of contact based upon this model; they showed that, when compared with the Greenwood and Williamson model[11] the introduction of the distribution of asperity curvatures markedly increases the contact pressures.

One further feature of the Whitehouse and Archard random signal model[13] is that the profile (and the surface, if isotropic) is defined by two parameters: the standard deviation of the height distribution, σ, and the correlation distance, β^*. It will be recalled that, in contrast, the Greenwood and Williamson model[11] is defined by three parameters (σ^*, R, η); both theoretical considerations and the results of digital analysis combine to suggest that these three parameters are inter-related by

$$\sigma^* R\eta = \text{constant} \simeq 0.05 \tag{9}$$

An alternative, and in many ways more rigorous, approach to the characterisation of random surfaces has been made by Nayak[15,16,17]. This includes a theory of contact based upon the assumption of plastic deformation. Many of the broad conclusions derived from these two alternative approaches[13,15,16,17] are similar.

6. SOME CONSEQUENCES FOR ENGINEERING PRACTICE

Common engineering practice and international agreements have resulted in the specification of surface finish in terms of the height distribution of the profile only. This is normally the RMS value or the centre line average (now specified as R_a). In addition normal practice usually specifies the method used to produce the finish.

In recent years some dissatisfaction with the use of a single parameter has combined with the theoretical arguments which have been outlined above to provoke discussion about improved methods for the specification of engineering surfaces. Thus, in their discussion of these problems Spragg and Whitehouse[18] include a proposal for a two parameter method for the specification of surface finish; the existing parameter of R_a would be combined with an average wavelength, λ_a. For random surfaces λ_a bears a simple relationship to the β^* of the Whitehouse and Archard model; similarly the practical conclusion to be drawn from the work of Nayak is that a parameter based upon wavelengths is an obvious requirement.

One can only wait for experience drawn from engineering practice to justify this type of approach.

7. PLASTIC FLOW AND PLASTICITY INDEX

Fig. 1 and the theoretical arguments associated with equation (3) show that the maximum Hertz pressure, p_0, in the elastic deformation of a spherical asperity depends only upon the elastic modulus and the shape of the deformed region. Thus

$$p_0 = \frac{2E'}{\pi}\left(\frac{w}{a}\right) = \frac{2E'}{\pi}\left(\frac{w}{R}\right)^{1/2} \tag{10}$$

Since the onset of plastic deformation occurs when $p_0 \simeq 0.6H$, where H is the hardness, the requirement for the avoidance of plastic deformation is

$$\frac{2E'}{\pi}\left(\frac{w}{a}\right) = \frac{2E'}{\pi}\left(\frac{w}{R}\right)^{1/2} < 0.6H \tag{11}$$

The application of these ideas to multiple contact models is obviously complex but it will be readily accepted that for the Greenwood and Williamson model[11], defined by σ^* and R, the probability of plastic flow will be related to a parameter

$$\psi = \left(\frac{E'}{H}\right)\left(\frac{\sigma^*}{R}\right)^{1/2} \tag{12a}$$

whilst for the Whitehouse and Archard model[13], defined by σ and β^* the appropriate parameter is

$$\psi^* = \left(\frac{E'}{H}\right)\left(\frac{\sigma}{\beta^*}\right) \tag{12b}$$

By a fortunate coincidence (despite the greater pressures expected from the latter model, as indicated above) it is found[14] that both parameters have the same significance. If ψ (or ψ^*) > 1 plastic flow will be significant, if ψ (or ψ^*) < 0.6 the chances of plastic flow are very small indeed.

One point about these calculations has escaped general comment. They are based upon the onset of plastic flow but, for a single contact, there is something like a 200-fold range of load (which implies, approximately, a 50-fold range of the deformation in depth, w) between *onset* and what Tabor[19] has termed '*full plasticity*'. Therefore if we are looking for severe plastic deformation it seems possible that much higher values of ψ or ψ^* may be needed.

An important investigation which shows how these considerations may be significant in Tribology has been reported by Hirst and Hollander[20]. Using a steel ball rubbing on stainless steel surfaces prepared with different values of σ and β^* they showed that the boundary between the presence and absence of severe damage, under conditions of boundary lubrication, was related to a parameter such as that given by equation (12b). More such work is needed which seeks a connection between tribological failure and these parameters of surface topography.

8. THE PROBLEM OF REAL SURFACES

The theories which have been outlined above have played a major role in shaping ideas about the nature of surface contact. But, at the same time, we must recognise their defects and their limitations. They make fairly specific assumptions about the character of the surface topography which are not always realised in practice. In particular, experience shows that surface topography is modified by running-in in the very same region of the height distribution which is crucial to these theories.

An alternative approach is to use profiles drawn from the surfaces used in the experiments and, with the aid of analogue/digital conversion, to simulate within the computer, the contact of the surfaces. Such an approach has been adopted by Williamson[21]. From a series of closely spaced profiles of a surface he created a map of the surface by a process which he termed 'microcartography'. He then performed a notional experiment in the computer by bringing two such surfaces into contact (although contact of the surfaces involved had not taken place).

In order to calculate areas of contact, certain simplifying assumptions had to be made. Contact was assumed to occur wherever the gap between the surfaces was less than zero. In our terminology, which will be used in what follows, the area of contact was assumed to be that given by the 'regions of interference'. These assumptions are probably a valid approximation if it be assumed that all deformation is plastic but if the deformation is elastic the true areas of contact are obviously less than the areas derived from the 'areas of interference'.

In what follows we shall pursue the idea of calculating the area of contact and the load borne by elastic deformation of the contacting asperities, it being assumed that the separation of the surfaces is known or can be deduced. It will be necessary to seek a relation between the area of interference, calculated from the 'regions of interference' and the true area of contact. To simplify the ideas so that the principles involved become more clearly apparent, it will be assumed, at the outset, that because the surfaces are isotropic, all the contacts can be regarded as the Hertzian contact of spheres. The later extension of the ideas to anisotropic contacts is a relatively simple matter. Secondly we shall assume that the information about each of the surfaces is reasonably complete and consists of a series of closely spaced parallel profiles suitable for microcartography. At a later stage it will be shown, somewhat surprisingly, that, for the calculation of the load borne by contact between the surfaces, a *very much smaller sample* of this total information may be quite adequate.

One feature of this procedure deserves comment. We are no longer faced with the problem (which is central in building theories of contact from the random signal model) of deciding what scale of size is appropriate. In the technique of direct computer simulation this problem is irrelevant; the appropriate scale of size is the scale of size revealed by the contact of the surfaces in the simulation.

Let us assume that we have a number of profiles taken from each of the surfaces with a spacing frequency of N profiles per unit distance in the y direction (normal to the profile as shown in Fig. 5). For later use we note that N is also the total length of profile which is used to represent unit area of the surface. Therefore in the computer simulation we have a number of 'profile pairs' which can be brought together as part of the simulation of the contact between the two surfaces. Interaction between the surfaces resulting in contact and load support is represented on these profile pairs by interference events of depth ω and width 2α as shown in Fig. 5. We seek to obtain the load supported by these interference events in terms of their dimensions and the elastic constants of the materials (since the deformation will be assumed to be elastic).

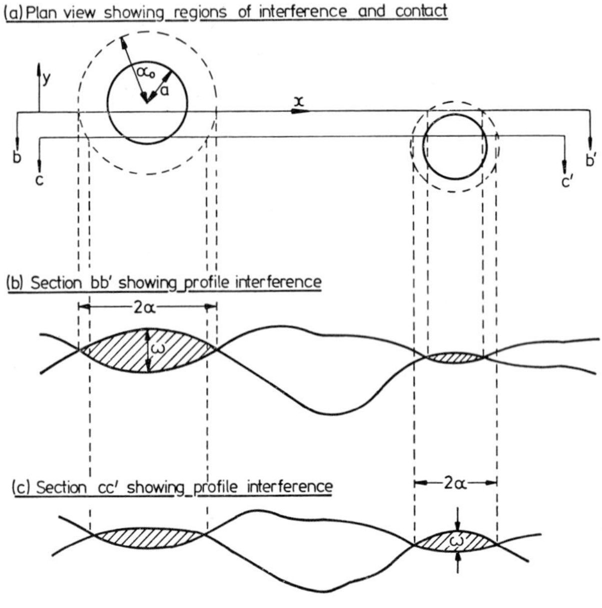

Fig. 5. Contact of surfaces as revealed by profiles

(a) Full circles represent the areas of contact; broken lines represent region of interference; (b) and (c) are two pairs of profile interactions; regions of interference (width: 2α, depth: ω) are revealed.

Consider one such region (Fig. 5) giving an area of contact, $\delta A = \pi a^2$, having a maximum depth of profile interference ω_0 and maximum width of profile interference $2\alpha_0$. If the local relative radius of curvature of the surfaces is R, geometric considerations and the Hertz equations give

$$\omega_0 = \frac{\alpha_0^2}{2R} = \frac{a^2}{R}, \quad \frac{\alpha^2}{\omega} = \frac{\alpha_0^2}{\omega_0} \tag{13}$$

Thus

$$\alpha_0 = a\sqrt{2} \tag{14}$$

This region of interference will appear on n profile pairs, where $n = 2\alpha_0 N$, and the mean width of profile interference, $2\alpha_{\text{mean}}$, and the mean depth of profile interference, ω_{mean}, observed on these profile pairs will be given by

$$\omega_{\text{mean}} = \tfrac{2}{3}\omega_0 \tag{15a}$$

$$\alpha_{\text{mean}} = \tfrac{1}{4}\pi\alpha_0 \tag{15b}$$

Hertz theory shows that the load δW supported by the area δA is given by

$$\delta W = \frac{4}{3}\frac{a^3 E'}{R} \tag{16}$$

where $1/E' = (1-\nu_1^2)/E_1 + (1-\nu_2^2/)E_2$ and using equations (13), (14) and (15)

$$\delta W = \frac{2\sqrt{2}}{3} E' \alpha_0 \omega_0 = \frac{4\sqrt{2}}{\pi} E' \alpha_{\text{mean}} \omega_{\text{mean}} \tag{17}$$

Suppose we now inspect the pairs of opposed profiles and, every time we detect an interference event, we note the interference depth, ω, and add it to a running total. We have noted that one interference region (associated with one area of contact) will appear on $2\alpha_0 N$ profile pairs so that the sum of interference depths recorded for this region will be

$$\Sigma\omega = 2\alpha_0 N \omega_{\text{mean}} = \frac{8}{\pi} N \alpha_{\text{mean}} \omega_{\text{mean}} \tag{18}$$

Comparison of equations (17) and (18) shows that the ratio of δW to $\Sigma\omega$ for one contact area is

$$\frac{\delta W}{\Sigma\omega} = \frac{E'}{N\sqrt{2}}$$

This ratio is independent of α_{mean}, ω_{mean} and R and will therefore apply to all other interference regions which are encountered. Thus the total load supported by the elastic contact of asperities is

$$W = \Sigma\delta W = \frac{E'}{N\sqrt{2}} \Sigma\omega,$$

where $\Sigma\omega$ is now the sum of all the interference depths on all the profile pairs which have been examined. As noted earlier, N is the total length of profile pairs analysed in the search of unit area of the surface.

Thus, if we examine a total length of profile pairs, l, and detect a sum of interference depths $\Sigma\omega$, the mean apparent pressure, $\bar{p}_a$ (i.e. the load supported by the elastic interaction of asperities per unit area) is

$$\bar{p}_a = \frac{E'}{\sqrt{2}} \frac{\Sigma\omega}{l}. \tag{19a}$$

The quantity $(\Sigma\omega/l)$ would be expected to reach a stable limit as we proceed with our inspection of the profile pairs. Therefore we do not need to inspect all possible combinations of profile pairs taken at very close intervals; in order to arrive at a reasonable estimate of the load we need only a limited sample of the data.

If the contact regions are ellipsoidal instead of spherical because, for example, the surface finish is anisotropic

$$\bar{p}_a = E' \frac{\Sigma\omega}{l} f_1(R_x/R_y) \tag{19b}$$

where $f_1(R_x/R_y)$ is a function of the assumed shape of the surfaces in the contact region as represented by their relative radii of curvature R_x, R_y parallel to and normal to the profile direction. $f_1(R_x/R_y)$ can be calculated[22,23] from the Hertz theory and over a range of values of (R_x/R_y) covering several orders of magnitude, is not very strongly dependent upon the assumed asperity shape.

Equations (19) provide the result which was sought. The load supported by asperity contacts can be calculated simply from the summation of the interference depths. It involves only the *depths* of the interference events and is independent of all other assumptions (α, R) about the nature of the regions of interference. Moreover, the result obtained is in a form readily available from a computer simulation using profiles in digital form.

10. APPLICATION OF THE THEORY

The use of the theory is most easily understood by the discussion of a specific problem. Consider two lubricated discs loaded together in rolling/sliding contact. The loaded region is Hertzian in shape but the surfaces are separated by a lubricant film whose thickness, h, can be calculated from the theory of elastohydrodynamic lubrication[24]; alternatively any other value of the film thickness can be assumed. The geometry of the main load bearing region is shown in Fig. 6 and consists of a rectangular area of dimensions $2L$ (the face width of the discs) by $2D$ (the width of the Hertzian band of contact). We consider how the theory of the last section can be used to derive the load borne by asperity contacts through the lubricant film.

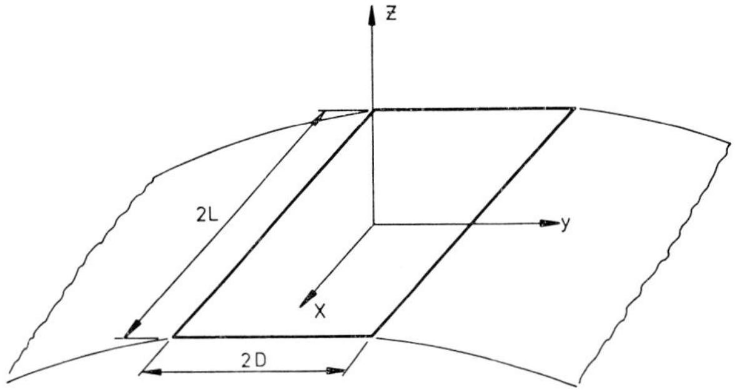

Fig. 6. Hertzian zone of an elastohydrodynamic contact between discs

Profiles from both discs are taken in the x direction (Fig. 6). If the rotations of the discs are unsynchronised, each 'profile pair', consisting of one profile from each disc, represents one of the many profile interactions which are possible. If m profiles are taken from each disc, and the effective length of each profile is l, a total length of profile interaction $m^2 l$ is available for study. The motion of the discs is in the y direction at right angles to the direction of the profiles.

It will be assumed that the deformations arising from the asperity contacts are independent of each other and that the deformation (that is the movement of the surfaces in the z direction) arising from any individual asperity contact is small compared with the deformation arising from the overall Hertzian pressure distribution imposed by the lubricant pressure. Therefore, as a first approximation, the analysis can proceed on the assumption that the asperity contacts are equivalent to those between two nominally flat surfaces of dimensions $2L \times 2D$ separated by a distance h.

Assuming, for the moment, that the asperities are spherical, equation (19a) gives the load per unit area. The total load which we wish to calculate is contained in a rectangle of area $4DL$ and we assume that a sum of interference depths, $\Sigma\omega$, is obtained from a total length of profile pairs $m^2 l$. It follows that the load, W, supported by asperity contacts within this area is

$$W = \frac{2\sqrt{2}L\,DE'}{m^2 l}\,\Sigma\omega \tag{20a}$$

An assumption implied in equation (20a) is that the total length of profile interaction ($m^2 l$) is sufficiently large to be representative of all the possible configurations of asperity interactions which can occur as the discs rotate.

Equation (20a) can also be modified, in the light of equation (19b) to take account of anisotropy of surface finish. Thus for a surface finish typical of discs ground circumferentially it can be shown[22,23] that

$$W = \frac{1.13\,DLE}{m^2\,l}\,\Sigma\omega \tag{20b}$$

In a similar fashion we can calculate the total area of contact A arising from asperity interactions. The result obtained[23] is

$$A = \frac{8\,DL}{m^2\,l}\,f_2(R_x/R_y)\,\Sigma\alpha \tag{21}$$

where, as in equation (19b), $f_2(R_x/R_y)$ can be calculated from Hertz theory; it has values between 0.38 and 0.50 for values of (R_x/R_y) between 10^{-3} and unity. In a similar fashion to equations (20) $\Sigma\alpha$ represents the summation of the half-widths of all the profile interference events observed in a total profile interaction of length $m^2\,l$.

11. PLASTIC FLOW IN COMPUTER SIMULATION

We can now consider the probability of plastic flow by a theory similar to that used in Section 7 above. Again we first consider the asperities as spherical. Equation (10) combined with equation (14) shows that the maximum Hertz pressure in contact simulation is given by

$$p_0 = \frac{2E'}{\pi}\left(\frac{w}{a}\right) = \frac{2\sqrt{2}E'}{\pi}\left(\frac{\omega_0}{\alpha_0}\right)$$

Then, as before putting $p_0 > 0.6H$ the criterion for the onset of plastic flow becomes

$$\frac{\omega_0}{\alpha_0} > \frac{0.6\pi}{2\sqrt{2}}\,\frac{H}{E'}. \tag{22}$$

Bearing in mind that for any contact

$$\frac{\omega_0}{\alpha_0} \geqslant \frac{\omega}{\alpha},$$

we shall adopt a criterion

$$\frac{\omega}{\alpha} > \frac{0.6\,\pi}{2\sqrt{2}}\,\frac{H}{E'} \tag{23a}$$

This criterion will give an indication of the contacts in which the elastic limit has been exceeded (compared with the exact statement of equation (22) it will underestimate the load involving plastic deformation). Thus, as in the derivation of equation (20a) we can write

$$W_p = \frac{2\sqrt{2}LDE'}{m^2 l} \Sigma\omega_p$$

where $\Sigma\omega_p$ indicates a summation only of those values of ω where equation (23a) is satisfied.

As in our earlier discussion it is a relatively straightforward extension of the theory to extend it to the analysis of ellipsoidal asperities. For this case equation (23a) is replaced by

$$\frac{\omega}{\alpha} > 0.6\, f_3(R_x/R_y)\, \frac{H}{E'} \tag{23b}$$

where $f_3(R_x/R_y)$ can be calculated[23] from Hertz theory and has values between 1.1 and 2.4 for values of (R_x/R_y) between 1 and 10^{-3}.

12. THE DISC MACHINE EXPERIMENTS

Discs employed in the experiments of Bell and Dyson[22,25] were used to illustrate some features of profile analysis using the computer and to test the theory of contact simulation which has been outlined above. Bell and Dyson interpreted their measurements of friction using these discs to illustrate their theory of mixed friction in an elastohydrodynamic system. This theory used the model of Greenwood and Williamson[11] to represent the contact of the surfaces through the lubricant film. A somewhat similar theoretical approach has been adopted by Johnson, Greenwood and Poon[26]. Therefore it seemed appropriate to use the Bell and Dyson discs with the eventual aim of comparing theories based on models with computer simulation using data derived directly from the surfaces.

A full account of the experimental conditions has been given by Bell and Dyson[22,25]. It will suffice to record that the discs, case hardened to a hardness of 870 V.p.n. had a diameter of 76.2 mm (3 in) and an effective contact width of 11.1 mm (7/16 in). They were run in a loading sequence up to a maximum load of 5.42 kN (1220 lbf) giving a mean Hertzian contact stress of 0.751 G Pa (1.09×10^5 lbf/in^2). The theoretical film thickness corresponding to the final running conditions was 0.14 μm. The discs were originally circumferentially ground to a finish of 0.406 ± 0.051 μm (16 ± 2 μin) R_a. In our examination of the discs it was apparent that their surface topography had been markedly modified by running-in. Therefore a similar profile analysis was conducted on freshly prepared discs which had been produced, as far as was possible, by identical methods.

A complete computing package has been devised which allows the computation of a whole range of data deduced from the profile. The print-out from this programme includes relevant tabulated data and, where appropriate, the results were also presented in graphical form. The calculated data included the height distributions of the ordinates and of the peaks and the valleys. Peaks and valleys were computed using three point analysis using a sampling interval of 2.5 μm. The slopes of the profile and the curvatures of the peaks were calculated in a similar fashion. The autocorrelation functions and power spectral densities were also computed. Some details of the methods of data handling and computing have been given elsewhere[13,27]. In what follows we shall give examples of the results of the profile analysis which seem of particular interest.

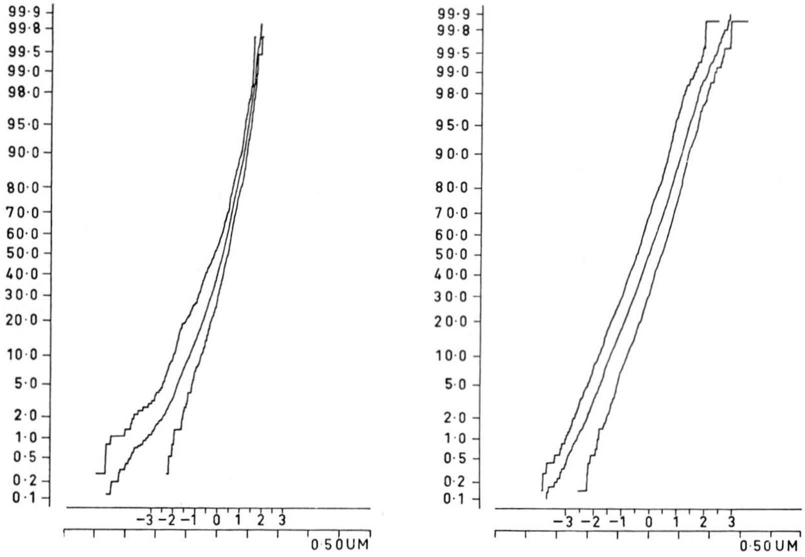

(a) *(b)*

Fig. 7. Computer print-out of cumulative height distribution from a) run-in, b) freshly prepared, disc

Distributions of ordinates, peaks, and valleys are shown. Upper horizontal scale is normalized by rms of ordinate height distribution.

Fig. 7a presents typical height distributions of ordinates, peaks and valleys from a disc used in the Bell and Dyson experiments and Fig. 7b presents similar distributions from a similar freshly prepared disc. It will be observed that the graphical plotting has been arranged to present the results in the usual form of a probability plot in which a Gaussian distribution appears as a straight line. The feature of interest in these plots lies the consequences of running-in. The freshly prepared surface has a distribution which is close to Gaussian. The distribution from the run-in surface suggests that two effects have occurred. The upper levels of the distribution have been depressed as expected. A less expected result is that the lower levels seem to have been raised. Further experiments are needed to confirm this result.

Fig. 8a shows the distribution of slopes from the run-in disc and Fig. 8b shows the corresponding distribution of slopes from the freshly prepared

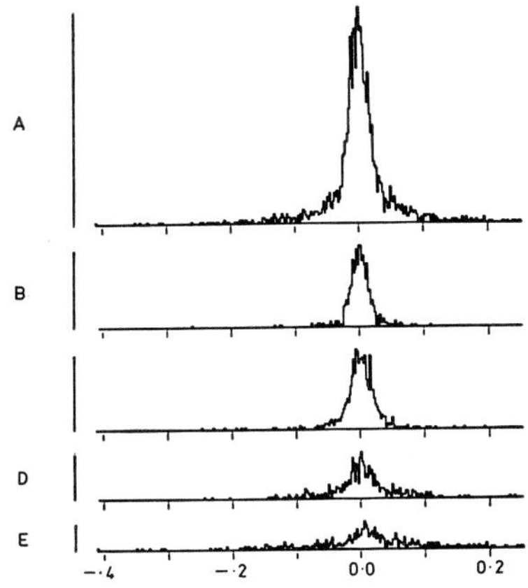

(a)

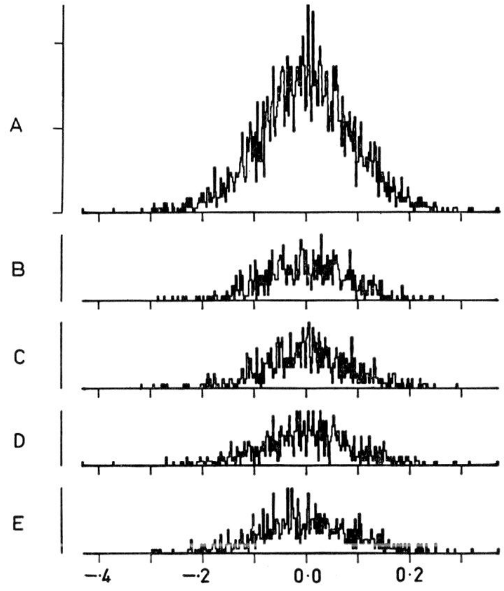

(b)

Fig. 8. Computer print-out of slopes from a) run-in, b) freshly prepared, disc

A, all slopes. B, C, D, E, slopes from four quartiles. B is highest quartile, E is lowest.

surface. In each example the upper plot shows the distribution of all slopes and the lower four plots show the slopes derived from the four quartiles of the profile. It will be seen that for the freshly prepared surface the slope distribution has a Gaussian shape and the slopes drawn from the four quartiles are essentially the same (within the limits of sampling errors). On the other hand the effect of running-in is shown to produce a significant reduction in the slopes and, as expected, this effect is most marked in the upper levels of the profile. The analysis similarly showed that running-in causes a marked reduction in the curvatures of the peaks. Similar consequences of running-in under slow speed boundary lubrication conditions have been demonstrated by Whitehouse[27].

These results and many others drawn from the analysis of surfaces used in experiments illustrate that such surfaces differ significantly from those which have been freshly prepared. Since many of the models for surface contact are based on data drawn from new surfaces we can see, very clearly, the limitations of theories based upon these models. These considerations also emphasize the importance of developing computer simulation methods based upon profiles taken from the surfaces used in experiments.

14. RESULTS OF COMPUTER SIMULATION

Twelve profiles were obtained from each disc in digital form. Some details of data handling and data processing methods have been given elsewhere[13,27]. A computer programme brought together any pair of profiles with their mean lines separated by any desired distance. The programme calculated any desired detail derived from the profile interference events including the details required in the theory which has been outlined above.

A typical length of profile interaction from any pair of profiles corresponded to approximately 5 mm on the disc surface. With 144 profile pairs in this programme the profile interaction employed in this study correspond to a distance of approximately 0.72 metres on the disc surfaces.

Results were calculated for each profile pair separately. With 144 such simulations available, our results could then be presented as a mean and standard deviation. In this way the internal evidence of the derived data gives an estimate of the extent to which the results were representative of the many contact situations which could occur. The results derived in this way are presented in Table 1. It is clear that with our sample of surface profiles (which represents a very small proportion of the total detail available on the surfaces of the discs) significant information can be obtained about the nature of the load borne by solid contacts and the way in which this varies with the separation. It is clear that with these run-in surfaces this load is almost entirely supported by elastic deformation and

Table 1. Computer simulation using discs from Bell and Dyson experiments

Separation (μm)	$\Sigma\omega$ (μm)	*Total solid contact load* (KN)	*Load borne by plastic contacts* (KN)
0.22	27.23 ± 1.85	1.278 ± 0.087	0.017 ± 0.017
0.45	16.68 ± 1.50	0.783 ± 0.070	0.008 ± 0.011
0.67	8.42 ± 1.15	0.395 ± 0.054	less than 1% of total
0.89	3.32 ± 0.79	0.156 ± 0.037	

Surface roughness of discs after running-in

Disc 1: $R_a = 0.29$ μm, $\sigma = 0.39$ μm
Disc 2: $R_a = 0.39$ μm. $\sigma = 0.50$ μm

Results are given as the mean value for one profile pair (approximately 5 mm in length on the disc surfaces) together with the standard deviation; these results were derived from 144 such profile pairs.

The Table also presents the values of the loads supported by solid contacts which have been deduced from the profile interference. The total applied load was 5.42 KN.

that plastic deformation is negligible. This last conclusion does not apply to a similar simulation of softer discs in their freshly prepared state where it was concluded that most of the deformation was plastic.

With the smaller separations of Table 1 a significant proportion of the total applied load is borne by solid contacts and some of the assumptions which are implicit in the work are no longer valid. An extension of the work might note the existence of contacts which are so close that they are no longer independent as a preliminary to the more exacting problem of considering their interaction.

As explained earlier, Table 1 gives deduced values of the load supported by asperity contacts within the main load bearing region. They are significantly greater than similar loads calculated by the methods of Bell and Dyson[22]. These authors used the difference between the frictional tractions observed with the rough discs and the corresponding values obtained with smooth discs as one element in their calculations; in using these measured frictional tractions it was assumed that the local coefficient of friction at the asperity contacts was 0.5. This latter assumption is a possible explanation of the large discrepancies between the loads calculated by the two methods. However, the contribution of the loads supported by asperity contacts in the curved portions of the discs, outside the main load bearing region, merits serious consideration. Using the results of the computer simulation reported here, Dyson[28] has provided a method for calculating these loads. In a typical situation they are 10 to 15% of the loads calculated for the main load bearing region such as those given in Table 1; therefore the loads supported by asperities outside the main load bearing region may not be negligible as was implicitly assumed in our calculations.

15. CONCLUSION

Some aspects of theories of surface contact have been reviewed. It has been shown that the stylus profilometer, in particular when linked with the digital computer, has played a significant role in the development of models to represent contact between surfaces. The major defect of these theories is that the models are mainly representations of freshly prepared surfaces.

The same techniques, using the stylus profilometer and the digital computer, can be used more directly in a study of the surfaces derived from tribological experiments. We have reported an exploration of some new techniques based upon these principles. It has been shown that significant information (for example, about the load support by asperity contacts through lubricant films) can be obtained from profilometer data which is a minute proportion of that required to represent all the surfaces.

ACKNOWLEDGMENTS

In the course of this work R.A. Onions was supported by a grant from the Shell International Petroleum Company and R.T. Hunt was supported by a grant from the Science Research Council to whom our thanks are offered. We are indebted to Mr. A. Dyson for much helpful discussion of this work.

REFERENCES

1. R. Holm, *Electrical Contacts*, H. Gerbers, Stockholm, 1946.
2. F.P. Bowden and D. Tabor, *Friction and Lubrication of Solids*, Clarendon Press, Oxford, 1954.
3. J.F. Archard, 'Contact and Rubbing of Flat Surfaces', *J. Appl. Phys.*, **24**, 1953, pp. 981–988.
4. J.T. Burwell and C.D. Strang, 'On the Empirical Law of Adhesive Wear', *J. Appl. Phys.*, **23**, 1952, pp. 18–28.
6. H. Blok, 'Comments on a Paper by R.W. Wilson', *Proc. Roy. Soc. Lond.*, A**212**, 1952, pp. 480–482.
7. J.S. Halliday, 'Surface Examination by Electron Microscopy', *Proc. Instn. Mech. Engrs.*, **169**, 1955, pp. 777–787.
8. J.S. Halliday, 'Application of Reflection Electron Microscopy to the Study of Wear', *Conference on Lubrication and Wear*, Instn. Mech. Engrs., 1957, pp. 647–651 and 867–869.
9. J.F. Archard, 'Elastic Deformation and the Laws of Friction', *Proc. Roy. Soc. Lond.*, A**243**, 1957, pp. 190–205.

10. J.F. Archard, 'Single Contacts and Multiple Encounters', *J. Appl. Phys.*, **32**, 1961, pp. 1420–1425.
11. J.A. Greenwood and J.B.P. Williamson, 'Contact of Nominally Flat Surfaces', *Proc. Roy. Soc. Lond.*, A**295**, pp. 300–319.
12. J. Peklenik, 'New Developments in Surface Characterisation and Measurement by Means of Random Process Analysis', *Proc. Instn. Mech. Engrs.*, **182**, Pt. 3K, Conference on the Properties and Metrology of Surfaces, 1967–'68, pp. 108–126.
13. D.J. Whitehouse and J.F. Archard, 'The Properties of Random Surfaces of Significance in their Contact', *Proc. Roy. Soc. London*, A**316**, pp. 97–121.
14. R.A. Onions and J.F. Archard, 'The Contact of Surfaces Having a Random Structure', *J. Phys. D. (Appl. Phys.)*, **6**, 1973, pp. 289–304.
15. P.R. Nayak, 'Random Process Model of Rough Surfaces', *Trans. ASME (J. Lubric. Technol.)*, **93**, 1971, pp. 398–407.
16. P.R. Nayak, 'Some Aspects of Surface Roughness Measurement', *Wear*, **26**, 1973, pp. 165–174.
17. P.R. Nayak, 'Random Process Model of Rough Surfaces in Plastic Contact', *Wear*, **26**, 1973, pp. 305–333.
18. R.C. Spragg and D.J. Whitehouse, 'A New Unified Approach to Surface Metrology', *Proc. Instn. Mech. Engrs.*, **185**, 1970–'71, pp. 697–707.
19. D. Tabor, *The Hardness of Metals*, Clarendon Press, Oxford, 1951.
20. W. Hirst and A.E. Hollander, 'Surface Finish and Damage in Sliding', *Proc. Roy. Soc. London*, A**337**, 1974, pp. 379–394.
21. J.P.B. Williamson, 'Topography of Solid Surfaces', in P.M. Ku (ed.), *Interdisciplinary Approach to Friction and Wear*, NASA, SP-181, NASA, Washington (D.C.), 1968, pp. 85–142.
22. J.C. Bell and A. Dyson, 'Mixed Friction in an Elastohydrodynamic System', *Symposium on Elastohydrodynamic Lubrication*, Instn. Mech. Engrs., 1972, pp. 68–76.
23. R.A. Onions, *Pitting in Gears*, Ph.D. Thesis, University of Leicester, 1973, Chapter 5.
24. D. Dowson and G.R. Higginson, 'New Roller Bearing Lubrication Formula', *Engineering Lond.*, **192**, 1961, pp. 158ff.
25. J.C. Bell and A. Dyson, 'The Effect of Some Operating Factors on the Scuffing of Hardened Steel Discs', *Symposium on Elastohydrodynamic Lubrication*, Instn. Mech. Engrs., 1972, pp. 61–67.
26. K.L. Johnson, J.A. Greenwood and S.Y. Poon, 'Asperity Contact in Elastohydrodynamic Lubrication', *Wear*, **19**, 1972, pp. 91ff.
27. D.J. Whitehouse, *The Properties of Random Surfaces of Significance in their Contact*, Ph.D. Thesis, University of Leicester, 1971.
28. A. Dyson, 'Thermal Stability of Models of Rough Elastohydrodynamic Systems' (to be published).

The interaction and lubrication of rough surfaces

*F. T. Barwell, M. H. Jones and S. D. Probert**

1. INTRODUCTION

Solid surfaces used in engineering are neither perfectly smooth nor flat so that, when two bodies are placed together, true contact only occurs over a small proportion ($\sim 0.1\%$) of the nominal contact area. The true area of contact affects the thermal and electrical resistances of the interface and is also of crucial importance with respect to the friction and wear behaviour if the surfaces are in relative motion.

The nature of the contact is governed by the extent to which the two bodies conform geometrically on a macro-scale as well as by the quality of their surface texture. As the materials involved suffer both elastic and plastic deformations, the contact pattern will be dependent upon the magnitude, direction and time dependence of any forces applied, as well as upon the geometries, stiffnesses and hardnesses of the materials composing them. In particular the stress distribution within the zone of contact between non-conforming bodies will be governed by Hertz's equations unless the elastic limit of the material has been exceeded. Even when the applied forces are sufficient to cause bulk deformation of the contacting solids, the microscopic features of the surface texture, although modified, are usually not completely suppressed.

The micro-texture of the surface of a component used in engineering is irregular and is best described in statistical terms. A convenient division between macroscopic and microscopic surface deviations has been suggested by Peklenik[1] based on the assumption that 'every surface profile may contain harmonics with at least one prevailing frequency and that the lowest estimated frequency determines the type of surface deviation'. The greatest estimated frequency f_s should be greater than 200 c/m for the feature to be considered microscopic or 'small scale'.

The most convenient method for studying a horizontal surface is by

* Now at School of Mechanical Engineering, Cranfield Institute of Technology, Bedford, U.K.

observation of the vertical displacement of a sharp stylus when traversed across the surface so as to obtain its 'profile'. Although this provides some qualitative information, the complete profile was not originally available for quantitative use and so some form of statistical assessment was required. The originators of the stylus method, Abbot and Firestone[2] introduced the concept of 'bearing area'. In this a diagram is prepared showing the area of surface intercepted by a series of imaginary planes parallel to the mean level of the surface and evenly spaced between the highest level at which a plane passes through continuous matter and that which passes through the highest peak of the surface.

The stylus instruments used to obtain profiles were also able to process the signal in an elementary way so as to provide a single number – a 'roughness' to describe a surface. This involved comparison of the elevation of each point on the path of the stylus with a computed mean height of the surface. The result was expressed either as the root mean square or the centre line average (CLA) and defined in Fig. 1. The finest surface attainable

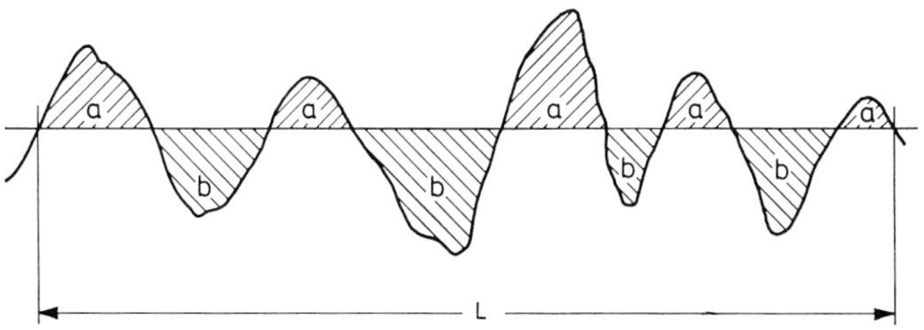

Fig. 1. Definition of centre-line average

$$\text{CLA of specimen surface} = \frac{\Sigma \text{ areas } a + \Sigma \text{ areas } b}{(\text{vertical magnification}) \times \text{L}}$$

Position of centre-line is such that Σ areas $a = \Sigma$ areas b.

in the laboratory usually has a roughness exceeding 3×10^{-2} μm CLA and the range of industrially prepared surfaces extends up to about 3 μm CLA.

Whilst representing a great advance, the single figure for roughness did not completely characterise a surface and provided little guidance with respect to the subsequent behaviour of the surface when placed in contact with another.

Research on contacts then advanced in two ways: one employed optics to provide a three-dimensional view of a surface and the other involved the connection of stylus instruments to computers so that both three-dimensional surface contours and complete statistical descriptions of surfaces could be obtained. Fig. 2 shows a model of a surface prepared by computer from stylus instrument input data and Fig. 3 shows a contour map derived therefrom[3].

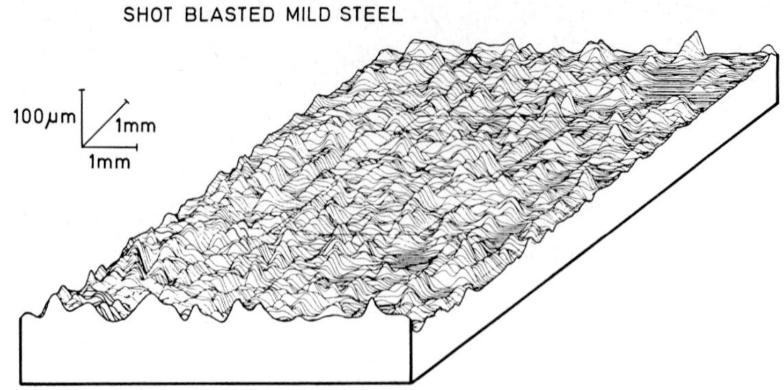

Fig. 2. Model of surface produced by computer

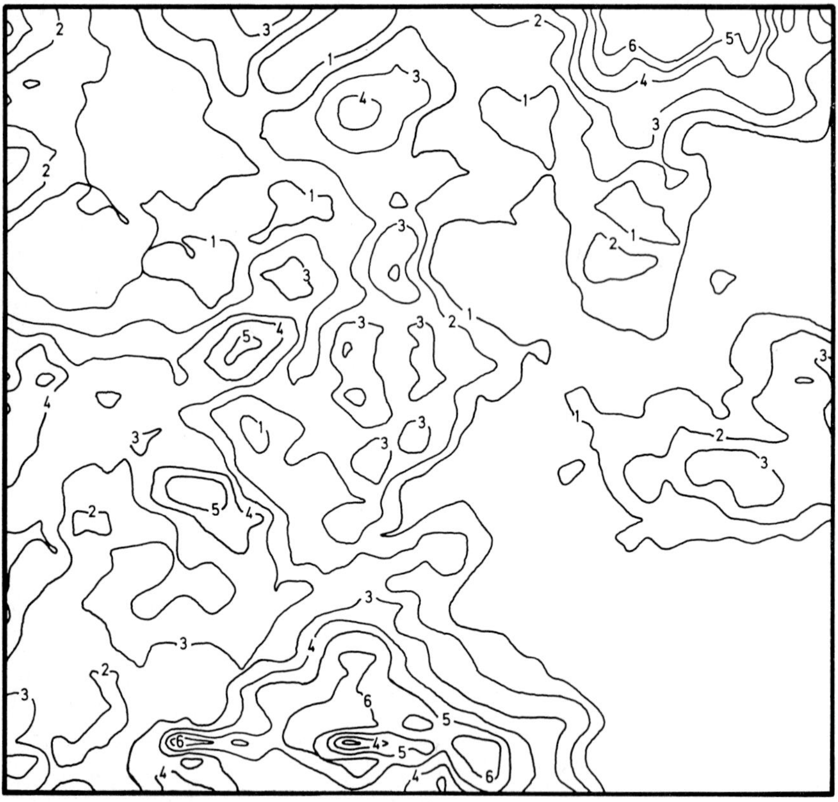

Fig. 3. Contours of shot blasted hot rolled sheet
Contour heights are: 1: 0.16 μ; 2: 0.41 μ; 3: 0.98 μ; 4: 0.98 μ; 4: 1.55 μ; 5: 2.12 μ; 6: 2.70 μ.

An important result[4] of computer analyses of stylus records is that the significant parameters which can be used to describe surfaces are usually distributed in a Gaussian manner. This implies that the Abbot curve should be sigmoid in shape to correspond with the ogive or cumulative distribution curve. This is indeed found to be the case. The dotted line on Fig. 4 shows

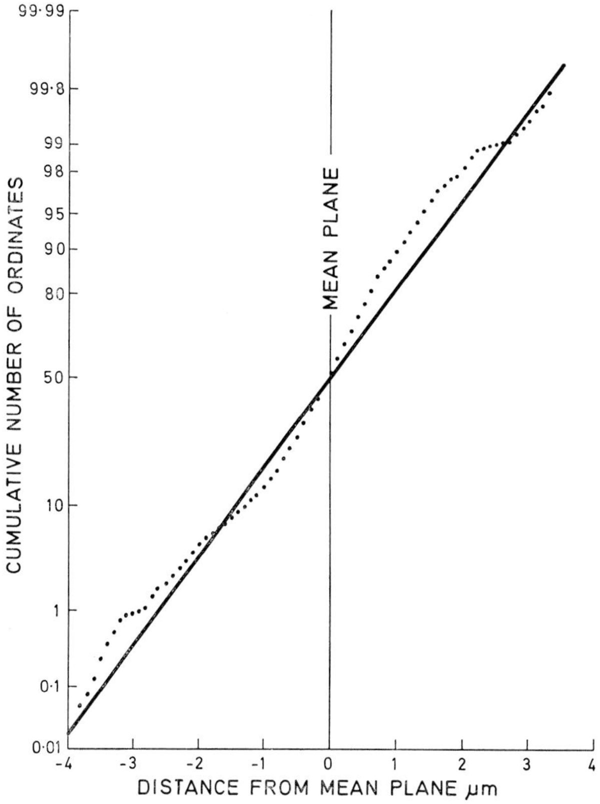

Fig. 4. Ordinate distribution for rough surface generated by bead blasting

the actual readings for comparison with the solid line which represents a perfect Gaussian distribution.

It will be noted that in the three cases described in Table 1, each of the factors which may be held to represent roughness, notably steepness of slope of asperities, and smoothness of peak or percentage of peaks above maximum peak height, increase together with the progression to rougher surfaces. Thus the use of the CLA value as an indication of surface roughness is justified. By restricting consideration to truly random surfaces, e.g. the bead-blasted surfaces described in reference[4], the CLA

Table 1. Characteristics of surfaces of hardened steel pistons (hardness 715 H_v)

Specimen	*CLA value* (10^{-2} μm)	*Estimated maximum peak ht* (10^{-2} μm)	*Average asperity slope (degrees)*	*Average asperity radius* (μm)	*Percentage of peaks above 2/3 of maximum peak height*
Finely finished as used in aircraft practice	3.56	10.2	0.16	508	1
Medium finish	13.5	81	3.1	40	1
Rough finish	17.8	250	7.35	18	7

value characterises the mean of the height or depth respectively of the peaks or valley bottoms from the central plane, but indicates nothing concerning the dispersion of these values, and so other parameters such as the standard deviation of peak heights for example may also be required.

Tsukizoe and Hisakado[5] proposed the use of the Gaussian probability function, $f(y)$ which they wrote as:

$$f(y) = \frac{1}{\sqrt{2\pi}} \exp(-y^2/2) \tag{1}$$

where $y = h/\sigma$, $h =$ deviation from median line of the profile curve and $\sigma =$ standard deviation. The measured profile trace of a 'random' surface is stationary and ergodic. The probability density can then be supplemented by the auto-correlation function or its power spectrum which is the Fourier transform of the auto-correlation function. A representative selection of these statistics is reproduced in Ref. [1].

The relationship between the cumulative height distribution curve derived from the surface profile and the bearing area curve for a random surface[2] requires some explanation. It appears intuitively that the bearing area is the product of the bearing line measured in two perpendicular directions. However, as pointed out by Williamson[4] the usual process of integration consists of summing an infinite number of closely-spaced parallel sections. If the surface is truly random and provided these sections are long enough, they will all contain the same information. Therefore statistics obtained from one trace are representative of the surface as a whole. If for example 10% of one profile lies above a certain height, 10% of the surface area will lie above that height and not one per cent which would have been the case had the product been the correct measure.

In order to permit the development of a theoretical treatment, Tsukizoe and Hisakado assumed that the asperities existing on a surface would take the form of cones having a base angle of θ. If such a surface existed and were examined with a stylus instrument the height 'h' of a point on a profile curve which traversed an asperity would be given by $\sqrt{(x^2+z^2)}\tan\theta$ where x and z are distances from the apex and that particular asperity measured in the horizontal plane. Then

$$\frac{\mathrm{d}h}{\mathrm{d}x} = \frac{x}{\sqrt{x^2+z^2}}\tan\theta = \cos\alpha\tan\theta \tag{2}$$

where $\alpha = \tan^{-1}(z/x)$.

The mean value of the profile slope as the stylus traverses the many asperities will be

$$\bar{\psi} = \frac{2}{\pi}\int\frac{\mathrm{d}h}{\mathrm{d}x}\,\mathrm{d}\alpha = \frac{2}{\pi}\tan\theta\int\cos\alpha\,\mathrm{d}\alpha$$

$$\therefore\ \theta \approx \frac{\pi}{2}|\bar{\psi}| \tag{3}$$

When two surfaces are brought together only very few asperities come into contact so that the applied load is concentrated thereon causing excessive stress locally. These asperities then deform so as to bring an increasing number into contact until the load can be supported without further deformation. The important questions that have to be answered in order to obtain a full understanding of the particular situation are as follows:

1. To what extent is the deformation elastic or plastic?
2. By what extent do individual asperities grow in size rather than increase in number?

If we consider a random surface to interact with a rigid plane which progressively approaches the surface and ignore the mechanism whereby the material of the intervening peaks is displaced, we may write,

$$A_R = L_x L_z\int_0^h f(h)\,\mathrm{d}h \tag{4}$$

where L_x and L_z represent lengths in the x and z directions respectively.

Holm[6] proposed a simple relationship, based on plastic flow, of the form

$$A_R = \frac{W}{\bar{p}} \tag{5}$$

where W = applied load and $\bar{p}$ denotes 'flow pressure' which is commonly assumed to correspond with the hardness of the material. This relationship has been used by Bowden[7] and others as an important component in the adhesion theory of friction.

$$\text{Nominal stress} = \frac{W}{A_n} = p$$

$$\text{then}\ \frac{A_R}{A_n} = \frac{p}{\bar{p}} \tag{6}$$

$$p = \frac{W}{L_x L_z} = \text{apparent compressive stress on surface} = \bar{p}\int_0^h f(h)\,\mathrm{d}h$$

$$\text{or } \frac{p}{\bar{p}} = \int_0^h f(h)\,\mathrm{d}h \tag{7}$$

The idealised depth of penetration $\bar{Y}$, i.e. the distance through which the hypothetical rigid flat moves into the surface by plastic deformation under the applied load W, is given by

$$\begin{aligned} \bar{Y} &= \gamma\sigma - h \\ &= (\gamma - y)\,\sigma \end{aligned}$$

where γ = a constant, so that $\gamma\sigma$ equals the distance from the highest peak to the median of the profile curve and $y = h/\sigma$. Let $H = j\sigma$ be the maximum peak height, then $\bar{Y} = (\gamma - y)(H/j)$.

If two rough surfaces are placed together, Tsukizoe and Hisakado suggest that the equivalent value of H is the root mean square of the values for each surface taken separately, thus

$$\bar{Y} = (\gamma - y)\,\frac{(H_1^2 + H_2^2)^{1/2}}{j\sqrt{2}} \tag{8}$$

A more convenient measure is however the separation, u, of the idealised flat surface from the mean level of the rough surface.

Thus if the rough surface has a Gaussian distribution of surface heights, then

$$f(u) = \frac{1}{\sqrt{2\pi\sigma^2}}\exp\{-\tfrac{1}{2}(u/\sigma)^2\}$$

or by writing $t = u/\sigma$,

$$\phi(t) = \frac{1}{\sqrt{2\pi}}\exp\{-t^2/2\} \tag{9}$$

When the separation is three times the standard deviation the Gaussian relationship gives a bearing area of less than 1%. We may then write $H \simeq 3\sigma$ or $t = 3$. However

$$\frac{\Delta L}{L} = \int_0^3 \phi(t)\,\mathrm{d}t - \int_0^t \phi(t)\,\mathrm{d}t \tag{10}$$

but erf $(3) = \frac{1}{2}$, following (9) we may write

$$\frac{\Delta L}{L} = \tfrac{1}{2} - \Phi(t) \tag{11}$$

where $\Phi(t)$ represents the probability integral $\int_0^t \phi(t)\,\mathrm{d}t$.

For fully-plastic deformation $\Delta L/L = p/\bar{p}$ from equation (5). Therefore this relationship between approach t of the surfaces and the applied normal pressure is given by $p/\bar{p} = 1/2 - \text{erf}\,(t)$. Uppal and Probert[8] observed that the initial two or three contact spots occur when $t = 3.5$ rather than 3.0. It was also deduced that, if the asperities are equivalent to cones, the number of contact points is determined by the degree of roughness and that over a wide range of applied load, the number of contact points increases much more rapidly than the average radius; factors of increase quoted being 42.2 and 1.6 respectively.

Whitehouse and Archard[9] confirmed that equation (2) applies to many surfaces and that the auto-correlation function of the profile

$$C(\beta) = \lim_{L\to\infty} \frac{1}{L} \int_{-1/2L}^{+1/2L} y(x)\, y(x+\beta)\, \mathrm{d}x \tag{12}$$

where $y(x)$ is the normalised height of the profile at a given coordinate x, and $y(x+\beta)$ is the height of an adjacent coordinate distance β from the first. For truly-random surfaces the correlation function takes the form

$$C(\beta) = \exp(-\beta/L) \tag{13}$$

where L is the correlation distance. When $\beta = 2.3L$, $C(\beta)$ has declined to 0.1. These authors regard this spacing as that at which two points on a profile can be regarded as independent.

Greenwood and Williamson[10] proposed that whether the asperities would be deformed elastically or plastically could be predicted by a 'plasticity index', given by

$$\lambda = \frac{E'}{h}\sqrt{\frac{\sigma}{r}} \tag{14}$$

where

$$\frac{1}{E'} = \frac{1-\gamma_1^2}{E_1} + \frac{1-\gamma_2^2}{E_2}$$

h = Hardness
σ = standard deviation of asperity height distribution, and
r = asperity radius

and γ_1 and γ_2 and E_1 and E_2 are the Poisson's ratios and Young's moduli of elasticity of the materials of the two contacting surfaces respectively.

The radii of curvature of the peaks are governed by the same probability considerations as the other surface parameters and Whitehouse and Archard were able to identify the relationship between r and L and to redefine the plasticity index as

$$\lambda = 0.6\left(\frac{E'}{h}\right)\left(\frac{\sigma}{L}\right) \tag{15}$$

The value of the constant was originally given as 0.3 but Hirst and Hollander[50] substituted the value 0.6. The most recent and probably the definitive model for rough surfaces is that proposed by Nayak[51] on the basis of random process theory. This method enables a distinction to be made between a 'peak' and a 'summit'. Thus when profiles are separated by a finite distance they will tend to pass over the shoulders of asperities rather than over the 'peak' or highest point. The shoulder will nevertheless appear as a peak on the profile though at a lower height than the true peak. Thus the summation of the peaks produces a lower number than actually exist on the surface. A similar error occurs in the determination of the mean slope as quantified in equation (3). All the necessary statistics for Gaussian isotropic surfaces can be derived from three parameters m_0, m_2 and m_4 obtained from a single profile.

By employing three non-parallel profiles the method enables a complete description to be given of a non-isotropic random surface and offers promise of extensions to industrial surfaces displaying a partially ordered structure.

Note:

$$m_0 = \sigma^2 = \int \xi(h)\, dh, \quad m_2 = \int h^2\, \xi(h)\, dh, \quad m_4 = \int h^4\, \xi(h)\, dh.$$

2. OPTICAL METHODS

Considering first the two methods for determining optically the distribution

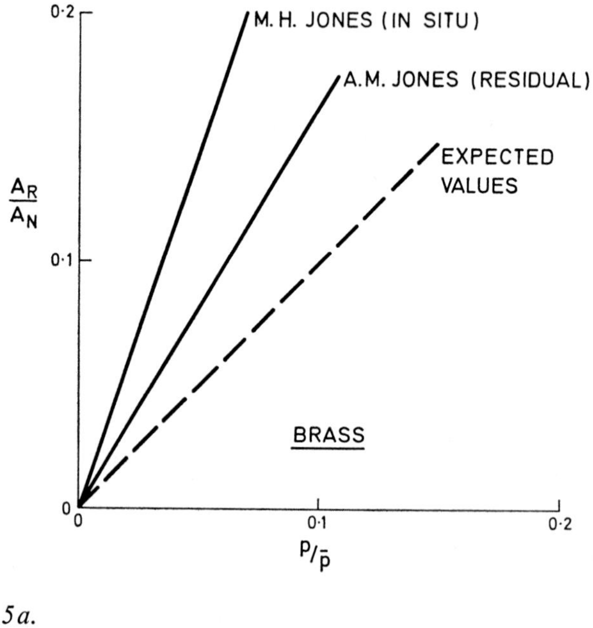

Fig. 5a.

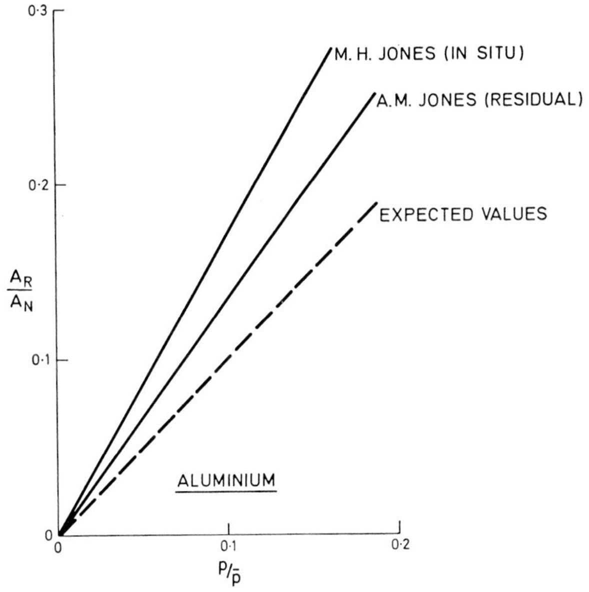

Fig. 5b.

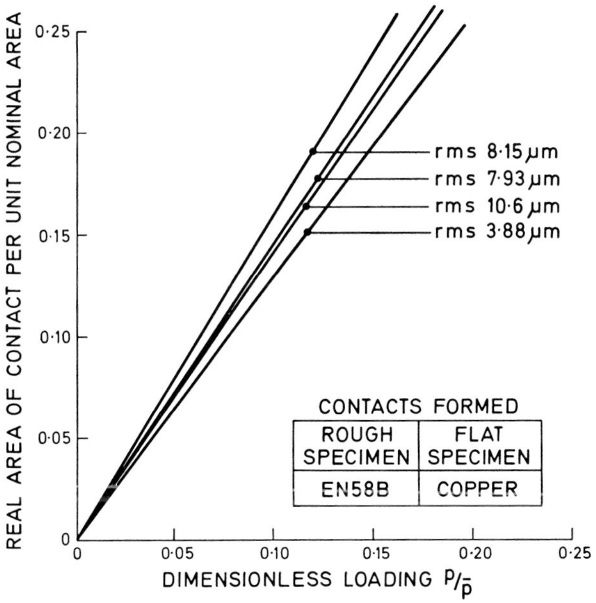

Fig. 5c. Real area of contact versus dimensionless loading pressure

and size of interacting asperities [13, 14, 15] the results of which are displayed in Fig. 5(a) and (b). In the first case the specimens were pressed against the silvered surface of a glass optical flat of a specially adapted microscope wherein the deformation at the contact spots was observed by Normarski phase-contrast illumination while the specimen was still under load. Thus elastic deformations were observed together with residual plastic deformations.

In the other case the specimen surface under study was polished to the quality necessary for an optical flat. Then a roughened steel anvil was imprinted onto the surface under a prescribed load and the resulting damaged optical flat was examined using a Quantimet 720 image analysing computer.

The flow pressures in each case were considerably lower than the hardnesses indicated by the Vickers Diamond method.

3. COMBINATION OF OPTICAL AND STYLUS METHODS

Four materials of differing hardnesses, as shown in Table 2, were polished to form optical flats and pressed against the flat surface of a hard stainless

Table 2. Vickers hardness and surface preparation of materials tested

Rough specimen		*Polished specimen*	
Material	*Vickers hardness*	*Material*	*Vickers hardness*
	kg/(mm)^2		kg/(mm)^2
EN58B Stainless steel	395	Copper (commercially pure)	99
		Aluminium (commercially pure)	114
		Brass	179
		Invar. Stainless steel	206

steel (EN 58B) specimen. The latter had been roughened by bead blasting to produce a Gaussian distribution of surface heights which was recorded using a Talysurf 4 surface analysis instrument coupled to a data logging system.

After separation, the imprinted micro-contact spots were visible as small indentations in the optical flat. These were counted and sized using the Quantimet image analyser. The indentations were visible as dark regions on a bright background. The system was set to measure the total number of features, the total area of those features and the size distribution of features with radii between 1.88 μm and 150 μm.

If the deformations were plastic and no interference between neighbouring asperities occurred, equation (5) applies. However the results were actually described by

$$\frac{A_R}{A_N} = C\left(\frac{p}{\bar{p}}\right) \tag{16}$$

where C took values of 1.6, 1.5, 1.41 and 1.33 for copper, brass, invar and aluminium respectively[14]. No systematic variation of C with surface roughness was observed as in Fig. 5(c).

Fig. 6 displays the probability of contacts of a given size on the basis that the surface roughness involves a Gaussian distribution of surface

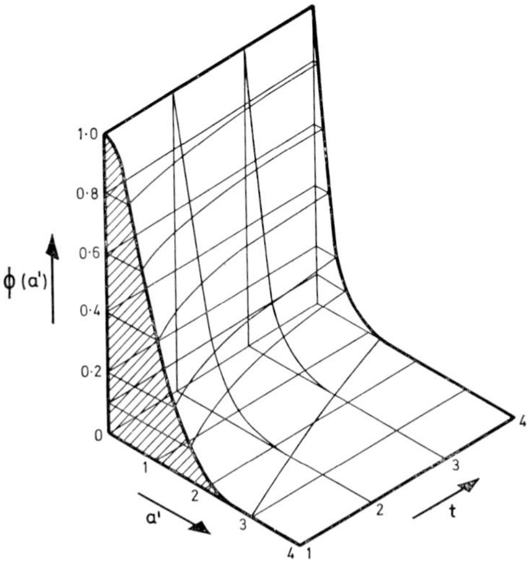

Fig. 6. Proportion of contacts with radius greater than given size

a' = dimensionless radius = $\pi\varphi a/2$

$\phi(a')$ = proportion of actual asperity radii greater than a'

t = dimensionless separation of planes = $\dfrac{\text{separation at mean planes}}{\text{RMS roughness}}$

ψ = mean slope/σ

heights, the individual asperities being conical in form having identical apex angles, in the manner postulated by Tsukizoe and Hisakado.

The contact spot size distribution for one contact between very rough stainless steel (RMS roughness = 3.88 μm) and a copper optical flat is shown in Fig. 7[14]. Agreement is good except for the range of contact sizes below 10 μm where it was increasingly difficult for the analyser to discriminate between contact regions and other features of the surface. Thus it can be asserted that there is good agreement between optical and stylus

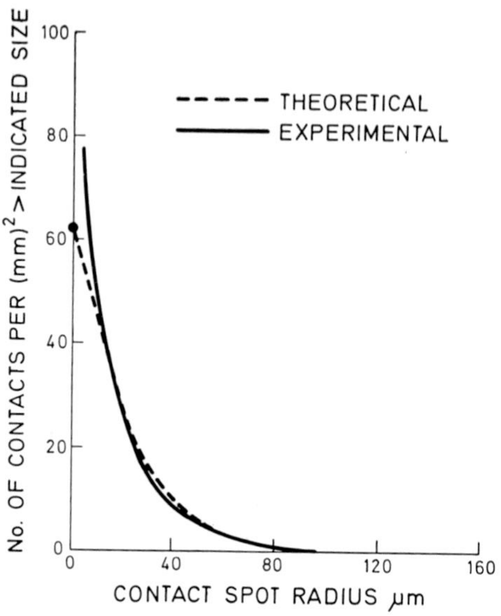

Fig. 7. Number of contact spots with radius greater than a given size

descriptions of a surface. Also although less theoretically convincing than the Whitehouse and Archard model, the Tsukizoe and Hisakado model provides a good working description of the surface.

4. CONTACT RESISTANCE

Problems involved in estimating the number and size of contacts from electrical and thermal resistance measurements arise from the fact that the constriction resistance is determined by the radius of each contact measured in the plane of the surface rather than by the actual contact area. Furthermore, additional resistance may be imposed by oxide and tarnish films, and conductive and radiative heat transfer through the interfacial fluid can occur over areas where the surfaces are not in contact[15, 16, 17, 18, 19]. Nevertheless it can be shown (see Fig. 8) from reference[20] that conduction is directly related to applied normal pressure between the surfaces.

Some account of the effect of oxide films on the contact may be taken by reference to results by Uppal et al.[21] (Table 3), where it is shown that electrical resistance varies inversely with temperature possibly due to

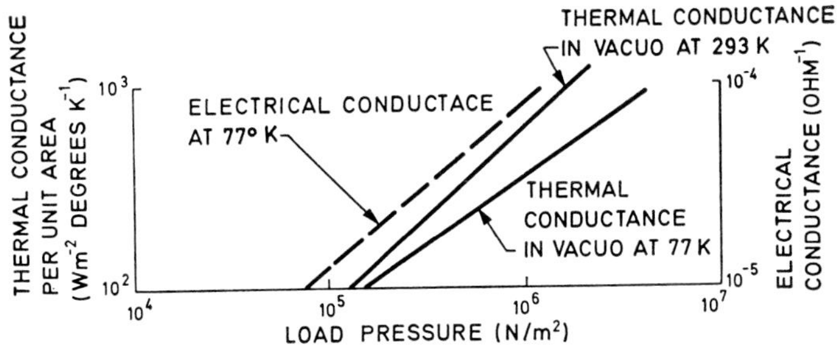

Fig. 8. Conductances of a contact between 14% *tungsten steel specimens (20)*

hardening of surface asperities at reduced temperatures and due to the resistivity of the oxide films increasing with decreasing temperature whereas the constriction resistance would be expected to increase with temperature.

Table 3. Effect of temperature on contact resistances

Material	*Thickness* (μm)	*Surface roughness CLA* (μm)	*Contact resistance* (mΩ)	
			Room temp. 290 K	*Liquid nitrogen temp.* (77 K)
18/8 Co Mo Steel	127	36.0×10^{-2}	2.16	25.0
13% Cr. Steel	330	38.1×10^{-2}	16.4	17.8
Brass	114	37.0×10^{-2}	2.6	3.1

High loadings

A number of departures from the simple theory may occur for contacts under high loadings due to such factors as the flow of displaced metal, work hardening or coalescing of asperities.

The failure of equation (5) to describe the situation even under light loads as revealed in Figs. 9 and 10 is somewhat mystifying and can only be attributed to the fact that the true flow stress and the Vickers hardness are not identical concepts. It would appear that the degree of support afforded to an asperity with relatively large apex angle is very different from that encountered by the steeply-sided diamond indentor. Flow stress will therefore be nearer to the compressive stress than to one-third of that value characteristic of hardness.

Some insight into such behaviour may be obtained by a drastic modification of the hardness test, such as that devised by Mallock and described by O'Neill[22]. Here cones of the material under investigation were pressed against a hard anvil under known loads. Thomas, Uppal and Probert[23] used cones of base diameter about 0.2 mm and examined the deformed cones by relocation profilometry and optical microscopy. The results were quoted in the form of 'deformation ratios' d/D where d is the diameter of the flat caused by deformation and D was the original base diameter. In the range $0.4 < d/D < 0.8$, the Mallock hardness – the ratio

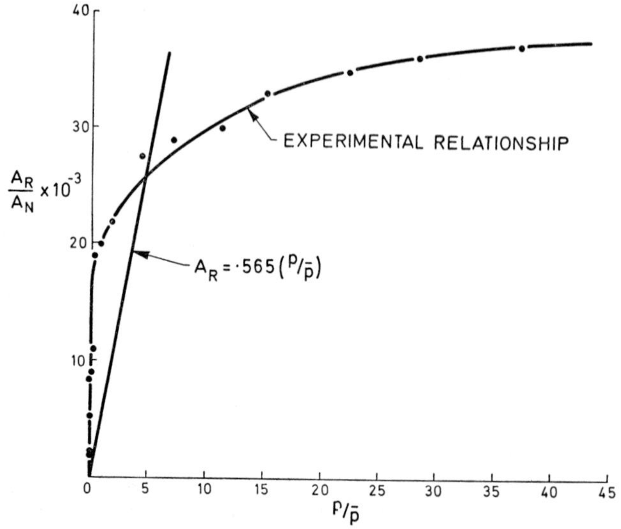

Fig. 9. Relationship between area of contact and applied pressure

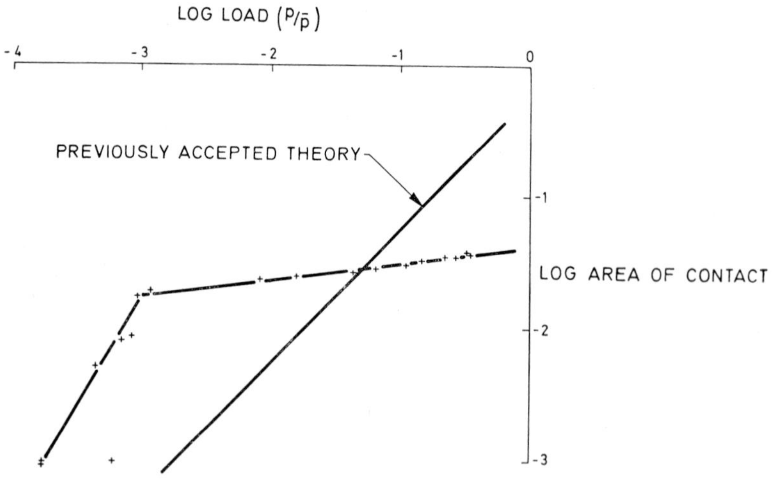

Fig. 10. Logarithmic presentation of area of contact against load

of the load on an asperity to the flattened area, was nearly constant and equal to about one-half of the Vickers micro-hardness measurement for the flats themselves. For values of d/D above this range, the Mallock hardness increased with deformation ratio until it reached several times the Vickers value. This is consistent with the fact that it is almost impossible to flatten asperities completely.

In all the experiments reported it appears that at low deformation ratios, the base diameter is unchanged, the displaced material being accommodated at the shoulders of the truncated asperities but at larger ratios the base of each cone begins to expand so that d may become greater than D. In a surface consisting of a Gaussian distribution of peaks it is likely that those of low deformation ratio will predominate in the initial stages so that the value of Mallock hardness will apply.

Another observation was that under high loads lateral displacement of surface material occurred at locations remote from the asperity. This reached a maximum at about twice and the perturbation extended to about three and a half times the initial base radius[24]. This effect may cause an array of asperities to behave differently from a single cone.

Further studies have been undertaken in which single asperities of tin, aluminium, silver, copper and silver steel have been progressively loaded and then studied by stylus instruments after precise re-location[24]. These confirmed that material from the apex was displaced to the shoulders and that, at higher loadings depending on the apex angle, material from the asperity base began to move outwards as shown in Fig. 13.

Multiple asperities are shown in Figs. 11 and 12 wherein it will be noted that at higher loads (~ 250 N on a circular area of 9.5 mm dia.) plastic flow resulted in the valleys becoming partially filled[25, 26, 27]. The surfaces were prepared by blasting small glass beads (2 μm to 3.75 μm average

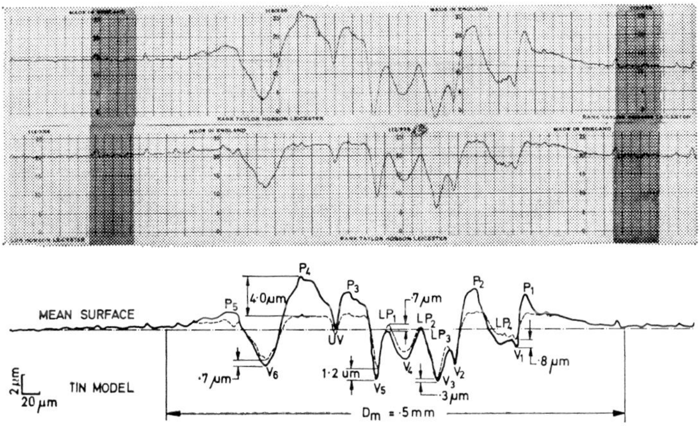

Fig. 11. Deformation of multiple asperities

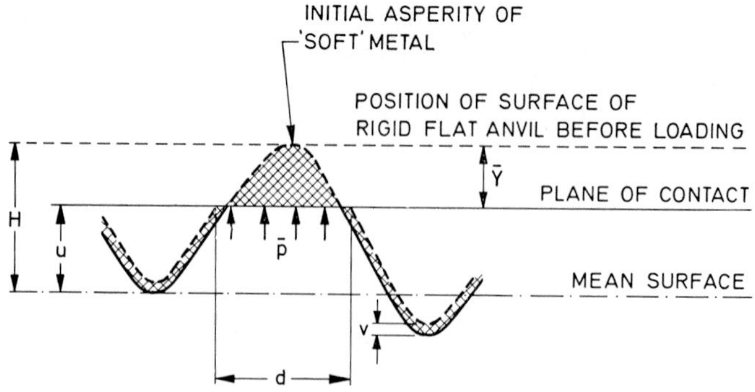

Fig. 12. Idealised asperity deformation

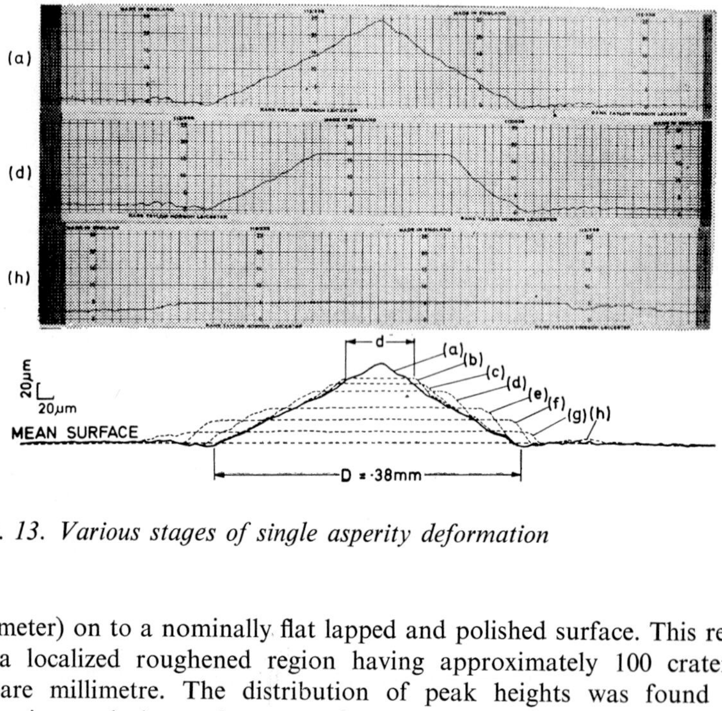

Fig. 13. Various stages of single asperity deformation

diameter) on to a nominally flat lapped and polished surface. This resulted in a localized roughened region having approximately 100 craters per square millimetre. The distribution of peak heights was found to be Gaussian and the surfaces appeared to be random and isotropic. The uniformly roughened zones were of limited area and a radial cross-section of these included about twenty peaks and valleys.

At high loads i.e. $p/\bar{p}$ approaching unity, there was a tendency for the bottoms of all valleys to rise. At light loads the deep valleys did not register any change but the shallower valleys rose.

Over a range of loading up to $p/\bar{p} = 0.025$[26], the area of contact was proportional to applied load with a mean flow pressure equal to the Mallock hardness (i.e. $\bar{p}/2$) but at higher loads A became proportional to $W^{0.67}$. The increasing resistance to further deformation was attributed to the interaction of zones of influence of the asperity bases.

Although it has been shown that the predictions of Tsukizoe and Hisakado regarding the mean radius and average concentration of contact areas are in qualitative agreement with experiment for moderate loads, an increasing disparity occurs at higher loads[27]. The maximum number of contact spots formed was about 100 per square millimetre of nominal area at a value of $p/\bar{p}$ of 0.05, as compared with the predictions of Tsukizoe and Hisakado of 380 spots per square millimetre at $p/\bar{p}$ equal to 0.15. This discrepancy is attributable in part to the failure of the theory to take into account the contribution of displaced material to the support of the applied load.

5. NUMBER OF CONTACT SPOTS

The relationship between the number of contact spots per unit length of a section and per unit surface area deserves some comment. At first sight the number per unit area might be assumed to be proportional to the square of the number per unit length, if the system were isotropic. Statistical arguments however point to a direct linear relation and it has been shown experimentally that:

$$\frac{\text{number of contact spots per unit area}}{\text{number of contact spots per unit length}} = \frac{1}{2\times \text{mean radius of contact spots}}$$

The relation between the RMS roughness σ_0 of the undeformed profile and the roughness σ_t of successive deformed profiles can be presented quite well by

$$\sigma_t = \sigma_0\left(\tfrac{1}{2} - \Phi(t)\right) \tag{17}$$

A comparison between the results for plastic contact at high load of Williamson[28], Uppal and Probert[29] and O'Callaghan and Probert[15, 16, 17, 18] is revealing. Qualitative agreement is reported at low and moderate loads but at a high load such that $p/\bar{p} > 0.5$ the flow pressure increases with load in accordance with

$$p = KW^{0.35} \tag{18}$$

This is again attributed to the interaction of asperities.

6. COMPLIANCE

The extent to which bodies in contact move towards each other under normal loads is as interesting technologically as the true area of contact. Some caution is however necessary regarding terminology because some workers in the field of surfaces employ the term 'compliance' in a different sense from the convention normally adhered to in engineering terminology. It will be recalled that the usual meaning is the reciprocal of stiffness. When applied to a particular component such as a spring it represents the amount of deformation divided by the applied load. With respect to the surface studies referred to here, it signifies the absolute value for the approach of two surfaces (usually as a result of irreversible plastic deformations) under a specified load. It is often more convenient to use the concept of 'separation' which is the distance between the mean planes of the two surfaces in contact. When normalised by dividing by the mean square of the RMS surface roughness it corresponds to t of equation (9).

When a hard flat surface is pressed against a relatively soft rough surface, subsequent examination of the latter surface by a stylus instrument reveals co-linear plateaux where asperities have been flattened and from which it is possible to measure the changes in separation[8].

Under low loads there is some correspondence between measured separations and the predictions of Tsukizoe and Hisakado, but a major disparity occurs at higher loads. The latter behaviour is more closely in accord with the prediction of Williamson et al.[28]. Uppal and Probert

Table 4. Transitions between regions of asperities in direction

Material	$p/\bar{p}$ *Upper limit of Region I*	$p/\bar{p}$ *Lower limit of Region III*
Tin	8.0×10^{-2}	3.4×10^{-1}
Aluminium	2.5×10^{-2}	4.4×10^{-1}
Silver	3.4×10^{-2}	2.0×10^{-1}
Copper	2.7×10^{-2}	—

were able to define three regions, summarised in Table 4. The transitions between regions would appear to be the results of increasing asperity interaction with loading.

7. SURFACES IN RELATIVE MOTION – UNLUBRICATED

In adhesion theory, it is assumed that welds occur at areas of intimate contact and these have to be sheared to permit relative sliding[31]. The coefficient of friction would then be an invariant function of $\bar{p}$. In fact this theory provides unrealistic predictions for the magnitude of the coefficient of friction and it is usually assumed that the area of the junctions grows with the application of tangential force until rupture ensues.

A simpler model can be chosen as the starting point for the descriptive treatment of kinetic friction. This gives a more realistic value for the coefficient of friction but its usefulness is limited to the somewhat artificial case where rupture is everywhere coincident with the direction of the original interface[32].

In reality, the presence of oxide and tarnish films as well as the heterogeneous strength properties of the asperity structures will moderate the degree of intimate contact between surfaces which occurs. Various models have been proposed to explain the interaction of asperities in relative lateral motion and although the coefficients of friction are predictable, at least within a single order of magnitude, wear rates may vary considerably. The energy expended in friction may be accounted for by shearing of the materials of the surfaces without the sheared material becoming separated from the bulk material. The most effective generalisation which can be drawn in our present state of ignorance is the distinction between 'mild' and 'severe' wear[33] or between 'tearing in depth' and 'polishing'[34].

It will be apparent that if the hardness or 'strength' in the bulk material decreases with increasing proximity to the interface the plane of rupture will tend to co-exist therewith so that the amount of disturbance will be limited and may even be confined to the oxide or the contaminating layer. If on the other hand the material at the interface is the strongest present, failure will take place on a plane of weakness remote therefrom so that a measurable thickness of material will be involved in the interaction with corresponding 'severe wear' or 'deep tearing'. Thus materials which tend to work harden may be difficult to lubricate satisfactorily. Confirmation of this has been provided by Jones and Stevens[12] who experimented with age-hardened copper-cobalt alloys ($2\frac{1}{2}$% Co). Severe wear could be changed to mild wear by annealing specimens so as to reverse the hardness gradient in the vicinity of the surface.

8. LUBRICATED SLIDING

Whereas, as is inferred earlier, chemical and metallurgical effects predominate over topological aspects when the contact is run 'dry', lubricated surface systems may or may not be affected by surface texture depending upon the thickness of the liquid lubricant layer. Tallian[35] suggests:

$$\xi = \frac{h}{\sqrt{\sigma_1^2+\sigma_2^2}}$$

where h is equivalent macroscopic film thickness and σ_1 and σ_2 are the RMS roughness of the two surfaces respectively. Assuming a Gaussian distribution of peak heights, it is to be expected that no appreciable asperity contact would occur if

$$h \geqslant 3.5\sqrt{\sigma_1^2+\sigma_2^2} \tag{19}$$

However Tallian reports continuing improvement of the fatigue life with h under elastohydrodynamic conditions at least up to $\bar{\xi} = 10$.

Dawson[36] also provides evidence of the beneficial effect of film thickness on pitting-fatigue life.

As the nominal film thickness becomes smaller, the surface texture will increase in importance and, in systems where Reynolds wedge action is absent, surface topography provides the sole mechanism for hydrodynamic pressure generation.

Consideration of a random surface as an agglomeration of micro-wedges is a possible approach but an alternative way of regarding surface topography which may merit attention for textured surfaces is to record a Fourier analysis of the surface, i.e. the sum of an infinite number of sinusoidal functions of frequency ω, 2ω, 3ω etc. and phases q_1, q_2, q_3 etc. By analogy with the 'describing function' used in control theory, only the fundamental is retained. This is useful for estimation of the hydrodynamic lubrication potential of lightly-loaded surfaces. Assuming the moving surface to be a sine wave of amplitude A and wave length l separated from an ideally flat surface at the point of closest approach by a minimum film thickness h_{min} then

$$h = h_{min} + A\left(1 - \cos\frac{2\pi x}{l}\right)$$

Possible values for A and l may be assumed from Peklenik[1] as shown in Table 5.

Table 5. Potential film forming power on various surfaces

Type of surface	A (μm)	l (mm)	$\frac{P}{\eta U}$
Spark erosion	13	0.45	1.4×10^6
Shaping	18	0.40	7×10^5
Milling	2.7	0.9	6.4×10^7
Fine turning	1.8	0.1	1.5×10^7
Surface grinding	1.3	0.2	6.2×10^7

As long ago as 1950, Salama[37] estimated the load carrying capacity of parallel surfaces due to macro-roughness on the assumption that cavitation occurred in those regions where the film profiles generated a diverging passage. His estimate of average pressure generated between parallel surfaces in our notation is

$$p = \frac{3}{2}\pi\eta U \frac{l}{A^2} S \tag{20}$$

where S was a function $(A + h_{\text{min}})/(A)$ which has a maximum value of 0.108 when $A = 1.7\,h_{\text{min}}$.

Substituting in (20) and re-arranging:

$$\frac{p}{\eta U} = 0.516 \frac{l}{A^2} \tag{21}$$

Values of $p/(\eta U)$ have been calculated from Peklenik's data and inserted in Table 5. Taking the example of a finely-ground finish, lubricated with oil of viscosity 10 CP (10^{-2} pascal seconds) and having a 2 m/s relative velocity, the pressure generated can be estimated to be 1.24×10^6 pascals.

The effect of surface texture as a random effect has been studied in association with bulk Reynolds action by Tzeng and Saibel. They used a Beta distribution to represent roughness as follows

$$f(e) = \frac{15}{16\,c}\left\{1 - \left(\frac{e}{c}\right)^2\right\} \tag{22}$$

where $-c \leq e \leq c$, and concluded that both the load-carrying capacity and the frictional forces are increased considerably when surface roughness is taken into account[38].

Similarly Christensen[39,40,41] viewed the film thickness itself as a stochastic variable to which Reynolds equation can be applied. In applying his treatment to bearings the film thickness was described by an equation having two terms, one describing the bulk geometry and any long wavelength disturbances and the other representing the randomly varying surface roughness. The load carrying capacity[42] estimated for corrugated slider bearings under various conditions tends towards that predicted by Christensen's statistical roughness theory.

It is probable that the ambient pressure surrounding the surfaces may be sufficiently high to prevent cavitation. However as Christopherson pointed out in the discussion of Salama's paper of 1950, the pressure may be much higher as the minimum film thickness is approached than in the divergent regions so that there will be a corresponding difference in viscosity. Thus there can be a net load-carrying capacity.

It is to be noted that this relates to the physical properties of the lubricant which contribute to the lubricating action under what are often referred to as boundary conditions. Indeed, as long ago as 1937 Bradford and Vandegriff[43] pointed out that the pressure-viscosity effect might be the basic cause of the mysterious property called 'oiliness'. More recently Fowless[44,45] applied elastohydrodynamic theory to the interaction of idealised surface asperities and he developed later a thermal elastohydrodynamic theory[46]. Fein and Kreutz[47] applied the Archard-Cowking formula to an idealized single asperity model and demonstrated the feasibility of micro-elastohydrodynamic film lubrication.

In contrast to Poon and Johnson[49] who regard the asperity contact and the elastohydrodynamic film as acting in parallel, Lee and Cheng[48] assumed that the presence of asperities enhanced the elastohydrodynamic action. They simplified the calculations by treating a single asperity on the face of one of two rollers in elastohydrodynamic contact and found that in general the overall film thickness profile increases with increasing asperity height.

The nature and intensity of wear may be expected to be determined by whether asperity deformation is plastic or elastic as determined by equation (15). Hirst and Hollander[50] carried out a series of experiments using a system of stainless steel lubricated by a solution of stearic acid in white oil. They quantified the wear process by determining the quantities, notably temperature and load pressure, which characterises the transition from moderate to high friction. Curves were plotted in which the transition boundaries were indicated on a graph of σ against L of equation (15). The principal boundary, approximately represented by a straight line sloping upwards from left to right is representative of a constant value of the plasticity index. This supports the view that wear is critically dependent on whether asperity deformation is elastic or plastic.

A horizontal line drawn at about $\sigma = 0.02$ μm forms the lower bound to the 'safe' region. This indicates that a surface may be too smooth to be effectively lubricated. This is consistent with the concept of micro-elastohydrodynamic lubrication as outlined for example in reference[40].

9. CONCLUSIONS

There is ample evidence to suggest that surface texture represents an important factor determining the interaction of lubricated surfaces. Caution is necessary in applying ideas developed on macroscopic experience to the microscopic scale represented by individual asperities. Thus Dowson, in a comprehensive review[52], states, that unwarranted idealization of surface topography or lubricant rheology can lead to erroneous predictions.

Archard, in a recent review article[53], envisages a situation in which the surfaces 'sit in a sea of classical hydrodynamic lubrication' with elastohydrodynamic action occurring at individual asperities and considers that surface profilometry particularly when used with relocation will continue to yield useful information.

The writers, in supporting these views, consider that further progress is dependent on a reliable way of characterising the surfaces which are encountered in practical engineering in contrast to the idealized concepts which were necessary to permit the early development of the subject.

ACKNOWLEDGMENTS

The authors wish to acknowledge the contributions of Dr. T.R. Thomas, now at Teesside Polytechnic who has been associated with the work at Swansea in various capacities for a number of years, Dr. P.W. O'Callaghan, now at Cranfield Institute of Technology and Dr. A.H. Uppal, now at the Engineering University Lahore, Pakistan. Dr. R. Stevens and Dr. A.M. Jones are also thanked for their major contributions to the work described.

The support of the Science Research Council by way of research assistants, studentships and finance for the purchase of apparatus is gratefully acknowledged.

Figs. 2 and 3 are based on the work of Thomas and Sayles whose permission to publish is acknowledged with thanks.

REFERENCES

1. J. Peklenik, 'New Developments in Surface Characterisation and Measurement by Process Analysis', *Proc. I. Mech. E.*, **182** Pt. 3K, 1968, pp. 108–126.
2. E.J. Abbot and F.A. Firestone, 'Specifying Surface Quality', *Mechanical Engineering*, **55**, 1933, pp. 569–572.
3. T.R. Thomas, private communication.
4. J.B.P. Williamson, 'Microtopography of Surfaces', *Proc. I. Mech. E.*, **182**, Pt. 3K, 1968, pp. 21–30.
5. T. Tsukizoe and T. Hisakado, 'On the Mechanism of Contact between Metal Surfaces —The Penetrating Depth and the Average Clearance', *Trans. ASME*, **87**, 1965, pp. 666–674 and Part II: 'The Real Area and the Number of the Contact Points', *Trans. ASME J. of Lubrication Technology*, **90**, 1968, pp. 81–88.
6. R. Holm, *Electric Contacts*, Hugo Gebers Forlag, Stockholm, 1946.
7. F.P. Bowden and D. Tabor, *The Friction and Lubrication of Solids*, Clarendon Press, Oxford, 1949 and Part II, 1964.
8. A.H. Uppal and S.D. Probert, 'Mean Separation and Real Contact Area between Surfaces Pressed together under High Static Loads', *Wear*, **23**, 1, 1973, pp. 39–53.
9. D.J. Whitehouse and J.F. Archard, 'Properties of Random Surfaces of Significance to their Contact', *Proc. Roy. Soc.*, **A316**, 1970, pp. 97–121.
10. J.A. Greenwood and J.B.P. Williamson, 'Contact of Nominally Flat Surfaces', *Proc. Roy. Soc.*, A**295**, 1966, pp. 300–319.
11. T.R. Thomas and S.D. Probert, 'Establishment of Contact Parameters from Surface Profiles', *J. Phys. D.*, **3**, 1970, pp. 277–289.
12. M.H. Jones and R. Stevens, 'The Thermal and Metallurgical Aspects of Wear', University College of Swansea, Department of Mechanical Engineering, Engineering Report No. MR/59/72, 1972.
13. A.M. Jones and F.T. Barwell, 'Prediction and Evaluation of Contact Parameters from Surface Analysis', SRC Grant, Report No. B/RG/2116, 1974.
14. A.M. Jones, P.W. O'Callaghan and S.D. Probert, 'Prediction of Contact Parameters

from the Topographics of Contacting Surfaces', to be published in *J. Mech. Eng. Sci.*
15. P.W. O'Callaghan and S.D. Probert, 'Real Area of Contact between a Rough Surface and a Softer Optically Flat Surface', *J. Mech. Eng. Sci.*, **12**, 4, 1970, pp. 259–267.
16. P.W. O'Callaghan and S.D. Probert, 'Prediction and Measurement of Thermal Contact Resistance', ARC Report, **33443** HMT, 1972, p. 296.
17. P.W. O'Callaghan and S.D. Probert, 'Correlations for Thermal Contact Conductance in Vacuo', *ASME, J. of Heat Transfer*, **95**C, **1**, 1973, pp. 141–142.
18. P.W. O'Callaghan and S.D. Probert, 'Thermal Resistance and Directional Index of Pressed Contact between Smooth, Non-Wavy Surfaces', *J. Mech. Eng. Sci.*, **16**, **1**, 1974, pp. 41–55.
19. T.R. Thomas and S.D. Probert, *Int. J. Heat Mass Transfer*, **9**, 1966, pp. 739–753.
20. T.R. Thomas and S.D. Probert, Unpublished paper presented at Conf. on Thermal Conductivity, NPL, London, 1964.
21. A.H. Uppal, S.D. Probert and M.C. Jones, 'Behaviour of Electrical Contact under Static and Dynamic Loads at Cryogenic and Room Temperatures', *Wear*, **13**, 1969, pp. 443–446.
22. H. O'Neill, *Hardness Measurement of Metals and Alloys*, Chapman and Hall, London, 1967.
23. T.R. Thomas, A.H. Uppal and S.D. Probert, 'Hardness of Rough Surfaces', *Nature—Physical Science*, **22**, **3**, 1971, pp. 86–87.
24. A.H. Uppal and S.D. Probert, 'Deformation of Single and Multiple Asperities on Metal Surfaces', *Wear*, **20**, 1972, pp. 381–400.
25. A.H. Uppal and S.D. Probert, 'Topography Changes Resulting from Surfaces being in Contact under Static and Dynamic Loads', *Wear*, **16**, 1970, pp. 261–271.
26. A.H. Uppal, S.D. Probert and T.R. Thomas, 'The Real Area of Contact between a Rough and a Flat Surface', *Wear*, **22**, 1972, pp. 164–183.
27. A.H. Uppal and S.D. Probert, 'Considerations Governing the Contact between a Rough and Flat Surface', *Wear*, **22**, 1972, pp. 215–234.
28. J.B.P. Williamson, J. Pullen and R.T. Hunt, 'Plastic Contact of Surfaces', *Burndy Res. Rep.*, **78** and **79**, 1970.
29. A.H. Uppal and S.D. Probert, 'The Plastic Contact between a Rough and Flat Surface', *Wear*, **23**, 1972, pp. 173–184.
30. P.W. O'Callaghan and S.D. Probert, 'Effects of Static Loading on Surface Parameters', *Wear*, **24**, 1973, pp. 133–145.
31. F.P. Bowden and D. Tabor, *Friction and Lubrication*, Methuen, London, 1956.
32. F.T. Barwell, 'Friction and Wear', *Proc. South Wales Institute of Engineers*, LXXXI, 1966, pp. 33–48.
33. W. Hirst, 'Wear of Unlubricated Metals', *Proc. I. Mech. E. Conf. Lub. and Wear*, 1957, pp. 674–681.
34. I.V. Kragelski, *Friction and Wear* (transl.), Butterworth, London, 1965.
35. T.E. Tallian, 'Rolling Contact Failure Control through Lubrication', *Proc. I. Mech. E.*, **182**, Part 3A, pp. 205–236.
36. P.H. Dawson, 'The Effect of Metallic Contact on the Pitting of Lubricated Rolling Surfaces', *Proc. I. Mech. E.*, **180**, Pt. 3B, 1965, pp. 95–100.
37. M.E. Salama, 'The Effect of Macro-Roughness on the Performance of Parallel Thrust Bearings', *Proc. I. Mech. E.*, **163** (K), 1950, pp. 149–161.
38. S.T. Tzeng and E. Saibel, 'Surface Roughness Effect on Slider Bearing Lubrication', *ASLE Trans.*, **10**, 1967, p. 334.
39. H. Christensen and K. Tonder, 'Tribology of Rough Surfaces—Stochastic Models of Hydrodynamic Lubrication', SINTEF Report No. 10/69–18, 1969.
40. As above 'Tribology of Rough Surfaces—Parametric Study and Comparison of Lubrication Models', SINTEF Report No. 22/69–18, 1969.
41. H. Christensen, 'Stochastic Models for Lubrication of Rough Surfaces', *Proc. I. Mech. E.*, **184**, Pt. 1, 1969–'70, pp. 1013–1022.

42. K. Tonder and H. Christensen, 'Waviness and Roughness in Hydrodynamic Lubrication', *Proc. I. Mech. E.*, **186**, 1972, pp. 807–812.
43. L.J. Bradford and C.G. Vandegriff, 'Relationship of the Pressure-Viscosity Effect to Bearing Performance', *I. Mech. E. Proc. General Discussion on Lubrication and Lubricants*, **1**, 1937, pp. 23–29.
44. P.E. Fowles, 'The Application of Elastohydrodynamic Lubrication Theory to Individual Asperity – Asperity Collisions', *Trans. ASME Series F, Journal of Lubrication Technology*, **91**, 1969, pp. 464–476.
45. P.E. Fowles, 'Extension of the Elastohydrodynamic Theory of Individual Asperity – Asperity Collisions to the Second Half of the Collision', *as above*, **93**, 1971, pp. 213.
46. P.E. Fowles, 'A Thermal Elastohydrodynamic Theory for Individual Asperity – Asperity Collisions', *as above*, **93**, pp. 383–397.
47. R.S. Fein and K.L. Kreutz, 'Contribution to Discussion, Interdisciplinary Approach to Friction and Wear', NASA, SP–181, 1972, pp. 358–376.
48. K. Lee and H.S. Cheng, 'Effect of Surface Asperity on Elastohydrodynamic Lubrication', NASA, CR 2595.
49. K.L. Johnson, J.A. Greenwood and S.Y. Poon, 'A Simple Theory of Asperity Contact in Elastohydrodynamic Lubrication', *Wear*, **19**, 1972, pp. 81–108.
50. W. Hirst and A.E. Hollander, 'Surface Finish and Damage in Sliding', *Proc. R. Soc. London A*, **337**, 1974, pp. 379–394.
51. P.R. Nayak, 'Random Process Model of Rough Surfaces', *Trans. ASME*, **93**F, 1970, pp. 393–407.
52. D. Dowson, 'Transition to Boundary Lubrication from Elastohydrodynamic Lubrication', *Boundary Lubrication, An Appraisal of World Literature*, American Society of Mechanical Engineers, 1969, pp. 229–240.
53. J.F. Archard, 'Elastohydrodynamic Lubrication of Real Surfaces', *Tribology*, **6**, February 1973, pp. 8–14.

Contact of rough surfaces of work-hardening materials

J. Halling and K. A. Nuri

1. INTRODUCTION

All solid surfaces are covered with asperities so that the nature of dry contact is intimately bound up with the behaviour of such deviations from flatness. Such behaviour is of primary importance in the study of electrical and thermal conductivity across interfaces, the stiffness and sealing efficiency of mechanical joints, and in all tribological situations. In particular, the relationship between the real area of contact and the applied load has been a continuing area of study. The laws of friction require a linear relationship between the load and the real area of contact even when the materials are deforming elastically. Archard[1] showed that such behaviour was possible when the nature of the asperities is considered, whilst Greenwood and Williamson[2] produced a classical analysis to this problem. They showed that the height distribution of the asperities was of primary importance in producing a linear load-real area relation for elastic deformation of the asperities. Recent investigations have produced experimental results which give general credence to these theoretical arguments[3].

To date such analyses have been treated by considering either elastic or perfectly plastic behaviour of the asperities. The following analysis shows how the effect of work-hardening may be introduced. The analysis requires a knowledge of certain physical constants which are chosen for the ease of their experimental determination.

2. ON THE CONTACT OF SPHERES

Well established solutions are already available for the elastic and perfectly plastic behaviour of spheres under load. For the present solution we require a more general understanding of this problem. To this end we shall assume a continuous function defining the stress/strain relationship of the material, namely:

$$\bar{\sigma} = B\bar{\varepsilon}^n \tag{1}$$

where $\bar{\sigma}$ and $\bar{\varepsilon}$ are the effective stress and strain values based on the reduced stress and strain increments, B is a scaling constant while the constant n defines the shape, i.e. the degree of work-hardening. It will be noted that when $n=1$ equation (1) represents the elastic behaviour with B representing the elastic constant, whilst when $n=0$ the behaviour is perfectly plastic with B now being the constant yield stress of the material.

In the contact of spheres, or a sphere on a plane, of such a material we shall assume that the state of stress may be characterised by a stress $\bar{\sigma}$ which is some linear function of the mean contact pressure, thus

$$\bar{\sigma} = P/c\pi a^2 \tag{2}$$

where P is the normal load and a is the radius of the current circular contact zone, $1/c$ being the constant of proportionality. This relationship is usually identified as the relation between current yield strength and current hardness with the constant c having a value of about 2.8 [4]. It will further be assumed that the state of strain may be characterised by a strain $\bar{\varepsilon}$ which is a linear function of a/β, where β is the radius of the sphere. This is justified by assuming that the strain is some function of a/β which may be expanded as a power series and since $a \ll \beta$, only the first term need be considered as significant. Hence

$$\bar{\varepsilon} = Da/\beta\,. \tag{3}$$

Combining equations (1), (2) and (3) gives:

$$P\beta^n = Ka^{(2+n)} \tag{4}$$

where

$$K = \pi BcD^n\,.$$

Since B, c, n and D are constants, we expect K to be a constant for a given material but only for particular values of B and n. When $n=1$ (elastic behaviour), equation (4) satisfies the known solution with $K_e = 4E'/3$, E' being the elastic modulus of the system given by:

$$\frac{1}{E'} = \frac{1-\nu_1^2}{E_1} + \frac{1-\nu_2^2}{E_2}$$

where E_1 and E_2 are the elastic moduli of the contacting materials and ν_1 and ν_2 are their values of Poisson's ratio.

For $n=0$ (perfectly plastic behaviour), equation (4) is again in agreement with the known solution with $K_p = \pi H$, H being the indentation hardness of the material.

From equation (4) the constant D of equation (3) now becomes

$$D = \left(\frac{K}{\pi Bc}\right)^{1/n}$$

Hence

$$\bar{\varepsilon} = \left(\frac{K}{\pi Bc}\right)^{1/n} \frac{a}{\beta}. \tag{5}$$

It is well established that the constant c has nearly the same value for most metals [4], therefore, we expect D to be a function of B and n. Since B is purely a scaling factor for the stress-strain curve we expect D to be a function of n only.

The advantage of equations (2) and (5) lies in the simplicity of determining the unknown constants. The following simple tests will provide the evaluation of all these constants.

1. A plane strain compression test will produce the stress/strain curve and thereby the most appropriate values of B and n. In deriving such curves it may be assumed that the strain ratios are constant and the principal axes do not rotate between successive strain increments, whence

$$\bar{\sigma} = [\tfrac{1}{2}(\sigma_1 - \sigma_2)^2 + (\sigma_2 - \sigma_3)^2 + (\sigma_3 - \sigma_1)^2]^{1/2}$$

and

$$\bar{\varepsilon} = \tfrac{1}{3}[2(\varepsilon_1 - \varepsilon_2)^2 + (\varepsilon_2 - \varepsilon_3) + (\varepsilon_3 - \varepsilon_1)^2]^{1/2}$$

During such a test the hardness may also be measured at each strain increment and there by produce a value for c.

2. A sphere pressed against the material at various loads will define the contact area and thereby produce a value for K. Moreover, using equations (2) and (5), these results may be used to redraw the stress/strain curve to obtain a check on experimental validity.

3. EXPERIMENTAL

Indentation experiments were conducted with materials of different degrees of work-hardening, namely, pure lead, aluminium, copper and mild steel. The aluminium and copper specimens, originally cold drawn, were first annealed at 340 °C and 450 °C respectively, for two hours and then furnace cooled. Stress/strain curves were obtained using the plane strain compression tests for all the materials used in these identation tests. A further sequence of compression tests were also performed for determining the indentation hardness at each increment of strain.

Steel balls mounted in special housings which may be easily fitted to the loading rig were used as indenting spheres. Five sizes of spheres were used: 0.635, 1.27, 1.905, 2.54, and 3.17 cm in diameter. Loading was effected by dead weights using a simple lever system for transmitting the load to the indenting sphere. The surfaces of the test specimens were first lapped flat, cleaned and then coated with a thin film of copper, using

Stead's reagent, to facilitate the measurement of the indentation area, particularly for low loads. The slightest pressure on the surface showed up as a bright spot. Measurement of the size of the indentation area was made with a low power travelling microscope. A further series of indentation tests were carried out to determine the normal approach between the indenting sphere and the specimen at each increment of load.

4. RESULTS AND DISCUSSION

The values of the index n for the various materials used were obtained by fitting theoretical curves to the stress/strain curves obtained from the compression tests. It will be appreciated that in practice only modest strains are involved during the process of contact between surfaces, and therefore only the initial part of the stress/strain curves will be of relevance to the present investigation. The following relationships were found to correlate with the experimental results up to 20% strain:

Pure lead	$\bar{\sigma} = 200 \quad \bar{\varepsilon}^{.15}$	(kg/cm^2)
Annealed aluminium	$\bar{\sigma} = 2010 \quad \bar{\varepsilon}^{.25}$	,,
Annealed copper	$\bar{\sigma} = 6220 \quad \bar{\varepsilon}^{.55}$	,,
Mild steel	$\bar{\sigma} = 715000\, \bar{\varepsilon}$	,,

Using the results of the indentation experiments, the values of the constant K of equation (4) were obtained by plotting $P\beta^n$ against $a^{(2+n)}$ (Fig. 1). The results clearly indicate a linear function as predicted by the simple argument used in deriving equation (4). It can also be seen that the results are almost unaffected by the size of the indenting sphere; however, with tests performed on brass specimens, varying the indenter's size produced considerable scatter in the results. This was attributed to the effects of grain size of this non-homogeneous alloy and consequently the tests with brass were disregarded.

Results for a variety of metals in the tests for the determination of c are shown in Fig. 2. The usual assumption that $c = 2.8$ is clearly validated in all cases.

The results of Fig. 1 can perhaps be more clearly demonstrated by plotting the values of the constant K in non-dimensional form against the index n. This is done in Fig. 3 where the dimensionless parameter K/B is plotted against n. The values of K/B in the extreme cases when $n = 0$ and $n = 1$ are theoretically determined as mentioned earlier. The results show clearly that K/B is linearly related to n. The significance of such a relationship will become more apparent when we refer to equation (5). Substituting the K/B function and the value of c in this equation yields a relationship between $\bar{\varepsilon}$ and a/β of the form

$$\bar{\varepsilon} = D\, a/\beta = (1 - .9\, n)^{1n/}\, a/\beta$$

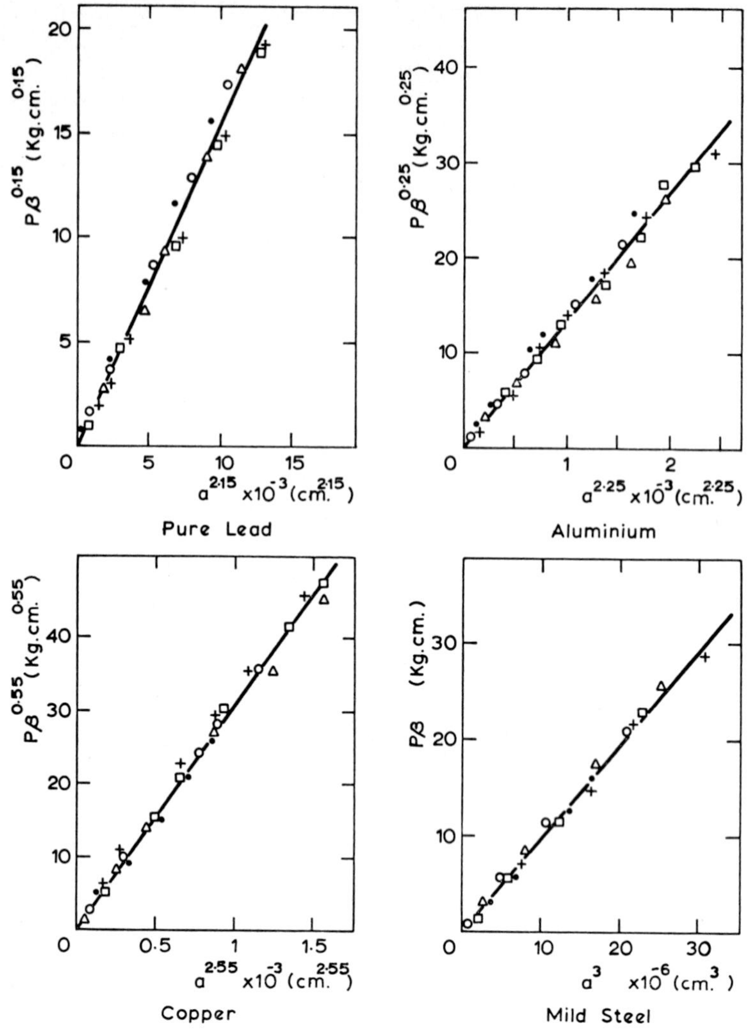

Fig. 1. The variation of $P\beta^n$ with $a^{(2+n)}$ obtained from indentation tests for various metals

● 0.635 ○ 1.27 △ 1.905 □ 2.54 + 3.17 cm diameter spheres

The relation between $\bar{\varepsilon}$ and a/β is often considered independent of the type of material used and empirical formulae are generally given; Tabor[4] gives $\bar{\varepsilon} = .2a/\beta$ while Richmond et al.[5] gives $\bar{\varepsilon} = .32a/\beta$. The result reported here, however, suggests a dependence of the $\bar{\varepsilon}/(a/\beta)$ relation on n. It is worth noting that for the range of values of n associated with ordinary metals, the values of D are comparable to those quoted by the previous

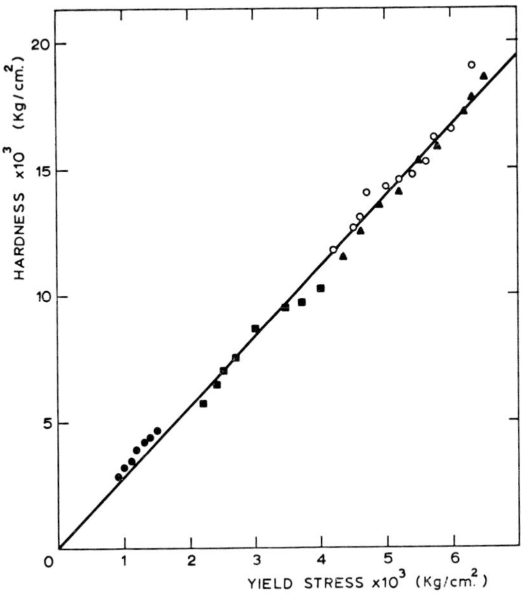

Fig. 2. Variation of the hardness with yield stress for various metals
● aluminium ■ copper ○ brass ▲ mild steel

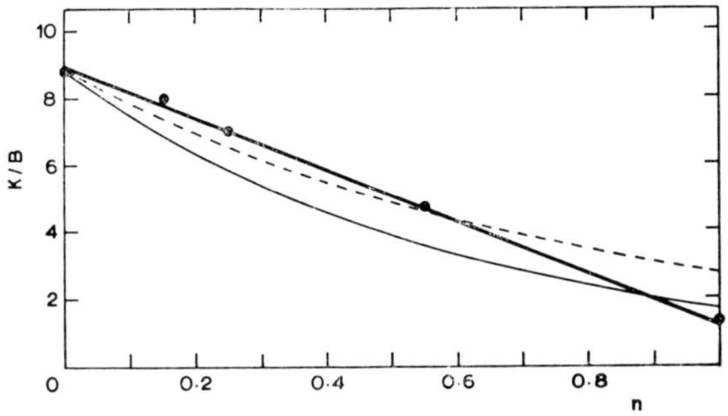

Fig. 3. The variation of K/B with degree of work-hardening
—●— experimental ——— after Tabor - - - - after Richmond et al.

investigators, e.g. for $n = .2$, $D = .37$ and for $n = .6$, $D = .274$. Fig. 3 also shows the variation of K/B with n using the assumption that $D = .2$ or $D = .32$ in equation (5). It is interesting to find that while the results are quantitatively comparable with those determined experimentally, the trends are clearly different.

In Fig. 4 a comparison is made between the stress/strain curves obtained from compression tests and from indentation experiments. In the latter case the effective stress is $p/2.8$, where p is the indentation pressure. Curves where the strain is determined by using Tabor's formula $\bar{\varepsilon} = .2a/\beta$ and Richmond's formula $\bar{\varepsilon} = .32a/\beta$ are also plotted. In general, we find that

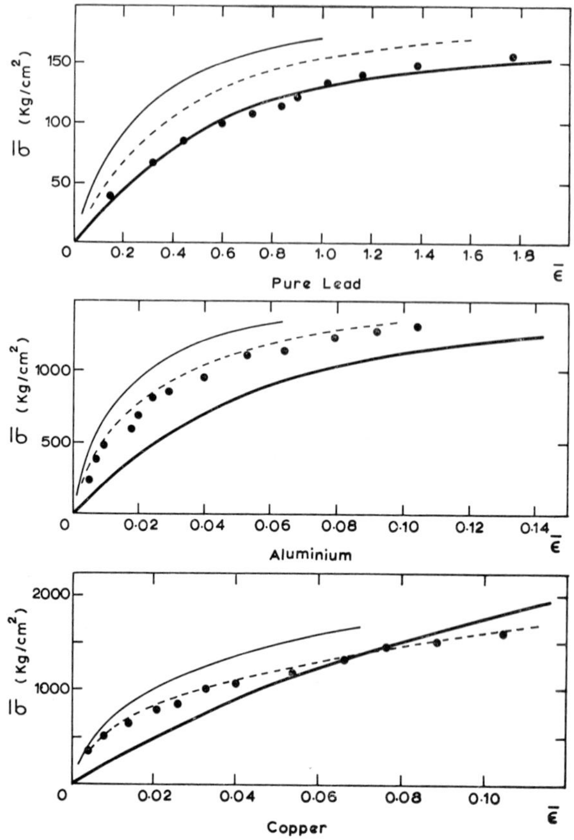

Fig. 4. Comparison of the stress/strain curves
——— compression tests ●●● indentation tests ——— Tabor's formula
- - - - Richmond's formula

the curves drawn for those values of D predicted by the present investigation are in better agreement with the compression tests' curves than the curves obtained by taking D as a constant. This emphasizes the validity of the $\bar{\varepsilon}/(a/\beta)$ relation presented here.

5. THE CONTACT OF ROUGH SURFACES

Consider a smooth flat plate pressed on to a rough flat surface to produce a number of discrete areas of contact at the tips of the asperities. The asperities will be assumed to be spherical in shape and for simplicity it will be considered that they are all of the same radius β. Following the arguments introduced by Greenwood and Williamson[2], but using the foregoing analysis for spherical contacts we may write for any single asperity

$$A_i = \pi a^2 = \lambda\pi\beta\delta \tag{6}$$

where A_i is the area of contact and δ is the normal approach between the asperity and the plate. λ is an area factor which depends on the material's degree of work-hardening. The two known solutions are for $n=0$ when $\lambda=2$ and for $n=1$ when $\lambda=1$.

Substituting for a in equation (4) gives the contact load P_i as

$$P_i = K\lambda^{\left(1+\frac{n}{2}\right)}\beta^{\left(1-\frac{n}{2}\right)}\delta^{\left(1+\frac{n}{2}\right)} \tag{7}$$

Consider the smooth flat surface to be loaded such that it is at distance d from some datum plane in the rough surface. For any contacting asperity of initial height z above this datum, the normal approach will be $(z-d)$, so that

$$A_i = \lambda\pi\beta(z-d)$$

and

$$P_i = K\lambda^{\left(1+\frac{n}{2}\right)}\beta^{\left(1-\frac{n}{2}\right)}(z-d)^{\left(1+\frac{n}{2}\right)}$$

The probability of making contact at any given asperity of hieght z is

$$\text{prob.}\ (z<d) = \int_d^\infty \phi(s)\,\mathrm{d}z$$

where $\phi(z)$ is the probability function of the asperity height distribution. The total real area of contact A where there are N asperities altogether thereby becomes

$$A = \lambda\pi N\beta\int_d^\infty (z-d)\,\phi(z)\,\mathrm{d}z \tag{8}$$

and the total load P will be

$$P = NK\lambda^{\left(1+\frac{n}{2}\right)}\beta^{\left(1-\frac{n}{2}\right)}\int_d^\infty (z-d)^{\left(1+\frac{n}{2}\right)}\phi(z)\,\mathrm{d}z\,. \tag{9}$$

It is usual practice to put $N=\eta\mathscr{A}$, where η is the surface asperity density and $\mathscr{A}$ is the nominal area, and to standardise the variables in terms of σ, the standard deviation of the height distribution. Thus, equations (8) and (9) become:

$$A = \lambda \pi \eta \mathscr{A} \beta \sigma F_1(h) \tag{10}$$

and

$$P = \eta \mathscr{A} K \lambda^{\left(1+\frac{n}{2}\right)} \beta^{\left(1-\frac{n}{2}\right)} \sigma^{\left(1+\frac{n}{2}\right)} F_{\left(1+\frac{n}{2}\right)}(h) \tag{11}$$

where

$$F_x = \int_h^\infty (s-h)^x \, \phi^*(s) \, \mathrm{d}s \, ,$$

$\phi^*(s)$ being the standardised height distribution and $h = d/\sigma$, $s = z/\sigma$.

For any given height distribution the integrals may be evaluated to obtain the appropriate values of A and P.

It is instructive to consider an exponential distribution; this readily provides an analytic solution since,

$$F_x(h) = x! \, \mathrm{e}^{-h} .$$

In this case equations (10) and (11) yield a load-area relation given by

$$\frac{P}{A} = \frac{K}{\pi} \left(\frac{\lambda \sigma}{\beta}\right)^{n/2} \left(1 + \frac{n}{2}\right)! \tag{12}$$

which for a given surface clearly indicates a linear relationship between the load and the real area of contact.

For any other distribution we have

$$\frac{P}{A} = \frac{K}{\pi} \left(\frac{\lambda \sigma}{\beta}\right)^{n/2} \frac{F_{(1+n/2)}(h)}{F_1(h)} \tag{13}$$

We shall now consider the form of equation (12) as n varies from 0 to 1, i.e. assuming an exponential distribution of asperities. It is immediately apparent that we must define values for σ/β for the surface and values of λ and K at each value of n. The results for the relation between the area of contact A and the normal approach δ for a single sphere, 1.905 cm in diameter, pressed against various materials are shown in Fig. 5. The anticipated linearity defined by λ in equation (6) is immediately apparent and Fig. 6 indicates the resulting variation of λ with n. A typical value for σ/β of 0.02 has been used and K/B has been used as specified by Fig. 3. These substitutions lead to a relationship between P/AB and n of the form shown in Fig. 7. The bounding curves for $\lambda = 1$ or $\lambda = 2$ at all values of n are also shown.

We first note that the results of Fig. 7 are only very marginally dependent on the variation of λ with n. It should however be stressed that for other

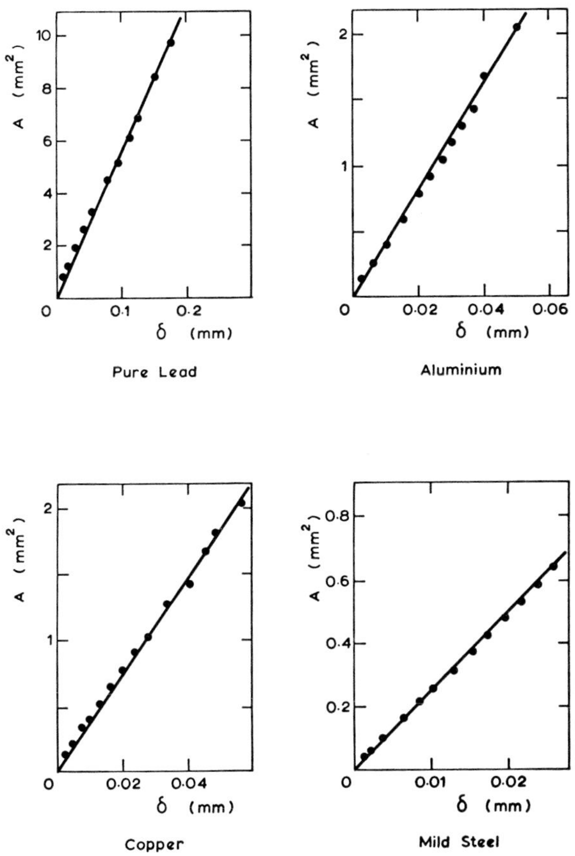

Fig. 5. Variation of the real area of contact with normal approach for various metals

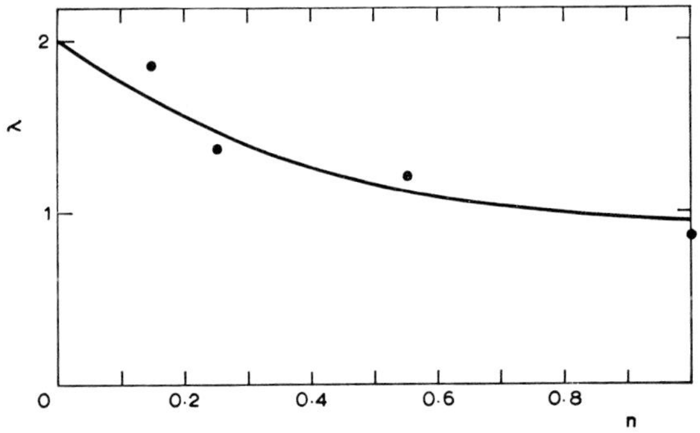

Fig. 6. Variation of the area factor with degree of work-hardening

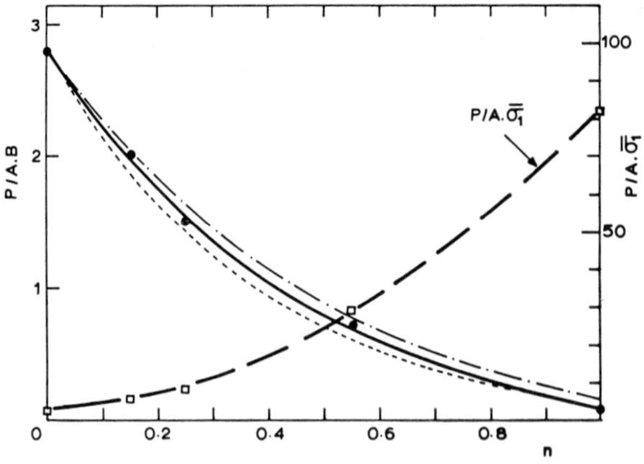

Fig. 7. Variation of the real pressure with n for exponential distribution of asperities
—□— for actual values of λ
– – – – for $\lambda = 1$ ——·—— for $\lambda = 2$

asperity distributions this may no longer be true, although it would be anticipated that many engineering surfaces would be reasonably represented by the general behaviour of an exponential type of distribution. Secondly, the form of Fig. 7 must be interpreted with some caution by a due recognition of the non-dimensional character of the ordinates. If one substitutes the values of B obtained for the materials studied in these tests, one finds that the P/A values show a progressive increase with increasing n, although at the lower values of n the increases are very modest.

A somewhat better physical interpretation of Fig. 7 may be obtained by considering a range of materials having various values of n but an identical value of 0.1% proof stress $\bar{\sigma}_1$. The value of B is then defined by

$$B = \bar{\sigma}_1/(0.001^n)$$

Substituting such values of B and plotting the $P/A\bar{\sigma}_1$ values against n yields the curve shown in Fig. 7 where the effect of an increasing work-hardening index is clearly indicated. It is immediately apparent that the work-hardening indices associated with common metals can produce an order of magnitude change in the stiffness of the texture.

REFERENCES

1. J.F. Archard, 'Elastic Deformation and the Laws of Friction', *Proc. Roy. Soc.*, **A243**, 1957, pp. 190ff.
2. J.A. Greenwood and J.B.P. Williamson, 'Contact of Nominally Flat Surfaces', *Proc. Roy. Soc.*, **A295**, 1966, pp. 300ff.

3. K.A. Nuri and J. Halling, 'The Normal Approach between Rough Flat Surfaces in Contact', Wear, 32, **81**, 1975.
4. D. Tabor. *The Hardness of Metals*, Oxford University Press, 1951.
5. O. Richmond, H.L. Morrison and M.L. Devenpeck, 'Sphere Indentation with Application to the Brinell Hardness Test', *Int. J. Mech. Sci.*, **16**, 1974, pp. 75ff.

Factors influencing the real trend of the coefficient of friction of two elastic bodies rolling over each other in the presence of dry friction

H. Krause and A. Halim Demirci

1. INTRODUCTION

One of the basic tasks of an engineer is to determine how the actual phenomena occurring in nature can be simplified for solving a certain problem. In technical mechanics, it has been found practical to consider the bodies in question as a continuum. The mechanical properties will, as a result, then be characterised by some few material characteristics such as the modulus of elasticity and Poisson's ratio. In this way, it has been possible to solve very many technical problems with sufficient accuracy. It is, however, necessary in the answering of certain questions to consider further properties of the bodies.

This is also the case in the friction system which will be dealt with in more detail here: two elastic bodies – cylindrical steel discs of the same size, driven

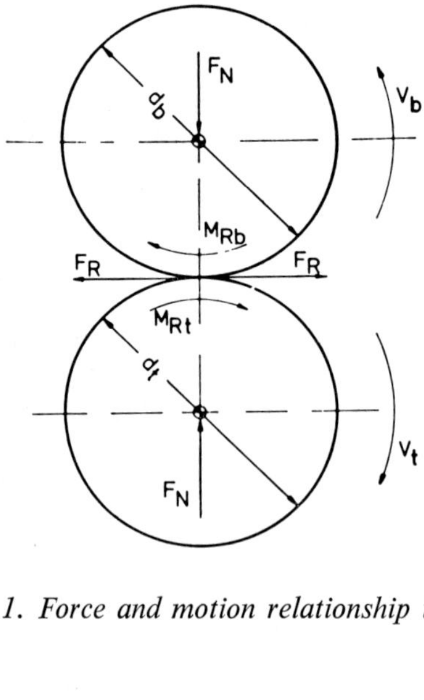

Fig. 1. Force and motion relationship in rolling friction

independently – roll over each other without lubrication. The rotational speed of both discs is constant but differs for each disc.

They are pressed together with a normal force F_N. As a result of the dry friction, the tangential or frictional force F_R acts in the common contact area of the bodies (Fig. 1). The frictional force F_R is proportional to the normal force F_N:

$$F_R = \mu F_N$$

The proportionality factor μ is generally known as the coefficient of friction.

According to Coulomb's laws, this coefficient of friction μ is a constant, the size of which is determined only by the material pairing. It is supposedly independent of the contact area, the normal force and the relative velocity or slip.

Although over the course of time many authors have been able to prove independently of each other that these postulates do not apply, Coulomb's laws are still quoted today as being generally valid. A further quantity which is also generally not taken into account is the stress duration or, in other words, the time over which the two bodies roll over each other in the presence of dry friction.

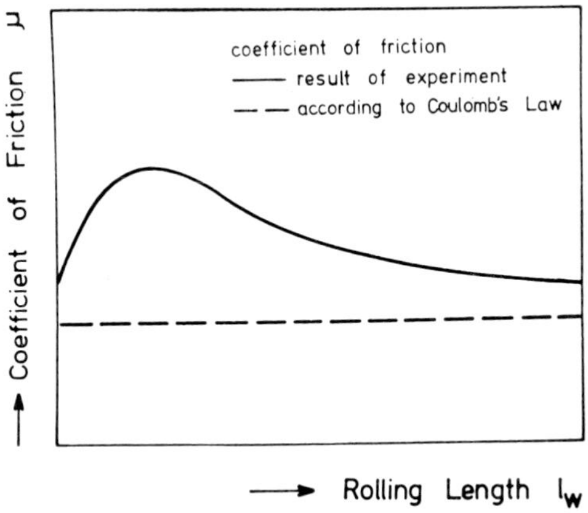

Fig. 2. Coefficient of friction as a function of rolling length l_w

Fig. 2 shows the trend, determined by experiment, of the coefficient of friction μ as a function of the stress duration, expressed here as rolling length l_w, compared with the constant coefficient of friction according to Coulomb. In order to explain the actual trend of the coefficient of friction, reference should be made to the quantities involved in the occurrence of friction and to the real material structure.

2. FACTORS INFLUENCING A FRICTION SYSTEM

A friction system is formed from the entirety of the quantities involved in the occurrence of friction. The plurality of these quantities can be seen in

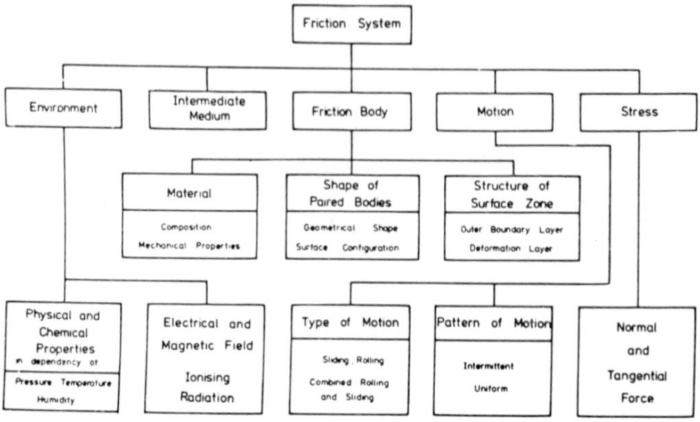

Fig. 3. Factors influencing a friction system

Fig. 3. In spite of their heterogenity, they can be divided into five main groups according to their effect:

1. Friction body.
2. Stress.
3. Motion.
4. Intermediate medium.
5. Environment.

The fact that the one or the other influencing factor can be allocated to more than just one of these groups only shows how complex the occurrence of friction is.

This diversity of the influencing factors makes it quite apparent that frictional force cannot be determined by a generally valid law.

3. REAL STRUCTURE OF THE SURFACE LAYERS OF METALLIC BODIES

The ferrous materials used in applied science do not have a pure metallic surface. They are usually covered by a thin absorption layer. This forms, together with the oxide or reaction layer, the outer boundary layer.

Mechanical stress creates additional property changes in the surface layers of the materials. The so-called inner boundary layer is formed (Fig. 4).

The physical and chemical properties of these surface layers deviate considerably from those of the metallic base material.

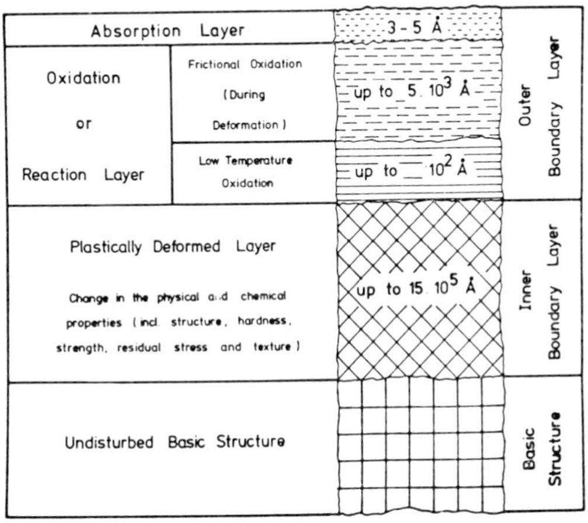

Fig. 4. Schematic representation of the structure of friction body surfaces in the zone of compact surface layers – with reference to[1,2,3] *(The layer thickness data are based on rolling friction investigations performed by the author on plain C steels)*

4. THE INFLUENCE OF THE SURFACE LAYERS ON FRICTIONAL BEHAVIOUR

4.1 *Test procedure*

The tests were performed on a rolling friction test machine of the type Bugarcic[4] which gives very good reproducibility of measured values. With this machine, the following influencing factors can be varied:

1. Normal force	F_N	= 250–700 N
2. Slip	s	= 0–1%
3. Air humidity	p_D	= 80–4000 Pa
4. Rolling element		
Diameter	d	= 52 mm
Width	b	= 8–15 mm

The two test rollers are driven separately. A rotational speed stability of ±0.01% is attained. The slip can be set to an accuracy of ±0.01%.

A range of various hypo-eutectic steels was used for the investigation.

4.2 *Influence of surface roughness on frictional and wear behaviour*

The actual contact area between two friction bodies reduces considerably as the roughness is increased. The greater the surface roughness, the greater are the specific surface pressure and the frictional force (Fig. 5).

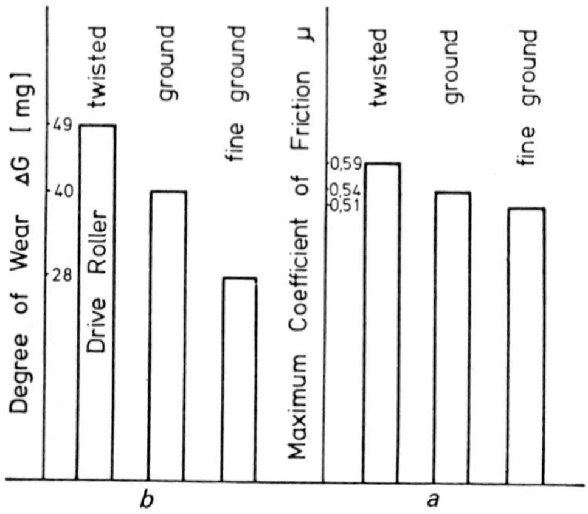

Fig. 5. Influence of surface roughness: — on the coefficient of friction (a) and — on the degree of wear (b)
F_N = 700 N; $p_D \leqq$ 80 Pa; s_m = 0,5%; l_w = 6800 m
Material: plain carbon steel $C = 0.76\%$ normally annealed

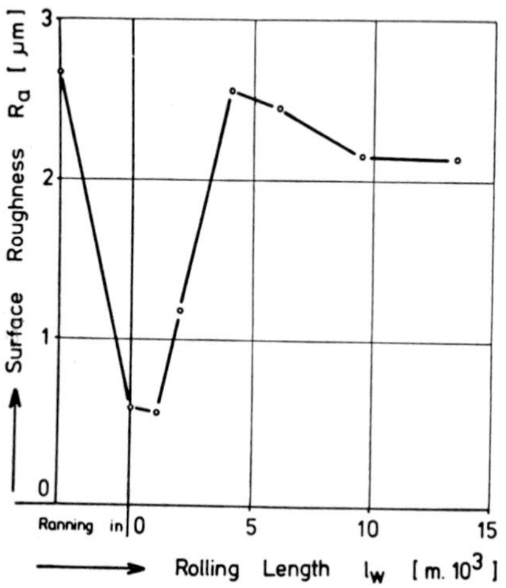

Fig. 6. Surface roughness as a function of rolling length
F_N = 700 N; $p_D \leqq$ 80 Pa; s_m = 0.5%
Material: Ck 45

The high specific surface pressure causes a high degree of plastic deformation of the small contact zones which, in turn, leads to a high intensity of frictional oxidation.

Because of the greater frictional force, rough rolling surfaces wear faster and to a higher degree than smooth ones (Fig. 5).

Independent of different initial roughnesses, the surface roughness of the two bodies will become equal after a certain rolling length.

All tests have shown that the roughness is initially considerably reduced with the formation of oxidic surface films. After the compact surface film has been destroyed, the roughness again increases rapidly, attains its maximum value and then tends towards a quasi-constant value (Fig. 6).

4.3 *Influence of the outer boundary layer on frictional behaviour*

As a result of rolling friction, the friction bodies are plastically deformed in the contact zone in the direction of the tangential stress. A strong chemical activation of the surface zones then occurs. The activated metallic zones create a spontaneous and visible oxide growth. Fink[5,6] calls this phenomenon 'frictional oxidation'.

Temperature and humidity, in this case, either promote or hinder oxidation in an extremely complex manner.

A visible frictional oxidation of the surfaces cannot be abserved during

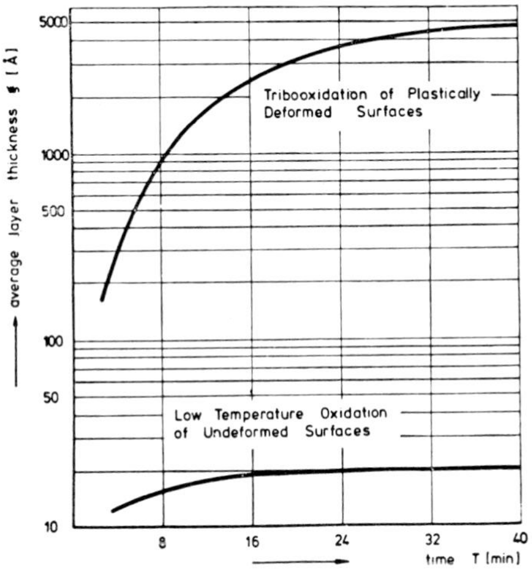

Fig. 7. Mean frictional oxidation layer thickness as a function of the test duration, compared with the low temperature oxidation of non-deformed surfaces

$F_N = 700$ N; $p_D \leqq 80$ Pa; $s_m = 0.5\%$

Material: Ck 45; braking roller [26]

pure rolling friction. When slip is present between the rollers, visible oxide layers are spontaneously formed, the adhesion of which appears to be dependent on the base metal.

The intensity of the chemical activation and thereby the thickness of the reaction layer (outer boundary layer) is strongly influenced by the plastic deformation (Fig. 7).

The dependency of the coefficients of friction on air humidity and slip in the compact surface zone can be seen from Fig. 8.

The air humidity is not the direct cause of the reduction in the coefficient

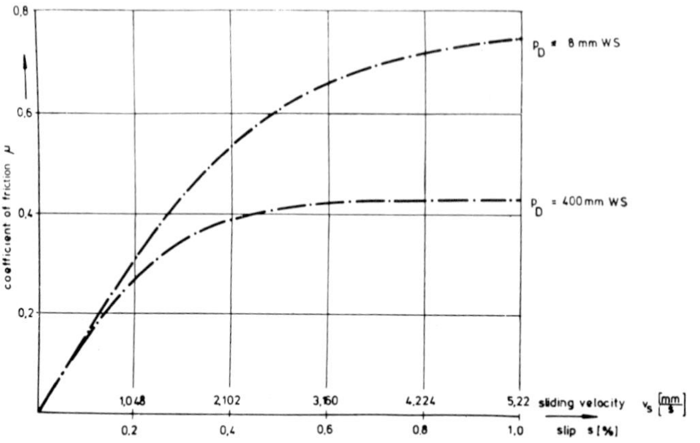

Fig. 8. Coefficient of friction as a function of slip and air humidity in the zone of compact oxidic surface layer [13]
Material: Ck 45; F_N = 500 N

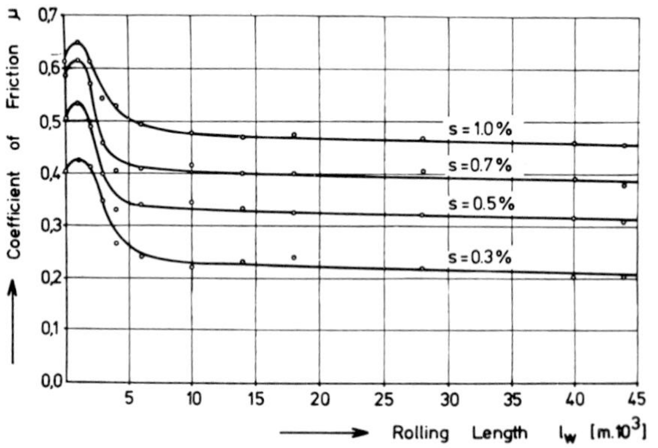

Fig. 9. Coefficient of friction as a function of rolling length
F_N = 700 N; $p_D \leqq$ 80 Pa
Material: plain carbon steel C = 0.5%

of friction. Investigations on steels which do not tend towards the formation of a reaction layer (e.g. V 2 A steel) showed that obsorbed water molecules do not have any influence worth mentioning on the frictional behaviour[7].

The different coefficients of friction can be explained by the dependency of the reaction layer formation on the air humidity.

As the surface layer is destroyed, the coefficient of friction reduces and tends towards a quasi-constant value (Fig. 9).

4.4 *Influence of the inner boundary layer on the frictional force*

It was only relatively recently that attention was given to the influence of the crystal orientation on the frictional behaviour. The investigations were performed on monocrystal specimens. Different values for the coefficient of friction were measured in various crystallographic directions and planes[8].

Alison and Wilman[9] determined that the coefficient of friction of metals with a cubic face centred lattice structure is greater than that of hexagonal structure metals.

Gwathmey, Leidheiser and Smith[10] were able to show on copper that the coefficient of friction also depends on the respective lattice plane of the single crystal. Thus the coefficient of friction of the dense (111) plane is smaller than that of the (100) plane.

Beyond this, investigations carried out by Buckley and Johnson[8] on various copper monocrystal orientations that the coefficient of friction in a given lattice plane is also dependent on the direction. The smallest value for the (111) plane was measured in the [110] direction.

It was shown in later tests, too, that the value of the coefficients of friction of hexagonal and cubic structure metals is smallest in privileged slip planes[11, 12].

These investigations were carried out under an ultra-high-vacuum on monocrystal specimens in order to bring the rolling surfaces into pure metallic contact with each other and to clearly be able to determine the influence of the orientation.

The metals used in applied science are not a continuum. They are neither monocrystals, nor do they have an ideal surface (Fig. 4).

The properties of the real crystal lattice deviate considerably from those of the ideal lattice.

When metallic materials are plastically deformed, a change of almost all physical and chemical properties occurs. This has a decisive influence on the frictional and wear behaviour of the materials. It can be seen that the hardness distribution, the apparent crystallite size and the lattice distortion, together with the crystal orientation (frictional texture) in the plastically deformed surface layers, all have to be known in order to clarify the frictional and wear behaviour.

4.4.1 *Hardness distribution in the surface layers of the rolling elements*

As a result of rolling friction, high degrees of deformation up to the limit malleability are reached in the metallic surface layers. The plastic deformations, with the simultaneous cold work-hardening, result in an increase in hardness of the surface layer.

The relationship between the hardness and the frictional force has already been determined[13].

The hardness distribution in the surface layer is shown in Fig. 10.

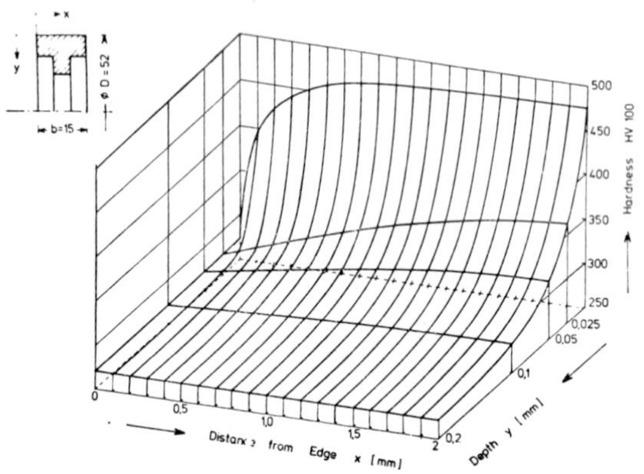

Fig. 10. Hardness distribution in the rolling surface zone
$F_N = 700$ N; $p_D \leqq 80$ Pa; $s_m = 0.5\%$; $l_w = 17000$ m

4.4.2 *Apparent crystallite size and lattice distortion from the spread of X-ray reflections*

Sharp interference in radiographic representations is a characteristic for crystallites which have uniform lattice constants and are to a large extent free from lattice dislocations. In the case of stress created by rolling with slip, lattice spaces are plastically deformed. The changes in the crystallite size and the changing of lattice parameters as a result of lattice distortion are the consequential effects of frictional forces. The lattice distortion and the change of crystallite size cause a spread of the *X*-ray reflections.

The detection of the interference spread can provide information on lattice dislocation and hence on the degree of plastic deformation and the depth of deformation.

According to Scherrer[15], the following equation is valid for the reflex spread as a result of crystallite size:

$$\beta_K = \frac{k \cdot \lambda}{D \cos \Theta}$$

β_K = reflex spread resulting from crystallite size
λ = wave length
Θ = diffraction angle
D = mean crystallite thickness at right angles to the reflected lattice plane
k = constant (at integral width $k = 1$).

The following is valid for reflex-spreading lattice dislocations:

$$\beta_v = \eta \operatorname{tg} \Theta$$

β_v = reflex spread resulting from lattice dislocations
η = apparent lattice distortion.

In the friction system considered here, the influences of the lattice distortion (η_v) and the crystallite size (D) are superimposed on the reflex spread (β_v and β_K). M. Trömel and H. Hinkel[16] have established an approximation function for separating the two influences:

$$\beta = \frac{\beta_K + \sqrt{\beta_K^2 + 4\beta_v^2}}{2}$$

The determination of D and η can be performed by calculation

$$D = \frac{k\lambda}{\beta_1^2 - q\beta_2^2}\left(\frac{\beta_1}{\cos\Theta_1} - q\,\frac{\beta_2}{\cos\Theta_2}\right)$$

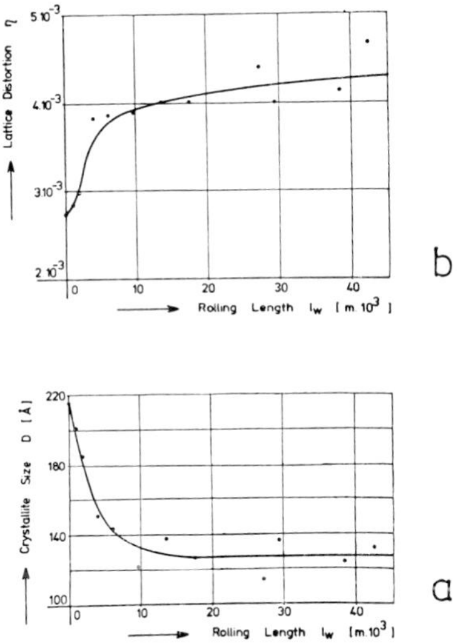

Fig. 11. Variation of the apparent crystallite size D(a) and of the lattice distortion η(b) under rolling stress
F_N = 700 N; $p_D \leqq 80$ Pa; s_m = 0.5%
Material: plain steel C = 0.5%

where

$$q = \frac{\mathrm{tg}^2\,\Theta_1}{\mathrm{tg}^2\,\Theta_2}.$$

$$\eta = \sqrt{\frac{\beta_1^2 \cdot \beta_2 \cdot \cos\Theta_1 - \beta_2^2 \cdot \beta_1 \cos\Theta_2}{\beta_2 \sin\Theta_1\,\mathrm{tg}\,\Theta_1 - \beta_1 \sin\Theta_2\,\mathrm{tg}\,\Theta_2}}$$

The graphs in Fig. 11 show the change of the apparent crystallite size and the apparent lattice distortion as a function of the rolling length. It can be seen that the lattice dislocations increase considerably with the stress while the apparent crystallite size asymptotically approaches a limit value.

If the graph in Fig. 11 is compared with the hardness trend or the texture change over the duration of stress, it can be seen that a final stage in the plastic deformation has been reached.

The relatively large scatter of the measured values is created by the influence of the measuring equipment and the anisotropy of the recorded (110) plane (texture).

4.4.3 *Crystal orientation (texture) in the surface layers of the rolling elements*

4.4.3.1 *Procedure for texture investigation.* For the texture investigations, the rolling elements were bored out to a thickness of 0.3 mm after the other tests had been completed. During the turning operation it was necessary to avoid the generation of heat by providing continuous cooling in order to prevent possible recrystallisation. The ring remaining after the boring out of the rolling element was cut open and flattened. It can be assumed that no additional texture changes occurred during flattening as the rollers were of sufficiently large diameter (52 mm) and were also adequately bored out (0.3 mm). This process has already been mentioned in technical publications. Stüwe uses it for investigations into the texture of wire of 4 to 5 mm diameter[17]. Specimens measuring 20 mm × 15 mm were manufactured from the cut and flattened metal ring. The investigations were carried out up to a polar angle of 80° whereby a correction of the measured intensities was provided with the aid of a texture-free specimen (Fe powder specimen). This correction is performed by inscribing polygonal figures through the recorded lines and determining the correction factors at intervals of 5° polar angle. Apart from this, the background radiation was subtracted.

In order to cover a greater area of the rolling surface, the specimen was moved to and fro.

The evaluation of the measured values was performed by means of a computer programme [18,19,20,21].

4.4.3.2 *Investigation results.* Fig. 12 shows (110) and (200) polar point diagrams of a burnished rolling element surface. They were recorded after a running-in length of 1450 m under the following test conditions:

$F_N = 550$ (N) (Normal force)
$s = 0.02$ (%) (Slip)
$p_D \leq 80$ (Pa) (Water vapour partial pressure).

The burnished rolling elements form the basic elements for further investigations.

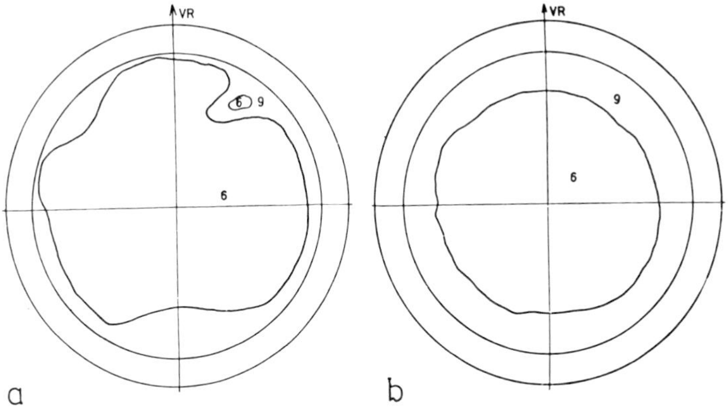

Fig. 12. (110)- and (200) polar point diagrams (a and b) of a burnished roller surface (the intensity increments in the polar point diagrams represent 20% of the mean intensity)

The (110) and (200) figures of the basic elements show, particularly in the middle of the polar point diagram (up to about 65° arc), an almost random orientation distribution in the surface zone (Fig. 12).

The increase of the density towards the edge can be partly attributed to correction influences and was not taken into consideration for the identification of the textures.

Already at the commencement of stress, the so-called frictional texture forms, even if only in part. Under further stress, the position of the maxima remains unchanged but their intensity increases as the deformation increases. Over a rolling length of approximately 17 000 m the intensity of the two maxima at position A and B attains its highest value (Fig. 13). The frictional texture has attained its ultimate condition. This ultimate stable condition remains unchanged under further stress. One can speak of an ultimate texture[22, 24] which is characterised by the ideal condition $(13\bar{2})$ [313].

As a deformation process simultaneously gives rise to texture changes, it can be assumed that this ultimate condition represents the maximum degree of deformation of constant stress.

The transitional forms of the frictional texture can, with respect to the two limit types—ultimate texture (Fig. 13) and initial texture (Fig. 12)—be quantitatively described.

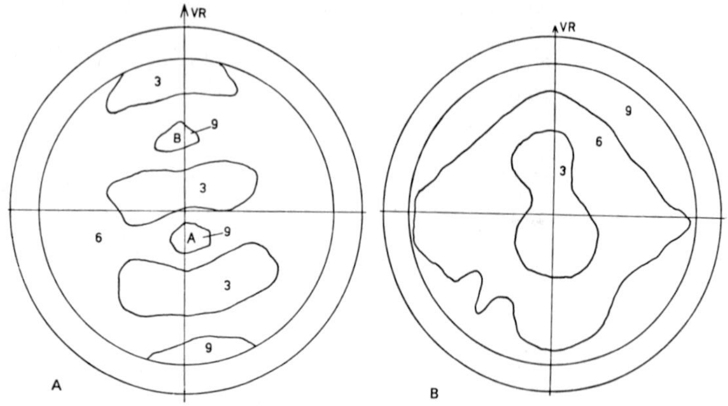

Fig. 13. (110)-(200) polar point diagrams (a and b) of a roller surface subjected to rolling friction
$F_N = 700$ N; $P_D \leqq 80$ Pa; $s_m = 0.5\%$; $l_w = 17000$ m

For this purpose, texture parameters are defined in accordance with [25]:

$$\alpha_A = \left(\frac{I_U - I_A}{I_E - I_A}\right) A$$

$$\alpha_B = \left(\frac{I_U - I_A}{I_E - I_A}\right) B$$

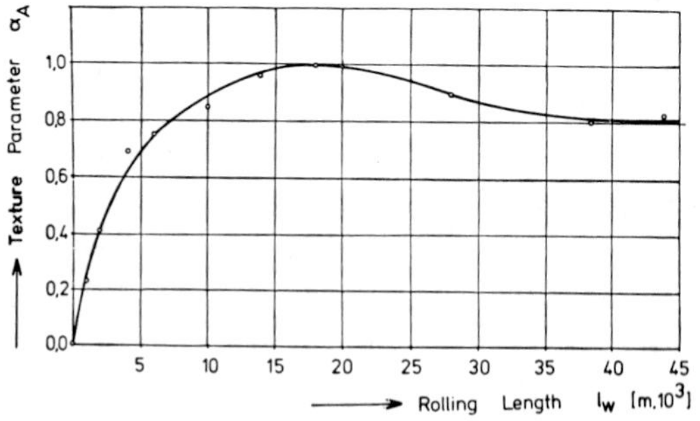

Fig. 14. Dependency of the texture parameter α_A on rolling length (Position A in the polar point diagram)
$F_N = 700$ N; $P_D \leqq 80$ Pa; $s_m = 0.5\%$

where

$I_{\dot{U}}$ = the relative intensity at selected and characteristic positions (Fig. 13) (with respect to the calculated mean intenstiy of the same figure)
I_A = the relative intensity of the initial texture, defined analogue $I_{\dot{U}}$
I_E = the relative intensity of the ultimate texture, defined analogue $I_{\dot{U}}$.

α_B is defined for position B as α_A for position A on the (110) polar point diagram.

The texture parameter (α_A) increases according to the degree of deforma-

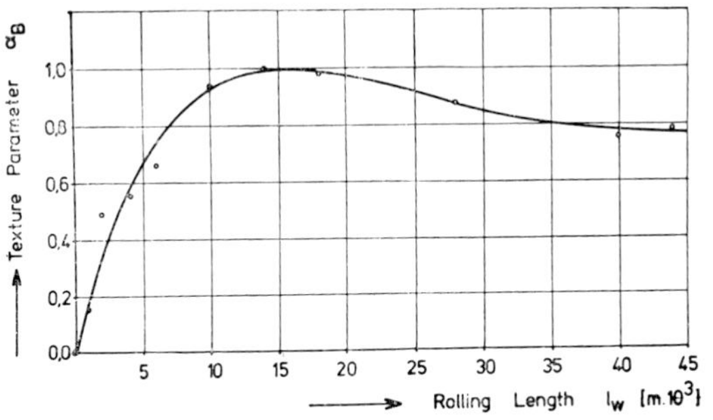

Fig. 15. Dependence of the texture parameter α_B on rolling length (Position B in the polar point diagram)
F_N = 700 N; $p_D \leqq$ 80 Pa; s_m = 0.5%

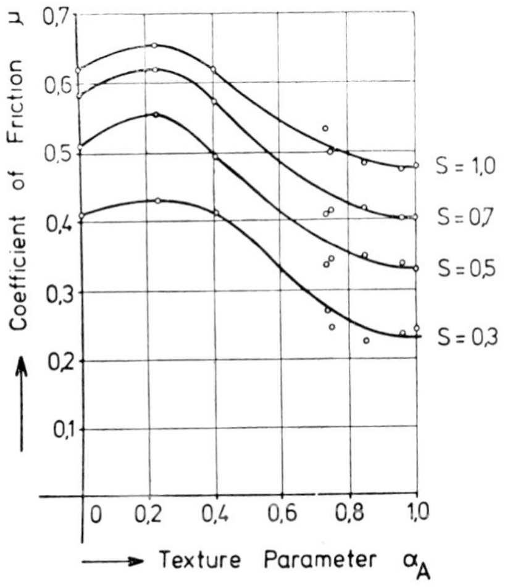

Fig. 16. Plotting of the texture parameter α_A at various degrees of slip againts the coefficient of friction
F_N = 700 N; $p_D \leqq$ 80 Pa

tion of the initial texture to the fritcional texture. (With the ultimate texture $\alpha_A = 1$ and with the initial texture $\alpha = 0$.)

Fig. 14 shows the change in texture as a func ion of the duration of stress.

As can be seen, α_A increases with the stress until it asymptotically assumes the value 1 (ultimate texture) at a rolling length of about 17 000 m. The drop of this parameter which occurs later can be attributed to the additional lattice distortion and the deviation of the surface roughness from the quasi-constant roughness. The drop only appears to be related to the rolling length. Obscured lattice distortion and surface roughness reduce the intensity of the X-ray reflections.

The values obtained from α_B agree with those from α_A (Fig. 15).

The change of the coefficient of friction as a function of the texture parameter for various slip values is shown in Fig. 16. At the beginning, the compact surface layer has a predominant influence (Fig. 16).

After the outer layer has been destroyed, the coefficient of friction reduces, while the frictional texture becomes more pronounced. When the ultimate texture is attained, the coefficient of friction assumes a minimum but constant value.

5. CONCLUSION

The trend shown in tests of the coefficient of friction μ of a friction system differs considerably from that which is to be expected according to Coulomb's laws of friction. The friction system consists of two separately driven, cylindrical steel discs which roll over each other under defined conditions.

To clarify the actual trend of the coefficient of friction, the quantities involved in the occurrence of friction have to be studied in detail together with the real material structure.

An attempt has been made to determine by technical tests the influence of the surface layers, surface roughness, hardness distribution, apparent crystallite size, lattice distortion and crystal orientation in order to explain the actual trend of the coefficient of friction.

ACKNOWLEDGEMENT

The authors like to thank J. Scholten for having carefully inspected the text of this report.

NOTATION

μ	coefficient of friction
ξ	layer thickness [Å]

T time [h, min]
s slip [%]
s_m average slip [%]
v_t rotational speed of the driven roll [m s^{-1}]
v_b rotational speed of the undriven roll [m s^{-1}]
F_N normal force [N]
F_R friction force [N]
p_D partial pressure of vapour [Pa]
M_{Rb} moment of friction of the undriven roll [Nm]
M_{Rt} moment of friction of the driven roll [Nm]
l_w rolling length [m].

REFERENCES

1. G. Schmaltz, *Technische Oberflächenkunde, Feingestalt und Eigenschaften von Grenzflächen technischer Körper, insbesondere der Maschinenteile*, Springer, Berlin, 1936.
2. E. Broszeit, *Modellverschleißuntersuchungen über den Einfluß des Zwischenmediums auf das Gleitverhalten unterschiedlicher Werkstoffpaarungen*, Dissertation, Darmstadt, 1972.
3. K.H. Kloos, 'Werkstoffpaarung und Gleitreibungsverhalten in Fertigung und Konstruktion', *Fortschritt-Berichte VDI-Z*, **2**, 25, 1972.
4. H. Bugarcic, *Einfluß der Feuchtigkeit auf mechanisch-chemische Vorgänge bei Reibungsbeanspruchungen von Armco-Eisen, Einsatz- und Radreifenstahl unter Verwendung einer neukonstruierten Reibungsprüfmashine*, Dissertation, Aachen, 1965.
5. M. Fink, 'Zur Theorie der Reiboxydation', *Archiv für Eisenhüttenswesen*, **6**, 1932/'33, 4, pp. 161–164.
6. M. Fink, 'Reiboxydation', *Zeitschrift VDI*, **77**, 1933, 37, pp. 997ff.
7. H. Krause, 'Tribochemische Reaktionen bei der Wälzreibung von Eisen', *Schmiertechnik und Tribologie*, **17**, 1970, 76–83.
8. R.L. Johnson and D.H. Buckley, 'Lubrication and Wear Fundamentals for High-Vacuum Applications', Conf. on Lubrication and Wear, London, 1967, Sess. 5.
9. J.P. Alison and H. Wilman, 'The Different Behaviour of Hexagonal and Cubic Metals in their Friction, Wear and Work-Hardening during Abration', *Brit. J. Appl. Phys.*, **15**, 1964, pp. 281–290.
10. A.T. Gwathmey, H. Leidheiser and G.P. Smith, 'Influence of Crystal Plane and Surrounding Atmosphere on some Types of Friction and Wear between Metals', *Proc. Roy. Soc.*, **A212**, 1952, pp. 464–467.
11. D.H. Buckley and R.L. Johnson, 'Friction, Wear and Adhesion Characteristics of Titanium-Aluminium Alloys in Vacuum', *NASA TN* ,D-**3235**, 1966.
12. D.H. Buckley, 'Adhäsion, Reibung und Verschleiß von Kobalt und Kobaltlegierungen,' *Kobalt*, **36**, 1968, pp. 17–24.
13. H. Krause, *Mechanisch-chemische Reaktionen bei der Abnutzung von St 60 VZA und Manganhartstahl*, Dissertation, Aachen, 1966.
14. H. Neff, *Grundlagen und Anwendung der Röntgen-Freinstukturanalyse*, Oldenbourg, Munich, 1962.
15. P. Scherrer, 'Bestimmung der Größe und der inneren Struktur von Kolloidteilchen mittels Röntgenstrahlen', *Göttinger Nachrichten, Math. Phys.*, 1918, pp. 98–100.
16. M. Trömel and H. Hinkel, 'Zur Methodik der Bestimmung von Teilchengrößen und Gittverzerrungen aus den Verbreiterungen der Röntgenreflexe', *Berichte der Bunsengesellschaft*, **69**, 1965, pp. 725–731.

17. G. Linßen, H. D. Mengelberg and H. P. Stüwe, 'Zyklische Texturen in Drähten kubisch flächenzentrierter Metalle', *Zeitschrift für Metallkunde*, **55**, 1964, pp. 600–604.
18. L.G. Schulz, 'A Direct Method of Determining Preferred Orientation of a Flat Reflection Sample Using a Geiger-Counter X-Ray Spectrometer', *J. Appl. Phys.*, **20**, 1949, pp. 1030–1036.
19. G. Wassermann and J. Grewen, *Texturen metallischer Werkstoffe*, Springer, Berlin 1962.
20. F. Haeßner, 'Zur Ermittlung von Texturen metallischer Werkstoffe mit Zählrohrverfahren, *Metallwirtschaft und Technik*, **12**, 1958, pp. 89–95, 1094–1101.
21. W. Bunk, K. Lücke and Masing, 'Zur Anwendung des Zählrohres für die Texturbestimmung, *Zeitschrift Metallkunde*, **45**, 1954, pp. 269–275.
22. E. Aernaudt and H.P. Stüwe, 'Die Endlagen der Verformungstextur insbesondere bei kubisch-flächenzentrierten Metallen', *Zeitschrift für Metallkunde*, **61**, 1970, pp. 128–136.
23. H. Siemes, 'Die Endlagen bei der Stauch- und Zugverformung von polykristallinen Metallen', *Zeitschrift für Metallkunde*, **58**, 1967, pp. 228–230.
24. W. Witzel, 'Die Endtextur bei der Scherverformung des Aluminiums, *Zeitschrift für Metallkunde*, **64**, 1973, pp. 813–817.
25. R. Alam, *Untersuchung der Texturänderung an Kupfer und α-Messing in Abhängigkeit von der Walztemperatur und dem ZN-Gehalt*, Dissertation, Aachen, 1967.
26. H. Krause, 'Tribochemical Reactions in the Friction and Wearing Process of Iron', *Wear*, **18**, 1971, pp. 403–412.

The frictional contact of rubber

A. Schallamach

1. INTRODUCTION

Conclusions on the frictional contact of rubber can be drawn from the load dependence of the friction force. If the friction force is taken to be proportional to the true area of contact, as in the theory of solid friction [1], simple assumptions on shape and deformation characteristics of the surface asperities on the rubber lead to a reasonable description of the load dependence of the coefficient of friction, but this approach cannot account for qualitative differences in temperature and velocity dependence between friction on smooth and on rough tracks. The reason is that track asperities not only affect the pressure distribution, and hence the true contact area but also produce mechanical energy losses during sliding which appear as friction.

2. FRICTION ON SMOOTH TRACKS

2.1 *Load dependence*

The coefficient of rubber friction on smooth surfaces, like plate glass, decreases with increasing normal pressure [2, 3]; similar results have recently been reported for friction on (cold) ice [4]. Rubber sliding on such tracks can develop large coefficients of friction without being abraded [3, 5]; this finding contrasts with solid friction in which abrasion is part and parcel of the friction process [1]. The bonds between rubber and another frictional member must be reversible.

If F is the friction force, and A the true area of contact resulting from the deformation of surface asperities, the basic assumption is that

$$F = \varphi A \tag{1}$$

where the proportionality factor φ will depend on velocity and temperature, and will also vary with the rubber compound. If the track is ideally flat, only the rubber asperities need be considered for the true contact area; their deformation is, however, elastic and not, as in solid friction, plastic. A further assumption has been that the rubber surface is covered by closely packed, identical asperities whose elastic behaviour can be described by a single elastic modulus E. The dependence of the coefficient of friction μ on the normal pressure p must then, for dimensional reasons, be given by the power series in equation (2) [6]:

$$\mu = (\varphi/E) \sum_n c_n (p/E)^{\beta n} \tag{2}$$

where the c_n and β_n are constants. The size of the asperities has dropped out of the equation. Its essential prediction is that the quantity $(\mu E/\varphi)$ should be a universal function of the ratio (p/E).

Equation (2) can be made more specific by assigning hemispherical shape to the asperities and using Hertz's equations for spherical contacts. This model gives [3]:

$$\mu = (\varphi/E)(p/E)^{-1/3} \tag{3}$$

The inverse cube root dependence of μ on p predicted by equation (3) has been verified for friction on glass [3] and on ice [4] at pressures between about 0.5 and 2.5 kgf/cm^2. At higher pressures, the hemispheres barrel out and jostle each other so that equation (3) loses its validity. In order to follow this process, a square array of 25 small rubber hemispheres was compressed against a glass plate in the model experiment of Fig. 1, and the contact area

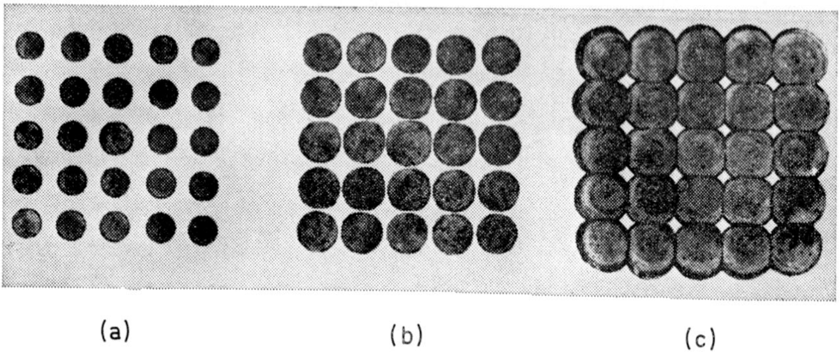

Fig. 1. Contact area of a model rubber surface under the pressures of: (a) 2.14; (b) 7.75; (c) 66.0 kgf/cm² (from [6])

was measured [6]. The gaps between the hemispheres close with increasing pressure, and the contact area approaches a constant value. According to equation (1), μ should then vary as $(1/p)$.

Fig. 2 shows the theoretical pressure dependence of μ as given by (3) at pressures $p < 0.1\,E$, and by the model experiment at higher pressures. Identifying E with Young's modulus, experimental data were plotted as

$E\mu = f(p/E)$ and superimposed on the theoretical curve; the necessary vertical displacement gives the factor φ in (1). The experimental points follow theory but fall on a flatter curve at pressures below about 0.04 E. This is

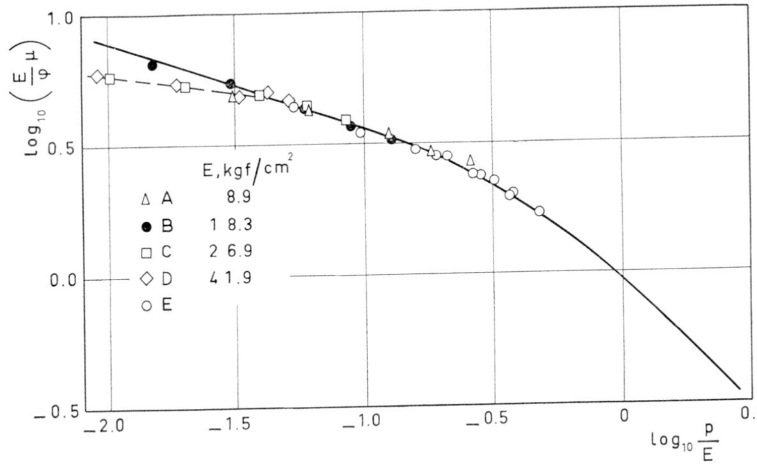

Fig. 2. Theoretical and experimental pressure dependence of the coefficient of friction on smooth tracks (from [6]*). Points A, B, C from*[3]*; E from*[2]*, unknown modulus, adjusted for best fit*

most probably due to a statistical distribution in the actual size of the asperities. Only the tallest asperities touch the track at low pressures, and the number of individual contacts increases with increasing pressure. Calculations show that in this case[6]:

$$\mu\,(\text{low } p) = \text{constant}\,(p/E)^{-1/7} \qquad (4)$$

The dotted line in Fig. 2 has the slope $-1/7$. Every deviation from a uniform surface topology tends to reduce the pressure dependence of the coefficient of friction.

The values of φ for the three unfilled vulcanizates of natural rubber in Fig. 2 are very nearly proportional to the modulus E.

2.2 *Surface condition*

It is evident from the derivation of equations (2) to (4) that the modulus E refers to the asperities in contact with the track rather than to the bulk of the sample. This distinction must be made because the asperities are deformed during sliding and lose part of their original stiffness because of stress softening[7]. Black-filled compounds simultaneously lose much of their initial electrical conductivity because of the break-down of an internal black-structure. This penomenon has served to demonstrate the existence of a surface layer on sliding rubber which differs in its physical properties from she bulk of the test piece[8]. Samples mounted on metal holders were pulled over a smooth metal track whilst the electrical impedance between holder and track was measured with a.c. of different frequencies. The impedance became frequency-dependent after previous abrasion, or after passage of a d.c. current through a new sample during sliding. The results suggested the presence of a surface layer of high resistance in which local deformation had practically destroyed the black structure. If a d.c. potential is applied, the voltage drop is concentrated in this layer and gives rise to a high field strength, resulting in electrostatic attraction between sample and track which operates like an additional normal load, with a consequent increase in frictional force. Fig. 3 is the result of such an experiment with an abraded

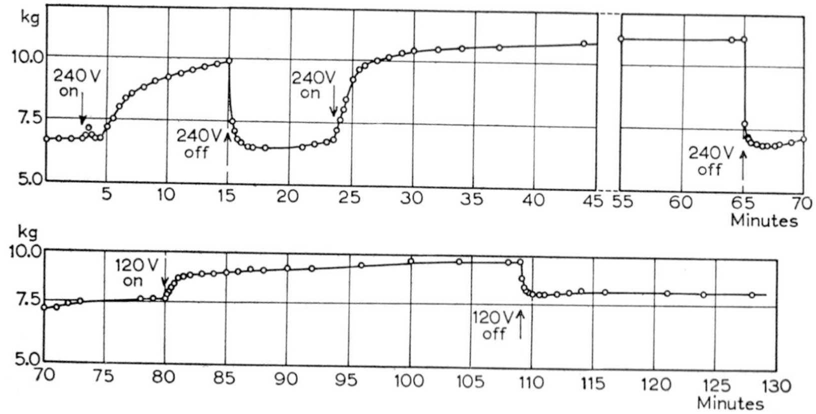

Fig. 3. Effect of a d.c. potential between sample and track on the friction of a natural rubber tyre tread sample (from[8])

tyre tread compound of natural rubber; a moderate voltage (240 V) increases the friction force by about 50%. We refer to the original publication[8] for a more detailed analysis of the results. The thickness of the softened layer could be estimated from the voltage dependence of the friction force as a few thousands of a centimetre.

2.3 *Temperature and velocity dependence*

The main outcome of work on the temperature and velocity dependence of rubber friction is that the velocity dependence at a temperature T_2 is obtained from the velocity dependence at a temperature T_1 by just multiplying the velocities by a factor $f(T_2)/f(T_1)$ in which the functions $f(T)$ depend on the temperature alone. If, therefore, μ is plotted as a function of the logarithm of the velocity v, with the temperature as parameter, the family of curves can be assembled into one 'mastercurve' by horizontal displacements of amount $[\log f(T_2) - \log f(T_1)]$ [9]. This interrelation between rate and temperature dependence is typical of visco-elastic processes, for which it has also been established that each elastomer has a characteristic reference temperature T_s such that the difference $\log f(T) - \log f(T_s)$ is the same for all elastomers. It is quantitatively given by[10]:

$$\begin{aligned}\log f(T) - \log f(T_s) &= \log a_T \\ &= -8.86\,(T-T_s)/(101.6+T-T_s)\end{aligned} \tag{5}$$

The characteristic temperature T_s lies about 50 °C above the glass transition temperature of the material.

The shift factors $\log a_T$ found in rubber friction follow equation (5), thus furnishing the strongest possible evidence for its viscoelastic nature[9]. The only exception appears to be natural rubber in which crystallization at times intervenes in the friction mechanism.

Fig. 4 shows mastercurves for the friction coefficient of four non-crystallizing, unfilled rubbers on various tracks. The 'pinhead' glass was smooth,

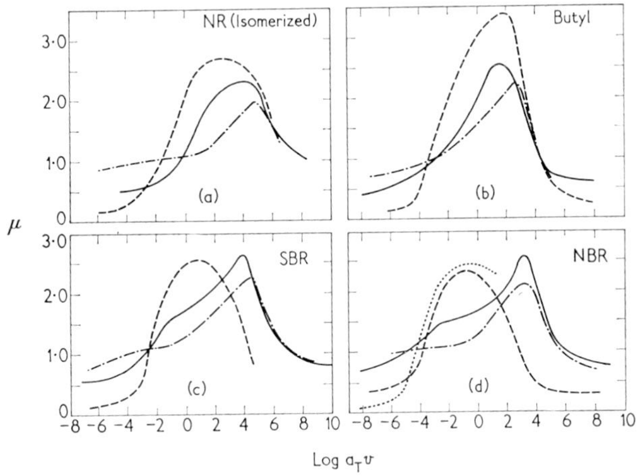

Fig. 4. Mastercurves for the velocity dependence of the friction of non-crystallizing
unfilled rubbers at 20 °C on pinhead glass – – – –; clean silicon carbide paper ————; silicon carbide paper dusted with magnesia –·–·–·–·–

with gentle protuberances, as used for bathroom windows; the coefficient of friction on such tracks is virtually independent of the normal load (see section 3.1). The mastercurves on this glass are all bell-shaped with a broad maximum at a velocity $v_{g,m}$ which could be related to the frequency $f_m(E'')$

at which the mechanical loss modulus of the rubber has a maximum[9]:

$$v_{g,m} = l_1 \times f_m(E^7) \tag{6}$$

with a length $l_1 = 6 \times 10^{-7}$ cm.

Anticipating here part of the discussion of friction on rough tracks, the principle of temperature-velocity equivalence holds also for friction on surfaces with asperities but the mastercurves look different from those determined on smooth tracks.

2.4 *The sliding process*

A simple interpretation of equation (6) is that rubber complexes of molecular size adhere to the track; the bonds are periodically broken by thermal agitation and by their share of the friction force, and are then re-formed in an advanced position. The loss modulus has its maximum at the circular frequency $\omega = 1/\tau$, where τ is the relaxation time of the rubber. Considering the breaking and making of a bond as a cyclic process, the length 6×10^{-7} cm in equation (6) is the forward jump of a bond during one period[9]. The model has been refined by several authors[11, 12, 13, 14]. Their theories, discussed in detail in[15], give a velocity dependence of friction broadly resembling the experimental data but it has been difficult to explain finite friction at very low velocities (static friction) and at very high velocities.

Recent work with transparent sliders or tracks has revealed the existence of a very different sliding mechanism[16]; these experiments were confined to relatively soft rubbers. Fig. 5 shows eight frames of a film of the contact

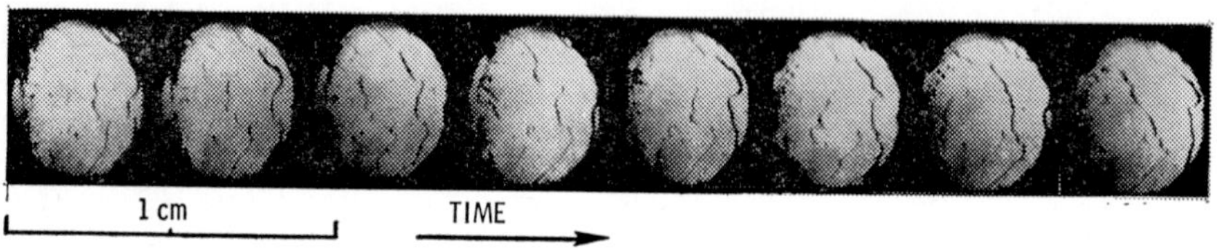

Fig. 5. Contact area of a hard, spherical slider moving to the right on a polyisoprene track at 1/32 sec intervals; v = 0.043 cm/sec (from[16]*)*

between a hard, spherical slider and a transparent track of poly-isoprene rubber. The black lines crossing the contact move from its front to its rear edge at speeds greatly exceeding the sliding velocity. Stereoscopic observation identifies them as folds in the rubber surface along which the rubber has lost contact with the track; complete adhesion is maintained between the moving folds which have been called 'waves of detachment'. The phenomenon has since been observed in this form by other workers[17].

Fig. 6 shows the contact area between a spherical slider of natural rubber and a perspex track; it resembles Fig. 5 but the waves move now from the rear to the front edge of the contact. Despite the outward similarity between the two sets of pictures, there are quantitative differences between them which

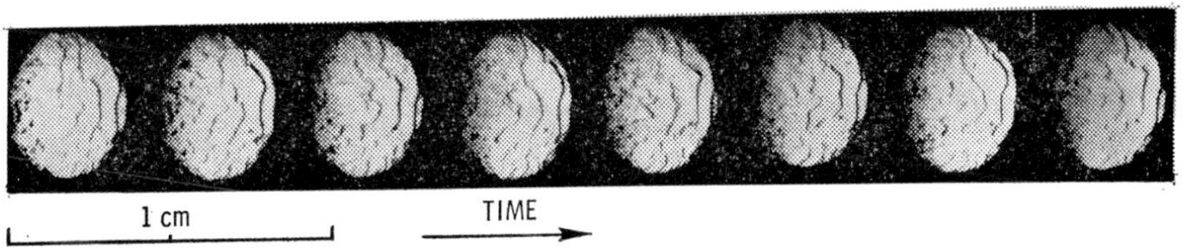

Fig. 6. Contact area of spherical natural rubber slider moving to the left on perspex at 1/32 sec intervals; $v = 0.043$ cm/sec (from[16]*)*

Table 1. Frequency of wave formation n (sec^{-1}) *and wave velocity w* ($cm\ sec^{-1}$) *at various sliding speeds v*($cm\ sec^{-1}$) *(from*[16]*)*

	Hard slider on polyisoprene rubber		*Natural rubber slider on perspex*	
v	n	w	n	w
0.024	5.6	0.86	5.2	0.36
0.043	11.7	1.07	7.1	0.76
0.093	23.6	1.45	9.2	1.34

emerge from Table 1 listing wave velocity w and frequency of wave formation n in the two cases at various sliding speeds. These data will be discussed presently. Waves of detachment form also on tracks of butyl rubber, which is very hysteretic, but sliders of butyl rubber on perspex do not move in this way. The contact area, Fig. 7, is reduced to parallel ridges normal to the sliding direction, and does not change during sliding.

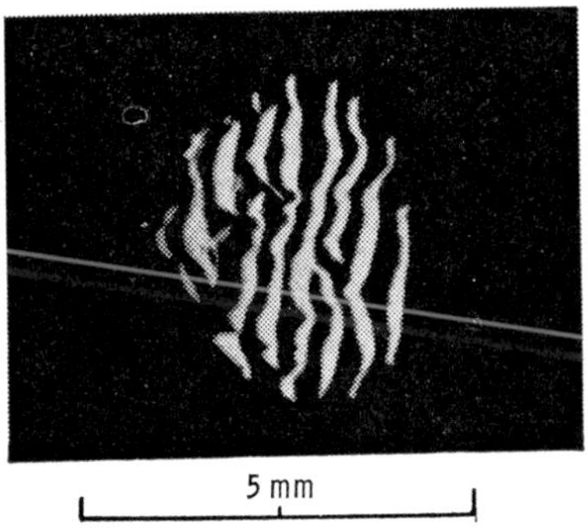

Fig. 7. Contact area of a spherical butyl rubber slider on perspex; $v =$ 0.043 cm/sec (from[16]*)*

The moving folds in the rubber surface constituting waves of detachment have been attributed to an elastic instability in the rubber adjacent to the other frictional member. The instability originates from compressive tangential stresses which produce buckling; Fig. 7 is a direct demonstration of this effect. A theory has been based on the model of the rubber surface in Fig. 8

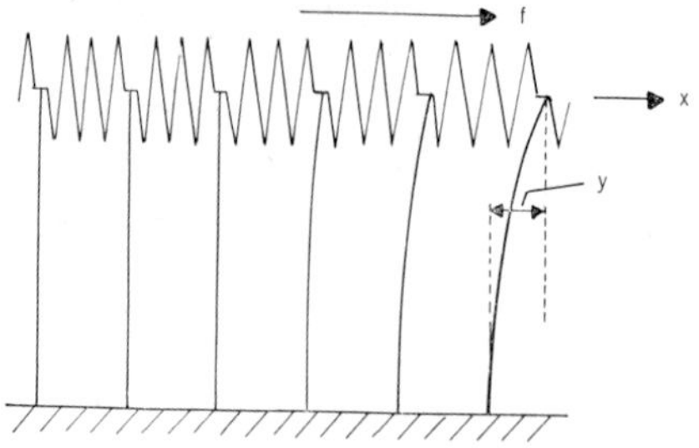

Fig. 8. Mathematical model for a rubber surface (from[18]*)*

which consists of a longitudinal spring supported by cantilever springs. If the force per unit strain in the longitudinal spring is E, and the force constant of the cantilevers is k per unit length of track, the tangential force per unit length f is given by

$$f = k(y - \sigma^2 d^2 y/dx^2) \tag{7}$$

where y is the tangential deflection, and σ stands for

$$\sigma = (E/k)^{1/2} \tag{8}$$

Taking Amontons' law to hold for the coefficient of friction, and assuming a normal pressure distribution of the 4th degree, f in equation (7) must equal[16,18]

$$f = \mu p_m (1 - x^4/a^4) \tag{9}$$

where the origin is in the centre of the contact of length $2a$, and p_m is the maximum pressure. The solution of equation (7) is:

$$y = (\mu p/k)\,[1 - 24\sigma^4/a^4 - 12\sigma^2 x^2/a^4 + 4\sigma/a(1 + 3\sigma/a + 6\sigma^2/a^2 + 6\sigma^3/a^3)\,\mathrm{e}^{-a/\sigma} \cosh x/\sigma] \tag{10}$$

The distribution of the deflection y along the contact according to equation (10) is shown in the upper part of Fig. 9, with $\sigma = a$ for the purpose of this illustration. The distribution of the surface strain is depicted in the lower part of the figure which shows a longitudinal strip of the track with initially equidistant lines when deformed by a slider moving to the right. The compression at the front edge is evident. For comparison with experiment,

Fig. 10 shows the deflection by a spherical slider of a transparent rubber sheet with a 2 mm square grid marked on it. The distortion of the grid in the

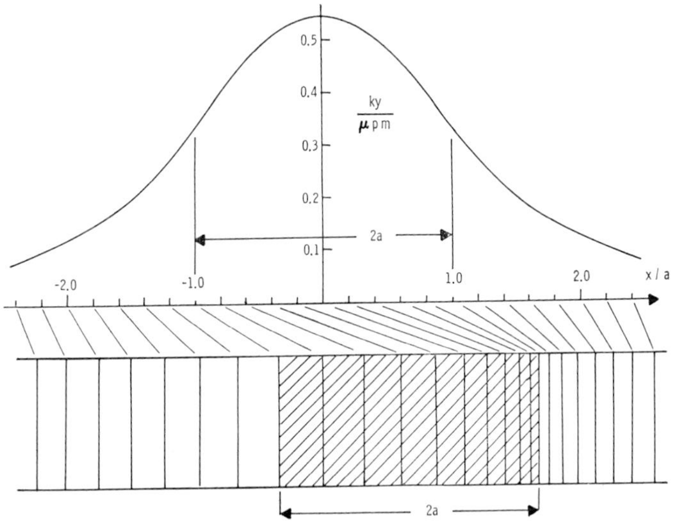

Fig. 9. Theoretical deflection of the model rubber surface by a slider moving to the right (from[18]*)*

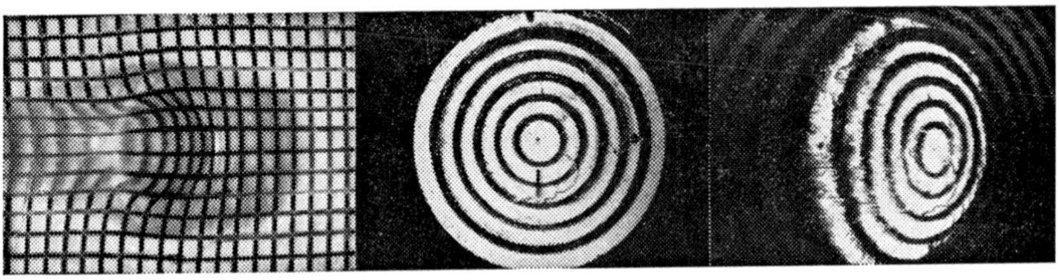

(a) (b)

Fig. 10. Distortion of a rubber surface by a hard, spherical slider (from[16]*)*

Fig. 11. Contact area of a spherical rubber slider on Perspex when (a) resting and (b) moving to the left (from[16]*)*

sliding direction strikingly resembles the theoretical deformation in Fig. 9. The theory also describes qualitatively the strains on a spherical rubber slider on a hard track. Fig. 11 (a) is a picture of the resting slider, and Fig. 11 (b) when moving to the left. The spacings of the circular reference marks, circumferentially 1 mm apart, become narrower at the rear of the contact, and widen at the front. The different number of rings passing through the centre line on either side of the apex show that the rubber sphere has rolled over to the left.

The mechanism of 'sliding' by means of waves of detachment is best understood as a process of stress relaxation. The rubber peeling off the track on one side of a fold is less compressed than the rubber contacting it on the other side. One must, however, differentiate between the cases of a hard slider on a rubber track, and a rubber slider on a hard track. A hard slider on a rubber track brings continuously new rubber into the contact so that tangential compression and the formation of a fold at the front edge are enforced by the sliding speed. Let the collapse of the rubber and the consequent stress relaxation due to the fold advance the sample by the amount δx; then

$$v = n\,\delta x \tag{11}$$

where n is the frequency of wave formation. Comparison between columns 1 and 2 in Table 1 confirm the proportionality between v and n. The fold is propagated as a wave with a velocity w expected to be proportional to the stress gradient, and hence proportional to the friction force. Data on the friction of similar compounds in[9] show that w in column 3 of the Table increases with speed at about the same rate as the friction.

This mechanism cannot operate on a rubber slider because the same rubber remains in contact with the track. Folds at the rear of the contact appear here most probably when the build-up in tangential compression has reached the critical value for buckling so that the frequency n should now depend on the friction force. This is borne out by the near proportionality between the data in columns 3 and 4 in Table 1. To maintain the imposed sliding speed, the wave velocity must have an enforced value which is given by[18]:

$$v = (w/2a)\,\Delta x \tag{12}$$

where Δx is the displacement of the slider during the passage of one wave through the contact; w in the 5th column of Table 1 is practically proportional to the velocity v.

Whether a molecular process or waves of detachment come into operation in rubber friction will depend on which mechanism is energetically more advantageous[16, 18]. The wave mechanism immediately explains the absence of abrasion and suggests a reason for static friction.

The limit of the tangential force below which no sliding occurs is reached when the tangential stress has become great enough to produce buckling of the rubber surface.

3. ROUGH TRACKS

3.1 *Load dependence*

The treatment of friction on smooth tracks in 2.1 can be extended to rough tracks by choosing track models suitable for the calculation of the true contact area. If the track asperities are approximated to hemispheres of much larger size than the rubber asperities, the coefficient of friction becomes[6]:

$$\mu = \text{const}\,(\varphi/E)(p/E)^{-1/9} \tag{13}$$

where the constant embodies the ratio between height and base diameter of the asperities. Equation (13) is seen to predict a far less pronounced pressure dependence of rubber friction on rough than on smooth tracks. Experimental data obtained, on silicon carbide paper (Fig. 12) show that μ follows equa-

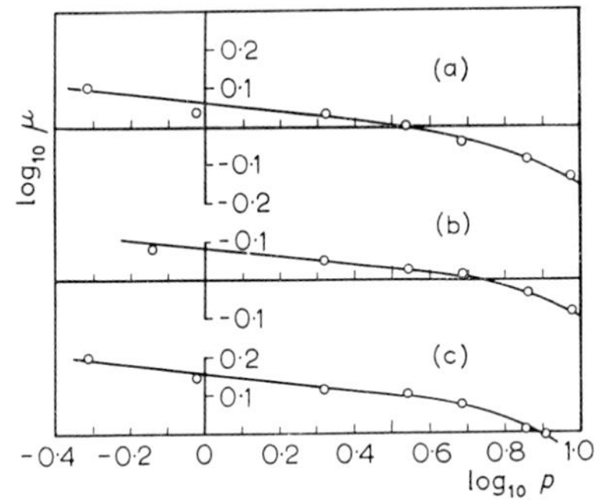

Fig. 12. Pressure dependence of the friction coefficient of rubber compounds on silicon carbide paper at 0.0021 cm/sec; (a) unfilled natural rubber; (b) natural rubber tyre tread; (c) styrene-butadiene tyre tread (from[6]*)*

tion (13) quite closely at the lower end of the pressure range but decreases faster than predicted at higher pressures. The reason for this deviation can lie in saturation effects of the type illustrated in Fig. 1; also, the interstices between track grains will fill with rubber under high loads.

As equation (13) does not contain the asperity size, μ should be independent of the roughness of the track as long as the grains remain geometrically similar. The friction coefficients of two tread mixes on three different grades of garnet paper in Table 2 confirm this; μ decreases at most by 15% for a

Table 2. Coefficients of friction on garnet paper of different coarseness at a sliding speed of 0.35 cm/sec *(from*[19]*)*

Compound		*Natural rubber tyre tread*		*Styrene-butadiene tyre tread*	
Pressure, kgf/cm²		0.53	1.84	0.53	1.84
Grain size, mm	0.13	1.42	1.40	1.41	1.41
	0.29	1.30	1.31	1.39	1.40
	0.63	1.21	1.27	1.29	1.36

nearly 5-fold increase in grain size, and hardly depends on the normal load[19].

The just quoted instances of agreement between theory and experiment may be fortuitous to a certain extent. If only the true contact area determined friction, the mastercurves for friction on smooth and rough tracks in Fig. 4 should not differ in shape.

3.2 *The effect of track asperities*

These differences are most obvious in the case of the two butadiene co-polymers. The mastercurves for friction on silicon carbide paper in Fig. 4 have an asymmetrical maximum at a higher velocity than the maximum on glass, but the curves obtained on clean abrasive have a hump at the velocity at which the maximum on glass is found. The hump vanishes when the track is dusted with magnesia. The composite nature of these mastercurves demonstrates that two friction mechanisms operate on a rough track: interfacial adhesion, and mechanical energy losses due to surface deformation of the rubber by the track asperities[9]. With magnesia between rubber and track, the dust apparently clings to the rubber so that magnesia, rather than rubber, sides on the track. This explains the absence of the hump and the almost constant friction at low velocities on a dusted track.

The velocity $v_{r,m}$ for maximum friction on silicon carbide paper could be empirically related to the frequency f_m (tan δ) at which the mechanical loss tangent of the rubber has a maximum[9]:

$$v_{r,m} = l_2 \times f_m(\tan \delta) \qquad (14)$$

with a length l_1 of 1.5×10^{-2} cm.

The frequency $f_m(\tan \delta)$ is always lower than $f_m(E'')$ in equation (6) but the difference is small for the butadiene co-polymers. As the length factors in equations (6) and (14) differ by several orders of magnitude, the two friction maxima are well separated. For butyl rubber, however, $f_m(E'')$ and $f_m(\tan \delta)$ differ by about the factor 3 000 so that the friction maxima can no longer be resolved. It must be admitted, though, that the attribution of the two friction mechanisms to different dynamic quantities rests on the dynamic peculiarities of butyl rubber; it would otherwise be difficult to decide in favour of E'' or tan δ in either case.

The length 1.5×10^{-2} cm in equation (14) very nearly equals the average distance between the silicon carbide grains on the abrasive track used in these experiments. The component of friction caused by gross mechanical energy loses appears therefore to be governed by the average frequency at which an element on the rubber surface encounters an asperity on the track.

The rubber surface is deformed both normally (indentation) and tangentially; the tangential deformation is thought to be the more important strain on dry tracks. The force exercised by a track asperity on the rubber should increase with the steepness of its leading flank. Experiments to elucidate this effect were carried out with the cylindrical and prismatic steel sliders shown in cross-section in Fig. 13 [20]. This form of slider was chosen to obviate difficulties with abrasion by pointed sliders. The chisel in Fig. 13(b) could be used in either direction so that it offered angles of attack of 25°

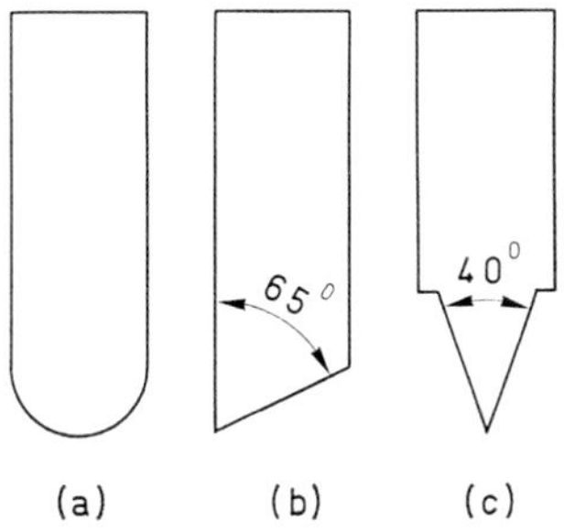

Fig. 13. Cross-section of metal sliders. The shanks are 1.25 cm square (from [20]*)*

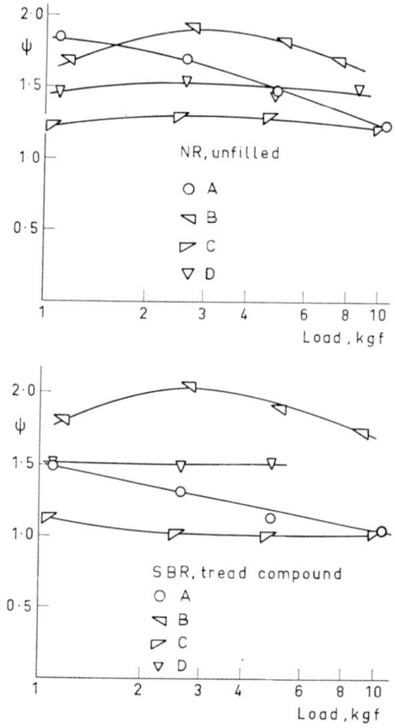

Fig. 14. Friction ratios of the sliders in Fig. 13 on two rubber tracks; A: cylinder; B: chisel with angle of attack of 90° and C: 25°; D: wedge; v = 0.046 cm/sec (from[20]*)*

and 90° at which to meet the rubber; the angle of attack was 70° for the wedge in Fig. 13(c).

Results obtained on two rubber tracks are shown in Fig. 14 as friction ratios ψ (= tractive force/normal force) to distinguish them from the friction coefficient μ between flat surfaces. Fig. 14, and other graphs in the original publication[20] confirm that ψ increases with increasing angle of attack but the effect is relatively small, an angle of 90° giving only about twice the friction ratio found at 25°. Naive reasoning suggests that ψ should become infinite when the cotangent of the angle of attack is equal to, or smaller than μ. What keeps ψ finite under these conditions is the extension of the rubber behind the slider. This extension has been demonstrated photo-elastically, and is responsible for abrasion[21]. The consequent tensile force helps to pull the rubber around the edge of the slider, and it prevents the friction ratio from becoming infinite. In the calculations, the edge of the slider was approximated to a cylindrical surface of small radius, and the friction coefficient μ was assumed to be constant. This approach, which is justified for a qualitative treatment, has the advantage that the radius of the cylindrical edge drops out of the final result. Fig. 15 gives the angles used to charac-

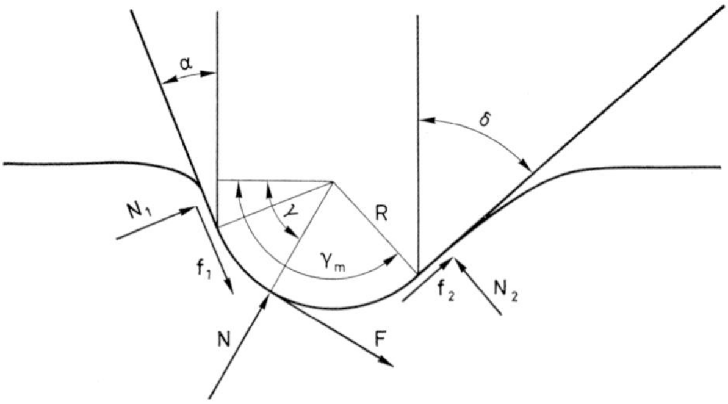

Fig. 15. Edge of prismatic slider moving to the left (from[20]*)*

terize a slider and the forces on its flanks and edge. The friction ratio becomes

$$\psi = \frac{\cos\alpha - (N_2/N_1)\cos\delta + \mu \sin\delta[N_2/N_1 + \exp\mu(\pi-\alpha-\delta)]}{\sin\alpha + (N_2/N_1)\sin\delta + \mu\cos\delta[N_2/N_1 + \exp\mu(\pi-\alpha-\delta)]} \tag{15}$$

ψ does not vanish when $\mu = 0$; its value gives then the friction on an ideally lubricated track which, in the theory by Greenwood and Tabor[22], is entirely due to deformation losses.

If μ is large, the back pressure N_2 in Fig. 15 becomes negligible, and the effective wrapping angle will be smaller than its theoretical value $(\pi-\alpha-\delta)$. It has been assumed that the rubber parts from the slider horizontally at an angle γ_e which defines an effective wrapping angle $\gamma_e - \alpha$. Equation (15) reduces then to

$$\psi = \frac{\mu \exp \mu(\pi - \alpha - \delta_e)}{\sin \alpha + \mu \cos \delta_e \exp \mu(\pi - \alpha - \delta_e)} \quad (16)$$

where δ_e is given by eqn (17):

$$\cos \alpha \exp[-\mu(\pi - \alpha)] = \mu(1 - \sin \delta_e) \exp(-\mu\delta_e) \quad (17)$$

Equation (16) contains neither the load nor the elastic properties of the track and can therefore only represent a rough approximation to the facts. Table 3 shows, nevertheless, that friction ratios calculated with $\mu = 1.0$ and

Table 3. Effective angles δ_e and friction ratios ψ at various angles of attack (from[20])

Angle of attack (°)	δ_e (°)		ψ	
	$\mu = 1.0$	$\mu = 1.5$	$\mu = 1.0$	$\mu = 1.5$
90	61.0	73.5	2.06	3.52
70	57.5	70.1	1.68	2.76
25	57.6	66.9	1.15	1.77

$\mu = 1.5$ are in the right ranking order, and agree with the experimental values in order of magnitude with $\mu = 1.0$.

3.3 *The frictional lift*

The vertical components of the tangential and radial forces on the edge of a moving slider create a normal pressure p_r in addition to the already present static pressure; p_r is theoretically given[23] by equation (18):

$$p_r = \text{constant}\,(1 - \mu \cot \gamma) \exp \mu(\gamma - \alpha) \quad (18)$$

and shown graphically in Fig. 16 for the chisel slider with an angle of attack of 90° and $\mu = 1.0$. The integral of p_r over the projected contact area of the edge is a force tending to raise the slider when it moves. This frictional lift has been confirmed experimentally. The records for the time dependence of the height of the slider in Fig. 17 show the effect on unfilled natural rubber. On releasing the tractive force, the slider drops immediately by a well defined amount and then sinks slowly because of creep.

As the lift develops predominantly at the rear of the edge (compare Fig. 16), the chisel is lifted more at an angle of attack of 25° than at 90°. The initial sinking of this slider at 90° comes from a transient pressure distribution in which the force on the leading vertical flank pulls the slider into the rubber before the tensile force has fully developed behind it. Apart from such

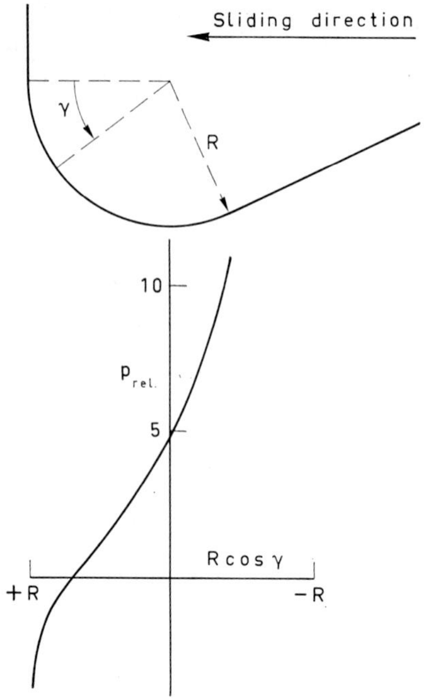

Fig. 16. Distribution of the additional pressure p_r on the edge of the moving chisel slider (from[23]*)*

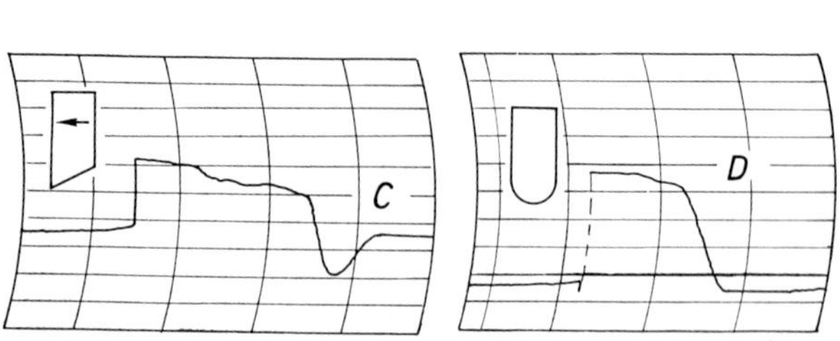

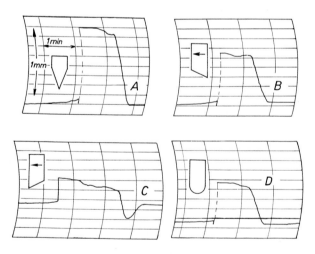

Fig. 17. Frictional lift of the various sliders on a track of unfilled natural rubber (from[20]*)*

general considerations, it is difficult to treat the frictional lift quantitatively on the basis of the present rudimentary theory. As it will depend on the stiffness of the track, the observed lifts in Table 4 are given as percentages of

Table 4. Frictional lift, in percent of the static identation (from [20]*)*

	Load, kgf	*NR, unfilled*	*NR, tread*	*SBR, tread*	*BR, tread*
cylinder	4.95	10.8	—	2.1	1.3
	10.55	14.7	2.9	1.9	—
wedge	4.95	17.2	12.1	15.2	6.9
	10.55	16.3	15.5	17.5	9.0
chisel, 25°	4.65	14.7	11.0	10.6	7.6
flank leading	9.91	15.2	15.3	10.9	6.9
chisel, 90°	5.29	4.2	—	9.0	0.2
flank leading	9.30	4.5	—	8.3	—
	11.28	—	9.0	—	1.6

NR = natural rubber; SBR = styrene butadiene rubber; BR = poly-butadiene rubber.

the static indentation. The cylindrical slider experiences an appreciable lift only on unfilled natural rubber because a sufficiently large wrapping angle can develop around its edge only on this relatively soft material. The lift is, understandably, smallest on poly-butadiene which has also the lowest coefficient of friction.

REFERENCES

1. F.P. Bowden and D. Tabor, *Friction and Lubrication of Solids*, Oxford Univ. Press, London, 1954, pp. 20–22, 30–32.
2. P. Thirion, 'Les Coefficients d'Ahérence du Caoutchouc', *Rev. Gén. Caout.*, **23**, 1946, pp. 101–106.
3. A. Schallamach, 'The Load Dependence of Rubber Friction', *Proc. Phys. Soc.*, **B65**, 1952, pp. 657–661.
4. E. Southern and R.W. Walker, 'A Laboratory Study of the Friction of Rubber on Ice', *ACS Internat. Symp. on Polymer Friction and Wear*, (Los Angeles, 1974). In press.
5. F.L. Roth, R.L. Driscoll and W.L. Holt, 'Frictional Properties of Rubber', *J. Res. Nat. Bur. Stand.*, **28**, 1942, pp. 439–462.
6. A. Schallamach, 'Friction and Abrasion of Rubber', *Wear*, **1**, 1958, pp. 384–417.
7. J.A.C. Harwood, L. Mullins and A.R. Payne, 'Stress Softening in Rubber', *J. Inst. Rubber Indus.*, **1**, 1967, pp. 17–25.
8. A. Schallamach, 'Surface Condition and Electrical Impedance in Rubber Friction', *Proc. Phys. Soc.*, **B66**, 1953, pp. 817–825.
9. K.A. Grosch, 'The Relation between the Friction and Viscoelastic Properties of Rubber', *Proc. Roy. Soc.*, **A274**, 1963, pp. 21–39.
10. M.L. Williams, R.F. Landel and J.D. Ferry, 'The Temperature Dependence of Relaxation Mechanisms in Amorphous Polymers and Other Glass-Forming Liquids', *J. Amer. Chem. Soc.*, **77**, 1955, pp. 3701–3707.

11. D. Bulgin, G.D. Hubbard, and M.H. Walters, 'Road and Laboratory Study of Friction of Polymers', *4th Rubber Technol. Conf.*, London, 1962, pp. 173–186.
12. A. Schallamach, 'A Theory of Dynamic Rubber Friction', *Wear*, **6**, 1963, pp. 375–382.
13. A.R. Savkoor, 'On the Friction of Rubber', *Wear*, **8**, 1965, pp. 222–237.
14. H. Rieger, 'Gedankenmodell zur Gummireibung', *Kautschuk und Gummi Kunstst.*, **20**, pp. 293–295.
15. A. Schallamach, 'Recent Advances in Knowledge of Rubber Friction and Tire Wear', *Rubber Chem. Technol.*, **41**, 1968, pp. 209–244.
16. A. Schallamach, 'How Does Rubber Slide?', *Wear*, **17**, 1971, pp. 301–312.
17. M. Barquins, R. Courtel and P. Thirion, 'Sur les Déformations Dynamiques Dues au Frottement du Caoutchouc', *C.R. Acad. Sc. Paris.*, **277**, 1973, pp. 479–481.
18. A. Schallamach, 'Elementary Effects in the Contact Area of Sliding Rubber', in D.F. Hays and A.L. Browne (eds.), *The Physics of Tire Traction*, Plenum Press, New York and London, 1974, pp. 167–177.
19. A. Schallamach, 'Abrasion and Tyre Wear', in L. Bateman (ed.), *The Chemistry and Physics of Rubberlike Substances*, Maclaren & Sons, Ltd., London, 1963, pp. 355–416.
20. A. Schallamach, 'Friction and Frictional Lift of Wedge Sliders on Rubber', *Wear*, **13**, 1969, pp. 13–25.
21. A. Schallamach, 'On the Abrasion of Rubber', *Proc. Phys. Soc.*, B**67**, 1954, pp. 883–891.
22. J.A. Greenwood and D. Tabor, 'The Friction of Hard Sliders on Lubricated Rubber', *Proc. Phys. Soc.*, **71**, 1958, pp. 989–1001.
23. A. Schallamach, 'Gummireibung', *Gummi Asbest Kunstst.*, **28,** 1975, pp. 142–155.

Applications for contact theories in nuclear reactor technology

L. A. Mitchell

1. INTRODUCTION

The nature of contact between two bodies is reflected in many phenomena. Friction, wear and adhesion are three such contact dependent phenomena that are of considerable importance in many branches of technology. Interest in contact mechanics in nuclear reactor technology is fairly broad; it includes those aspects relevant to the three phenomena mentioned above and also those influencing thermal contact resistance of interfaces and the determination of the characteristics of static seals. Additionally, there is a need to consider how contact conditions can change in corrosive environments, when chemically formed films achieve dimensions comparable to, or greater than, those of the original surface roughness.

The aim of this paper is to outline briefly the importance of these contact phenomena in nuclear reactor technology, to demonstrate how an appreciation of contact mechanics can assist in understanding the true causes of practical problems and to suggest aspects of contact theories that could be developed to advantage. The arguments are based to a large extent on work which has been carried out by the author and his colleagues at the Berkeley Nuclear Laboratories of the Central Electricity Generating Board. For this reason the validity of some of the comments may appear to be restricted to the gas-cooled reactor systems developed in Britain. However, this would be a false impression, as in general terms, the problems on all reactor systems have much in common. Indeed many of the problems discussed are not peculiar to the nuclear industry at all but derive solely from a hostile atmosphere and/or high temperatures.

2. THE STATISTICAL NATURE OF FRICTION

Wherever mechanisms are required to operate in environments in which friction coefficients are high, difficulties are encountered in obtaining geometric configurations for the mechanisms that will function reliably. This is because, for most mechanisms, there is a critical coefficient of friction at which the reactive forces in the system become indeterminate and the mechanism cannot transmit power. In lubricated situations it is relatively easy to avoid this condition, but it becomes progressively more difficult as the level of friction rises. As the margins between the allowable coefficient of friction and the mean value decrease, it becomes important to pay attention to the constancy of the coefficient of friction.

There are severe experimental difficulties in studying the statistical fluctuations in the coefficients of friction because a measuring system with a very high natural frequency is required. Nevertheless, with care, it can be done and Fig. 1 shows the probability distribution of the coefficient of

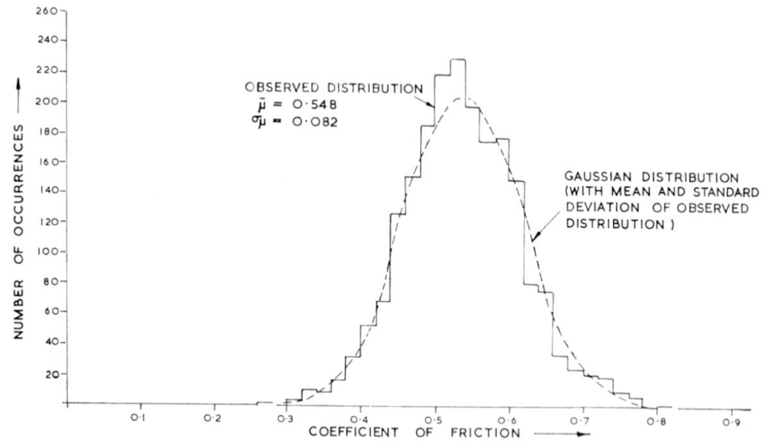

Fig. 1. A probability distribution of the coefficient of friction of stainless steel sliding on itself in air

friction obtained for stainless steel sliding on itself unlubricated in air. It is obvious from this figure that the variance of the friction coefficient is significant compared with its mean and, where this is true, the variability of friction can be a determining factor in the reliability of mechanisms.

Mitchell and Osgood [1] have considered the failure probability of a simple mechanism, assuming that the fluctuating friction forces can be described by a normal probability distribution together with an exponential autocorrelation function. For a typical mechanism they concluded that, to achieve a failure rate lower than 1 in 1000, it would be necessary that the critical coefficient of friction should be at least 6 standard deviations above the mean. A reliability of this order is not a demanding requirement from a mechanism, but accurate specification of the coefficient of friction at this level is by no means simple.

From contact theory it is clear that the instantaneous friction force is the sum of the tangential forces resisting motion at a number of interacting surface asperities. Thus it should not be surprising that friction is a fluctuating quantity and it might be surmised that the size of fluctuations should be strongly dependent upon the average number of contacts on the surface. On this simple argument the deviations from the mean would be expected to decrease if the number of contacts increased. Such a conclusion would have important consequences for the application of experimental friction data to practical problems because the probability distribution of the friction coefficient would depend on how the contact conditions had been 'scaled'.

In an attempt to formulate a theory for the statistical nature of contact Mitchell and Osgood[2] have developed a modification to the adhesive theory of friction, in which it is assumed that there are only two types of asperity interaction, namely welded and non-welded. In their model the sensitivity of the coefficient of friction to environment is supposed to result from the mean proportion of welded contacts changing rather than from a change in the mean strength of the contacts as supposed by Bowden and Tabor in their original theory[3]. On this basis they were able to develop expressions for the ratio of the standard deviation of friction coefficient to the mean value. For a massive slider, the ratio is given by the expression

$$\frac{\sigma_\mu}{\bar{\mu}} = \frac{1}{\bar{m}^{1/2}}\left[\frac{\{\bar{p}+(1-\bar{p})\tan^2\phi\}^{1/2}}{\bar{p}+(1-\bar{p})\tan\phi}\right] \tag{1}$$

where

$\bar{m}$ is the mean number of contacts,
$\bar{p}$ is the probability that the contact will be welded,
ϕ is the mean absolute surface slope.

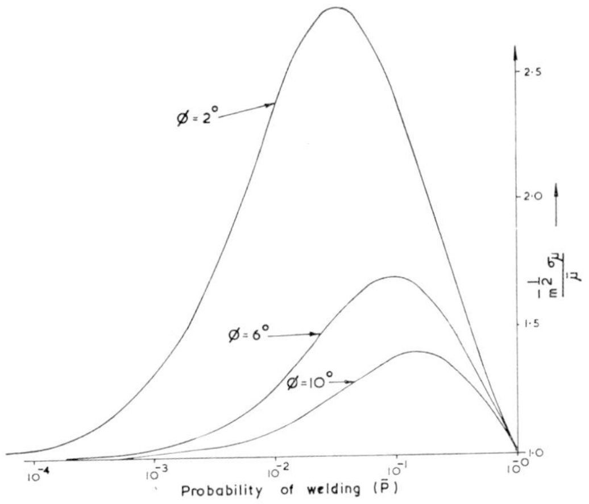

Fig. 2. $\bar{m}^{\frac{1}{2}}\sigma_\mu/\bar{\mu}$ *as a function of the probability of welding*

Fig. 2 shows a plot of $\overline{m}^{\frac{1}{2}}\sigma_\mu/\bar{\mu}$ against $\bar{p}$ for various values of surface slope.

For a sliding system where the mean number of contacts is 100 and the value of ϕ is 2°, $\sigma_\mu/\bar{\mu}$ assumes values between 0.1 and 0.27 depending on $\bar{p}$. Thus for a mechanism of the type considered by Mitchell and Osgood, operating in the most adverse conditions, the critical friction would have to be some 2.6 times the mean value if a reliability of 1 in a 1000 is to be achieved.

The validity of this general model can be checked by conducting experiments to determine the probability distribution of friction in a controlled environment with varying contact stress. In these circumstances, $\bar{p}$ and ϕ in equation (1) may be assumed to be fixed and $\sigma_\mu/\bar{\mu}$ becomes proportional to $1/\overline{m}^{\frac{1}{2}}$. Results of this kind have been obtained by Rabinowicz et al. [4] and by the author. Both supply support for a relationship of this general type. However, there is clearly considerable scope for further experimental work in this area but it is unlikely that it will prove possible to determine the extremes of the probability distribution of the friction coefficient with adequate accuracy for predicting machine reliability. Calculations of this kind will need to rely, therefore, on theoretical arguments about the form of the distribution. Thus it may be expected that there will be a need for a more sophisticated contact model. For example, in its present form, the theory takes no account of the variation in tangential force during the life of an individual contact. Effects of this kind should become important when there are very small numbers of contact on the surface, a situation which is not unlikely when severe wear prevails.

Apart from concern over the form of the probability distribution of the friction coefficient, the application of reliability theories will rely heavily on estimates of the mean number of contacts on real bearing surfaces. This implies a need for an improved understanding of the nature of contacts between 'engineering' surfaces with 'engineering' properties. It is reassuring that there are at least two papers in this Symposium which deal with these subjects.

3. CRITERIA FOR WEAR

The most demanding problems in nuclear technology arising from wear are a consequence of the need to ensure that it will not prove a life limiting process on reactor components which should not require replacement. In most cases the concern results from unintended movements due to vibration or thermal effects and the materials in contact are frequently not a 'compatible sliding pair'. As a reactor is generally designed for a life of 30 years it is clearly impossible to demonstrate the adequacy of components by simulation experiments, accurate in all respects. There is a requirement, therefore, for a criterion for wear which allows data from short term experiments to be extrapolated in time. Ideally, to avoid the need for expensive full scale experiments, it is desirable that this criterion should be capable also of relating wear measurements on small scale apparatus to practical situations.

Archard[5] proposed a model of surface contact which allowed him to formulate a theory of adhesive wear. His basic equation was modified by Rabinowicz[6] to give an equation for the volume of material removed

$$V = kLx/3H \tag{2}$$

where

k is the wear coefficient (and is equal to the probability of forming a detached wear fragment in an asperity interaction),

L is the applied load,

x is the sliding distance,

H is the material hardness.

It might be expected that the probability of forming a detached wear fragment would be a function of the environment alone. Thus, for a given material pair in a given environment, equation (2) suggests that the quantity V/Lx should be a constant. This grouping of parameters is commonly referred to as the specific wear rate and would appear to satisfy the need for a prediction criterion.

Before the specific wear rate can be applied to practical problems with confidence it is necessary to consider the experimental evidence for its validity. To demonstrate its value in extrapolation with respect to time it is sufficient to establish its constancy during wear processes. However, to enable it to be used to relate data from different sliding configurations, it is necessary to establish that it is a quantity which is dependent upon materials and environment alone.

There is a considerable body of evidence that, after an initial phase, the specific wear rate remains constant in a particular experiment. It would clearly be desirable to improve the understanding of the early stages of wear but, in many cases, the duration of this period is small in comparison to the total time of sliding. Thus in many practical situations it is found that the specific wear rate may be an adequate criterion for the restricted exercise of extrapolating in time only.

With regard to the effects of contact geometry, it is well known that many materials can exhibit a transition in wear behaviour from mild wear to severe wear, and that this is associated with large changes in the rate of material removal. The parameters influencing this transition have been shown to include contact stress, temperature and sliding speed[7,8]. It is likely therefore, that superficially similar experiments and apparatus with different geometric configurations could lead to different results if conditions are close to the transition. An individual laboratory cannot easily test the sensitivity of wear to experimental configuration because it will be limited by the apparatus available. However, the OECD organised a series of experiments in which prepared specimens in standard materials were issued to a number of laboratories, who were given identical instructions to carry out wear tests on their own equipment. The results have been reported by Begelinger and De Gee[9]. If equation (2) is employed to interpret their data, values of the probability of forming a detached fragment in an asperity collision may be deduced. A histogram may then be plotted showing the number of laboratories recording the same results. Such a histogram is plotted in Fig. 3 for copper sliding on mild steel. It may be seen that the

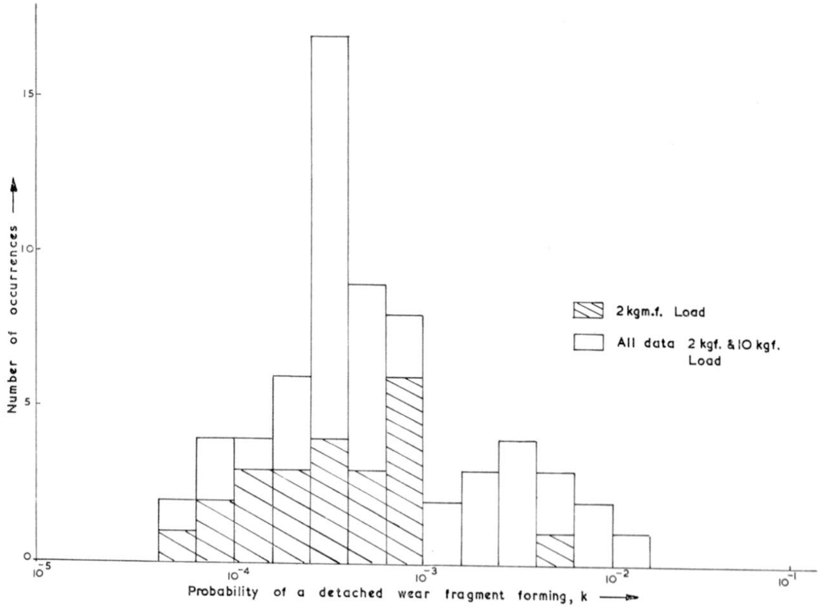

Fig. 3. Histogram of wear results obtained by different laboratories for copper sliding on mild steel

scatter in results is very large indeed and that the histogram is bi-modal. Other results for more compatible materials were, in comparison, repeatable.

This observation of large discrepancies in experimental wear data for poor bearing combinations is in general accordance with the experience of the author, although, because it is from limited sources, this experience cannot be described quantitatively. From a practical viewpoint, the implication of these observations is clear; data from experimental apparatus cannot be applied to practical problems involving unintended sliding of incompatible materials unless very large margins are allowed. Frequently the need for introduction of adequate margins is a cause of embarrassment.

The bi-modal character of the histogram in Fig. 3 suggests that one of the contributing causes of the scatter in results could be a mild to severe wear transition. If this is so, a marked improvement in the uncertainty in the application of such data could be achieved if the conditions in which such a transition occurs could be better identified. It has already been said that the number of parameters affecting wear transition is large and it would seem to be fortuitous if this issue could be resolved by a completely empirical approach. There is a strong need, therefore, for a theory of contact to be extended to give a better appreciation of the factors controlling wear.

4. THERMAL CONTACT RESISTANCE OF INTERFACES

Interest in the thermal contact resistance of interfaces derives from the need to minimise the temperature difference between nuclear fuel and coolant. As the fuel is clad with a protective sheath, one of the barriers to the heat flow is the interface between the fuel and its clad.

If the sole means of heat transport is conduction through metal, the problem of heat flow at a contact becomes analogous to electrical conduction. This subject has received considerable attention, the most notable early work being that of Holm[10]. Greenwood[11] made a most pertinent contribution when he extended the theory from consideration of individual contacts. He considered the resistance of a number of contacts, and developed an equation for the total contact resistance of the form

$$R \approx \frac{\rho}{2\Sigma a_i} + \frac{\rho}{\pi m^2} \sum_{i \neq j}\sum \frac{1}{s_{ij}}$$

where

a_i is the radius of the ith contact,
s_{ij} is the distance between the ith and jth contacts,
m is the number of contacts,
ρ is the resistivity.

The first term on the right-hand side is the usual constriction resistance, due to the channelling of the current through contact points, and the second term is an additional resistance due to the interaction of the potential field of neighbouring contacts. In certain circumstances the latter term can have a very considerable effect.

In practice the problem of heat conduction is not analogous to that of electrical conduction because heat can be conducted through the gas in the voids between the surfaces. Other analyses, for example that of Henry and Rohsenow[12], have attempted to model the overall heat flow problem. Their contact model was, however, grossly idealised and had a completely uniform distribution of contacts.

Experimentally it is very difficult to obtain repeatable results for the thermal contact resistance of an interface. The reason for this becomes obvious when the surface profile of a fuel pellet is examined. Fig. 4 shows such a profile for a fuel compact proposed for use in the high temperature reactor.

It can be seen from the axial profile of the circumferential surface, which would be in contact with the cladding material, that it is concave and tapered

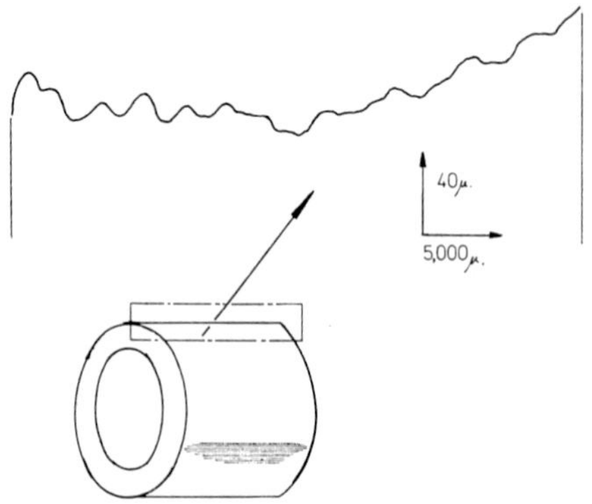

Fig. 4. The surface profile of a fuel compact for a high temperature reactor

and that the deviation from linearity associated with macroscopic distortion is 5 times the scale of the surface roughness. This distortion is a systematic effect caused by the manufacturing process. The contact between such a compact and a cylindrical surface would be restricted to a region close to its end. In this case the interaction term in the electrical conduction model would become very high and the effect would be that the predominant heat transfer mode would be conduction through the gas gap rather than the solid/solid contact.

Clearly a theoretical understanding of this problem will not emerge from the more conventional analyses of surface profiles. In general the sample considered is too small and the longer wavelengths which are so important for this problem are overlooked. Accompanying analyses of the behaviour of contacts under load should consider the elastic or creep distortion of these long wavelength features.

5. THE CHARACTERISTICS OF STATIC GAS SEALS

A reactor vessel is a large engineering structure containing the coolant, from which only low leakage is allowed. The permissible leakage rates may be low because of the need to contain any active fission products in the circuit or, in some reactor systems, because the coolant is expensive or chemically reactive. There are numerous penetrations through the basic shell of the vessel to permit operations such as refuelling and the removal of equipment for maintenance. Most leakage occurs through the static seals associated with these penetrations; diffusion losses through the vessel itself are, in comparison, negligible. The seals themselves usually consist of a deformable element, or gasket, trapped between two rigid mating components. The gasket material may be either metallic or elastomeric depending on the temperature seen during service. The technical arguments described below refer specifically to metallic gaskets when the flow is plastic, but the general conclusions drawn apply to all seals.

To obtain a good seal it is obviously necessary to force surfaces into intimate contact to exclude the voidage which provides the leakage path. The difficulty of achieving complete contact was demonstrated in the classic experiment of Moore[13] when he showed that the surface roughness could persist even when the substrate material had undergone plastic deformation. A similar result is illustrated in Fig. 5, which shows a surface profilometer

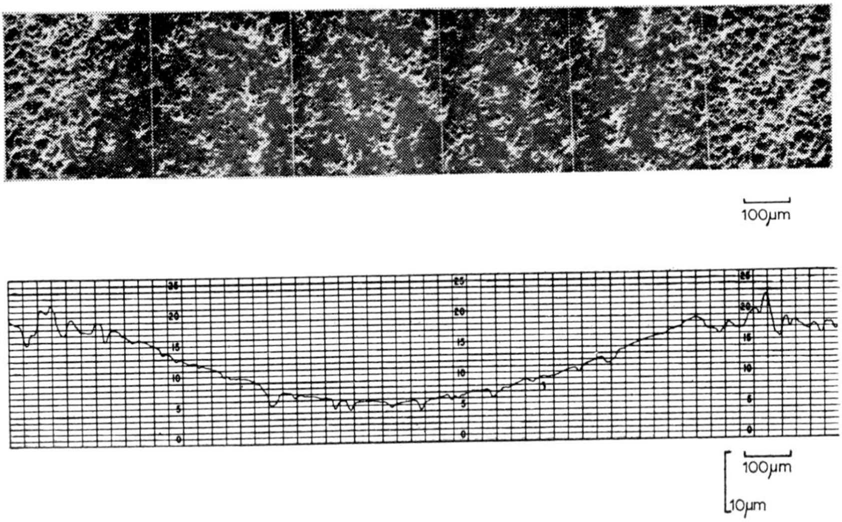

Fig. 5. Profilometer trace and scanning electron micrograph of the surface below a spherical indentation

trace and scanning electron micrograph of the surface below a spherical indentation.

This difficulty in obliterating surface roughness is not surprising from plastic flow considerations, because the pressures required to promote flow are strongly dependent upon geometric constraints. The use of a constant microscopic hardness to relate real area of contact to normal load is only valid for plastic contact when the points of contact are discrete, the plastic field is entirely contained by an asperity and geometric similarity is preserved during the deformation. At high contact stresses these conditions will be violated and plastic flow in individual asperities will be impeded by the proximity of their neighbours, where-upon it becomes impossible to maintain geometric similarity. Initial stages of compression of an asperity can be treated as a quasi steady state problem and the deformation pressure is then constant. When plastic fields in adjacent asperities begin to interact, the deformation pressure begins to rise and the plasticity solutions are only true instantaneously. Approximate solutions for the compression of an array of uniform wedges have been obtained by Mitchell and Rowe[14] and their results are reproduced in Fig. 6. Recently slip-line field solutions for

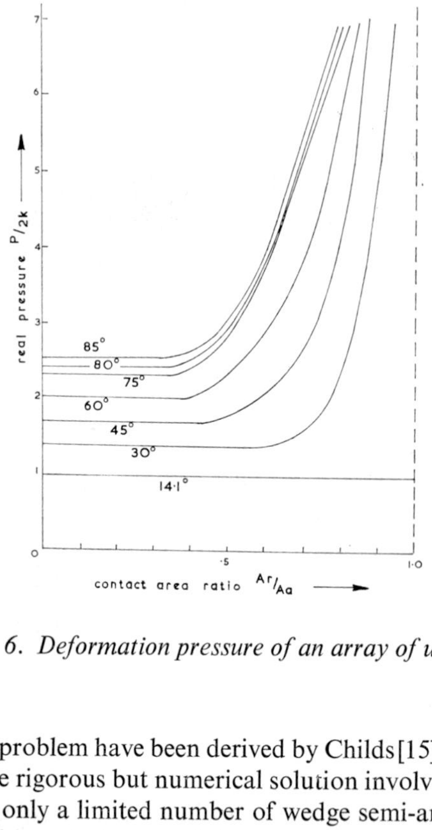

Fig. 6. Deformation pressure of an array of uniform wedges

this problem have been derived by Childs[15]. This method is mathematically more rigorous but numerical solution involves a lengthy graphical procedure and only a limited number of wedge semi-angles have been considered.

Although these analytical developments were primarily directed towards the development of theories of real contact area under high contact stresses it is clearly a simple matter to determine the variation of residual voidage as a function of contact stress on such a basis. Mitchell and Rowe have also considered this aspect and by assuming laminar, incompressible flow in the residual voidage obtained the results shown in Fig. 7. This graph shows a plot of a non-dimensional leakage term, $Q\mu(1/h^3)\{p_2/(p_1^2-p_2^2)\}$, against the non-dimensional contact stress $p/2k$ for various wedge semi-angles. In these non-dimensional groups p_1 and p_2 are the gas pressures on the two sides of the seal, l is the seal width, h is the wedge height and K is the yield shear stress of the gasket material. In these non-dimensional groups p_1 and p_2 are the

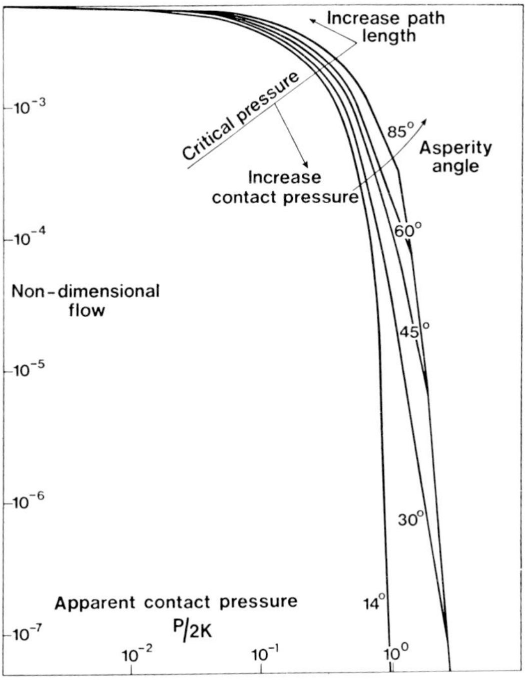

Fig. 7. The variation of leakage with apparent contact stress

gas pressures on the two sides of the seal, l is the seal width, h is the height, k is the yield shear stress of the gasket material, μ is the viscosity of the gas and Q the volumetric leakage rate.

There is usually a practical limit to the load available to compress the gasket and an important decision in seal design is the selection of the gasket width. If the gasket is wide the contact stress will be low but the leakage paths long, whereas for narrow gaskets high contact stresses can be obtained but the leakage paths will be short. It can be shown that a worst gasket width exists and that this corresponds to a point where the gradient of the curves in Fig. 7 is equal to -1. Sealing performance will improve continuously with changing contact stress in either direction from this point. Low contact stresses are employed in large flanged connections (e.g. turbine casings) but most high integrity gaskets aim to provide maximum contact stress. In such high integrity seals, the essence of good seal design lies therefore in providing macroscopic constraints to inhibit the flow of bulk material, thereby maximising the contact stresses.

Chivers, Mitchell and Rowe[16] have made measurements of the real area of contact under high stresses and compared their results to the theoretical predictions of Mitchell and Rowe. They found the results to be in agreement with the theoretical predictions, provided corrections were made for the changes in surface hardness resulting from strain of the surface layers. Indeed, the effects of strain hardening were of such a magnitude that they concluded that it is more important to refine the knowledge of material properties than the analytical method for treating the plastic flow.

This work provides a valuable insight to the mechanics of sealing and enables the potential of a particular seal design to be assessed. One significant limitation of the current theory is the inability to cope with combined normal and tangential loading.

6. ADHESION AND FRICTION OF CORRODING CONTACTS

When surfaces remain in contact for long periods at high temperatures in a chemically reactive atmosphere, it becomes necessary to consider how the properties of the junction might change with time. There are basically two situations of interest:

1. where the surface films are protective in nature and the rate of increase of film thickness decreases with time, and
2. when the surface film grows at an approximately constant rate unaffected by mechanical stresses.

Protective films become of concern when they can achieve dimensions comparable with the surface roughness. In these circumstances, their role can extend beyond that of a simple barrier to adhesion at the points of intimate contact between a pair of metals. By intergrowth of the oxide on the two surfaces, the film can develop strength of its own and this essentially a 'junction growth' mechanism which can develop until the real contact area becomes equal to the total apparent area. At this stage shear strength of the contact would be approximately equal to the shear strength of a thin oxide film and any dependency upon the normal flow would be expected to be weak.

For this situation, the criteria for assessing the strength of junctions would appear to be clear. However, intergrowth of chemical films on surfaces is not the only mechanism by which adhesion can develop at high temperatures. The temperature of many components is such that creep of metals can be a significant factor, and thus the possibility exists that the initial true contact points will grow under the high local stresses. A simplified contact model suggests that, in a fixed time, the contact area would increase by a constant factor independent of the contact stress. If the mean shear strength of the junctions remains constant, the adhesive forces would change by an equal factor.

Experimental results for stainless steels in carbon dioxide indicate that the two processes described above can occur in parallel. Fig. 8 shows a plot of the shear strength against contact stress. The two lower lines on the plot bound the friction coefficients measured immediately after the bond devel-

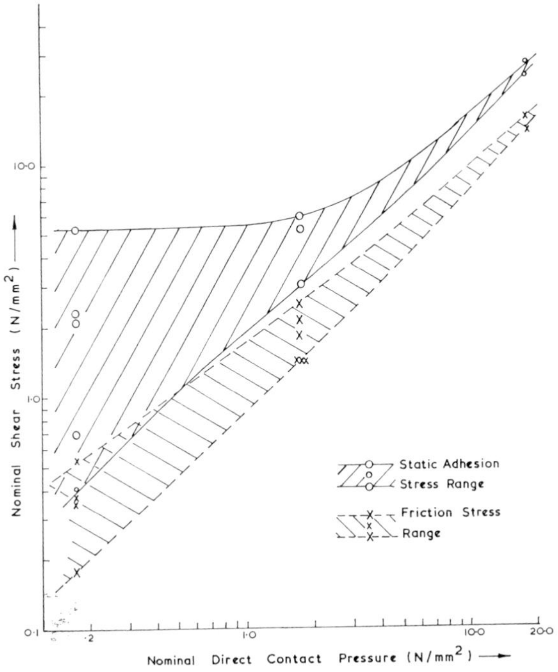

Fig. 8. The effect of contact stress on the shear strength of adhesive bonds developed in an oxidising environment

oped during exposure had been broken. The total adhesive forces can be represented as the sum of the friction force, a constant multiple increase due to creep (a line parallel to the friction line on the log log plot) and an additional constant term attributable to the oxide strength.

This model is rather tentative and requires considerable development, but the general conclusions are of considerable practical importance in relating experimental results to real situations. They suggest that, at low contact stresses, adhesive forces will be largely determined by the oxide cementing process and almost proportional to contact area. However, at high stresses, the dominant term would be the junction growth through creep and the controlling factor would be the normal load. However, as the effects are the combination of two time-rate processes of different activation energies, the relative importance of the two mechanisms would change with temperature.

It would also be expected that it would be very sensitive to the chemical potential of the atmosphere.

When the surface films are not of a protective character, the corrosion can continue until films are large compared with surface roughness features and their thickness may exceed the original mechanical clearances between components. Such is the case with 'breakaway corrosion' of mild steel in carbon dioxide[17].

The high corrosion rates are maintained because gas diffuses through cracks in the oxide layer, forming fresh oxide at the metal/oxide interface. Thus conditions at the original interface between two corroding surfaces are equivalent to the contact of two oxide surfaces, formed relatively early in time, being pressed together under an increasing load. By conducting experiments with simplified geometry selected such that the normal stress is proportional to the extent of oxide interference, it was shown that a conventional frictional relationship applies between the normal load and the tangential load required to produce relative movement of the surfaces. Applying this result to more complex geometries, it is obvious that the problem is one of determination of the interactive forces normal to the plane of the interface.

This point may be illustrated by considering a pin-in-bush geometry, which is a reasonable representation of many practical interfaces. Shear strength against time plots are shown in Fig. 9 for such geometries; two sets of data are given with different ratios of radii. The strains in the bush are well above the elastic limit and at the temperature of oxidation the straining mechanism is through creep. Thus the interface stress continues to grow and approaches asymptotically an equilibrium level at which the creep relaxation

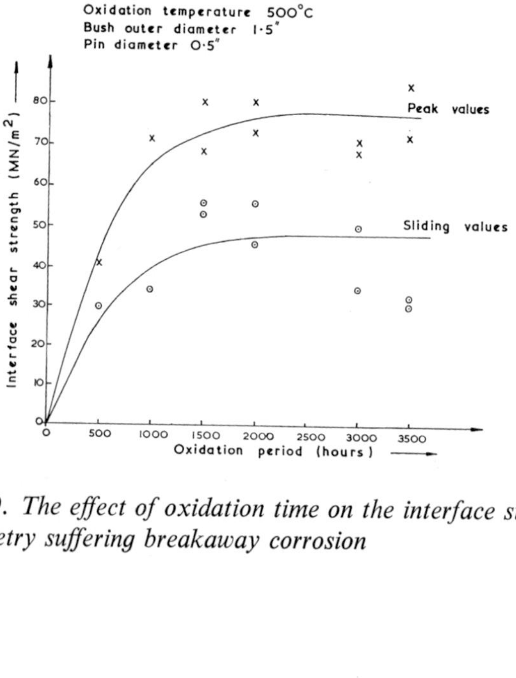

Fig. 9. The effect of oxidation time on the interface strength of a pin-in-bush geometry suffering breakaway corrosion

rate exactly offsets the rate of oxide growth. It can be seen from Fig. 9 that the equilibrium level is dependent upon geometry and it can be shown that the differences are explicable by steady-state creep analyses assuming secondary creep.

The problems remaining in this area relate to more complex geometries and to obtaining solutions for the transient situation whilst the stress is growing. In the latter case, the analysis needs to consider both primary and secondary creep behaviour but in both cases, the general requirement is for improved methods of creep analysis for deformation caused by an expanding interface.

7. CONCLUSIONS

The value of contact theories in seeking solutions to practical problems in nuclear reactor technology associated with contact dependent phenomena has been reviewed. Generally, the theoretical methods are shown to have considerable value, but areas where further developments would be advantageous have been suggested. Of the practical problems considered, contact theories have most deficiencies in relation to the subject of wear. Here theories have almost no predictive value and, at a lower level, cannot be relied upon to improve perspective when examining experimental data. This is a very important problem area in practical terms and any developments of theory would be of immense value.

ACKNOWLEDGEMENTS

This paper is published by permission of the Central Electricity Generating Board.

REFERENCES

1. L.A. Mitchell and C. Osgood, 'Prediction of the Reliability of Mechanisms from Friction Measurements, *Proc. of First European Tribology Congress, I. Mech. E.*, London, 1973, pp. 63–70.
2. L.A. Mitchell and C. Osgood, to be published.
3. F.P. Bowden and D. Tabor, *The Friction and Lubrication of Solids*, Clarendon Press, Oxford, 1950.
4. E. Rabinowicz, B.G. Rightmore, C.E. Tedholm and R.D. Williams, 'The Statistical Nature of Friction', *Trans. ASME*, **77**, 1955, pp. 981–984.
5. J.F. Archard, 'Contact and Rubbing of Flat Surfaces', *J. Appl. Phys.*, **24**, 1953, pp. 981–988.
6. E. Rabinowicz, *Friction and Wear of Materials*, Wiley, New York, 1965, pp. 125–166.
7. J.K. Lancaster, 'The Formation of Surface Films at the Transition between Mild and Severe Wear', *Proc. Roy. Soc.*, **A273**, 1963, pp. 466–483.

8. J.F. Archard and W. Hirst, 'The Wear of Metals under Unlubricated Conditions', *Proc. Roy. Soc.*, A**236**. 1956, pp. 397–410.
9. A. Begelinger and A.W.J. De Gee, 'Synopsis of the Results from an International Co-operative Wear Programme', *Lubrication Engineering*, 1970, pp. 56–63.
10. R. Holm, *Electric Constants*, Springer, Berlin, 1967.
11. J.A. Greenwood, 'Constriction Resistance and the Rael Area of Contact', *Br. J. Appl. Phys.*, **17**, 1966, pp. 1621–1632.
12. W.M. Rohsenow, *Developments in Heat Transfer*, Edward Arnold, London, 1964, pp. 354–370.
13. A.J.W. Moore, 'Deformation of Metals in Static and Sliding Contact, *Proc. Roy. Soc.*, A**195**, 1948, pp. 231ff.
14. L.A. Mitchell and M.D. Rowe, 'Influence of Asperity Deformation Mode on Gas Leakage between Contacting Surfaces', *J. Mech. Eng. Sci.*, **11**, 1969, pp. 534–545.
15. T.H.C. Childs, 'The Persistence of Asperities in Indentation Experiments', *Wear*, **25**, 1973, pp. 3–16.
16. T.C. Chivers, L.A. Mitchell and M.D. Rowe, 'The Variation of Real Contact Area between Surfaces with Contact Pressure and Material Hardness', *Wear*, **28**, 1974, pp. 171–185.
17. J.E. Antill, K.A. Peakall and J.B. Warburton, 'Oxidation of Mild and Low Alloy Steels in CO_2 based Atmospheres', *Corrosion Science*, **8**, 1968, pp. 689–701.

Linearized contact vibration analysis

P. R. Nayak

1. INTRODUCTION

In the Hertzian contact of two solid elastic bodies the relative displacement normal to the contact surface of points distant from the contact point is a nonlinear function of the compressive force. When the two bodies are further made to roll or slide over one another, the waviness of the contacting surfaces causes the bodies to vibrate. In the range of relatively low forcing frequencies, a linear dynamic analysis neglecting the Hertzian displacement is adequate. In the region of relatively high frequencies, however, the nonlinear contact spring causes a 'contact resonance' phenomenon to appear. With usual damping levels, the vibration levels around the contact resonance frequency can be quite high.

Interest in contact vibrations is exemplified by the recent appearance of four publications[1–4], in which experimental and analytical support is provided for the claim that these vibrations are instrumental in causing surface fatigue, surface corrugations, loss of tractive capacity, and high radiated noise levels.

A nonlinear analysis of random contact vibrations of two rolling discs is presented in[2]. However, this analysis is valid only for certain types of surface waviness spectra. Moreover, it is difficult to see how such an analysis could be extended to more complex systems. In response to the need for an alternative analytical technique, Gray and Johnson[3] developed a linearized analysis of the rolling/sliding contact of two discs, and demonstrated reasonable agreement between analysis and experiment. This technique was then used by the present author to investigate the rolling contact of a wheel and a rail[4]. Neither of these two publications, however, suggests a general linearization technique for the analysis of the contact of arbitrary linear elastic solids.

Such a general linearized theory is presented here, with the two attendant assumptions that the input waviness be either a sinusoidal or a stationary Gaussian random process, and that the contacting solids be individually linear. The theory involves replacing the nonlinear contact spring by an equivalent linear spring with precompression, the stiffness and precompression being determined so as to minimize the error in linearization[3]. The response of the contacting bodies is then determined from their point impedances. Tabulations of impedances are available for a wide variety of systems[5, 6].

In section 2, the general nonlinear problem is formulated in terms of the point impedances of the contacting bodies. In section 3, the linearized formulation is developed, and the general solution is reduced to the simultaneous solution of three nonlinear algebraic equations, each involving an integration. Problems involving either a sinusoidal or a random waviness are shown to fit into the framework of the general solution, and for these two cases, specific (but still general) expressions are derived for the amplitude of the dynamic load. For the case of a random Gaussian waviness, general expressions for various statistics of the dynamic load are also derived.

It is shown in section 4 that the formulation of the two problems studied in detail previously[3, 4] follows in a straightforward fashion from the general analysis developed in section 3. For the case of two discs rolling against each other, a detailed comparison is made between the results of the linearized analysis and the nonlinear analysis given in[2]. Important differences between the results of the two analyses are highlighted.

Finally, some concluding thoughts and comments are offered in section 5.

2. FORMULATION OF THE NONLINEAR PROBLEM

Two specific examples of contact vibrations have been discussed in the literature in great detail[2, 3, 4], so that the physics of the problem needs no further detailed exposition. Here, a brief review and generalization is presented.

Let two bodies in point contact roll over each other at a steady velocity V. If their surface waviness be denoted by $h_1(x)$ and $h_2(x)$, x being a spatial coordinate measured along the line of rolling, the total input roughness is given by

$$h(x) = h_1(x) + h_2(x). \tag{1}$$

This input roughness may also be considered to be a function of the time $t = x/V$.

Let $y_i(t)$ $(i = 1, 2)$ be the deflections of reference points within the bodies. These points are assumed to be sufficiently remote from the contact zone so that their relative displacement is given with adequate accuracy by Hertz' expression (see equation (3) below) for the displacement of points 'at infinity'. The displacements y_i are normal to the contact zone and are positive for displacements away from the contact zone. Furthermore, these displacements are measured from the configuration in which the bodies are just touching, under no load, and in the absence of any waviness. Then the elastic Hertzian deflection of the two bodies is

$$z_H = h - (y_1 + y_2), \tag{2}$$

and the contact force given by [7]

$$P_c = C z_H^{3/2} H(z_H) \tag{3}$$

where C is a constant dependent on the curvatures near the contact point and the elastic constants of the two bodies, and $H(z)$ is the Heaviside unit-step function, defined by

$$H(z) = \begin{cases} 1, & z \geq 0 \\ 0, & z < 0. \end{cases} \tag{4}$$

In the following, it is generally assumed that C is time-invariant, that is, that the geometrical and elastic properties of the bodies around the contact point are constant over time.

The contact force will, in general, have a nonzero mean, equal to the static contact force, P_0:

$$\overline{P_c(t)} = P_0. \tag{5}$$

This static contact force is an externally applied, 'given' force, i.e. it is an independent variable.

Thus, the dynamic contact force may be defined by

$$P_d = P_c - P_0. \tag{6}$$

Now if the equations of motion of the individual contacting bodies are linear, a point impedance $Z_i(\omega)$ $(i = 1, 2)$ may be defined for each of the two bodies:

$$Z_i(\omega) = P_d/(\mathrm{d}y_i/dt),$$

where P_d is taken to have a harmonic temporal variation with a circular frequency $\omega = 2\pi f$:

$$P_d = P_{d0} \exp(-i\omega t).$$

Then the deflections $y_j(t)$ may be written in the Fourier form

$$y_j(t) - \bar{y}_j = \frac{i}{(2\pi)^{1/2}} \int_{-\infty}^{\infty} \frac{v_j(\omega)}{\omega} \mathrm{e}^{-i\omega t} \, \mathrm{d}\omega = \frac{i}{(2\pi)^{1/2}} \times$$

$$\times \int_{-\infty}^{\infty} \frac{P_d(\omega)}{\omega Z_j(\omega)} \mathrm{e}^{-i\omega t} \, \mathrm{d}\omega, \tag{7}$$

where $\bar{y}_j$ is the mean value of $y_j(t)$, $v_j(\omega)$ is the Fourier transform of

$$v_j(t) = \mathrm{d}y_i/\mathrm{d}t, \tag{8}$$

and $P_d(\omega)$ is the Fourier transform of $P_d(t)$. Equation (7) thus holds for any temporal variation of P_d.

Combining equations (2), (3), (6) and (7), one obtains the following equation of motion

$$P_d(t) = C\left\{h(t) - (\bar{y}_1+\bar{y}_2) - (2\pi)^{-1/2}\int_{-\infty}^{\infty}\left(\frac{1}{Z_1(\omega)}+\frac{1}{Z_2(\omega)}\right)\times \right.$$

$$\left. \times \frac{P_d}{\omega}\,e^{-i\omega t}\,d\omega\right\}^{3/2} H - P_0. \tag{9}$$

the argument of H being the expression appearing within the braces.

Thus the problem is reduced to solving equation (9) given Z_1, Z_2, P_0 and $h(t)$, subject to the condition that $\bar{P}_d(t)=0$.

This nonlinear problem cannot in general be solved exactly by analytical means. In the sequel, a linearized approximation to the problem is developed, and a general linearized solution presented.

Before proceeding to the solution, it is worth noting that equation (9) is valid only for frequencies such that the contact zone and the 'distant' reference points (with displacement y_i) are moving in phase. Speaking generally, the wavelength λ_E of elastic waves propagating into the bodies must be large compared to the distance from the contact zone to the reference points. Approximately, the requirement is $\lambda_E \gg \kappa^{-1}$ where κ is the mean curvature of the surface of the contacting body at the contact point. Since $\lambda_E = 2\pi C_E/\omega$ where C_E is the speed of propagation of the elastic waves, an equivalent requirement is $\omega \ll 2\pi C_E \kappa$.

3. THE LINEARIZED SOLUTION

The linearization used here was first applied to the contact vibration problem by Gray and Johnson[3] and later by Nayak[4]. It involves replacing equation (3) by the linear approximation

$$P_c = P_0 + K_L(z_H - \bar{z}_H), \tag{10}$$

where K_L and $\bar{z}_H$ are constants to be determined in such a way as to minimize the error involved in linearization. K_L is an equivalent linear contact stiffness, and $\bar{z}_H$ the mean (linearized) Hertzian deflection. When $\bar{z}_h=0$, $P_c=0$ according to equation (3), but $P_c = P_0 - K_L\bar{z}_h$ according to equation (10). The latter value, is a pre-compression force in the linearized problem.

Since, by definition, $\bar{h}(t)=0$, equation (2) yields

$$\bar{z}_H = -(\bar{y}_1+\bar{y}_2), \tag{11}$$

and equation (10) may therefore be written in the form

$$P_d = P_c - P_0 = K_L\{h - (y_1-\bar{y}_1) - (y_2-\bar{y}_2)\}. \tag{12}$$

Equations (7) and (12) may now be combined, resulting in

$$P_d(t) = K_L\, h(t) - \frac{iK_L}{(2\pi)^{1/2}} \int_{-\infty}^{\infty} \frac{P_d(\omega)}{\omega Z_c(\omega)}\, \mathrm{e}^{-i\omega t}\, \mathrm{d}\omega\,, \tag{13}$$

where $Z_C(\omega)$ is a composite impedance defined by

$$Z_c(\omega) = \frac{Z_1(\omega) + Z_2(\omega)}{Z_1(\omega)\, Z_2(\omega)}. \tag{14}$$

The Fourier transform of equation (13) is

$$P_d(\omega) = \frac{K_L\, h(\omega)}{1 + iK_L/\omega Z_c(\omega)}, \tag{15}$$

where $h(\omega)$ is the transform of $h(t)$.

It immediately follows from equation (15) that within the approximations made, $P_d(t)$ has a power spectral density (PSD)

$$\Phi_{Pd\,Pd}(\omega) = \frac{K_L^2}{|1 + iK_L/\omega Z_c|^2}\, \Phi_{hh}(\omega)\,, \tag{16}$$

where $\Phi_{hh}(\omega)$ is the PSD of $h(t)$, related to the PSD (in the spatial domain) of $h(x)$ by

$$\Phi_{hh}(\omega) = \Phi_{hh}(k) \left| \frac{\mathrm{d}k}{\mathrm{d}\omega} \right|, \tag{17}$$

where $k = 2\pi/\lambda$ is the wavenumber, λ being the wavelength.

Since

$$\omega = 2\pi f = 2\pi V/\lambda = Vk\,,$$

equation (17) becomes

$$\Phi_{hh}(\omega) = \frac{1}{V}\, \Phi_{hh}(k = \omega/V)\,, \tag{18}$$

and equation (16) may be written in the form

$$\Phi_{Pd\,Pd}(\omega) = \frac{K_L^2}{V} \frac{\Phi_{hh}(k)}{|1 + iK_L/\omega Z_c(\omega)|^2}. \tag{19}$$

The variance of P_d is given by

$$\sigma_{Pd}^2 = \int_{-\infty}^{\infty} \Phi_{Pd\,Pd}(\omega)\, \mathrm{d}\omega\,, \tag{20}$$

and that of z_H by

$$\sigma_{zH}^2 = (\sigma_{Pd}/K_L)^2\,. \tag{21}$$

Now suppose the probability density of z_H is denoted by $p(z_H)$; then it follows that the probability density of the dynamic load P_d is given by

$$p(P_d) = p[K_L(z_H - \bar{z}_H)]. \tag{22}$$

To complete the general analysis, one must determine $\bar{z}_H$, K_L, $\bar{y}_1$, and $\bar{y}_2$. The constants $\bar{z}_H$ and K_L may be determined by developing to constraints minimizing the error in linearization. One is a physical constraint, that the mean value of the contact load should be P_0. Thus, since $P_c = Cz_H^{3/2} H(z_H)$, one must have

$$C \int_0^\infty z_H^{3/2}\, p(z_H)\, \mathrm{d}z_H = P_0. \tag{23}$$

A second constraint is obtained by minimizing the error

$$E = \int_{-\infty}^\infty [Cz_H^{3/2} H(z_H) - P_0 - K_L(z_H - \bar{z}_H)]^2\, p(z_H)\, \mathrm{d}z_H$$

with respect to K_L, resulting in

$$K_L \int_{-\infty}^\infty (z_H - \bar{z}_H)^2\, p(z_H)\, \mathrm{d}z_H = \int_0^\infty [Cz_H^{3/2} - P_0]\, (z_H - \bar{z}_H)\, p(z_H)\, \mathrm{d}z_H,$$

or, from the definitions of σ_{zH}^2 and $\bar{z}_H$,

$$K_L \sigma_{zH}^2 = C \int_0^\infty z_H^{3/2} (z_H - \bar{z}_H)\, p(z_H)\, \mathrm{d}z_H. \tag{24}$$

There are now three equations relating $\bar{z}_H$, σ_{zH} and K_L: equations (21), (23) and (24). (Note that $p(z_H)$ contains $\bar{z}_H$ and σ_{zH} as parameters.) If these three equations are solved simultaneously, $\bar{z}_H$, σ_{zH} and K_L may simultaneously be determined. How this is to be done is demonstrated below for sinusoidal and random waviness. Before this, $\bar{y}_1$ and $\bar{y}_2$ must be determined. This is simple, since they are simply the static values of y_1 and y_2. Suppose $\bar{y}_1$ is determined from the static problem for body 1 relative to some arbitrary datum. Then $\bar{y}_2 = -(\bar{y}_1 + \bar{z}_H)$, from equation (11).

3.1 *Sinusoidal waviness*

If the surface waviness may be represented by

$$h = h_0 \sin(2\pi x/\lambda) = h_0 \sin(\omega_0 t),$$

where $\omega_0 = 2\pi V/\lambda$, then

$$\Phi_{hh}(\omega) = \frac{h_0^2}{4} [\delta(\omega - \omega_0) + \delta(\omega + \omega_0)],$$

and, from equations (16), (20) and (21),

$$\sigma_{zH}^2 = \frac{h_0^2 |Z_c(\omega_0)|^2}{2|Z_c(\omega_0) + iK_L/\omega_0|^2}. \tag{25}$$

Now, for the linearized system, when the input h is sinusoidal, so will be the response z_H:

$$z_H = \bar{z}_H + \sqrt{2}\,\sigma_{zH} \sin(\omega_0 t + \theta), \tag{26}$$

where θ is a phase lag, given by

$$\tan\theta = -\left\{\frac{1 - \text{Imaginary part of } [K_L/\omega_0\, Z_c(\omega_0)]}{\text{Real Part of } [K_L/\omega_0\, Z_c(\omega_0)]}\right\}. \tag{27}$$

Thus, the probability density of z_H is given by

$$p(z_H) = \begin{cases} \dfrac{1}{\pi}\,[2\sigma_{zH}^2 - (z_H - \bar{z}_H)^2]^{-1/2}, & |z_H - z_H| \le \sqrt{2}\,\sigma_{zH} \\ 0 \quad \text{otherwise}. \end{cases} \tag{28}$$

Introducing this expression into equation (23) and using the transformation $(z_H - \bar{z}_H) = \sqrt{2}\sigma_{zH} \sin\varphi$, one obtains

$$P_0 = \frac{C}{\pi}\,(\sqrt{2}\,\sigma_{zH})^{3/2} \int_{\varphi_{\min}}^{\pi/2} (u_0 + \sin\varphi)^{3/2}\, \mathrm{d}\varphi, \tag{29}$$

where

$$u_0 = \bar{z}_H/\sqrt{2}\,\sigma_{zH} \tag{30}$$

and

$$\varphi_{\min} = \begin{cases} -\pi/2, & \bar{z}_H \ge \sqrt{2}\,\sigma_{zH} \\ -\sin^{-1}(\bar{z}_H/\sqrt{2}\,\sigma_{zH}) & \text{otherwise}. \end{cases} \tag{31}$$

One similarly obtains from equation (24) the following constraint:

$$K_L = \frac{2^{5/4}\,C}{\pi}\,\sigma_{zH}^{1/2} \int_{\varphi_{\min}}^{\pi/2} \sin\varphi\,(u_0 + \sin\varphi)^{3/2}\, \mathrm{d}\varphi. \tag{32}$$

From equations (29) and (31), one obtains (by numerical methods) a relationship between K_L/K_{0H} and σ_{zH}/z_{0H}, where

$$K_{0H} = 1.5\,P_0^{1/3}\,C^{2/3} \tag{33}$$

is the small-amplitude stiffness of the Hertzian contact, and

$$z_{0H} = (P_0/C)^{2/3} \tag{34}$$

is the static Hertzian deflection. This relationship is shown in Fig. 1(a).

The solution may now be completed as follows. For any given value of h_0, superpose on Fig. 1(a) the relationship between σ_{zH}/z_{0H} and K_L/K_{0H} given by equation (25). Intersections of the two curves give possible solutions, with values of σ_{zH} and K_L given simultaneously. The mean deflection $\bar{z}_H$ may then be obtained from equation (29), whose graph is shown in Fig. 1(b). By varying ω_0, the complete response diagram may be obtained. Questions of

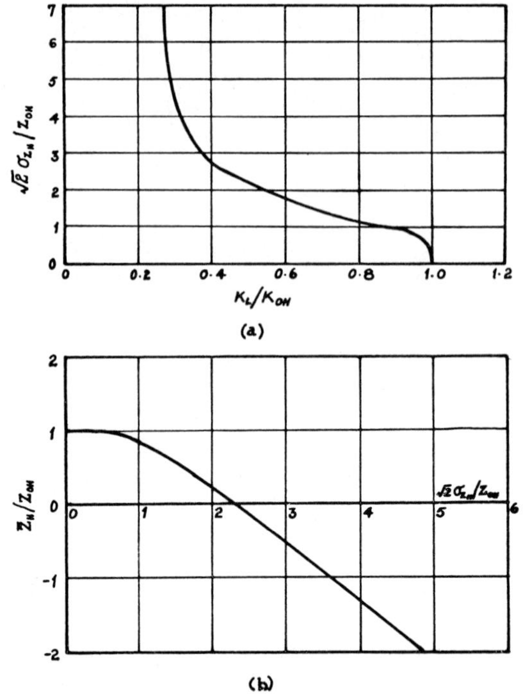

Fig. 1. Constraints obtained by minimizing the linearization error for sinusoidal inputs. (a) Relation between the standard deviation of the Hertzian deflection and the equivalent linear stiffness K_L. (b) Relation between the mean value $\bar{z}_H$ of the Hertzian deflection and σ_{zH}.

stability of the response do remain, but will not be dealt with here. A brief discussion of stability may be found in[2].

3.2 *Random waviness*

If the surface waviness is a stationary, Gaussian random function with a PSD $\Phi_{hh}(k)$, then its PSD in the frequency domain is given by equation (18). Furthermore, in the linearized solution, z_H will also be a Gaussian random function, with a PSD given by equation (16). Thus, the probability density of z_H is

$$p(z_H) = \frac{1}{\sigma_{zH}(2\pi)^{1/2}} \exp\left[-\frac{1}{2}\left(\frac{z_H - \bar{z}_H}{\sigma_{zH}}\right)^2\right] \tag{35}$$

Introducing this expression into equation (23), one obtains[3]

$$P_0 = \frac{C\sigma_{zH}^{3/2}}{(2\pi)^{1/2}} \int_{-u_0}^{\infty} (u+u_0)^{3/2} \exp(-u^2/2)\, du\,, \tag{36}$$

where

$$u_0 = \bar{z}_H/\sigma_{zH}\,. \tag{37}$$

Similarly, equation (24) yields

$$K_L = \frac{C\sigma_{zH}^{1/2}}{(2\pi)^{1/2}} \int_{-u_0}^{\infty} u(u+u_0)^{3/2} \exp(-u^2/2)\, du\,. \tag{38}$$

Equations (36) and (38) may be manipulated to yield two graphs, one showing (σ_{zH}/z_{0H}) as a function of (K_L/K_{0H}), the other showing $(\bar{z}_H/z_{0H})$ as

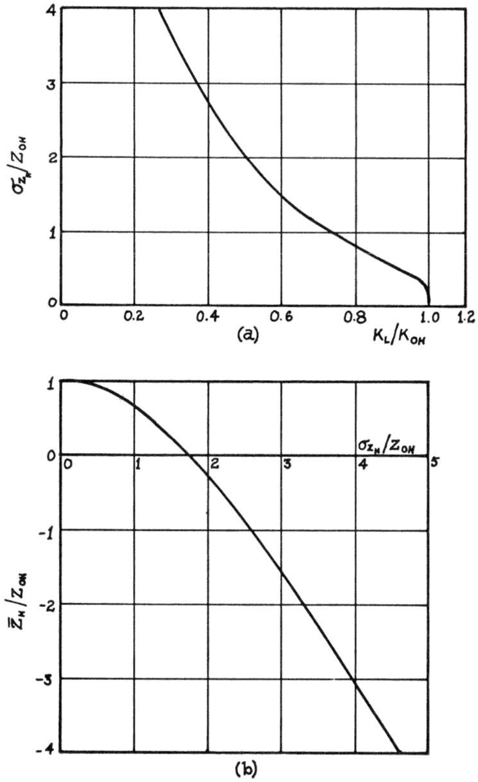

Fig. 2. Constraints obtained by minimizing the linearization error for random Gaussian inputs. (a) Relation between the standard deviation σ_{zH} of the Hertzian deflection and the equivalent linear stiffness K_L. (b) Relation between the mean value z_H of the Hertzian deflection and σ_{zH}.

a function of (σ_{zH}/z_{0H}). These graphs are in Figs. 2(a) and 2(b) respectively. Equations (16), (20) and (21) may be solved to obtain (σ_{zH}/z_{0H}) as another function of K_L/K_{0H}; this function may then be superposed on Fig. 2(a) to obtain the solution, which is the intersection of the two graphs. Once σ_{zH} and K_L are obtained, $\bar{z}_H$ may be obtained from Fig. 2(b). The spectra of the dynamic displacements y_{di} follow from equation (7):

$$\Phi_{ydi\,ydi}(\omega) = \Phi_{Pd\,Pd}(\omega)/\omega^2\,|Z_i(\omega)|^2 \quad (i = 1, 2). \tag{39}$$

For the sake of clarity it may be noted that once σ_{zH} and K_L are obtained, σ_{Pd} is obtained from $\sigma_{Pd} = K_L \sigma_{zH}$. The probability density of the dynamic contact load P_d is then given by

$$p(P_d) = \frac{1}{\sigma_{Pd}(2\pi)^{1/2}} \exp[-\tfrac{1}{2}(P_d/\sigma_{Pd})^2], \tag{40}$$

and of the total contact load $P_c = P_0 + P_d$ by

$$p(P_c) = \frac{1}{\sigma_{Pd}(2\pi)^{1/2}} \exp\left[-\frac{1}{}\left(\frac{P_c - P_0}{\sigma_{Pd}}\right)^2\right]. \tag{41}$$

3.3 *Statistics of the contact load*

Only the case of random waviness need be dealt with here. Typical statistics of interest are:

a. The probability of loss of contact.
b. The probability of plastic indentation of the contacting surfaces.
c. The frequency of plastic indentations.
d. The average length of these indentations.
e. The probability distribution of the depth of the indentations.
f. The mean 'clump size' of indentations, which is an indicator of the tendency of indentations to be grouped together, rather than being uncorrelated spatially.

3.3.1 *The probability of loss of contact*

In the nonlinear analysis, there is no ambiguity in the definition of loss of contact: it corresponds to both $P_c = P_0 + P_d = 0$ and $z_H \leq 0$. In the linearized analysis, however, when $P_C = 0$, $z_H = -P_0/K_L + \bar{z}_H$, and when $z_H = 0$, $P_C = P_0 - K_L \bar{z}_H$. Since there appears to be no physical argument to guide one in the choice of definition, both definitions will be used here, a choice being made in section 4 on the basis of a comparison with the nonlinear theory.

For $P_C \leq 0$, the probability is found from equation (41):

$$p(P_c \leq 0) = \tfrac{1}{2}\operatorname{erfc}(P_0/\sigma_{Pd}\sqrt{2}) = \tfrac{1}{2}\operatorname{erfc}\left(\frac{2}{3}\frac{K_{0H}}{K_L}\frac{z_{0H}}{\sigma_{zH}}\right). \tag{42}$$

the latter expression resulting from the definitions of σ_{zH}, K_{0H} and z_{0H} in equations (21), (33) and (34).

For $z_H \leq 0$, the probability is found from equation (35):

$$p(z_H \leq 0) = \tfrac{1}{2}\,\mathrm{erfc}(\bar{z}_H\,\sigma_{zH}\sqrt{2}) = \tfrac{1}{2}\,\mathrm{erfc}\left(\frac{1}{\sqrt{2}}\,\frac{\bar{z}_H}{z_{0H}}\,\frac{z_{0H}}{\sigma_{zH}}\right). \tag{43}$$

Only when $K_L\,\bar{z}_H = P_0$ will equations (42) and (43) yield identical results.

3.3.2 *The probability of plastic indentations*

Plastic indentation occurs when the contact load P_c exceeds some plastic load P_p. This plastic load depends, among other things, on the curvatures of the contacting surfaces, the yield strengths of the bodies, and on whether or not work-hardening and shake-down have occurred. These factors will not be considered here. The probability of plastic indentation is

$$p(P_c \geq P_P) = \tfrac{1}{2}\,\mathrm{erfc}\left(\frac{P_P - P_0}{\sigma_{Pd}\sqrt{2}}\right). \tag{44}$$

3.3.3 *The frequency of plastic indentations*

The frequency ν_P with which indentations occur is given by[2]:

$$\nu_P = \frac{1}{2\pi}\,\frac{\sigma_{vH}}{\sigma_{zH}}\,\exp\left[-\frac{1}{2}\left(\frac{P_P - P_0}{\sigma_{Pd}}\right)^2\right], \tag{45}$$

where σ_{vH} is the standard deviation of $v_H \equiv \mathrm{d}z_H/\mathrm{d}t$; it is given by

$$\sigma_{vH}^2 = \frac{1}{K_L^2}\int_{-\infty}^{\infty} \omega^2\,\Phi_{Pd\,Pd}(\omega)\,\mathrm{d}\omega\,. \tag{46}$$

3.3.4 *The average indentation length*

When an indentation occurs, its length along the line of rolling will at least equal the length of the contact zone, denoted here by 'a'. If the contact load P_c exceeds the plastic load P_p for a finite length of time t_0, the length of the indentation will be greater than a by an amount Vt_0, V being the rolling velocity (strain rate effects being ignored here). The average value of t_0 is

$$\bar{t}_0 = p(P_c \geq P_p)/\nu_p,$$

so that the average indentation length is

$$\bar{l}_i = a + V p(P_c \geq P_p)/\nu_P\,. \tag{47}$$

3.3.5 *The indentation depth*

The Hertzian deflection corresponding to the plastic load P_p is

$$z_{HP} = (P_p - P_0)/K_L + \bar{z}_H\,. \tag{48}$$

When $z_H \geq z_{HP}$, an indentation of depth $(z_H - z_{HP})$ occurs. The mean depth of all indentations is thus given by

$$\bar{d}_i = \int_{zHP}^{\infty} (z_H - z_{HP})\, p(z_H)\, \mathrm{d}z_H$$

$$= \sigma_{zH} \left\{ \frac{1}{\sqrt{2\pi}} \exp(-q^2) - \frac{q}{\sqrt{2}} \operatorname{erfc}(q) \right\}, \quad q \equiv (z_{HP} - \bar{z}_H)/\sigma_{zH}\sqrt{2}. \tag{49}$$

3.3.6 *The clump size of indentations*

Depending on whether the spectrum $\Phi_{PdPd}(\omega)$ is narrow-band or not, indentations will or will not tend to be clustered together. For a narrow-band spectrum, $P_d(t)$ will resemble a free, undamped oscillation, the amplitude varying slowly with time. If the amplitude modulation is slow compared to the oscillation frequency, one indentation will generally be followed by many more, all occurring in the period when the envelope is above the level P_p. The average number of indentations in a cluster is termed the 'clump size', and is given by [4]:

$$\bar{n}_i = \frac{1}{(2\pi)^{1/2}} (1 - \lambda_1^2/\sigma_{zH}^2 \sigma_{vH}^2)^{-1/2} \left(\frac{K_L \sigma_{zH}}{P_P - P_0} \right), \tag{50}$$

where

$$\lambda_1 = \frac{2}{K_L^2} \int_0^{\infty} \omega \Phi_{PdPd}(\omega)\, \mathrm{d}\omega. \tag{51}$$

For a broad-band contact force spectrum, $\bar{n}_i$ tends to be a small number – often less than one – and loses its physical significance. Alternative estimates of $\bar{n}_i$ not having this deficiency are available [8], but they differ little from equation (50) for narrow-band spectra, on which most interest is likely to be concentrated.

4. EXAMPLES

Two examples will be considered:

a. a wheel on a rail;

b. one disc against another.

Both have been studied before; the first one will be used only to demonstrate the application of the general theory of section 3. The second example – that of two discs – will be considered in detail in order to explore the accuracy of the linearized solution.

The random waviness spectrum will be assumed in both cases to have the form [9]:

$$\Phi_{hh}(k) = 1/Lk^4, \tag{52}$$

where L is a constant indicating the magnitude of the waviness; large values of L indicate smooth surfaces. In using this spectrum, one precaution must be observed. Equation (52) indicates a singularity at $k=0$ (i.e., wavelength $=\infty$), which in fact does not exist in nature. Equation (52) is simply a good representation at the frequencies one is normally concerned with in contact vibration analysis. Thus, in various integrals over the range $(-\infty, +\infty)$, the singularity at $k=0$ (or $\omega=0$) will be ignored.

From equation (18), one obtains

$$\Phi_{hh}(\omega) = V^3/L\omega^4 . \tag{53}$$

4.1 *Wheel on a rail*

If the wheel be modeled as a lumped mass M, its impedance is

$$Z_1 = -i\omega M . \tag{54}$$

If the rail be modeled as an infinite beam (with mass m per unit length and bending rigidity EI) on a point-reacting foundation (stiffness K_f), its impedance is

$$Z_2 = \frac{2\sqrt{2}K_{EQ}}{\omega} i(1-\Omega^2)^{3/4} , \tag{55}$$

where

$$K_{EQ} = (EI)^{1/4} (K_f)^{3/4}, \quad \Omega = \omega/\omega_{nr}, \quad \text{and} \quad \omega_{nr} = (K_f/m)^{1/2}.$$

For $\Omega > 1$, that branch of the function is to be chosen which yields

$$(1-\Omega^2)^{3/4} = -(\Omega^2-1)^{3/4} (1+i)/\sqrt{2}.$$

4.1.1 *Sinusoidal waviness*

Using equations (14), (54) and (55) in equation (25), one obtains

$$\sigma_{zH}^2 = \frac{h_0^2}{2} \left| 1 - \frac{K_L}{M\omega_{nr}^2\Omega_0^2} + \frac{(K_L/K_{EQ})}{2\sqrt{2}(1-\Omega_0^2)^{3/4}} \right|^{-2} , \tag{56}$$

where $\Omega_0 = \omega_0/\omega_{nr}$.

When this function $\sigma_{zH}(K_L)$ is superposed on Fig. 1(a), the solution for K_L and σ_{zH} is immediately obtained. The mean deflection $\bar{z}_H$ may then be obtained from Fig. 1(b).

4.1.2 *Random waviness*

By combining equations (14), (19) to (21), and (53) to (55), one obtains

$$\sigma_{zH}^2 = \frac{1}{L}\left(\frac{V}{\omega_{nr}}\right)^3 \int_{-\infty}^{\infty} \left| 1 - \frac{K_L}{M\omega_{nr}^2\Omega^2} + \frac{(K_L/K_{EQ})}{2\sqrt{2}(1-\Omega^2)^{3/4}} \right|^{-2} \frac{d\Omega}{\Omega^4} . \tag{57}$$

The procedure is to numerically obtain σ_{zH} as a function of K_L and superpose this function on Fig. 2(a), thus obtaining the solution values of K_L and σ_{zH}. $\bar{z}_H$ may then be obtained from Fig. 2(b).

If one assumes that $K_L \ll \omega^2 M$ in the frequency range where the spectral representation of equation (52) is correct, and furthermore confines one's attention to the region $\Omega \geq 1$, where the integrand will be large, one obtains

$$\sigma_{zH}^2 \approx \frac{1}{L}(V/\omega_{nr})^3 \int_1^\infty \frac{d\Omega}{\Omega^4\{[(\Omega^2-1)^{3/4}-r_K]^2 + r_K^2\}}, \tag{58}$$

where $r_K \equiv K_L/4k_{EQ}$.

Equation (58) is identical to that obtained in [4] but for a factor 2, which arises from the fact that the waviness spectrum is taken to be two-sided here, whereas it was taken to be one-sided in [4].

4.2 *Two rolling discs*

For the sake of simplicity, it is assumed that both disc-systems are identical, and may be modeled by a mass M with an attached spring of stiffness K_S and a dashpot with velocity coefficient C_D. The impedances Z_1, Z_2 and Z_C are then given by

$$Z_1 = Z_2 = 2Z_c = C_D - i(\omega M - K_s/\omega). \tag{59}$$

Again, for the sake of simplicity, it is assumed that in the frequency region of interest (i.e., $\omega^2 \approx 2K_{0H}/M$), $K_s \ll \omega^2 M$, i.e. that the 'suspension' spring stiffness is small compared to the contact stiffness. Then

$$Z_c \approx \tfrac{1}{2}(C_D - i\omega M). \tag{60}$$

4.2.1 *Sinusoidal waviness*

The procedure of section 4.1.1 yields

$$\left(\frac{\sigma_{zH}}{z_{0H}}\right)^2 = \frac{1}{2}\frac{\Omega_0^2(\Omega_0^2+4\zeta^2)}{(K_L/K_{0H}-\Omega_0^2)^2 + 4\zeta^2\Omega_0^2}\left(\frac{h_0}{z_{0H}}\right)^2, \tag{61}$$

where

$$\Omega_0 = (\omega_0/\omega_{0c}), \quad \zeta = C_D\omega_{0c}/4K_{0H}, \tag{62}$$

ω_{0c} being the small-amplitude contact resonance frequency, given by

$$\omega_{0c} = (2K_{0H}/M)^{1/2}. \tag{63}$$

The constants K_{0H} and $\bar{z}_{0H}$ are defined in equations (33) and (34). As in section 3, ω_0 is the frequency of the input waviness.

When the graph of equation (61) is superposed on Fig. 1(a) for given values of Ω_0, ζ and $(h_0/\bar{z}_{0H})$, three outcomes are possible. One, two or three intersections may occur. A single intersection yields the high and low frequency asymptotes of the response σ_{zH} as a function of Ω_0. Two intersections indicate 'jump' frequencies, the response jumping from a low to a a high value when sweeping the frequency Ω_0 upwards, at an 'upward jump' frequency $\hat{\Omega}$, and from a high to a low value when sweeping the frequency Ω_0 downwards, at a 'downward jump' frequency $\breve{\Omega}$. The frequency $\hat{\Omega}$ is mainly a function of (h_0/z_{0H}). The frequency $\breve{\Omega}$ is mainly a function of $(h_0/z_{0H}\zeta)$. When three intersections occur, the solution corresponding to the intermediate value of σ_{zH} is unstable. Which of the two stable solutions

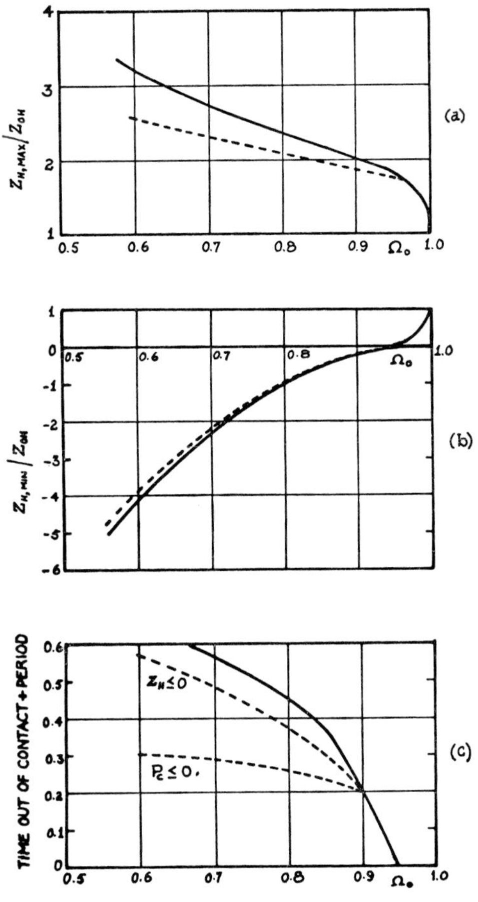

Fig. 3. Comparison of the linearized and nonlinear solutions for the free undamped vibrations of two discs. (a) The maximum value of the Hertzian deflection. (b) The minimum value of the Hertzian deflection. (c) The fraction of a period that the discs are out of contact. Ω_0 is the normalized vibration frequency

——— nonlear; – – – – linearized.

will occur depends on the history of the system. A detailed discussion of all these phenomena can be found in [2].

A comparison of the linearized and the nonlinear analyses is facilitated if one confines one's attention to the problem of free undamped vibrations. From this comparison, much may be learnt about the problem of forced, damped vibrations. The latter problem contains too many variables to allow a comparison brief enough to be accommodated here.

For free undamped oscillations, $h_0 = \zeta = 0$ in equation (61). Thus, a non-zero value of σ_{zH} can result only if $K_L/K_{0H} = \Omega_0^2$. The graph of Fig. 1(a) therefore represents the solution if Ω_0^2 is substituted for K_L. In [2], a complete solution to the nonlinear problem may be found. Fig. 3 compares the linearized and nonlinear results.

In Fig. 3(a), the values of the maximum deflection during the oscillation, $z_{H,\max}$ are compared. It may be seen that except near $\Omega_0 = 1$, the value from the linearized analysis is considerably lower than that from the nonlinear analysis.

In Fig. 3(b), the values of the minimum deflection during the oscillation, $z_{H,\min}$, are compared. Here too, the modulus of the value is larger for the nonlinear analysis than for the linearized analysis, but not by much.

In Fig. 3(c), the fraction of a period for which the bodies are out of contact are compared. In the linearized analysis, two definitions of loss of contact are possible, as was pointed out in section 3.3.1: one corresponds to $z_H \leq 0$, the other to $P_c \leq 0$. From Fig. 3(c), it is apparent that the more accurate result is obtained by the definition $z_H \leq 0$. Even with this definition, however, the linearized analysis results in a considerable underestimate of the time spent out of contact, except near $\Omega_0 = 1$.

From Fig. 3, one may conclude that for the case of harmonic inputs with relatively low damping, both the likelihood of high stresses (i.e. high z_H) as well as the time spent out of contact are likely to be underestimated by the linearized analysis. The degree of underestimation depends on the input (h_0/z_{0H}), the damping ζ and the forcing frequency Ω_0. In general, the error will be significant only when analysing the large-amplitude stable solution.

4.2.2 *Random waviness*

Here, one obtains

$$\left(\frac{\sigma_{zH}}{z_{0H}}\right)^2 = \frac{(V/\omega_{0c})^3}{L z_{0H}^2} \int_{-\infty}^{\infty} \frac{\Omega^2 + 4\zeta^2}{\Omega^2\{(K_{0H}/K_L - \Omega^2)^2 + 4\zeta^2\Omega^2\}} \, d\Omega . \tag{64}$$

Ignoring the multiple pole at $\Omega = 0$ (see the beginning of section 4), one obtains the following expression by contour integration:

$$(\sigma_{zH}/z_{0H}) = \left[\frac{\pi(V/\omega_{0c})^3}{2\zeta L z_{0H}^2}\right]^{1/2} (K_{0H}/K_L)^{1/2}, \quad \zeta \ll 1 . \tag{65}$$

In [2] it has been shown that the exact value (i.e. the value obtained from the nonlinear analysis) of the standard deviation of the normalized Hertzian velocity $\dot{\eta} = v_H/z_{0H}\omega_{0c}$ is given by

$$\sigma_{\dot{\eta}}^* = \left[\frac{\pi(V/\omega_{0c})^3}{2\zeta L z_{0H}^2}\right]^{1/2}, \tag{66}$$

so that equation (65) becomes

$$(\sigma_{zH}/z_{0H}) = \sigma_{\dot{\eta}}^*(K_{0H}/K_L)^{1/2}. \tag{67}$$

For each value of $\sigma_{\dot{\eta}}^*$, the graph of equation (66) may be superposed on Fig. 2(a) to obtain solution values of (K_L/K_{0H}) and (σ_{zH}/z_{0H}). The solution value of $(\bar{z}_H/z_{0H})$ may then be obtained from Fig. 2(b). These solutions are shown in Fig. 4.

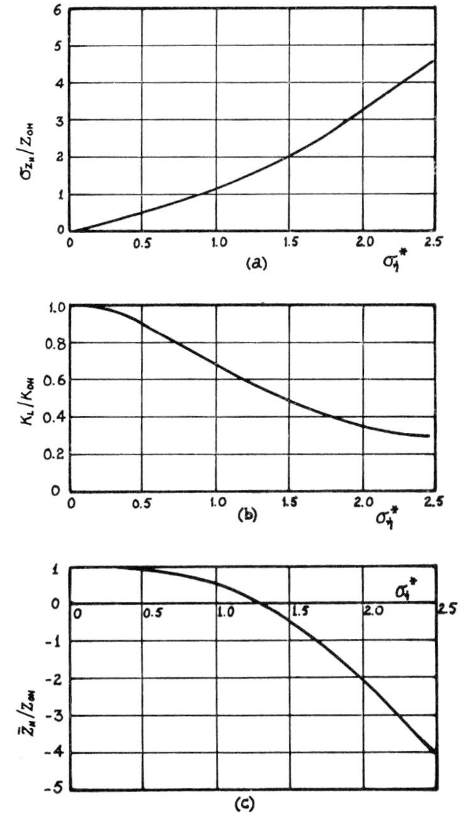

Fig. 4. The linearized solution for the contact of two discs with a random surface roughness. $\sigma_{\dot{\eta}}^*$ is an input variable defined in equation 66. (a) The standard deviation of the Hertzian deflection. (b) The equivalent linearized stiffness. (c) The mean value of the Hertzian deflection

The linearized probability density for z_H is given by equation (35). The exact (nonlinear) density was found in [2] to be given by

$$p(z_H) = G \exp\left\{-\frac{2}{3(\sigma_{\dot{\eta}}^*)^2}\left[0.4(z_H/z_{OH})^{5/2}\, H(z_H) - (z_H/z_{OH})\right]\right\}, \qquad (58)$$

where G is a normalizing constant.

The probability densities of equations (35) and (68) are compared in Figs. 5 and 6. It is apparent that only for low values of $\sigma_{\dot{\eta}}^*$ – say $\sigma_{\dot{\eta}}^* \leq 0.5$ – is

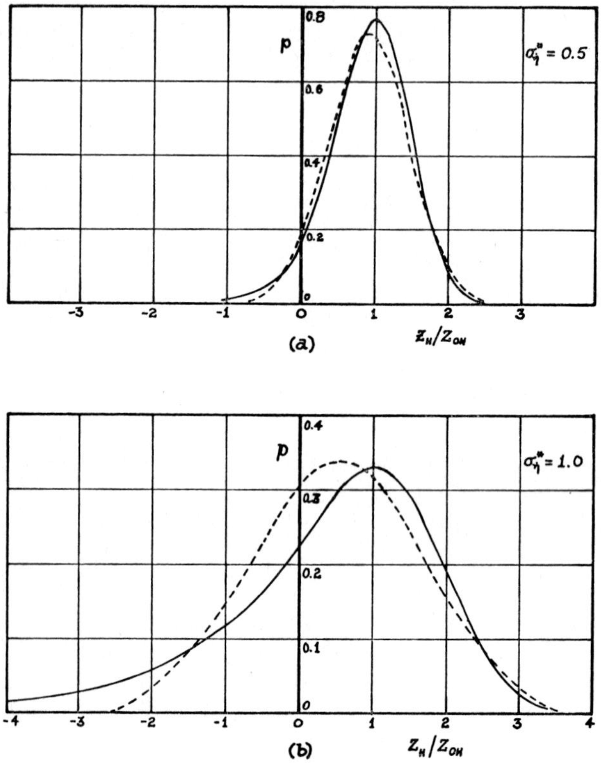

Fig. 5. Comparison of the nonlinear and linearized probability densities of the Hertzian deflection z_H, for the case of two discs with a random roughness. $\sigma_{\dot{\eta}}^*$ is an input variable defined in equation 66. (a) For $\sigma_{\dot{\eta}}^* = 0.5$. (b) For $\sigma_{\dot{\eta}}^* = 1.0$
——— nonlinear; – – – – linearized.

the linearized probability density reasonably accurate. The following general observations may be made.

First, the linearized analysis overestimates the probability of exceeding high values of z_H, for example $z_H \geq z_{OH}(1+2\sigma_{\dot{\eta}}^*)$. This is because the linearized density varies as $\exp(-z_H^2)$, whereas the nonlinear density varies as $\exp(-z_H^{5/2})$ for high values of z_H.

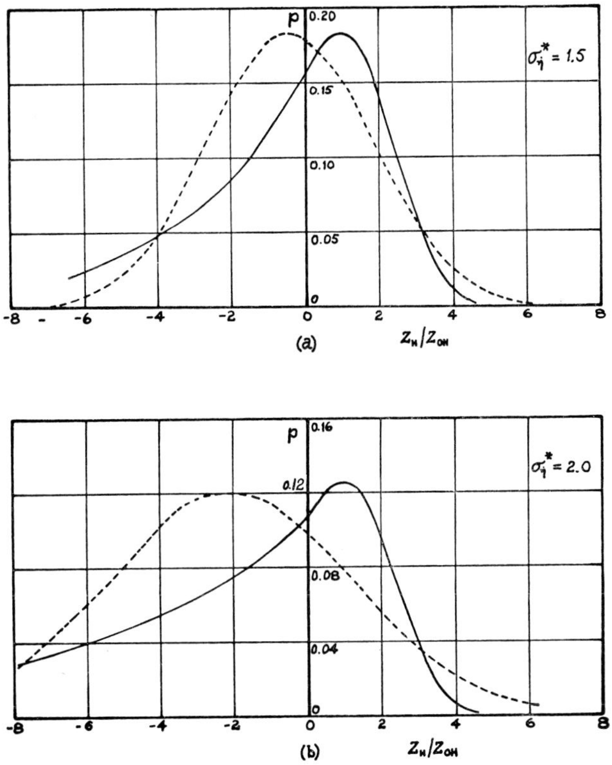

Fig. 6. Comparison of the nonlinear and linearized probability densities of the Hertzian deflection z_H, for the case of two discs with a random roughness. $\sigma_{\dot\eta}^*$ is an input variable defined in equation 66. (a) For $\sigma_{\dot\eta}^* = 1.5$. (b) For $\sigma_{\dot\eta}^* = 2.0$
——— nonlinear; – – – – linearized.

Second, the linearized analysis underestimates the probability of exceedance of intermediate values of z_H. As an example, Fig. 7(*a*) compares the estimate of the probability of exceeding the static value z_{0H} according to the linearized and nonlinear analyses.

Finally, the linearized analysis again *overestimates* the probability of loss of contact, if one defines loss of contact by $z_H \leq 0$. The definition $P_c \leq 0$ leads to a gross understimate, on the other hand, as is shown in Fig. 7(b). It is worth recalling from Fig. 3(c) that for sinusoidal waviness, the definition $z_H \leq 0$ for loss of contact led to more accurate results.

One thus observes a curious twist: for sinusoidal waviness, the linearized analysis *underestimates* both the maximum Hertzian deflections as well as the time spent out of contact; for random waviness, the linearized analysis *overestimates* both the probability of high deflections as well as the probability of loss of contact.

May one conclude from Figs. 5 and 6 that the probability of exceeding a high load is overestimated by the linearized analysis? In general, yes. However, care must be exercised: for a deflection $z_H = rz_{0H}$, with $r \geq 1$ (say), the loads are $P_c = P_0 r^{3/2}$ for the nonlinear analysis and $P_c = P_0[1 + 1.5(K_L/K_{0H}) \times (r - \bar{z}_H/z_{0H})]$ for the linearized analysis. In general, the latter value will be smaller than the former.

Before concluding, a few brief remarks will be made about two other

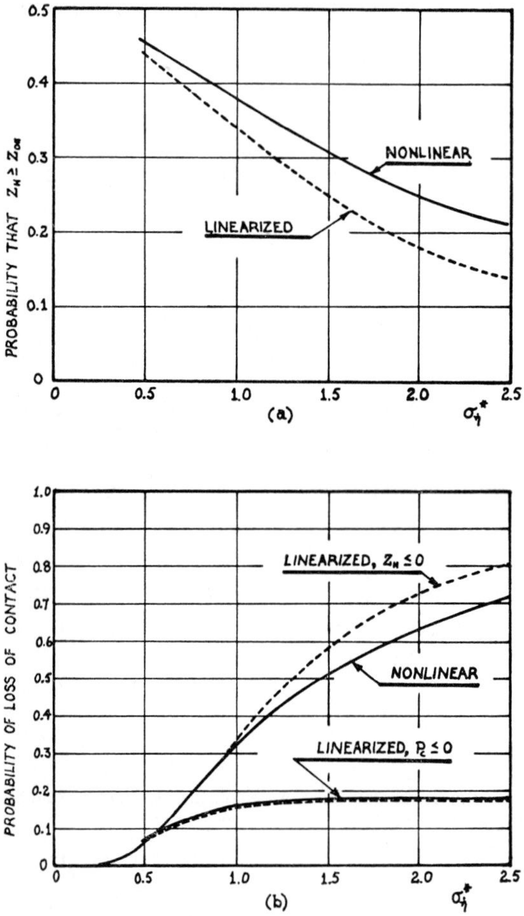

Fig. 7. Comparison of nonlinear and linearized response statistics for the case of two discs with a random roughness. (a) The probability of exceeding the static Hertzian deflection z_{0H}. (b) The probability of loss of contact. σ_{η}^{*} is an input variable defined in equation 66

statistics: the frequency of exceedances and their clump size (see sections 3.3.3 and 3.3.6). First, an examination of equations (45) and (46) and of the 'exact' results of [2] will indicate that the linearized analysis again overestimates the exceedance frequency for relatively high exceedance levels. Lack of space precludes a detailed examination here. Second, it is worth noting that at present no clump size estimates are forthcoming from the nonlinear analysis, since spectrum estimates are not available. From the linearized analysis one obtains

$$\bar{n}_i \approx (8\zeta)^{-1/2}\left(\frac{K_L\sigma_{zH}}{P_P-P_0}\right). \tag{69}$$

5. CONCLUSION

Linearization is a powerful method of analysis for contact vibration analysis – as it is for most nonlinear problems. It has been demonstrated that the method can be applied to a wide variety of contact configurations. Like any mathematical technique, however, linearization has its attendant pitfalls, the major one being how to interpret the results. There are two ways in which the accuracy of the technique can be checked. One is to compare the results of linearized and nonlinear analyses for some simple case (for which a nonlinear analysis is possible). This has been done here. The general conclusions that can be drawn are straightforward enough:

1. The linearized solution is accurate for relatively small values of the dynamic Hertzian deflection. This requires either a low magnitude of input roughness or a high level of damping. Specific criteria are given in the text.
2. With large inputs or low damping, the error due to linearization – stated qualitatively – is as follows:
   a. For a sinusoidal roughness, the linearized solution underestimates the maxima and minima of response, as also the time spent out of contact.
   b. For a random roughness, the linearized solution overestimates the probability of extreme values and of loss of contact, but underestimates the probability of exceeding intermediate values of the Hertzian deflection z_H.
3. In general, all criteria should first be developed for the nonlinear problem in terms of the Hertzian displacement and applied in this form to the linearized problem. For example, loss of contact should be defined by $z_H \leq 0$ rather than $P_c \leq 0$. Similarly, a plastic indentation [defined in the nonlinear problem by $P_c \geq P_p$ or $z_H \geq z_{HP} = (P_p/C)^{2/3}$] should be defined by $z_H \geq z_{HP}$ rather than $P_c \geq P_p$.

However, this comparison with a nonlinear analysis is by itself not adequate. The question arises: is the nonlinear analysis accurate? This question has been partially tackled by Gray and Johnson[3], who, in an experimental study of rolling discs, found that the linearized analysis – and therefore, *a fortiori*, the nonlinear analysis – was sufficiently accurate for the roughness and damping obtaining in their experiments. But there are aspects of nature that make analyses difficult and results wrong, among which might be mentioned the following:

1. nonlinear damping, for example from hysteresis losses;
2. plastic flow;
3. roughness patterns that are neither random nor sinusoidal (i.e. periodic with more than one component);
4. non-Hertzian contact conditions, including tractive forces, non-localized contact and simply bodies which cannot quite be approximated by half spaces (e.g. the contact of gear teeth); and
5. non-stationary effects wherein the contact geometry varies with time, as in a cam-and-follower.

How these phenomena are to be dealt with is a question looking for an answer.

One other curious phenomenon concerning contact vibrations is worth mentioning. Gray and Johnson [3] have demonstrated that small-wavelength components of the roughness are attenuated by the finite width of the contact patch, so that dynamic load spectral levels at high frequencies ($f \gg V/a$, where V is the rolling velocity and 'a' the length of the contact patch) should be extremely low. It is a well-known fact, however, that rolling and sliding contacts generate plenty of high-frequency noise. These two findings are by no means incompatible; but the mechanism of high-frequency noise generation remains a mystery.

REFERENCES

1. R.M. Carson and K.L. Johnson, 'Surface Corrugations Spontaneously Generated in a Rolling Contact Disc Machine', *Wear*, **17**, 1971, pp. 59–72.
2. P.R. Nayak, 'Contact Vibrations', *J. Sound Vib.*, **22**, 1972, pp. 297–322.
3. G.G. Gray and K.L. Johnson, 'The Dynamic Response of Elastic Bodies in Rolling Contact to Random Roughness of Their Surfaces', *J. Sound Vib.*, **22**, 1972, pp. 323–342.
4. P.R. Nayak, 'Contact Vibrations of a Wheel on a Rail', *J. Sound Vib.*, **28**, 1973, pp. 277–293.
5. M.A. Heckl, 'Compendium of Impedance Formulas', Bolt Beranek and Newman Inc., Report No. 774, May 1961.
6. L. Cremer and M. Heckl, *Körperschall*, Springer, New York. 1967.
7. S. Timoshenko, *Theory of Elasticity*, McGraw-Hill Book Co., Inc., New York, 1951.
8. S.H. Crandall, 'First Crossing Probabilities of the Linear Oscillator', *J. Sound Vib.*, **12**, 1970, pp. 285–299.
9. T.G. Pearce and B.J. May, 'A Study of the Lateral Stability, Curving and Dynamic Response of the Linear Induction Moter Test Vehicle', Report to the U.S. Department of Transportation, under Contract No. 3–0261, 1969.